T0252100

JOHN DEERE

SHOP MANUAL JD-202

Models ■ 2510 ■ 2520

Model ■ 2040

Models ■ 2240 ■ 2440 ■ 2630 ■ 2640

Models ■ 4040 ■ 4240 ■ 4440 ■ 4640 ■ 4840

I&T

SHOP MANUALS

Information and Instructions

This shop manual contains several sections each covering a specific group of wheel type tractors. The Tab Index on the preceding page can be used to locate the section pertaining to each group of tractors. Each section contains the necessary specifications and the brief but terse procedural data needed by a mechanic when repairing a tractor on which he has had no previous actual experience.

Within each section, the material is arranged in a systematic order beginning with an index which is followed immediately by a Table of Condensed Service Specifications. These specifications include dimensions, fits, clearances and timing instructions. Next in order of arrangement is the procedures paragraphs.

In the procedures paragraphs, the order of presentation starts with the front axle system and steering and proceeding toward the rear axle. The last paragraphs are devoted to the power take-off and power lift systems. Interspersed where needed are additional tabular specifications pertaining to wear limits, torquing, etc.

HOW TO USE THE INDEX

Suppose you want to know the procedure for R&R (remove and reinstall) of the engine camshaft. Your first step is to look in the index under the main heading of ENGINE until you find the entry "Camshaft." Now read to the right where under the column covering the tractor you are repairing, you will find a number which indicates the beginning paragraph pertaining to the camshaft. To locate this wanted paragraph in the manual, turn the pages until the running index appearing on the top outside corner of each page contains the number you are seeking. In this paragraph you will find the information concerning the removal of the camshaft.

More information available at haynes.com
Phone: 805-498-6703

J H Haynes & Co. Ltd.
Haynes North America, Inc.

ISBN-10: 0-87288-366-3
ISBN-13: 978-0-87288-366-6

Disclaimer

There are risks associated with automotive repairs. The ability to make repairs depends on the individual's skill, experience and proper tools. Individuals should act with due care and acknowledge and assume the risk of performing automotive repairs.

The purpose of this manual is to provide comprehensive, useful and accessible automotive repair information, to help you get the best value from your vehicle. However, this manual is not a substitute for a professional certified technician or mechanic.

This repair manual is produced by a third party and is not associated with an individual vehicle manufacturer. If there is any doubt or discrepancy between this manual and the owner's manual or the factory service manual, please refer to the factory service manual or seek assistance from a professional certified technician or mechanic.

Even though we have prepared this manual with extreme care and every attempt is made to ensure that the information in this manual is correct, neither the publisher nor the author can accept responsibility for loss, damage or injury caused by any errors in, or omissions from, the information given.

JOHN DEERE

Models ■ 2510 ■ 2520

Previously contained in I&T Shop Manual No. JD-33

SHOP MANUAL

JOHN DEERE

SERIES 2510—2520

Tractor serial number is located on right rear of transmission case. Engine serial number is stamped on a plate on right side of engine cylinder block.

INDEX (By Starting Paragraph)

CONDENSED SERVICE DATA

GENERAL	2510 Gasoline	2510 Diesel	2520 Gasoline	2520 Diesel
Engine Make	Own			
No. of Cylinders	4			
Bore-Inches	3.86	3.86	3.86	4.02
Stroke-Inches	3.86	4.33	4.33	4.33
Displacement-Cubic Inches	180	202	202	219
Compression Ratio	7.5:1	16.3:1	7.8:1	16.3:1
Main Bearings, No. of	3	5	3	5
Cylinder Sleeves-Type	Wet			
Battery Terminal Grounded	Negative			

TUNE-UP

	2510 Gasoline	2510 Diesel	2520 Gasoline	2520 Diesel
Firing Order	1-3-4-2			
Compression Pressure @ Cranking speed (PSI)	120	300	105	325
Valve Tappet Gap:				
Intake	0.014	0.014	0.014	0.014
Exhaust	0.022	0.018	0.022	0.018
Ignition Distributor:				
Make	Prestolite	......	Prestolite	......
Model	IBT-4101S	......	IBT-4101S or IBT-4101U	......
Breaker Contact Gap	0.020	......	0.020	......
Ignition Timing	"S" Mark @ 2500 RPM	......	"S" Mark @ 2500 RPM	......
Plug Electrode Gap	0.025	......	0.025	......
Engine Low Idle-RPM	800	800	800	800
Engine High Idle-RPM	2680	2650	2680	2650
Engine Full Power-RPM	2500	2500	2500	2500
PTO High Idle-RPM	689-1275	680-1260	689-1275	680-1260
PTO Full Power-RPM	643-1190	643-1190	643-1190	643-1190
Horsepower at PTO (Manufacturer's Observed Rating at ASAE 540-1000 RPM PTO Speed)				
Syncro-Range	48.2	49.2	53.2	55.8
Power Shift	45.0	45.8	51.9	51.9

SIZES-CAPACITIES-CLEARANCES

	2510 Gasoline	2510 Diesel	2520 Gasoline	2520 Diesel
Crankshaft Journal Diam.	3.123-3.124			
Crankpin Diameter	2.308-2.309	2.748-2.749	2.308-2.309	2.748-2.749
Camshaft Journal Diam.	2.1997-2.2007			
Piston Pin Diameter	1.1875-1.1879			
Valve Stem Diameter	0.3715-0.3725			
Main Bearing Clearance	0.0016-0.0046			
Rod Bearing Clearance	0.0014-0.0044	0.0012-0.0042	0.0014-0.0044	0.0016-0.0046
Camshaft Journal Clearance	0.0035-0.0055			
Camshaft End Play	0.0025-0.0085			
Crankshaft End Play	0.002-0.008			
Piston Skirt Clearance (Bottom)	See Paragraph 35			
Cooling System—Qts.	14			
Crankcase (With Filter) Qts. (Dry)	7			
Fuel Tank—Gallons	26			
Transmission & Hydraulic System—Gallons:				
Syncro-Range-Dry	11			
(Refill)	8			
Power Shift-Dry	14			
(Refill)	11			

TIGHTENING TORQUES-FT.-LBS.

	2510 Gasoline	2510 Diesel	2520 Gasoline	2520 Diesel
Cylinder Head	110			
Main Bearings	85			
Con. Rod Bearings (Oiled)	40-45	60-70	40-45	60-70
Rocker Arm Assembly	35			
Flywheel	85			

FRONT SYSTEM

Tractors are available with three tricycle type front end units and two axle types. Refer to Figs. 1 through 7 for exploded views. Tricycle systems consist of a fork mounted single wheel, dual wheel tricycle, or "Roll-O-Matic" dual wheel tricycle, which attach directly to the center steering spindle. Axle types consist of a narrow width fixed tread or adjustable axle which may be either standard or high clearance.

AXLE AND SUPPORT
Models So Equipped

1. **HOUSING & PIVOT BRACKET.** The front axle attaches to tractor frame by pivot bracket (1—Fig. 1 or 2—Fig. 2). Axle pivot bushing clearance should be 0.009-0.012. Tighten pivot bracket to frame bolts to a torque of 150 ft.-lbs. and pivot bracket to support bolts to 300 ft.-lbs.

2. **SPINDLES & BUSHINGS.** Steering arm (2—Fig. 3 or 4) is splined to spindle (7) and retained by a bolt. Spindle bushings are pre-sized. Thrust washers (6) are positively located on spindle and axle extension by dowels. When renewing, be sure ground surfaces of thrust washers are together. Maximum allowable end play of spindle is 0.036. Adjustment is made by means of shim washers (8) which are 0.036 thick.

On all models, tie rods should be adjusted equally to provide ⅛-⅜ inch toe-in. On adjustable axle models, tighten clamp bolts to a torque of 35 ft.-lbs.

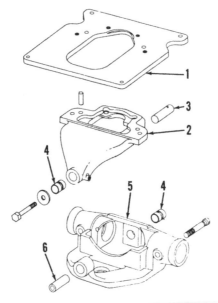

Fig. 2—Exploded view of front axle housing, pivot bracket and associated parts used on models with fixed tread.

1. Adapter
2. Pivot bracket
3. Pivot pin
4. Bushing
5. Axle housing
6. Pivot pin

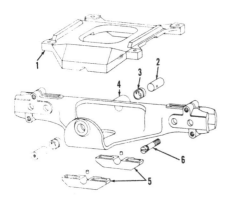

Fig. 1—Exploded view of front axle housing, pivot bracket and associated parts used on adjustable axle models.

1. Pivot bracket
2. Pivot pin
3. Bushing
4. Axle housing
5. Pivot clamps
6. Lock bolt

using the special John Deere lifting plate (JDG-3) or other suitable support attached to tractor front support. Tighten the retaining cap screws to a torque of 300 ft.-lbs. when unit is reinstalled.

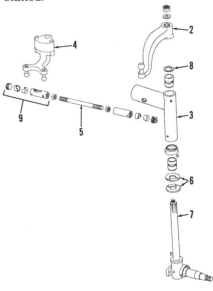

Fig. 4—Exploded view of fixed tread side axle. Refer to Fig. 3 for parts identification except for (9) which is tie rod end.

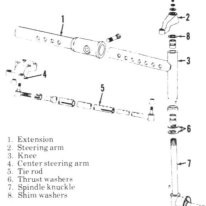

Fig. 3—Exploded view of adjustable extension, steering spindle and associated parts.

1. Extension
2. Steering arm
3. Knee
4. Center steering arm
5. Tie rod
6. Thrust washers
7. Spindle knuckle
8. Shim washers

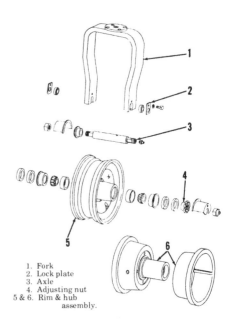

Fig. 5—Exploded view of front wheel fork and axle assembly used on single wheel tricycle models. The one-piece wheel (5) is for 7.50–16-inch tire, the two-piece rim (6) for 9.00–10-inch tire.

1. Fork
2. Lock plate
3. Axle
4. Adjusting nut
5 & 6. Rim & hub assembly.

TIE RODS & TOE IN
All Axle Models

3. The socket type tie rod ends (9—Fig. 4) used on narrow tread models can be adjusted by removing the cotter pin and tightening plug to remove all end play without binding. Ball sockets are not spring loaded.

SPINDLE EXTENSION (PEDESTAL)
Tricycle Models

4. **REMOVE & REINSTALL.** The spindle extension (pedestal) attaches directly to steering motor spindle by four cap screws. To remove the unit, support front of tractor with a hoist

5. **SINGLE WHEEL TRICYCLE.** The fork mounted single wheel is supported on taper roller bearings as shown in Fig. 5. Bearings should be adjusted to provide a slight rotational drag by means of adjusting nut (4). The one-piece wheel & rim assembly (5) accommodates a 7.50"-16" tire, the two-

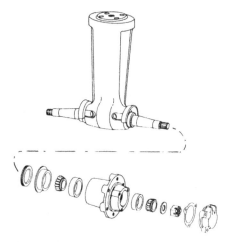

Fig. 6–Dual wheel tricycle pedestal showing one wheel hub and associated parts.

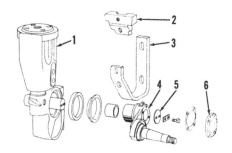

Fig. 7–"Roll-O-Matic" spindle extension showing component parts. Lock (2) and support (3) may be installed to provide rigidity.

1. Pedestal extension
2. Lock
3. Support
4. Knuckle
5. Thrust washer
6. Cap

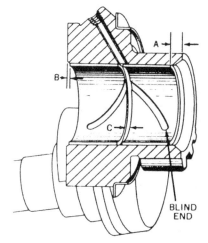

Fig. 8–Cross sectional view of knuckle showing details of bushing installation. Refer also to paragraph 7.

piece rim (6) a 9.00"-10" tire. Different wheel fork (1) is required for the two tire sizes.

6. **DUAL WHEEL TRICYCLE.** An exploded view of the dual wheel tricycle pedestal and wheel is shown in Fig. 6. Horizontal axles are not renewable. Service consists of renewing the complete pedestal assembly.

7. **ROLL-O-MATIC UNIT.** The "Roll-O-Matic" front wheel pedestal and associated parts are shown exploded in Fig. 7. The unit can be overhauled without removing the assembly from tractor.

Support front of tractor and remove wheel and hub units. Remove knuckle caps (6) and thrust washers (5), then pull knuckle and gear units (4) from housing.

Tractors may be equipped with regular or heavy duty "Roll-O-Matic" units. The regular unit is equipped with a lock (2) and lock support (3) which may be installed for rigidity

when desired. Check the removed parts against the values which follow:

Regular "Roll-O-Matic"
Knuckle Bushing ID . . .1.873-1.875
Knuckle Shaft OD 1.870-1.872
Thrust Washer Thickness 0.156

Heavy Duty "Roll-O-Matic"
Knuckle Bushing ID . . .2.127-2.129
Knuckle Shaft OD 2.124-2.126
Thrust Washer Thickness 0.187

Bushings are pre-sized and contain a spiral oil groove which extends to one edge of bushing. When installing new bushings, use a piloted arbor and press bushings into knuckle arm so that OPEN end of spiral grooves are together as shown in Fig. 8. Bushing at spindle end of Regular "Roll-O-Matic" unit should be pressed into arm so that outer edge (B) is flush with machined surface. On Heavy Duty "Roll-O-Matic", distance (B) should measure 1/32-inch. There should be a gap (C) of 1/32-1/16 inch between bushings, and distance (A) from edge of inner bushing to edge of bore should measure ¼ inch on Regular "Roll-O-Matic" or 3/16 inch on

Fig. 9–Make sure timing marks (M) are aligned when installing knuckles in "Roll-O-Matic" unit.

Heavy Duty unit. Soak felt washers in engine oil prior to installation. Install one of the knuckles so that wheel spindle extends behind vertical steering spindle. Pack the "Roll-O-Matic" unit with wheel bearing grease and install the other knuckle so that timing marks on gears are in register as shown at (M—Fig. 9). Tighten the thrust washer attaching screws to a torque of 55 ft.-lbs.

POWER STEERING SYSTEM

All models are equipped with full power steering. No mechanical linkage exists between steering wheel and front unit; however, steering can be manually accomplished by hydraulic pressure when tractor hydraulic unit is inactive. Power is supplied by the same hydraulic pump which supplies power for the brake system and hydraulic lift. A pressure control (priority) valve is located in outlet line from main hydraulic pump. Valve gives steering and brakes first priority on hydraulic flow.

OPERATION
All Models

8. The power steering system consists of the tractor hydraulic system described in paragraph 153, plus the steering control unit and motor described in this section. Refer to Fig. 10 for a schematic view of steering control unit and motor.

The control unit contains a double acting slave cylinder (2) on steering wheel shaft which is of approximately equal displacement to the cylinders in

steering motor (9). Valve actuating levers (5) are in contact with operating collar (4) also mounted on steering shaft.

When the control unit is in neutral position, there is no fluid flow but oil at pump pressure is available at pressure line (6). When steering wheel is turned for a right or left turn, the steering shaft screw (3) will meet the resistance of slave cylinder (2) and shaft will move collar (4) and levers (5) to open one pressure and one return needle valve. Fluid at pump pressure will then

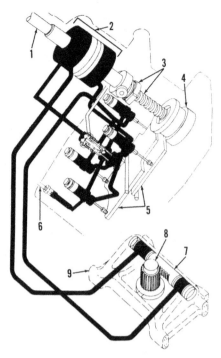

Fig. 10–Schematic view of power steering system showing operating parts. Refer to paragraph 8 for description.

1. Steering shaft
2. Slave cylinder
3. Actuating screw
4. Actuating collar
5. Operating levers
6. Pressure line
7. Operating piston
8. Steering spindle
9. Steering motor

enter the control valve, lines and cylinders.

On a right hand turn, pressurized fluid is allowed to enter the bottom part of slave cylinder (2), fluid from top of cylinder is forced to left side of steering motor (9) and fluid from right side of steering motor (9) returns to sump. On a left hand turn, fluid at pump pressure is allowed to enter right cylinder of steering motor (9) and fluid from left cylinder is forced to top of slave cylinder (2). Fluid from bottom part of slave cylinder returns to sump through the open return valve. When turning in either direction, the slave cylinder meters the extent and speed of fluid flow and turning action, and is directly controlled by the steering wheel.

When no pressure is present at inlet pressure line (6), a check valve prevents the entry and discharge of fluid from steering lines. The piston in slave cylinder is manually moved by the steering shaft worm (3) and manual steering is accomplished by exchange of trapped fluid between slave cylinder (2) and steering motor (9).

TROUBLE SHOOTING

NO POWER STEERING
    Low transmission oil level
    Oil filter plugged

    Filter bypass valve open
    Cooler relief valve stuck open
    Transmission pump failure
    Transmission pump suction screen
      plugged
    PTO valve leaking
    Differential lock valve sealing rings
      failed
NO POWER STEERING WHILE OPERATING FUNCTION
    Low transmission oil level
    Oil filter plugged
    Filter bypass valve stuck open
    Transmission pump suction screen
      plugged
    Transmission pump failure
    Pressure control valve stuck open or
      adjusted incorrectly
    Brake inlet valve leaking
NO OR POOR POWER STEERING TO LEFT
    Leaking right pressure or right
      return valve
    Inlet check valve failure
NO OR POOR POWER STEERING TO RIGHT
    Upper or lower steering valve piston
      rod packing failure
    Left return or left pressure valve
      leaking
    Steering check valve piston packing
      failure
    Steering motor piston seal failure
    Inlet check valve failure
NO OR POOR MANUAL STEERING TO LEFT
    Steering valve cylinder piston seal
      failure
    Synchronizing valve failure
    Lower steering valve piston rod
      packing failure
    Inlet check valve seat or packing
      failure
    Steering check valve ball seat failure
    Steering check valve piston stuck
    Steering motor piston seals failure

NO OR POOR MANUAL STEERING TO RIGHT
    Upper steering valve piston rod
      packing failure
    Steering valve cylinder piston
      packing failure
    Synchronizing valve leaking
    Steering check valve piston packing
      failure
    Steering motor piston seal failure
STEERING WANDERS TO LEFT OR RIGHT
    Upper steering piston rod packing
    Steering valve cylinder piston
      packing
    Steering valve operating shaft collar
      assembly loose
    Left pressure and return valve
      leaking
    Steering check valve piston packing
      failure
    Improper valve adjustment
    Steering motor piston seals failure

FREQUENT SYNCHRONIZATION
    Upper steering valve piston rod
      packing failure
    Steering check valve piston packing
      failure
    Steering motor piston seals failure
    Synchronizing valve failure
EXCESSIVE STEERING WHEEL FREE PLAY
    Steering valve shaft nut and rod
      loose
    Collar loose
    Air in steering system

BLEEDING
All Models

9. To bleed the system first remove the cowling and the small machine screw from bleed screw (Fig. 11). Attach a hose to bleed screw and run free end back to reservoir. Start engine and run at slow idle speed. Turn steering wheel to full right, then to full left, to allow steering valve to synchronize with front wheels.

Leave steering wheel and front wheels in full left turn position, loosen bleed valve lock nut and back out bleed valve approximately ½ turn. With engine at slow idle and without moving front wheels, turn steering wheel very slowly to full right. Close bleed valve and allow front wheels to turn full right.

Repeat the procedure, if necessary, until air-free fluid is being returned to reservoir.

STEERING CONTROL UNIT
All Models

10. **REMOVE AND REINSTALL.** To remove the steering control unit, first remove steering wheel and cowl; and on diesel models, remove the hood.

Disconnect electrical connections to instruments. Loosen gage clamps and slip heat gages out of instrument panel. Remove adjusting screw from

Fig. 11–To bleed the steering system, remove the cowl and attach bleed line to bleed screw, then refer to paragraph 9 for procedure.

fast idle stop plate and dog point set screw from throttle lever housing, lift off throttle lever; then remove instrument panel.

Disconnect steering fluid lines, being sure to cap all fittings, then unbolt and remove the complete steering unit.

When reinstalling, bleed steering system as outlined in paragraph 9 and adjust throttle linkage as in paragraph 60, 64 or 66. Tighten steering wheel nut to a torque of 50 ft.-lbs.

11. **OVERHAUL.** To disassemble the removed steering control unit, first remove lower cover (20—Fig. 13). Remove cotter pin and nut (12) from lower end of steering shaft, then unbolt and withdraw valve housing (18) and operating collar (10). Check valve stop; spring and shaft (16) will fall out as housing is removed, and must be recovered and saved.

Remove nut (9), spring (8) and collar (7). The rollers and pins in collar must be removed before collar can be withdrawn. Temporarily install steering wheel and turn wheel counter-clockwise to force cylinder cover (6) from cylinder housing (1), then withdraw cover, piston and lower steering shaft from cylinder housing. Upper steering shaft is retained in steering column by a snap ring. Remove shaft if service is required on oil seal, bushing, shaft or housing. Withdraw springs from operating valves in valve housing, remove valve balls and inspect balls and seat for line contact. Note: If inlet check valve spring or guide damage is noted, there has been excessive flow through the assembly. Check for foreign particles in a control valve orifice or for particles holding the control valve or control valve ball off its seat. Check for seat damage due to foreign material. Check synchronizing valve assembly for sticking. This can prevent the proper synchronizing between valve and motor. Renew any parts that are damaged or worn.

Clean all parts by washing in clean

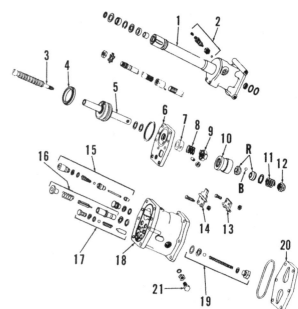

Fig. 13–Exploded view of steering valve housing and associated parts.

1. Cylinder housing
2. Bleed screw
3. Steering shaft
4. Seal
5. Piston & rod assy.
6. Cylinder cover
7. Collar
8. Spring
9. Nut
10. Operating collar
11. Spring
12. Nut
13. Operating lever
14. Operating lever
15. Operating valve
16. Check valve
17. Inlet check valve
18. Valve housing
19. Return valve
20. Cover
21. Lever plug.

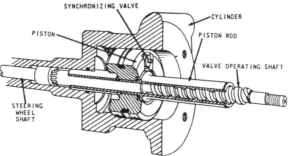

Fig. 14–Cross sectional view of steering valve operating piston, cylinder and steering shaft. Piston is moved up or down in cylinder by helical thread on steering shaft. The synchronizing valve which corrects for internal leaks is shown in Fig. 15.

solvent and immerse all parts including O-rings and backup washers in clean hydraulic fluid before assembly.

When reassembling, tighten spring loaded nut (9) on operating shaft until a gap of approximately 5/16-inch exists between nut (9) and collar (7). This tension provides the friction which gives a feeling of stability to the steering effort. Lever plugs (21) control the end play of pivot shafts on operating levers (13 & 14). Adjust the plugs, if necessary, until levers turn freely but end play is limited to a maximum of 0.003. The ball races (R)

for operating collar (10) contain 13 loose bearing balls (B). Tighten nut

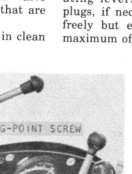

Fig. 12–Dog point set screw and screw retaining fast idle stop plate must be removed to remove hand throttle lever.

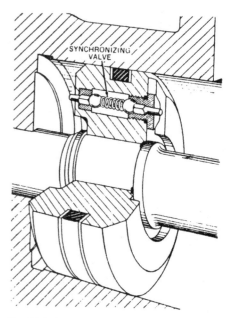

Fig. 15–Steering valve piston must be synchronized with steering motor for full turning action. Synchronization is automatically accomplished. When control valve piston reaches end of its stroke, the extended rod unseats the ball check valve allowing pressurized fluid to flow through piston until motor and valve are synchronized.

(12) to 5 ft.-lbs., loosen to nearest castellation and install cotter pin. Operating collar (10) must turn by hand on shaft. Tighten the cap screws retaining valve housing to cylinder cover to a torque of 85 ft.-lbs. Adjust the valve levers as in paragraph 12.

12. **ADJUSTMENT.** Clamp steering control unit in a vise as shown in Fig. 16, and install the special neutral stops (JDH-3C) and dial indicator. Turn steering wheel to full right and attach a weight to steering wheel rim as shown. Loosen the locknuts on adjusting screws (A & B—Fig. 17). With dial indicator pointer contacting operating levers midway between the two adjusting screws, adjust one 'A' and one 'B' screw to allow a total free movement of operating lever of 0.003 when measured with dial indicator. Adjust the two remaining screws to reduce the movement to 0.001 after the first two screws have been adjusted. Install lower cover and tighten retaining cap screws to a torque of 20 ft.-lbs.

STEERING MOTOR
All Models

13. **REMOVE AND REINSTALL.** To remove the steering motor, first remove cowl and hood and support front of tractor from a hoist by installing engine sling. Remove fuel tank and right hand side frame. On tricycle models, remove wheels and pedestal assembly. On axle models, remove front axle and support as a unit.

Place a rolling floor jack under steering motor to support the motor. Disconnect and plug fluid lines and vent line. Remove the cap screws securing steering motor to left side frame and front frame, lower the jack and roll steering motor away from tractor.

To install the steering motor, reverse the removal procedure and bleed system as outlined in paragraph 9. Tightening torques are as follows:

Motor to side frames . . . . 250 ft.-lbs.
Motor to front frame . . . . . . 150 ft.-lbs.
Steering spindle
 flange bolts . . . . 300 ft.-lbs.

14. **OVERHAUL.** To disassemble the removed steering motor, turn unit upside down on bench and remove the cap screws securing spindle retainer (15—Fig. 18) to motor housing. Tap spindle retainer from its doweled position and withdraw spindle (16), retainer (15) and bearing (13) as a unit. Spindle bearing is retained by snap ring (12) and can be removed if service is indicated on any of the components. Install spindle bushing (11) in housing with split to rear if bushing is removed.

To remove the piston (4), first remove the spindle. Remove outer washer (9) and snap ring (8), then reinstall cap screw to assist in pulling end plug (7). Remove opposite end plug in same manner, then push piston from cylinder bore.

When reassembling the steering motor, reinstall either piston end plug,

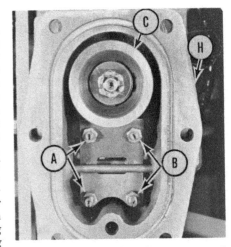

Fig. 17—Front view of valve with cover removed. Correct positioning of operating collar (C) with relation to housing face (H) is accomplished by the two blocks JDH-3C shown in Fig. 16.

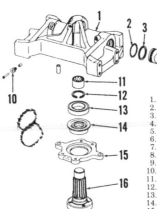

1. Housing
2. O-ring
3. Backup ring
4. Rack & piston
5. O-ring
6. Backup ring
7. Piston plug
8. Snap ring
9. Washer
10. Vent tube
11. Bushing
12. Snap ring
13. Bearing
14. Oil seal
15. Retainer
16. Spindle

Fig. 18—Exploded view of steering motor showing component parts.

Fig. 19—V-mark on spindle flange should align with mark on housing at either end of piston stroke.

Fig. 16—Steering valve positioned for adjustment. Refer to paragraph 12.

then install piston (4), pushing it into bore until it contacts the installed plug. Install spindle and retainer, with V-mark on spindle flange aligned with housing scribe mark on same side as the installed end plug. Complete the assembly by reversing the disassembly procedure and install the assembled unit as outlined in paragraph 13.

ENGINE AND COMPONENTS

REMOVE AND REINSTALL
All Models

15. To remove engine and clutch as a unit, first drain cooling system and if engine is to be disassembled, drain oil pan. Remove front weights if any are installed. Remove side grille screens,

hood and cowl.

Disconnect tachometer cable, battery ground cable, oil pressure switch and throttle linkage. Remove hydraulic line clips and radiator brace rods and disconnect wire from fuel gage sending unit. Disconnect hydraulic fluid lines at front junction and at firewall, remove hydraulic line spacer clamps then remove the lines.

Shut off fuel and remove fuel line. On diesel models, remove leak-off line leading to fuel tank, and disconnect ether starting aid pipe. On all models, disconnect wiring harness, air cleaner and coolant hoses; and detach heat indicator sending unit from cylinder head.

NOTE: Before splitting a tractor that has an accumulator, discharge the accumulator by opening the right hand brake bleed screw, and holding right brake pedal down for a few seconds.

Remove the cap screw which fastens accumulator support bracket to the engine, if so equipped. On Power Shift models, remove left tractor step and disconnect hydraulic pump inlet pipe.

Attach tractor split stands or support rear of tractor on a rolling floor jack and front half from a hoist. Remove the cap screws securing clutch housing to engine and roll rear half of tractor back.

Support engine in a hoist (or block up beneath side frames). Disconnect hydraulic pump drive coupler and remove cap screws securing engine block to front support and side frames and slide engine to the rear.

CAUTION: If fuel tank is full or front end is weighted, front half may tip forward when engine is removed. Check stability of front frame before removing engine.

Reassemble tractor by reversing the disassembly procedure. Tightening torques are as follows:

Clutch housing to engine . . 170 ft.-lbs.
Cylinder block to
    front support . . . . 170 ft.-lbs.
Cylinder block to
    side frames . . . . 250 ft.-lbs.
Hydraulic pump
    drive coupler . . . . 20 ft.-lbs.

CYLINDER HEAD
All Models

16. To remove the cylinder head, first drain coolant and remove hood. Disconnect battery ground straps. Remove air cleaner tube and water outlet elbow.

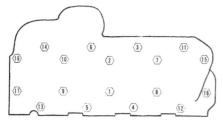

Fig. 20—When installing cylinder head, tighten the cap screws to a torque of 110 ft.-lbs. using the sequence shown.

On gasoline tractors, disconnect fuel line and throttle linkage from carburetor and unbolt and remove the manifold. Disconnect coil wires and spark plug wires and remove coil and spark plugs.

On diesel models, disconnect injector leak-off pipe from fuel tank, injectors and injection pump and remove the pipe. Disconnect fuel lines from filters and unbolt and remove filters. Disconnect high pressure lines and remove injector clamps and injectors.

On all models, remove crankcase vent pipe from rocker arm cover and unbolt and remove the cover. Remove rocker arms and shaft assembly, push rods and, on diesel models, remove valve stem caps. Disconnect coolant temperature bulb and ventilator pump air intake tube, then unbolt and remove the cylinder head.

CAUTION: Cylinder sleeves are slip fit in block bores. Do not turn crankshaft with head removed, unless cylinder sleeves are secured with washers and short cap screws.

When installing cylinder head, be sure that all cap screws have a flat washer, and use a thin coat of non-hardening sealer on both sides of head gasket. Install and tighten cylinder head cap screws to a torque of 110 ft.-lbs. using the sequence shown in Fig. 20. Be sure oil holes in rear rocker arm bracket and cylinder head are open and clean. Rear oil passage provides lubrication for rocker arms.

Head bolts should be rechecked after engine has been run in for about an hour at 2500 rpm at half load. Loosen head bolts about 1/6-turn then retighten to 110 ft.-lbs. Adjust valve tappet gap hot or cold, using the procedures outlined in paragraph 18.

VALVES AND SEATS
All Models

17. Intake and exhaust valves for all engines seat directly in the cylinder head, and the valve guides are an integral part of the cylinder head, and are non-renewable. Exhaust valves of gasoline engines are fitted with "Rotocaps" while intake valves are equipped with an O-ring seal. All valves in diesel engines are equipped with renewable, hardened stem caps.

Intake and exhaust valve stem diameter is 0.3715-0.3725 for all models, with a recommended operating clearance of 0.002-0.004 in stem bores. Valves are available with 0.003, 0.015 and 0.030 oversize stems for installation in reamed bores if clearance is excessive. If valve stems do not show any measurable wear, the valve guides can be knurled and reamed to the standard clearances. If necessary to knurl

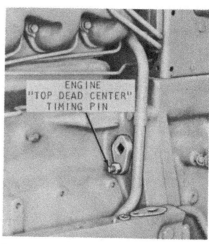

Fig. 21—View of engine showing timing hole cover and timing pin.

guides, use the tool exactly as recommended by the manufacturer.

Recommended valve seat angle is 45° for all models, with a suggested face angle of 43½° on diesel engines or 44° on gasoline engines to establish the recommended interference angle. Suggested valve seat width is 1/16 inch for diesel engines or 5/64 inch for gasoline models. Seats can be narrowed using 20 and 70 degree stones. Valve tappet gap should be adjusted using the procedure outlined in paragraph 18.

18. **TAPPET GAP ADJUSTMENT.** On all models, the two-position method of valve tappet gap adjustment is recommended. Refer to Figs. 22 and 23 and proceed as follows:

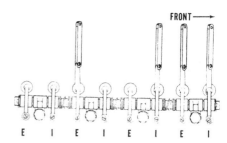

Fig. 22—With No. 1 piston at TDC on compression stroke, adjust the indicated valves to clearance given in paragraph 18. Refer also to Fig. 23.

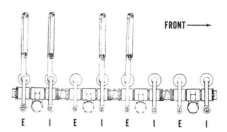

Fig. 23—With No. 4 piston at TDC on compression stroke, adjust the indicated valves to clearance given in paragraph 18. Refer also to Fig. 22.

Turn engine crankshaft by hand until "TDC" timing pin can be inserted in hole in flywheel as shown in Fig. 21. Check the No. 2 exhaust valve (fourth valve from front) to determine whether No. 1 or No. 4 piston is on compression stroke. If No. 2 exhaust valve is partly open, No. 1 cylinder is on compression stroke; adjust the four tappets shown in Fig. 22. If valve is closed, adjust the tappets shown in Fig. 23. When first adjustment is completed, turn crankshaft one complete revolution until "TDC" timing pin can again be inserted, then adjust remainder of valves using the other diagram.

Recommended valve tappet gaps are as follows:

Diesel Models
Intake valves .... 0.014 inch
Exhaust valves .... 0.018 inch

Gasoline Models
Intake valves .... 0.014 inch
Exhaust valves .... 0.022 inch

VALVE GUIDES
All Models
19. Valve guides are integral with cylinder head and have an inside diameter of 0.3745-0.3755. Normal operating clearance for valve stems is 0.002-0.004, with a maximum allowable clearance of 0.006.

When excessive clearance cannot be corrected by installing new valves, guides may be knurled, using procedures outlined in paragraph 17, or ream guide to fit the next available oversize valve stem. Valves are available with stems in oversizes of 0.003, 0.015 and 0.030.

VALVE SPRINGS
All Models
20. Inlet and exhaust valve springs

are interchangeable. The same spring is used for gasoline and diesel models. Springs may be installed either end up. Renew any spring which is distorted, discolored, rusted, or does not meet the specifications which follow:
Free length (approx.) .... 2⅛ in.
Lbs. test @ 1 13/16 in. 52-64
Lbs. test @ 1 23/64 in. 133-153

VALVE ROTATORS
All Gasoline Models
21. Positive type valve rotators are used on exhaust valves of gasoline engines. Normal servicing of the rotators consists of renewing the units. It is important, however, to check operation of the rotators. If rotator is removed, see that it turns freely in one direction only. If rotator is installed, be sure valve rotates a slight amount each time it opens.

ROCKER ARMS AND SHAFT
All Models
22. Rocker arms are interchangeable and bushings are not available. Inside diameter of shaft bore in rocker arm is 0.790-0.792. Outside diameter of shaft is 0.787-0.788. Normal operating clearance between rocker arm and shaft is 0.002-0.005, renew rocker arm and/or shaft if clearance is excessive.

Valve stem contacting surface of rocker arm may be refaced but original radius must be maintained.

When reinstalling rocker arm assembly, be sure oil holes and passages are open and clean, and plugs in each end of shaft are tight. Pay particular attention to the rear mounting bracket as lubrication is fed to rocker arm shaft through passages at this point. Oil hole in rocker arm shaft must face downward when installed on cylinder head.

CAM FOLLOWERS
All Models
23. The cylindrical type cam followers (tappets) can be removed from below after camshaft has been removed. If necessary, followers can also be removed from above after cylinder head, rocker arms assembly and push rods are removed. Cam followers are available in standard size only and operate directly in machined bores in cylinder block.

It is recommended that new cam followers be installed whenever camshaft is renewed.

VALVE TIMING
All Models
24. Valves are correctly timed when timing mark on camshaft gear aligns with timing tool (JOHN DEERE JD-

Fig. 25–Diesel engine timing gear train with timing gear cover removed.

B. Balance shaft gear
C. Crankshaft gear
G. Camshaft gear
L. Lower idler gear
O. Oil pump drive gear

P. Injection pump drive gear
T. Thrust screw
U. Upper idler gear

254) and No. 1 piston is at TDC on compression stroke. Refer to Fig. 26.

TIMING GEAR COVER
All Models
25. To remove the timing gear cover, first remove hood and grille screens. Drain radiator and remove radiator hoses. Remove any front end weights. Disconnect tachometer cable, oil pressure switch and throttle linkage. Remove hydraulic line clips and radiator brace rods, and disconnect wire from fuel gage sending unit. Disconnect hydraulic fluid lines at front junction and at firewall, remove line spacer clamps then remove the lines.

Shut off fuel and remove fuel line. On diesel models, remove leak-off line leading to fuel tank. On all models, disconnect wiring harness, air cleaner and coolant hoses; and detach heat indicator sending unit from cylinder head.

Fig. 24–Oil pressure relief valve (R) is located in timing gear cover as shown.

Fig. 26–With No. 1 piston at TDC on compression stroke and timing tool (TT) positioned as shown, camshaft gear timing mark (TM) will be directly under edge of timing tool.

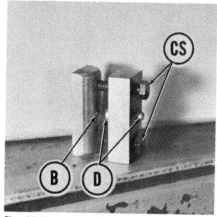

Fig. 27–View of tool (JD-255) used to stake balance shaft bushings. Half-round (left) part of tool contains staking ball (B) and is used inside bushing. Dowel (D) positively locates the tool. Cap screws (CS) should be tightened evenly.

Support rear of tractor on a rolling floor jack or from an engine sling. Attach front half of split stand to side frames or suitably support front unit. Disconnect hydraulic pump coupler. Remove the cap screws securing engine block to front support and side frames; then roll side frames, front support, front axle and associated parts away from engine.

Remove fan, fan belt, alternator and water pump. Remove crankshaft pulley using a suitable puller. Remove the oil pressure regulating plug, spring and valve as shown in Fig. 24. Drain and remove oil pan, then unbolt and remove timing gear cover.

Crankshaft front seal can be renewed at this time. To renew the seal, coat outside of seal with a suitable sealing compound and install seal from outside of cover with lip toward inside. Cover should be supported around seal area and seal should bottom in its bore.

On gasoline engines, the governor shaft front bushing is also located in timing gear cover. Renew the bushing if necessary.

Install timing gear cover by reversing the removal procedure, tightening retaining cap screws to a torque of 30 ft.-lbs.

CAMSHAFT
All Models

26. To remove the camshaft, first remove timing gear cover as outlined in paragraph 25. Remove vent tube, rocker arm cover, rocker arms assembly, push rods and fuel pump. On gasoline engines, remove distributor. Use wires with a 90° bend in end and push into push rod bore of tappet to hold tappet away from camshaft lobes. (Wood dowels of proper size, and spring clothes pins or rubber bands can also be used). Turn engine until thrust plate

retaining cap screws can be reached through holes in camshaft gear, then remove cap screws and pull camshaft and thrust plate from cylinder block.

Support camshaft gear, press camshaft from gear and remove Woodruff key. If tachometer drive worm in rear of camshaft must be renewed, thread exposed end and install a nut. Attach puller to nut to remove the drive.

The camshaft is carried in three unbushed bores in cylinder block. When checking camshaft, also inspect journal bores using the following data:

Camshaft journal OD .. 2.1997-2.2007
Camshaft bore ID 2.2042-2.2052
Diametral clearance ... 0.0035-0.0055
Maximum allowable
  clearance 0.007
Camshaft end play 0.0025-0.0085
Maximum allowable end
  play 0.015
Thrust plate thickness
  (new) 0.156-0.158

When installing new camshaft gear, be sure timing mark is toward front and support camshaft under front bearing journal. When installing new tachometer drive worm, support camshaft under rear journal, and press worm gear on shaft with slot in gear facing 180 degrees from camshaft drive gear keyway.

When installing camshaft in cylinder block, be sure timing mark is aligned as shown in Fig. 26. Tighten thrust plate cap screws to a torque of 35 ft.-lbs.

BALANCER SHAFTS
All Models

All models are equipped with two balancer shafts located below and parallel with crankshaft at each side of cylinder block. The right hand balancer shaft is driven by the lower idler gear and left hand balancer shaft by oil pump drive gear. Fig. 25 shows timing gear train with cover removed. Shafts rotate in opposite directions at twice crankshaft speed and are designed to dampen the vibration inherent in four cylinder engines.

27. To remove the balancer shafts, first remove timing gear cover as outlined in paragraph 25, then remove lower idler gear and oil pump gear. Check end play of balancer shafts. End play should be 0.002-0.008, if it exceeds 0.015 renew thrust plates while shafts are removed.

Identify balancer shafts as to right and left, unbolt and remove thrust plates and carefully withdraw balancer shafts from cylinder block. Specifications are as follows:

Shaft journal OD 1.4995-1.5005
Bushing ID 1.502-1.504
Shaft operating
  clearance 0.0015-0.0045

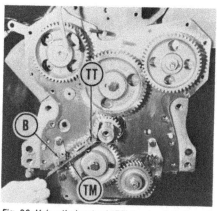

Fig. 28–Using timing tool (TT) to check timing of right balance shaft (B). Timing mark (TM) on shaft gear should align with tool when No. 1 piston is at TDC on either stroke.

Maximum allowable
  clearance 0.006
Shaft end play 0.002-0.008
Maximum allowable end
  play 0.015
Thrust plate thickness 0.117-0.119

Renew any parts which fail to meet specifications. The two front balancer shaft bushings for either shaft can be renewed with engine in tractor. If either rear bushing is to be renewed, remove engine, flywheel and flywheel housing to permit staking the bushing.

When installing bushings, use a piloted driver (JD-249 or equivalent) and install bushings from front so that front of bushing is flush with chamfer at front of bore and oil holes are aligned with holes in cylinder block. After installation, stake the bushings as follows:

Use John Deere Tool JD-255 shown in Fig. 27. Place half-round portion of tool in ID of bushing so that staking ball (B) is in round relief in bushing groove directly opposite to bushing oil hole. Turn square half of tool so the correct size dowel (D) fits in lower hole at bushing base. Install cap screws (CS) then recheck to be sure staking ball (B)

Fig. 29–Timing left balance shaft. Procedure is the same as that outlined in Fig. 28.

Fig. 30–Timing the injection pump drive gear. No. 1 piston must be at TDC on compression stroke.

is still in relief area. If it is not, remove half-round portion, turn it end-for-end and reinstall. Tighten cap screws evenly until half-round portion of tool is drawn tight against the bushing, to indent bushing into dowel hole and lock bushing in bore.

If new gears are being installed on balancer shafts, be sure timing mark is toward front and support shaft on both sides of front journal with tool JD-247 or equivalent. Press gear on shaft until flush at forward end, within 0.001. Position of gear controls shaft end play.

Reinstall balance shafts by reversing removal procedure. Before installing lower idler gear, set No. 1 piston on TDC and align balancer shaft gear timing marks using timing tool JD-254 or equivalent as shown in Figs. 28 and 29.

NOTE: When shafts are properly timed and No. 1 piston is at TDC, keyways in both balancer shaft gears should be pointing straight up. If timing marks are aligned and keyways are incorrectly positioned, balance shafts are installed in wrong bores and must be interchanged.

IDLER GEARS
All Models

28. Upper and lower idler gears (U & L—Fig. 25) are bushed and operate on stationary shafts attached to engine front plate with cap screws. Idler gear end play is controlled by thrust washers. Both idler gears are driven by the crankshaft timing gear. The upper idler drives camshaft gear and injection pump drive gear on diesel models; or camshaft and governor gear on gasoline models. The lower idler gear drives the right hand balancer shaft gear and oil pump drive gear on all models, and ventilator pump gear on models so equipped.

To remove either idler gear, remove oil pan and timing gear cover, remove retaining cap screw and pull gear and

thrust washer from shaft. Idler gear shaft can be removed after gear is off. Specifications are as follows:

Shaft OD 1.7495-1.7505
Bushing ID 1.7515-1.7535
Operating clearance 0.001-0.004
Maximum allowable
　clearance 0.006
End play 0.001-0.007
Maximum allowable end
　play 0.015

Reinstall by reversing the removal procedure. Be sure all gears are timed as shown in Figs. 26, 28, 29 and 30. Tighten shaft cap screws to a torque of 85 ft.-lbs.

TIMING GEARS
All Models

29. **CAMSHAFT GEAR.** The camshaft gear (G—Fig. 25) is keyed and pressed on shaft. When removing the gear, it is first recommended that shaft be removed as outlined in paragraph 26. Camshaft is correctly timed when timing mark on gear aligns with centerline of camshaft and crankshaft with No. 1 piston at TDC. Refer to Fig. 26 for recommended timing procedure.

30. **CRANKSHAFT GEAR.** Renewal of crankshaft gear requires removal of crankshaft as outlined in paragraph 38. Gear is keyed and pressed on crankshaft. Support crankshaft under first throw when installing new gear. Heat gear in oil for easier installation.

Crankshaft gear has no timing marks but keyway in crankshaft will be straight up when No. 1 piston is at TDC.

31. **INJECTION PUMP GEAR AND SHAFT.** To renew the injection pump drive gear on tractors with Model DB injection pump, first remove thrust plunger and spring from end of shaft. Wedge gears with a clean cloth, then back off the retaining nut until it is flush with end of shaft. Attach a puller to flat bottomed puller holes in gear web and pull gear from shaft, being careful not to pull shaft from injection pump.

To renew injection pump shaft and/or seals on Model DB pumps, remove thrust plunger and spring, then pull gear and shaft assembly from injection pump. Loosen gear retaining nut until nut is flush with end of shaft, then press shaft from gear. Note the dimple in tang at rear of shaft. This dimple must mate with a similar dimple in pump rotor when shaft is installed, to insure proper timing.

New seals are installed with lips opposed. (facing away from center) Use JD256 or no. 13371 drive shaft installation tool to install seals on shaft and shaft in pump. Coat seals and the area

between seals with Lubriplate prior to installation. New seals can also be installed after pump is removed as outlined in paragraph 57.

To renew drive gear on tractors equipped with Model C injection pump, remove gear mounting screw, screw retainer and the three gear retaining cap screws.

Reinstall by reversing the removal procedure. Make sure the correct timing mark is aligned as shown in Fig. 30.

NOTE: The same injection pump drive gear is used on some three cylinder diesel engines. Use the mark having a "4" stamped beside it when timing the gear.

32. **GOVERNOR GEAR.** Refer to paragraph 67 for information on the gasoline engine governor assembly. Governor gear has no timing marks and need not be timed.

TIMING GEAR BACKLASH
All Models

33. Use the following information when checking backlash of gears in timing gear train. Excessive gear backlash is corrected by renewing gears as necessary.

Crankshaft gear
　to upper idler 0.003-0.012
Camshaft gear to
　upper idler gear 0.003-0.014
Injection pump gear
　to upper idler 0.003-0.014
Crankshaft gear
　to lower idler 0.003-0.014
Lower idler
　to balance shaft 0.002-0.016
Lower idler to
　oil pump gear 0.002-0.015
Oil pump gear to
　balancer shaft 0.002-0.014
Upper idler gear
　to governor gear 0.002-0.013

ROD AND PISTON UNITS
All Models

34. Piston and connecting rod assemblies can be removed from above after removing cylinder head, oil pan and connecting rod caps. Secure cylinder sleeves in block using short cap screws and flat washers to prevent sleeves from moving when crankshaft is turned.

All pistons have word "FRONT" stamped on head of piston. Diesel engine connecting rods also have the word "FRONT" cast into web of connecting rod. Gasoline engine connecting rods have small raised "pip" marks. These marks should be in register and toward camshaft side of engine when rods are installed. Replacement rods are not numbered and should be stamped with correct cylinder number. When installing rod and piston units, lubricate rod cap screws and tighten to a torque of 40-45 ft.-lbs. on gasoline engines or 60-70 ft.-lbs. on diesel engines.

PISTONS, RINGS AND SLEEVES
All Models

35. All pistons are cam ground, forged aluminum alloy, and are fitted with three rings located above the piston pin. All pistons have the word "FRONT" stamped on piston head and are available in standard size only.

Piston rings are available in standard size only and are furnished with correct end gap. Renew pistons if side clearance of new rings exceed 0.008 in any groove. Top compression ring is chrome plated. Compression rings are marked "TOP" for correct installation.

The oil control ring for gasoline engines is not marked and can be installed either side up. Oil control ring for diesel is marked "TOP" and uses an expander which should be installed with plastic sleeve on expander under end gap of ring.

Cylinder block is fitted with renewable wet-type sleeves available in standard size only. Inside and outside diameters of gasoline and 2510 diesel engine sleeves are the same but 2510 diesel sleeves are longer because of the longer stroke of diesel engines. 2520 gasoline and diesel engines have the same 4.33 inch stroke, but the diesel has a larger bore. Lower ends of sleeves are sealed by a rubber ring and upper end by cylinder head gasket.

After cylinder head is removed and before turning crankshaft, secure sleeves from moving up in bores by using flat washers and short cap screws. If out-of-round or taper exceeds 0.005, renew sleeve, piston and rings. If sleeve is reusable, deglaze sleeve bore using a 45° cross-hatch pattern. Sleeve standout should be 0.000-0.004 above top face of cylinder block measured at outer edge (not inner step) as shown in Fig. 31. Lower packing must be removed when taking the measurement. If standout is excessive, check for foreign material in sleeve bore and counterbore. It may be necessary to select another sleeve. Specifications are as follows:

Sleeve Bore
   180 & 202 Engine ... 3.8576-3.8590
   219 Engine 4.0150-4.0164
Piston Skirt Diameter
   Gasoline 3.8543-3.8553
   Diesel 3.8548-3.8558
   219 Engine 4.0120-4.0140
Piston Skirt Clearance
   Gasoline 0.000-0.0034
   Diesel 0.001-0.0044

NOTE: Manufacturer indicates that zero clearance is apparent, rather than actual, as piston skirt will conform within narrow limits to cylinder size.

PISTON PINS AND BUSHINGS
All Models

36. The 1.1875-1.1879 full floating piston pins are retained in piston by snap rings. A pin bushing in upper end of connecting rod must be reamed after installation to provide a thumb press fit for piston pin. Pin to piston clearance is 0.0001-0.0008.

CONNECTING RODS AND BEARINGS
All Models

37. The steel-backed, aluminum lined bearings can be renewed from below after removing oil pan and rod bearing caps. Rod and cap are numbered and number should be transferred to replacement connecting rod. Gasoline engine connecting rods and caps have small raised "pip" marks which should be in register and face toward camshaft side of engine. Diesel engine connecting rods are marked "FRONT" and parting line of rod and cap are milled to prevent cap from being reversed.

Connecting rod bearings are available in standard size and undersizes of 0.002, 0.003, 0.010, 0.012, 0.020, 0.022, 0.030 and 0.032. Refer to the following specifications.

Diesel Engines
   Crankpin diameter 2.748-2.749
   Bearing Clearance
     2510 .... 0.0012-0.0042
     2520 .... 0.0016-0.0046
   Cap screw torque ...... 60-70 ft.-lbs.

Gasoline Engines
   Crankpin diameter .. 2.3085-2.3095
   Bearing clearance ... 0.0014-0.0044
   Cap screw torque ..... 40-45 ft.-lbs.

CRANKSHAFT AND BEARINGS
All Models

38. Crankshaft in gasoline engines is supported in three main bearings while diesel engines use five bearings. Main bearing inserts may be removed from below after removing oil pan, oil pump and main bearing caps. All caps are interchangeable except rear cap, and should be identified. Rear main bearing insert is flanged to control crankshaft end play which should be 0.002-0.008.

To remove the crankshaft, engine must first be removed from tractor as outlined in paragraph 15 and completely disassembled as follows: Remove oil pan, oil pump, cylinder head

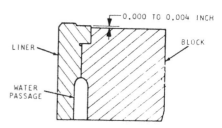

Fig. 31–Sleeve stand-out without packing should be 0.000-0.004 above cylinder block as shown.

and rod and piston units. Remove clutch, flywheel and flywheel housing. Remove timing gear cover, both idler gears, camshaft, injection pump drive gear & shaft, balancer gears & shafts and engine front plate. Remove main bearing caps and lift out the crankshaft. Specifications are as follows:
Main journal diameter . 3.1235-3.1245
Gasoline crankpin 2.3085-2.3095
Diesel crankpin 2.748-2.749
Recommended end play ...0.002-0.008
Maximum allowable end play .. 0.015
Main bearing diametral
   clearance 0.0016-0.0046
Maximum allowable
   clearance 0.006

Renew or regrind crankshaft if any journal is out-of-round more than 0.003 or tapered more than 0.001, or is otherwise damaged. Main bearing inserts are available in standard size and undersizes of 0.002, 0.003, 0.010, 0.012, 0.020, 0.022, 0.030 and 0.032.

CRANKSHAFT REAR OIL SEAL
All Models

39. The lip type crankshaft rear oil seal is contained in flywheel housing and seals on a wear ring pressed on flywheel mounting flange of crankshaft.

To renew the seal, first remove the clutch, flywheel and flywheel housing. If wear ring on crankshaft is damaged, spread the ring using a dull chisel on flat seal surface and withdraw the ring. Install new ring with rounded edge to rear, being careful not to cock the ring. Ring should start by hand and can be seated using a suitable driver such as JD-251. Install seal in flywheel housing working from rear with seal lip to front. When properly installed, seal should be flush with rear of housing bore. Use care to avoid folding or curling lip of seal under while installing housing.

FLYWHEEL
All Models

40. Flywheel is positioned by a locating dowel and secured by four cap screws. Two threaded holes in flywheel hub are provided for easier removal.

Flywheel ring gear is a shrink fit on front of flywheel and must be heated for installation. Install chamfered ends of ring gear teeth to front.

When installing flywheel, tighten retaining cap screws to a torque of 85 ft.-lbs.

FLYWHEEL HOUSING
All Models

41. The cast iron flywheel housing (engine adapter) is secured to rear face of engine block by eight small and one large cap screw. Flywheel housing contains the crankshaft rear oil seal and tachometer drive. The housing can be removed after removing clutch and flywheel.

Use a suitable seal protecter when installing flywheel housing. Tighten the small retaining cap screws to a torque of 35 ft.-lbs. and the large cap screw to 170 ft.-lbs.

OIL PUMP AND RELIEF VALVE
All Models

42. To remove the oil pump, first remove timing gear cover as outlined in paragraph 25. Turn crankshaft until No. 1 piston is at TDC and lock in place with timing pin. Remove oil pump drive gear retaining nut, attach a puller and remove drive gear; then unbolt and remove the oil pump.

To overhaul the removed pump, use Fig. 32 as a guide and proceed as follows: Remove gears (6) and drive shaft (4) from pump housing (3). Check to see that groove pin retaining pump gear (6) to shaft (4) is tight. Bearing OD of drive shaft should be 0.6295-0.6305 and shaft should have a normal operating clearance of 0.001-0.004 in bushing (2). Renew bushing and/or shaft if clearance is excessive. Idler gear shaft (5) can be pressed from housing if renewal is necessary. Radial clearance between ends of pump gear teeth and body bore should be 0.001-0.004 and end clearance of gears should be 0.001-0.006 when measured with a straight edge and feeler gage.

When installing the oil pump, tighten the drive gear nut and the four pump retaining cap screws to a torque of 35 ft.-lbs., then stake drive gear nut to shaft.

On early models, a relief valve (11) and spring (12) were located in pump outlet tube (10), and a pressure regulating valve entered engine oil gallery through timing gear cover as shown in Fig. 33. On late models, the relief valve in pump outlet tube has been eliminated. On all models, pressure is adjustable by adding or removing shims (3) in regulating valve plug (1). Regulated pressure should be 30-40 psi on early models or 50-60 psi on late models, with engine at operating temperature and running at a speed of 2500 rpm. The number of shim washers (3) must not exceed four. Regulating valve seat in front face of engine block can be renewed after removing timing gear cover. Use Tool JD-248 or equivalent and press the seat into block until outer recessed edge is flush with bottom of counterbore. Do not press on or otherwise damage the raised inner rim of valve seat.

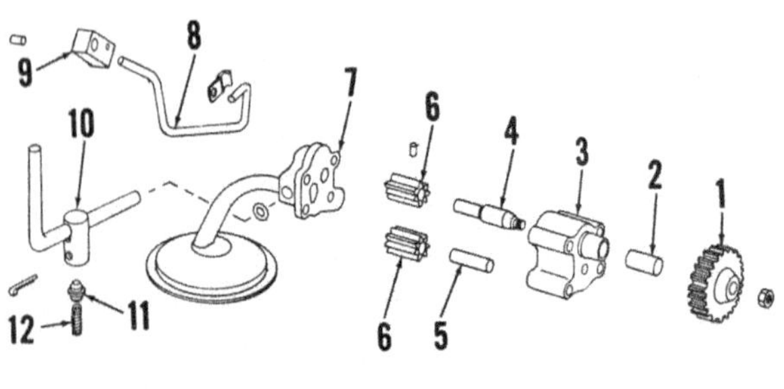

VENTILATOR PUMP
Models So Equipped

43. Some models may be equipped with an engine driven ventilator pump which provides forced ventilation to crankcase. The pump is driven by the timing gear lower idler. The air supply must first pass through the engine air cleaner and is therefore filtered of dust and dirt.

The pump impeller rotates at 1⅓ engine speed and revolves in a housing bore which is offset in relation to impeller hub and partially filled with engine oil. In operation, the oil is picked up and forced to outer wall of casing by centrifugal force where it rotates with the impeller. Air is drawn into the ventilator pump from the inlet tube and is trapped between the impeller blades and the rotating stream of oil. As the impeller turns, the air and a small amount of oil is discharged into the crankcase where it scavenges the gases and is discharged through the outlet tube in rocker arm cover. Approximately 35 cubic feet of air per hour is circulated by the ventilator pump.

To remove the ventilator pump, first drain and remove the oil pan. Remove oil pump auxiliary oil lines, then unbolt and remove oil pump cover with

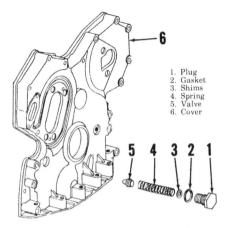

1. Plug
2. Gasket
3. Shims
4. Spring
5. Valve
6. Cover

Fig. 33–Timing gear cover showing relief valve exploded.

intake and pressure lines attached. Disconnect air intake line from ventilator pump. Remove ventilator pump drive gear nut and gear, using a suitable puller. Remove the pump cover retaining cap screws, lift off the cover; then remove ventilator pump from front plate.

Withdraw impeller and shaft assembly from housing and examine shaft and bore in housing for wear. Shaft and impeller are available only as an assembly, all other parts are available individually.

When reassembling, tighten oil pump and ventilator pump cover retaining cap screws to a torque of 35 ft.-lbs. Tighten ventilator pump drive gear retaining nut to 35 ft.-lbs. and stake in place. Complete the assembly by reversing the disassembly procedure.

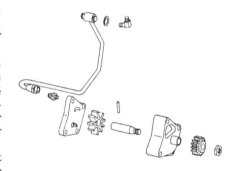

Fig. 34–Exploded view of crankcase ventilator pump showing component parts.

CARBURETOR

All models are equipped with a Marvel-Schebler TSX type carburetor. Late units use an electrically actuated fuel shut-off solenoid.

Fig. 32–Exploded view of engine oil pump. Outlet relief valve (11 & 12) was not used on late tractors and tube (10) is plain.

1. Gear		7. Cover	
2. Bushing		8. Tube	
3. Body		9. Block	
4. Shaft		10. Tube	
5. Shaft		11. Valve	
6. Gears		12. Spring	

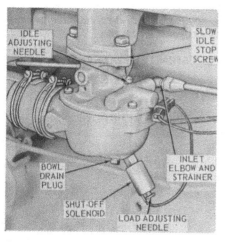

Fig. 35–View of carburetor showing points of adjustments.

ADJUSTMENT
All Models

44. Initial settings are 1½ turns open for idle mixture adjustment screw and 2⅛ turns open for load adjustment screw. Adjustment should be re-set for best performance with engine at operating temperature. Refer to Fig. 35 for installed view of late carburetor showing points of adjustment.

OVERHAUL
All Models

45. Recommended float setting is ¼-inch measured from nearest edge of float to gasket surface with throttle body inverted. Both halves of float must be adjusted alike and float must not rub or bind on castings when carburetor is reassembled. If adjustment is required, carefully bend float arms using a bending tool or needle nose pliers.

Carburetor must be disassembled for float adjustment. First clean outside of carburetor with a suitable solvent and remove main and idle adjustment needles. Remove the four screws which retain throttle body to fuel bowl and lift off throttle body, gasket, float and venturi as a unit. Remove float shaft and float, gasket, venturi and inlet needle valve. Withdraw venturi from gasket and discard the gasket. Remove fuel shut-off solenoid if carburetor is so equipped. Remove inlet needle valve seat, idle jet, power jet and main nozzle. Discard the gaskets from nozzle and needle valve seat.

Remove throttle and choke valves, shafts and packing. Bushings are serviced for throttle shaft. Clean all parts except shut-off solenoid in a suitable carburetor cleaner and rinse in clean mineral solvent. Discard gaskets and packing. Reassemble by reversing the disassembly procedure, using Fig. 36 as a guide, and adjust the assembled unit as outlined in paragraph 44.

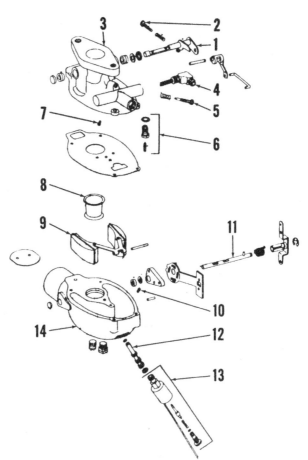

Fig. 36–Exploded view of Marvel-Schebler carburetor showing main components.

1. Throttle shaft
2. Idle speed screw
3. Throttle body
4. Inlet elbow
5. Idle mixture needle
6. Inlet needle & seat
7. Idle jet
8. Venturi
9. Float
10. Power jet
11. Choke shaft
12. Main nozzle
13. Shut-off solenoid
14. Fuel bowl

DIESEL FUEL SYSTEM

FILTERS AND BLEEDING
All Models

46. **FILTERS.** The renewable element type fuel filter attaches to right side of cylinder head. Renew filter element every 500 hours of operation or whenever system is contaminated. A glass sediment bowl and strainer is located on top of primary fuel pump and another sediment bowl underneath the fuel filter.

47. **BLEEDING.** Whenever fuel system has been run dry or a line has been disconnected, system must be air bled as follows:

Be sure there is sufficient fuel in tank and that shut-off valve is open. Loosen bleed screw on top of filter and actuate primer lever on fuel transfer pump until a solid, bubblefree stream of fuel emerges from bleed screw. Pump primer lever several extra strokes to prime the injection pump which is self-bleeding. Crank the engine. If engine will not start, or misses, loosen pressure line connections at injector assem-

blies and, with throttle open, turn engine over with starter until fuel is being pumped from injector lines. Tighten the connections and start engine. If engine still will not start, repeat the procedure until system is free of trapped air.

NOTE: If primer lever will not pump fuel and no resistance is felt, rocker arm may be setting on high point of cam lobe. Turn engine with starter to reposition camshaft, then try again.

FUEL LIFT PUMP
All Models

48. The fuel lift pump is shown exploded in Fig. 37. Most parts are available individually and overhaul kits are provided. Fuel pump can be tested on tractor by disconnecting outlet line and turning engine over with starter. NOTE: Throttle lever should be in "STOP" position to keep engine from starting. If fuel spurts from the line with force, fuel pump is probably satisfactory.

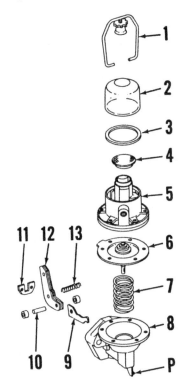

Fig. 37–Exploded view of diaphragm type fuel pump showing component parts.

1. Ball	8. Body
2. Fuel bowl	9. Link
3. Gasket	10. Pivot
4. Screen	11. Retainer
5. Cover (with valves)	12. Lever
6. Diaphragm	13. Spring
7. Spring	P. Primer lever

If fuel flow is unsatisfactory, check to be sure that primer lever (P) is in operating position and that lines and connections look good before removing fuel pump.

To disassemble the removed fuel pump, first remove bail, fuel bowl, gasket and screen. Mark body (8) and cover (5) to save time in reassembly, remove body screws and lift off the cover. Remove lever spring (13) and retainer (11). Push down on diaphragm (6) to overcome spring pressure, then unhook and withdraw rocker arm (12), link (9) and associated parts as an assembly.

Examine diaphragm for cracks or deterioration and slot in pull rod for wear. Examine the valves in pump cover (5). (Valves are not available separately).

Assemble by reversing the disassembly procedure, being sure rocker arm link (9) is hooked in slot in diaphragm pull rod. Align the previously scribed assembly marks on cover (5) and body (8) and install the screws loosely. Actuate rocker arm vigorously to be sure diaphragm has enough slack, then tighten the screws evenly. Install and bleed system as outlined in paragraph 47.

INJECTOR NOZZLES

WARNING: Fuel leaves the injectors with sufficient force to penetrate the skin. When testing, keep clear of the spray.

All Models

49. TESTING AND LOCATING A FAULTY NOZZLE. If one engine cylinder is misfiring, it is reasonable to suspect a faulty injector. Generally, the faulty unit can be located by running engine at a slow idle speed and loosening, one at a time, each high pressure line at injector connection. As in checking spark plugs in a spark ignition engine, the faulty unit is the one that least affects the running of the engine when its line is loosened.

Remove the suspected injector unit from engine as outlined in paragraph 50. If a suitable nozzle tester is available, test the injector as outlined in paragraphs 51 through 55. If a tester is not available, reconnect pressure line to injector and with nozzle tip directed where it will do no harm, crank engine and observe spray pattern. The conical spray pattern should be symmetrical and finely atomized. If pattern is ragged or wet, or if nozzle dribbles, the nozzle valve is not seating properly and unit should be overhauled or renewed.

50. REMOVE AND REINSTALL. To remove an injector, first remove hood and wash injector, lines and surrounding area with clean diesel fuel or solvent. Use hose clamp pliers to expand clamp and pull leak-off boot from injector. Disconnect high pressure line, then cap all openings. Remove cap screw from nozzle clamp and spacer. Pull injector from cylinder head.

NOTE: Unless the carbon stop seal has failed causing injector to stick, the injectors can be easily removed by hand. If injectors cannot be pulled by hand, use John Deere Nozzle Puller JDE-38 and be sure to pull injector straight out of bore. DO NOT attempt to pry injector from cylinder head or damage to injector could result.

When installing injector, be sure nozzle bore and seal washer seat are clean and free of carbon or other foreign material. Install new seal washer and carbon seal on injector and insert injector using a slight twisting motion. Install and align locating clamp then install hold-down clamp and spacer. Tighten retaining cap screw to a torque of 20 ft.-lbs.

51. TESTING. A complete job of nozzle testing and adjusting requires the use of an approved nozzle tester. Only clean, approved testing oil should be used in the tester tank. The nozzle should be tested for spray pattern, opening pressure, seat leakage and back leakage. Injector should produce a

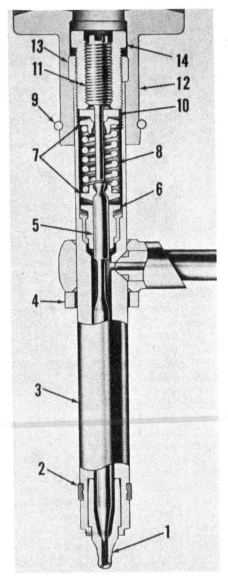

Fig. 38–Cross sectional view of Roosa-Master injector. Nozzle tip (1) and valve guide (6) are parts of finished body and are not serviced separately.

1. Nozzle tip	8. Pressure spring
2. Carbon seal	9. Boot clamp
3. Nozzle body	10. Ball washer
4. Seal washer	11. Lift adjusting screw
5. Nozzle valve	12. Boot
6. Valve guide	13. Lock nut
7. Spring seat	14. Pressure adjusting screw

distinct audible chatter when tested, and cut off quickly at end of injection with a minimum of seat leakage.

NOTE: When checking spray pattern, turn nozzle about 30 degrees from vertical position. Spray is emitted from nozzle tip at an angle to centerline of nozzle body. Unless injector is angled, the spray may not be completely contained by the glass test jar.

52. SPRAY PATTERN. Attach injector to tester and operate tester at approximately 60 strokes per minute. A finely atomized spray should emerge at each of the four nozzle holes and a distinct chatter should be heard as tester is operated. If spray is not symmetrical or not finely atomized, or if injector does not chatter, overhaul the injector as outlined in paragraph 56.

53. **OPENING PRESSURE.** There should not be more than 100 psi difference between the four nozzles in any engine except when intermixing new and used nozzle units.

To check nozzle opening pressure, attach injector to tester and pump tester several times to free the nozzle valve. Now observe tester gage pressure at which nozzle operates.

The opening pressures are as follows:

Part No. AT18064 (2510 before 092279)
Opening pressure
  New nozzle spring . . . . 2750-2850 psi
  Used nozzle spring . . . 2550-2650 psi
Nozzle valve lift adjustment, ¾ turn from bottomed.

AT30299 (2510 after 092280 and 2520 before 177720)
Opening pressure
  New nozzle spring . . . . 2950-3050 psi
  Used nozzle spring . . . 2750-2850 psi
Nozzle valve lift adjustment, ½ turn from bottomed.

AR49877 (2520 after 177721)
Opening pressure
  New nozzle spring . . . . 3150-3250 psi
  Used nozzle spring . . . 2950-3050 psi
Nozzle valve lift adjustment, ½ turn from bottomed.

If opening pressure is not correct but nozzle will pass other tests, refer to Fig. 38 and proceed as follows: Loosen locknut (13), then hold pressure adjusting screw (14) from turning and back out valve lift adjusting screw (11) at least one full turn. Adjust the opening pressure as required by turning pressure adjusting screw (14).

When the nozzle opening pressure has been correctly set, hold pressure adjusting screw (14) from turning and gently turn valve lift adjusting screw (11) clockwise until it bottoms, then back it out specified amount to provide the correct valve lift. Hold pressure adjusting screw and tighten locknut to a torque of 110-115 inch-pounds for early type adjusting screws, or 70-75 inch-pounds for later type adjusting screws as shown in Fig. 39.

NOTE: A positive check can be made to be sure lift adjusting screw is bottomed by actuating tester lever while screw is bottomed. If valve does not open at a pressure 250 psi above opening pressure, lift adjusting screw is bottomed.

54. **SEAT LEAKAGE.** To check nozzle seat leakage, raise pressure to approximately 2400 psi and hold the pressure for ten seconds. Check nozzle

tip. A slight dampness is permissible, but if a drop forms on tip, renew the injector or disassemble and clean as outlined in paragraph 56.

55. **BACK LEAKAGE.** Turn injector on tester until tip is slightly above horizontal. Raise and hold pressure at 1500 psi and observe leakage from return (top) end of injector. After the first drop falls, back leakage should be 5-8 drops every thirty seconds. If back leakage is excessive, renew the injector.

NOTE: If the nozzle performs properly in all tests, no further service is necessary and it can be installed in the engine.

56. **OVERHAUL.** First clean outside of injector thoroughly. Place nozzle in a holding fixture and clamp fixture in a vise. Loosen locknut (13—Fig. 38) and back out pressure adjusting screw (14). Invert the nozzle body and allow ball washer (10), spring seats (7) and spring (8) to fall from nozzle body into your hand. Catch nozzle valve (5) by its stem as it slides from body. If nozzle valve will not slide from body, use the special retractor or discard the injector assembly.

Nozzle valve and body are a matched set and should never be intermixed. Keep parts for each injector separate and immerse in clean diesel fuel in a compartmented pan as injector is disassembled.

Clean all parts thoroughly in clean diesel fuel using a brass wire brush and lint-free wiping towels. Hard carbon or varnish can be loosened with a suitable non-corrosive solvent.

Clean the spray tip orifices first with an 0.008 cleaning needle held in a pin vise; then with a 0.012 needle. Clean the valve seat using a Valve Tip Scraper and light pressure while rotating scraper. Use a Sac Hole Drill to remove carbon from inside of tip.

Piston area of valve can be lightly polished by hand, if necessary, using Roosa Master No. 16489 lapping compound. Use the valve retractor to turn valve. Move valve in and out slightly while rotating, but do not apply down pressure while valve tip is in contact with seat.

Valve and seat are ground to a slight interference angle. Seating areas may be cleaned up, if necessary, using a small amount of 16489 lapping compound, very light pressure and no more than 3 to 5 turns of valve on seat. Thoroughly flush all compound from valve body after polishing.

When assembling the nozzle, back lift adjusting screw (11) several turns out of pressure adjusting screw (14), and reverse disassembly procedure using Fig. 38 as a guide.

Adjust the opening pressure and

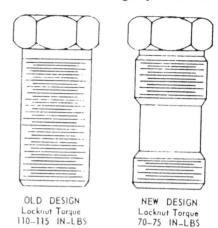

OLD DESIGN
Locknut Torque
110-115 IN-LBS

NEW DESIGN
Locknut Torque
70-75 IN-LBS

Fig. 39–Before tightening lock nut on injector pressure adjusting screw, determine which type is used and tighten to the recommended torque.

valve lift as outlined in paragraph 53, after valve is assembled.

INJECTION PUMP

All diesel tractors before Serial No. 014297 were equipped with a Roosa-Master Model DBG injection pump. Later models use Roosa-Master Model CDC, CBC or JDB pump. All pumps are flange mounted on left side of engine front plate and driven by upper idler gear of timing gear train. JDB or DBG pumps are not interchangeable with C model and procedures for removal, installation and adjustment differ. Refer to the appropriate following paragraphs for procedures.

Model DBG and JDB Pump

57. **REMOVE AND REINSTALL.** To remove the injection pump, first close fuel shut-off valve on fuel tank, then clean injection pump, line connections and surrounding area with clean diesel fuel. Turn engine crankshaft until No. 1 piston is at TDC on compression stroke.

NOTE: Pump can be removed and reinstalled without regard to crankshaft timing position, however, positioning crankshaft at TDC is recommended so timing can be properly checked and/or adjusted when pump is reinstalled. If timing is not to be checked, scribe timing marks on injection pump mounting flange and engine front plate which can be aligned when pump is reinstalled.

Disconnect fuel inlet line, fuel return line and throttle link rod from injection pump. Disconnect pressure lines from injectors and pump and remove the lines. Remove pump mounting stud nuts and pull pump straight to rear off pump shaft.

To install the pump without changing the timing, refer to Fig. 40. Punch mark on tang of exposed end of pump drive shaft must align with sim-

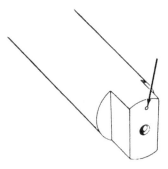

Fig. 40—End view of pump drive shaft. Arrow points to timing mark on tang.

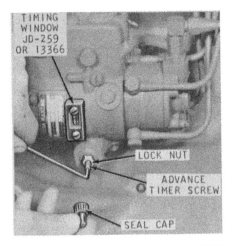

Fig. 42—Adjusting advance timing on Model DBG and JDB injection pump. Refer to Fig. 41 for timing positions.

ilar mark on pump rotor. Align marks and, using a seal compressor, carefully install the injection pump. Make sure scribe lines on pump flange and engine front plate are aligned, then tighten pump mounting studs securely.

To install and time the injection pump, recheck to be sure that No. 1 piston is at TDC on compression stroke. Remove timing hole cover from side of injection pump and turn pump rotor, if necessary, until timing scribe line on governor weight retainer aligns with scribe line on pump cam. Install the pump using a seal compressor. Recheck alignment of scribe lines as stud nuts are tightened.

Reinstall timing hole cover and complete pump installation by reversing the removal procedure. Bleed pump and lines as outlined in paragraph 47 and adjust pump linkage, if necessary as outlined in paragraph 60.

58. **TIMING.** To check injection pump timing without removing injection pump, first turn engine until No. 1 piston is starting compression stroke, then remove timing pin and cover, and reinsert timing pin into threaded hole in flywheel housing. Continue to turn crankshaft until timing pin slides into timing hole in flywheel.

Shut off fuel and remove timing hold cover from injection pump. The timing lines on governor weight retainer and cam should be in perfect alignment. If

they are not, loosen pump mounting stud nuts and move pump housing until alignment is obtained. Hold pump in this position and tighten mounting stud nuts securely.

59. **ADVANCE TIMING.** The injection pump is provided with automatic speed advance which is factory set and will not normally need to be checked or reset. Minor adjustments can, however, be made without removal or disassembly of the pump. To check the advance mechanism, proceed as follows:

Shut off fuel, remove pump timing hole cover and install timing window as shown in Fig. 42. Shift the window

when installing, until one of the scribe lines on window is positioned as nearly as possible over scribe line on pump cam ring. Refer to Fig. 41. Turn on fuel and bleed fuel system, then start and run engine. Maximum advance should be 6° (three scribe lines) for some early model DBG pumps or 8° (four scribe lines) for late DBG or all JDB pumps.

NOTE: Model DBG injection pumps having two different advance curves have been used. Early models (before Pump Serial Number 700034) had a 6° maximum advance unless later modified. Later pumps (or modified early pumps) have a maximum advance of 8 degrees.

If maximum advance is not correct when tested at 2400 engine rpm, renew or overhaul the pump. If maximum advance was correct, check the intermediate advance at the engine speed and advance setting indicated in the following table. To adjust the intermediate advance, remove seal cap as shown in Fig. 42, loosen locknut and, with engine running at specified speed, turn adjusting screw until intermediate advance setting is correct.

60. **LINKAGE AND SPEED ADJUSTMENT (DBG and JDB models.)** To adjust the throttle linkage, first start and warm engine. Refer to Fig. 43 and Fig. 44. Disconnect control rod (3) from injection pump and move pump throttle arm to high idle

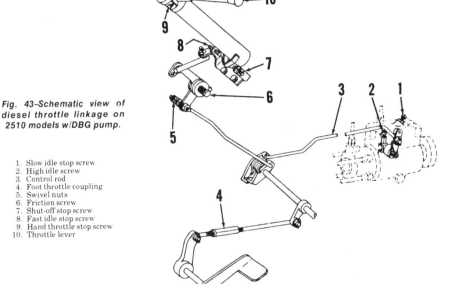

Fig. 43—Schematic view of diesel throttle linkage on 2510 models w/DBG pump.

1. Slow idle stop screw
2. High idle screw
3. Control rod
4. Foot throttle coupling
5. Swivel nuts
6. Friction screw
7. Shut-off stop screw
8. Fast idle stop screw
9. Hand throttle stop screw
10. Throttle lever

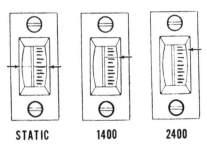

STATIC   **1400**   **2400**

Fig. 41—Injection pump timing window installed showing advance timing positions. Refer to paragraph 59.

Pump Model	Engine RPM	Intermediate Advance
Early DBG	1400 rpm	3° (1½ marks)
Modified DBG	1400 rpm	5° (2½ marks)
Late DBG	1400 rpm	5° (2½ marks)
All JDB	1900 rpm	6° (3 marks)

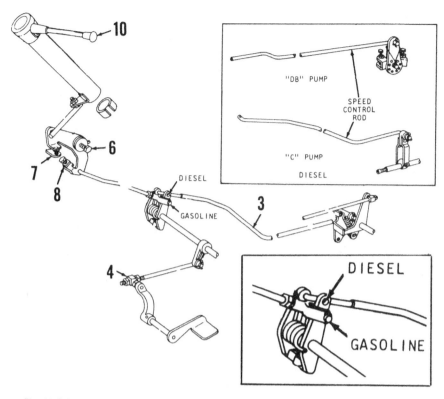

Fig. 44—Schematic view of 2520 throttle linkage. Note choice of diesel or gasoline settings.

3. Control rod	6. Friction screw	8. Fast idle stop screw
4. Foot throttle coupling	7. Slow idle stop screw	10. Throttle lever

position. Check engine high idle speed which should be 2650 rpm. If adjustment is required, break seal and readjust high idle screw (2—Fig. 43) as required. Move pump throttle arm to slow idle.

Throttle arm also controls shut-off linkage on 2510 models. A small amount of free travel exists between slow idle setting and shut-off position.

Slow idle engine speed of 800 rpm can be adjusted if necessary, by turning slow idle stop screw (1—Fig. 43) on governor housing cover. Reconnect control rod (3) after checking settings.

Move hand throttle lever (10) up to first stop (without pulling out on knob) and adjust swivel nuts (5—Fig. 43), or slow idle stop screw (7—Fig. 44) if necessary, until engine speed is 800 rpm.

Pull hand throttle down to first stop (without pulling out on knob) and check rated setting which should be 2330 rpm (2100 rpm load speed). If setting is incorrect, remove the cover and loosen hand throttle stop screw (9), reset the stop and tighten screw.

Pull out on throttle knob and move hand throttle down carefully until pump throttle arm contacts high idle screw (2—Fig. 43) for 2510, and engine speed is 2650 rpm; then readjust fast idle stop screw (8) if necessary until it contacts welded stop on throttle tube.

Adjust friction screw (6) as required so that lever will maintain any setting

without creeping, but can be easily moved (about 10-15 lbs. pull at throttle knob).

Adjust foot throttle coupling (4), to obtain high idle engine speed of 2650 rpm as pedal contacts platform.

Model CDC and CBC Pump

61. **REMOVE AND REINSTALL.** To remove the injection pump, first close shut-off valve on fuel tank, then clean injection pump, line connections and surrounding area with clean diesel fuel.

Turn engine crankshaft until No. 1 piston is at TDC on compression stroke. Drain radiator and remove lower radiator hose, then remove access plate from front of timing gear cover. Refer to Fig. 45. Remove pump gear mounting screw and the three gear retaining screws. Disconnect throttle rod, shut-off solenoid wiring lead, fuel supply line and fuel return line from injection pump. Remove injector lines and cap or plug all fuel openings. Support the pump and remove retaining stud nuts and washers, then pull injection pump from engine front plate. Pump drive gear will be retained by timing gear cover.

When reinstalling pump, be sure crankshaft is positioned with No. 1 piston at TDC. Remove injection pump timing pin (Fig. 46) from drive housing, reverse pin and reinsert it in threaded hole, pointed end down. Turn

pump drive shaft until timing pin drops (about 1/16-inch) into timing hole in pump drive shaft, then install pump on engine leaving stud nuts loose. Install pump drive gear, tightening retaining screws to a torque of 15 ft.-lbs. Rock pump slightly on mounting studs until timing pin is indexed in timing hole, then tighten stud nuts securely. Complete the assembly by reversing the removal procedure.

Bleed fuel system as outlined in paragraph 47 and, if necessary, check and/or adjust throttle linkage as in paragraph 64.

62. **TIMING.** To check the timing (static) without removing injection pump, first remove engine timing pin from flywheel housing and pump timing pin from pump housing. Turn crankshaft until No. 1 piston is at TDC on compression stroke and BOTH timing pins should drop into their respective timing holes. Adjust by loosening the injection pump mounting stud nuts and shifting the pump on engine front plate.

63. **ADVANCE TIMING.** The injection pump is provided with an automatic speed advance which is factory set and will not normally need to be checked or reset. Minor adjustments can, however, be made without removal or disassembly of pump, proceed as follows:

Shut off fuel and remove advance cam hole plug (Fig. 46); then install No. 17180 timing window as shown in Fig. 47. The timing window contains a series of concentric circles designed so that all lines are two pump degrees apart. Sight through the window and align the pump cam pin with any of the lines, then observe advance movement of pin as gaged by the lines.

Five different Model "C" pumps have

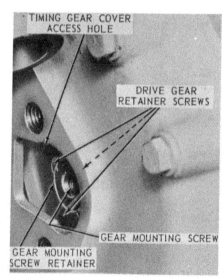

Fig. 45—View of timing gear cover with pump cover removed to prepare for removal of Model CBC and CDC injection pump.

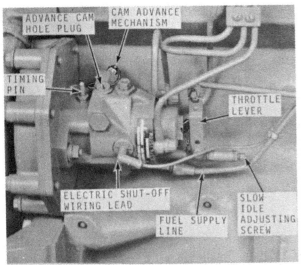

Fig. 46–Installed view of Roosa-Master Model CBC and CDC injection pump.

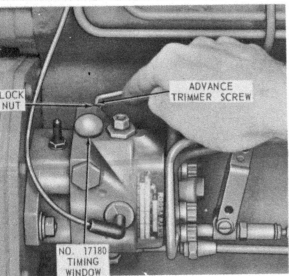

Fig. 47–Adjusting the advance timing on Model CBC and CDC pump. Refer to paragraph 63 for advance settings.

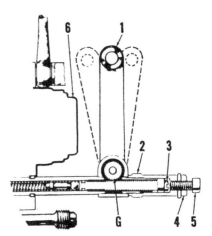

Fig. 48–Schematic view of Model CDC throttle arm showing points of adjustment.

1. Throttle arm
2. Control cap
3. Adjusting screw
4. Jam nut
5. Stop screw
6. Pump body
G. Throttle gear

On CBC pump, to make fast idle adjustment, remove throttle control cap (Fig. 49) and turn fast idle adjusting nut out to increase speed or in to decrease speed to 2650 rpm. Install throttle cap and tighten.

On CDC and CBC pumps, to make the slow idle adjustment (Figs. 48 and 49), move the throttle lever forward and adjust slow idle screw to 800 rpm. Turn screw in to increase speed, or out to decrease speed.

Reconnect control rod to throttle arm. On Model 2510, move hand throttle lever (1-Fig. 50) to first idle stop without pulling out on throttle knob. Adjust swivel nuts (6) if necessary until injection pump throttle arm clears pump body by about ⅛ inch.

Be sure correct holes (5 & 7) in bellcrank are used. The other holes are for gasoline tractors. If used with diesel linkage, too little movement of hand throttle lever will result causing difficulty in selecting intermediate speeds and making stop adjustments impossible.

Model CB Injection Pump

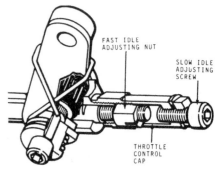

Model CB Pump Idle Adjusting Points

Fig. 49–Schematic view of CBC pump idle adjusting points (Model 2520).

been used and the advance curve differs slightly between each of the pumps. Pump model numbers and advance data are as shown in table.

Pump models CDC 431-6DG, 431-9DG, and 431-14DG were used on 2510 models, and CDC 431-21DG and CBC 431-16AL were used on 2520 models.

If intermediate advance is not as indicated, remove cap seal and loosen locknut on advance trimmer screw, then adjust the intermediate advance as shown in Fig. 47. Check maximum advance which should be as stated. If it is not, pump must be overhauled or renewed.

64. LINKAGE AND SPEED ADJUSTMENT. CBC and CDC Models. First start and warm engine. Disconnect speed control rod from injection pump throttle arm (1-Fig. 48). Check engine high idle speed which should be 2650 rpm. If adjustment is required, proceed as follows:

On the CDC pump, remove slow idle stop screw (5). NOTE: Never remove the throttle control cap (2) from CDC pump sleeve while making any adjustments. Insert small bladed screwdriver and turn fast idle adjusting screw (3) to obtain 2650 rpm. Turn screw in to increase, or out to decrease speed. Reinstall slow idle screw.

Pump Models CDC 431-	6 DG	9 DG	14 DG	21 DG
Maximum advance (@ 2400 RPM)	6°	8°	8°	8° (@ 2250 RPM)
Intermediate speed @ (no load RPM)	1800 4°	1800 6°	1600 5°	1600 5°
CBC 431-16AL Maximum advance (@ 2400)		8°	(@ 1900 no load) 6°	

With engine running, pull hand throttle down to first stop (without pulling out on knob) and check rated speed setting which should be 2330 rpm (2100 rpm rated load speed). If setting is incorrect, remove the cover, loosen hand throttle stop screw (2), reset stop and retighten screw. Pull out on hand throttle knob and pull lever down until welded stop contacts screw (4), and adjust maximum high idle speed to 2650 rpm.

On 2520 models, the above procedure can be followed using Fig. 44 as a guide for linkage adjusting points.

65. INJECTION PUMP SHUT-OFF. The Model C injection pump is equipped with a fuel solenoid which must be energized to provide fuel flow. The shut-off valve is spring loaded in the closed position and is opened when key switch is turned to "ON" or "START".

If tractor will not start, turn switch to "ON" position and check continuity of solenoid lead.

GASOLINE

ENGINE

GOVERNOR

The centrifugal, flyweight type gasoline engine governor is mounted on left side of engine front plate and driven from the timing gear upper idler. The forward end of governor shaft is supported in a bushing in engine timing gear cover.

LINKAGE ADJUSTMENT
All Models

66. Refer to Fig. 50 for an exploded view of governor and throttle linkage and to Fig. 51 for an exploded view of governor assembly. To adjust the linkage, proceed as follows:

Disconnect carburetor rod (5—Fig. 51) from throttle arm. Hold both the governor arm and throttle arm in wide open position and adjust length of rod (5) until it will just enter hole in throttle arm. Shorten rod one turn and reconnect. Start and warm the engine. Move hand throttle lever up to first idle stop (without pulling out on knob) and adjust the two locknuts on swivel (6—Fig. 50, or 7—Fig. 44) to obtain an engine idle speed of 800 rpm.

Refer to Fig. 50 for 2510 models and move hand lever down to high speed stop without pulling out on lever knob,

then check idle speed which should be 2300 rpm. If adjustment is required, remove cover at (2), loosen clamp screw and reposition high idle stop to obtain the desired speed. Pull out on knob (1) and pull hand throttle down beyond rated speed stop as far as it will go and adjust fast idle stop screw (4) to obtain a fast idle speed of 2680 rpm. Move hand throttle up and adjust linkage on foot throttle so that it contacts platform at an engine high idle speed of 2680 rpm.

Refer to Fig. 44 to make above adjustments on 2520 models. Make sure that proper hole in bell crank is used for gasoline linkage.

On models without shut-off solenoid on carburetor, shut-off set screw (3) should be adjusted to obtain an engine shut-off speed of 425 rpm when hand throttle is moved to slow idle position with hand knob (1) pulled. Adjust friction bolt (8) so hand throttle will maintain any setting without creep, but will still move easily.

GOVERNOR
All Models

67. REMOVE AND REINSTALL. To remove the governor assembly, first loosen clamps on inlet hose to carburetor, disconnect the hose and move air cleaner inlet pipe down out of the way. Disconnect clevis clip from speed change lever. Remove the two retaining cap screws then lift off governor housing (1—Fig. 51) and associated parts. Weight shaft (15) and thrust sleeve (13) will usually remain with engine, but can be withdrawn.

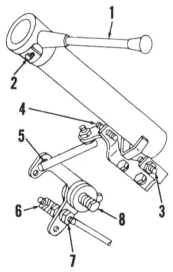

Fig. 50—Schematic view of throttle linkage used on 2510 models with CDC injection pump.

1. Throttle lever	5. Bellcrank attaching hole
2. Hand throttle stop screw	6. Swivel nuts
3. Shut-off stop screw	7. Bellcrank attaching hole
4. Fast idle stop screw	8. Friction screw

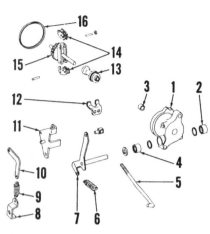

Fig. 51—Exploded view of gasoline engine governor and associated parts.

1. Housing	9. Spring
2. Bearing	10. Counterbalance arm
3. Bushing	11. Speed change lever
4. Bearing	12. Fork
5. Throttle rod	13. Sleeve
6. Spring	14. Weights
7. Lever	15. Shaft and gear
8. Bracket	16. Seal

When installing the governor assembly, position weight & gear assembly in tractor, then install housing and associated parts over weight unit. Tighten the retaining cap screws to a torque of 20-25 ft.-lbs., install the clevis clip and adjust the governor linkage as outlined in paragraph 66.

68. OVERHAUL. Governor weights (14—Fig. 51) are a free fit on weight pins, but weight pins should fit tight in carrier. Examine rounded ends of weight feet which contact sleeve (13) for flat spots, and examine surface of sleeve for wear or ridging. Weights are available individually, but should be renewed in pairs. Bushing (3) should be threaded for pulling and should be bottomed when installing. Press on closed end of bearing (2) or lettered end of bearing (4) when installing. Refer to paragraph 67 for governor installation procedure and to paragraph 66 for linkage adjustment.

COOLING

SYSTEM

RADIATOR
All Models

69. To remove the radiator, drain cooling system and remove side shields, grille side screens, cowl and hood. Remove upper and lower radiator hoses and front air intake hose.

Remove the sheet metal screws securing fan shroud to radiator and move shroud back against front of engine. Remove radiator brace rod and mounting cap screws. On power shift models, disconnect oil cooler from radiator frame. On all models, slide radiator to left and lift from tractor. Install by reversing the removal procedure.

FAN AND WATER PUMP
All Models

70. **REMOVE AND REINSTALL.** To remove fan and/or water pump, drain cooling system and remove side shields, grille side screens, cowl and hood. Remove alternator adjusting screw. Disengage fan belt from alternator pulley and allow alternator to tip to outside, then remove fan belt. Remove the cap screws securing fan blades to hub and lay blades forward against radiator. Disconnect lower radiator hose and bypass hose from water pump, remove the attaching cap screws and withdraw water pump assembly from left side of tractor.

Install by reversing the removal procedure. Adjust fan belt to ¾-inch deflection midway between alternator and crankshaft pulleys.

71. **OVERHAUL.** To disassemble the pump, support fan pulley in belt groove and, using a suitable mandrel, press shaft from pulley hub. Suitably support water pump housing and working from front end, press shaft & bearing assembly, seal and impeller as a unit from housing.

Bearing outer race is a tight press fit in housing bore and is not otherwise secured. It is important therefore, that reasonable precautions be taken during assembly to prevent bearing

movement during installation of impeller or pulley hub. Refer to Fig. 52 and assemble as follows:

Coat outer edge of seal (5) with sealant and install in housing (8), using a socket or other driver which contacts only the outer flange of seal. Lubricate lip of seal, then insert long end of bearing (6) through front of housing bore. Use Tool No. JD-262 or other similar tool which contacts only

outer race of bearing and press bearing into housing bore until outer race is flush with front of bore. Support front end of shaft and press impeller (4) on rear of shaft until rear of fins are flush with gasket flange of housing. Support impeller end of shaft and press pulley (9) on front of shaft until pulley is flush with end of shaft. Complete the assembly by reversing the disassembly procedure.

IGNITION AND ELECTRICAL SYSTEM

DISTRIBUTOR
All Models

72. **TIMING.** A "TDC" timing hole is located in flywheel but ignition is timed with an "S" mark on crankshaft pulley which aligns with a scribed line on right side of timing gear cover when advance spark should occur. Suggested timing method is by using a power timing light and timing with marks aligned at 2500 engine rpm.

Emergency static timing can be made by turning crankshaft until No. 1 piston is on compression stroke, then continuing to turn shaft until "S" mark is aligned. Remove distributor cap and hold rotor counter-clockwise to full advanced position; then turn distributor body until points just begin to break. Recommended breaker point gap is 0.020.

73. **REMOVE AND REINSTALL.** Before removing the distributor assembly, turn crankshaft until No. 1 piston is at TDC on compression stroke. Remove distributor cap and disconnect coil lead. Remove distributor clamp screw, then withdraw distributor assembly from cylinder block. When installing the distributor, refer to Fig. 54 and proceed as follows:

Turn distributor housing until primary lead is pointing directly toward engine, then turn distributor shaft until rotor is pointing about 75 degrees clockwise from primary lead as shown.

Fig. 53—Distributor is correctly timed if spark occurs when "S" mark on crankshaft pulley aligns with timing mark on engine casting.

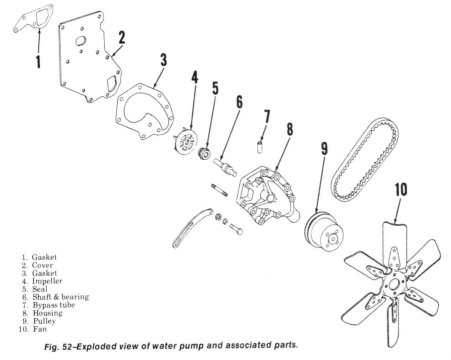

1. Gasket
2. Cover
3. Gasket
4. Impeller
5. Seal
6. Shaft & bearing
7. Bypass tube
8. Housing
9. Pulley
10. Fan

Fig. 52—Exploded view of water pump and associated parts.

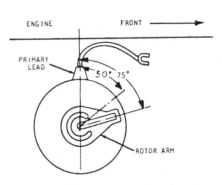

Fig. 54—With "S" timing mark aligned as shown in Fig. 53 and No. 1 piston on compression stroke, install distributor with rotor arm and body positioned as shown.

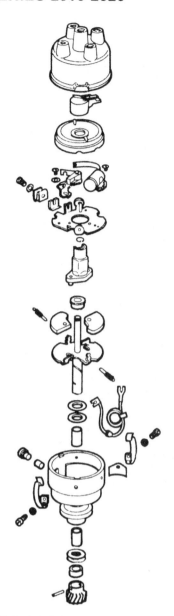

Fig. 55—Exploded view of distributor of the type used.

With No. 1 piston at TDC on compression stroke, install the distributor. With gears fully meshed, rotor should point about 50 degrees clockwise from primary lead. Complete the assembly, start engine and time with a timing light as outlined in paragraph 72.

NOTE: An error of one tooth in installation timing will make 30° difference in rotor position, and is readily evident after refering to Fig. 54.

Firing order is 1-3-4-2 and recommended distributor point gap is 0.020.

74. **OVERHAUL.** A PRESTOLITE Model IBT-4101S or IBT-4101U distributor is used. Breaker point gap is 0.020 and breaker point spring tension should be 17-22 oz. measured at center of contact. Tension can be adjusted by sliding the spring in or out on at-

taching screw. Direction of rotation is counter-clockwise, viewed from rotor end of shaft. Distributor shaft end play is 0.002-0.010. Distributor shaft bushings may be renewed in housing and both bushings should be installed 3/32-inch below flush with housing bore. Grease hole must be drilled in upper bushing after installation, by first removing the grease hole plug. When the distributor is assembled at the factory, the spring pin (roll pin) which retains the drive gear is left protruding about 1/16-inch on the end which aligns with rotor. An exploded view of distributor is shown in Fig. 55. Cam dwell angle is 66-72 degrees and advance data are as follows, given in distributor degrees and distributor rpm:

Start advance .... 0° @ 200 rpm
Intermediate advance . 4° @ 400 rpm
Maximum advance ... 15° @ 1200 rpm

ALTERNATOR AND REGULATOR
All Models

75. All models are equipped with a 12 volt electrical system. 2510 diesel tractors use two 12 volt batteries which are connected in parallel. 2520 Diesel tractors use two 6 volt batteries hooked in series to obtain 12 volts, which increases the capacity, or reserve. The negative battery post is grounded on all batteries.

The transistorized regulator and an isolation diode are mounted on rear end frame of alternator. The isolation diode (See Fig. 56) is mounted on one accessory terminal (a) and one ground terminal (b), but is insulated from ground terminal by sleeve (S) and washer. The primary purpose of the isolation diode is to permit the use of a charging indicator lamp. The isolation diode and brush holder can be renewed without removal or disassembly of alternator. All other service requires removal and disassembly.

The regulator is fully transistorized and consists of only a voltage regulator. Current regulation and reverse current protection are provided in the design of the alternator unit.

Failure of the isolation diode is usually indicated by the charging indicator lamp, which glows with key switch off and engine stopped if diode is shorted; or with engine running if diode is open.

Failure of a rectifying diode is usually indicated by a humming noise when alternator is running and diode is shorted; or by a steady flicker of charge indicator light at slow idle when diode is open. Either fault will reduce alternator output.

To check the charging system, refer to Fig. 56 and proceed as follows:

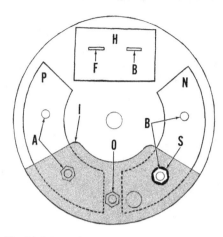

Fig. 56—Schematic view of alternator rear end frame. Isolation diode (I) is mounted on one accessory terminal (A) and one battery return terminal (B), but is insulated from (B) terminal by sleeve (S) and washer. The regulator mounts on alternator frame at brush holder (H).

A. Accessory terminals
B. Battery return terminals
F. Field terminal
H. Brush holder
I. Isolation diode
N. Negative heat sink
O. Output terminal
P. Positive heat sink

(1). With key switch and all accessories off and engine not running, connect a low reading voltmeter to terminals A-B. Reading should be 0.1 volt or less. A higher reading would indicate a short in isolation diode (I), key switch or wiring. Find and correct the trouble, then proceed to Test 2.

(2). With a suitable voltmeter connected to terminals A-B, start engine and gradually increase engine speed to approximately 1300 rpm. Reading should be 15 volts. If a lower reading is obtained, proceed as outlined in Test 4; if reading is 15 volts or better, proceed to Test 3.

(3). If a 15 volt reading was obtained in test 2, move voltmeter lead from Terminal A to Terminal 0. Reading should drop one volt from the reading obtained in Test 2, reflecting the resistance built into the isolation diode. If battery voltage (12 volts) is obtained, isolation diode is open and must be renewed.

(4). If a reading lower than the recommended 15 volts was obtained when alternator is checked as in Test 2, stop the engine and remove the voltage regulator. Connect a jumper wire from terminal (A) on positive heat sink to terminal (F) on brush holder (H). Connect a suitable voltmeter to terminals A-B. Start engine and slowly increase engine speed while watching voltmeter. If a reading of 15 volts can be obtained with an engine speed of 1300 rpm or less, renew the regulator. If a 15 volt reading cannot be obtained, renew or overhaul the alternator.

1. Spacer
2. Front end frame
3. Bearing
4. Snap ring
5. Rotor
6. Bearing
7. Stator
8. Insulated sleeve
9. Rear end frame
10. Brush holder
11. Brush return wire
12. Isolation diode
13. Insulated sleeve
N. Negative heat sink
P. Positive heat sink

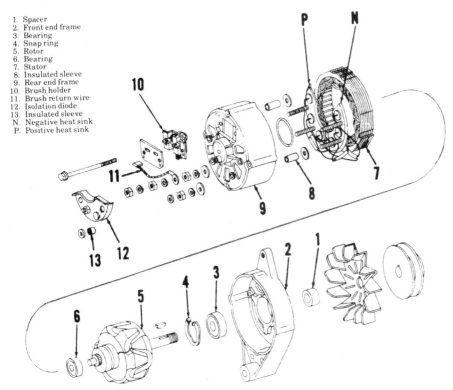

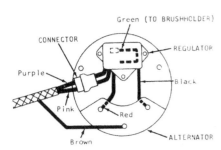

Fig. 57–Exploded view of Motorola Alternator of the type used.

DR-1107339
Brush Spring Tension .... 35 oz.
No-Load Test
    Volts 10.6
    Amperes 65-100
    RPM 3600-5100
DR-1107577
Brush Spring Tension .... 35 oz.
No-Load Test
    Volts 10.6
    Amperes 105-200
    RPM 6500-14000
DR. 1108319
Brush spring tension .... 35 oz.
No-load Test
    Volts 9.0
    Amperes 55-80
    RPM 3400-6000
DR 1107863
Brush spring tension .... 35 oz.
No-load Test
    Volts 9.0
    Amperes 40-140
    RPM 8000-13000
DR 1107871
Brush spring tension .... 40 oz.
No-load Test
    Volts 9.0
    Amperes 40-140
    RPM 8000-13000

CAUTION. DO NOT allow voltage to rise above 16.5 volts when making Test 4, or alternator may be damaged.

Refer to Fig. 57 for an exploded view of alternator and to Fig. 58 for a view of alternator wiring connections for 2510. 2520 is the same except the pink wire is orange, and the brown wire is red.

Fig. 58–Alternator wiring connections correctly made for 2510. See text for 2520 Color Code.

STARTING MOTOR
All Models
76. Delco-Remy starting motors are used. 2510 Gasoline tractors use Model 1107339 motor, diesel tractors use Model 1107577. 2520 gasoline tractors use model 1108319 motor, diesel tractors use model 1107863, or 1107871. Test specifications include solenoid. Specifications are as follows:

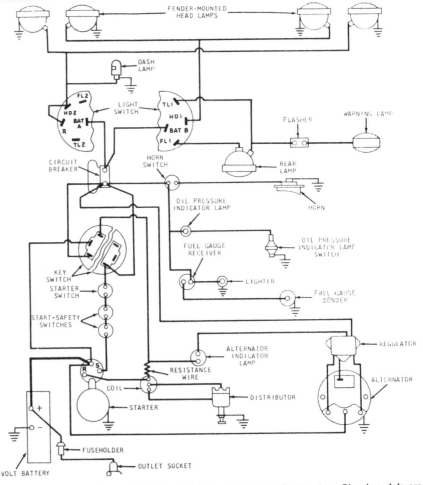

Fig. 59–Schematic view of a typical wiring diagram used on gasoline tractors. Diesel models are similar.

CIRCUIT DESCRIPTION
All Models

77. All models are equipped with a 12 volt electrical system with the negative battery post grounded. 2510 diesel models use two 12 volt batteries connected in parallel. 2520 diesel models use two 6 volt batteries in series. Refer to Fig. 59 for a schematic view of a typical wiring diagram used on Gasoline tractors. Diesel models are similar except for omission of ignition components and the use of two batteries.

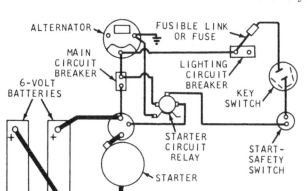

Fig. 60–Schematic view of starting circuit on 2520 diesel tractors. Gasoline models use one 12 volt battery.

ENGINE CLUTCH

NOTE: This section covers tractors equipped with Syncro-Range, manual shift transmission only. For models equipped with power shift transmission, refer to paragraph 109.

LINKAGE ADJUSTMENT
All Models

78. **TRANSMISSION CLUTCH.** To check the transmission clutch pedal free play, place transmission lever in "PARK" position and have engine running at 1900 rpm. Free play measured

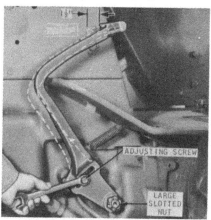

Fig. 61–Loosen the large slotted nut and adjusting clamp screw to adjust clutch pedal free play to 1½ inches as shown.

at pedal pad should be 1½ inches as shown in Fig. 61.

Linkage should be readjusted whenever free play measures less than ¾ inch. To adjust the linkage, loosen the large slotted nut and adjusting clamp cap screw and reposition pedal on clutch fork shaft arm.

As an emergency measure when insufficient adjustment remains, temporary adjustment can be made through clutch housing bottom opening by loosening all three clutch finger adjusting nuts equally ½-turn. Adjustment must be made only once and clutch must be

overhauled as soon as possible after making emergency adjustment.

79. **POWER TAKE-OFF CLUTCH.** The power take-off clutch is not adjustable on Syncro-Range models. The early clutch is over-center type and spring loaded in the engaged position. The late type is hydraulically applied and held in the engaged position.

When pto clutch operating lever is pulled to rearmost position, clutch is released and a brake is applied which stops pto shaft rotation.

TRACTOR SPLIT
All Models

80. To detach (split) engine from clutch housing, proceed as follows:

Disconnect the batteries and remove cowl and hood. Disconnect tachometer cable and unplug wiring harness connections at rear of engine block. Remove choke cable (gasoline) and speed control rod. On Power Shift, remove left tractor step and disconnect hydraulic pump inlet pipe. Relieve any pressure on hydraulic system components by actuating the control levers, remove the hydraulic fluid line clamps and spacers, then disconnect, plug and remove interfering hydraulic fluid lines. On tractors with a power brake accumulator, remove the capscrew fastening the accumulator support bracket to the engine.

NOTE: Open right hand brake bleed screw and hold the right brake pedal down for a few seconds to discharge accumulator before separating tractor.

If tractor is equipped with front end weights, remove the weights.

Support engine and transmission separately, remove the connecting cap screws and roll transmission assembly rearward away from engine.

To connect, reverse the above procedure.

R&R AND OVERHAUL
All Models

81. To remove the clutch assembly after engine has been separated from clutch housing, remove the six retaining cap screws and lift off the assembly. Transmission clutch disc (1—Fig. 62) will be free when clutch is removed, and may be renewed without disassembly of clutch unit.

1. Transmission clutch disc
2. Operating bolt
3. Transmission clutch plate
4. Spring
5. Inner spring
6. Outer spring
7. Retainer
8. PTO clutch plate
9. PTO clutch disc
10. Spring

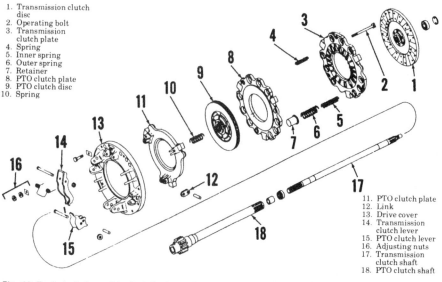

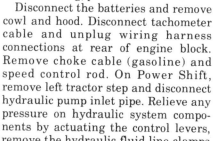

11. PTO clutch plate
12. Link
13. Drive cover
14. Transmission clutch lever
15. PTO clutch lever
16. Adjusting nuts
17. Transmission clutch shaft
18. PTO clutch shaft

Fig. 62–Exploded view of typical dual type transmission clutch of the type used on Syncro Range Models.

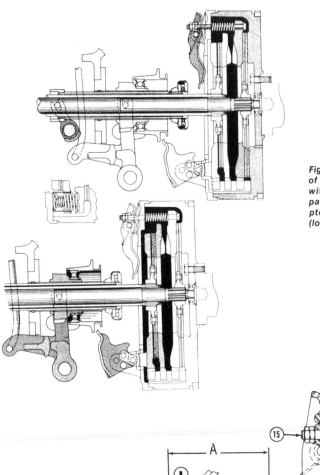

Fig. 63–Cross sectional view of early clutch and flywheel with transmission drive parts shaded (top view) and pto drive parts shaded (lower view). Spring detail is shown in center inset.

Disassembly and reassembly of the removed clutch unit will be made easier by removal of the flywheel (or use of a spare flywheel) for use as a holding jig. To disassemble the clutch, place flywheel front side down on a bench. Place transmission clutch disc and clutch assembly in operating position on flywheel and secure with three alternately spaced jack screws with nuts and washers, as shown in Fig. 64. Tighten jack screw nuts evenly to take pressure from clutch fingers. Remove jam nuts, adjusting nuts and operating bars from transmission clutch fingers, then back off jack screw nuts evenly until spring pressure is released. Remove jack screws then remove the clutch components.

Outer clutch springs (6—Fig. 62) should have a free length of approximately 3¼ inches and should test 105-129 lbs. when compressed to installed length of 1¾ inches. Inner springs (5) should have a free length of approximately 3 5/16 inches and should test 79-97 lbs. when compressed to 1¾ inches. Springs (4) should have a free length of 2¼ inches, a compressed length of 1 25/32 inches and test pressure of 63.5-77.5 lbs. All parts are available individually.

Install flywheel (if removed) using new cap screws and tighten to a torque of 85 ft.-lbs. Long hub of transmission clutch disc should face forward. Long hub of pto clutch disc should face rearward. Refer to Fig. 63. Install clutch assembly using Special Tools JDE-52 and JDE-52-2 or other suitable pilot tool. Tighten cover cap screws to a torque of 35 ft.-lbs. and bend lock plates. Use Special Tool JDE-51 or equivalent for early type with overcenter pto clutch fingers, or JDE59 for late type, and adjust each clutch finger as shown in Fig. 65 or 65-A. Lock by tightening the jam nuts, then remeasure.

Place screwdrivers under each of the pto clutch operating levers as shown in fig. 65-B to hold the pto clutch disc in solid contact with clutch pressure plates. Use special gage JDE-59 and adjust each lever to just touch gage, tighten jam nuts, and recheck. On late models, be sure installed length of pto clutch return spring is adjusted to 3.91″ ± .010. See Fig. 63-A and 66-A.

Reconnect the tractor as outlined in paragraph 80. Check and adjust operating linkage as outlined in paragraph 78.

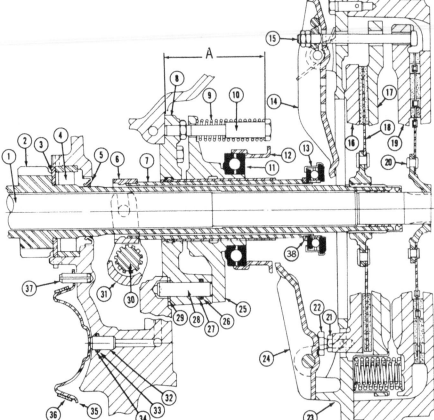

Fig. 63A–Cross sectional view of late clutch and flywheel, with hydraulic PTO components. See Fig. 66A for dimension "A".

1. Transmission clutch shaft	12. PTO clutch collar	27. Sealing ring
2. PTO shaft and gear	13. Clutch release bearing	28. PTO clutch piston
6. Clutch collar	14. Clutch release lever	29. Sealing ring
7. Clutch collar tube	16. PTO clutch rear plate	30. Clutch fork shaft
8. Sleeve support	17. PTO clutch front plate	31. Clutch fork
9. Return spring	18. PTO clutch disc	32. PTO brake piston
10. Special capscrew	19. Transmission clutch plate	33. Sealing ring
11. PTO clutch bearing	20. Transmission clutch disc	34. Backing ring
	23. Clutch cover	35. PTO brake shoe
	24. PTO clutch levers	
	25. PTO clutch sleeve	
	26. Backing ring	

CLUTCH SHAFT
All Models

82. To renew either the transmission or pto clutch shaft, it is first necessary to detach (split) tractor between the

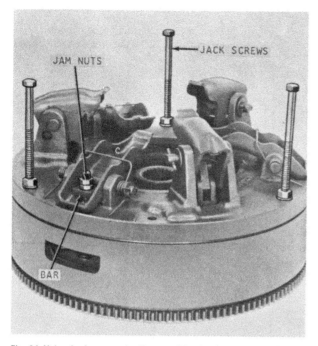

Fig. 64–Using jack screws to disassemble clutch as outlined in paragraph 81.

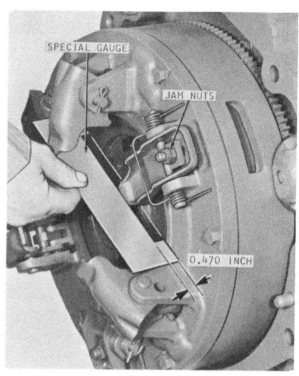

Fig. 65–Using special gage to adjust transmission clutch lever height on early models. PTO clutch levers are not adjusted.

clutch housing and transmission case as follows:

83. **TRACTOR SPLIT.** First drain the transmission and hydraulic system fluid and remove batteries and battery boxes. Disconnect transmission shifter linkage and clutch pedal return spring. Discharge brake accumulator, if so equipped.

Disconnect and cap the main hydraulic pump inlet line, rockshaft cylinder inlet line and the hydraulic brake lines. Disconnect pressure pipe from rockshaft housing, and the hydraulic pump seal drainpipe. Move tractor seat to extreme rearward position. Loosen the control support cover knob, raise support cover to clear platform, then remove operator's platform. Remove differential lock pedal pivot pin and lock control link, if so equipped.

Disconnect the rockshaft push-pull cable or rod. Remove the remote cylinder hydraulic lines, hydraulic brake lines and wiring shields. Disconnect and pull wiring forward to clear transmission case and remove transmission top cover. Remove front pto guard and bearing quill being careful not to damage seal. (Catch the trapped oil).

Support both halves of tractor separately, remove connecting cap screws and separate the two units.

When reassembling, tighten the clutch housing to engine retaining cap screws to a torque of 170 ft.-lbs.

84. **R&R AND OVERHAUL.** After splitting tractor as outlined in para-

graph 83, clutch shafts may be withdrawn rearward out of clutch housing. To prevent damage to inner seal, remove both shafts as a unit. Withdraw transmission clutch shaft carefully from pto clutch shaft after both shafts are removed from housing.

Both oil seals can be renewed at this time. Outer seal is located in clutch housing in front of pto shaft roller bearing which must first be removed for access to the seal. If bushing inside pto clutch shaft is renewed, install bushing from front with closed ends of oil groove toward the front next to oil seal. Installed ID of pto clutch shaft bushing is 1.082-1.083.

CONTROL LINKAGE
All Models

85. To overhaul the clutch control linkage it is necessary to remove the clutch housing from tractor. To remove the housing, first detach engine from clutch housing as outlined in paragraph 80. Attach a chain hoist to clutch housing, then detach from transmission as outlined in paragraph 83. Refer to Fig. 66 for an exploded view of linkage and associated parts.

Clutch release bearings (4 & 6) and pto operating fork and linkage can be removed from front of housing without making rear split. The pivot shaft for pto operating fork (10) is a tight press fit in right side of housing and must be driven out toward the left.

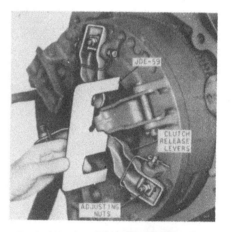

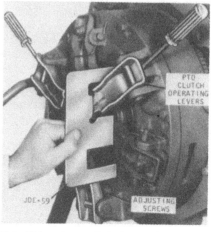

Figs. 65A and 65B–Method of adjusting late transmission clutch release levers and PTO clutch operating levers using special tool JDE-59.

Transmission clutch fork (16) and arm (18) are splined to operating shaft (17). Correct position of installation is indicated by index marks on shaft, fork and arm.

When reassembling, adjust linkage as outlined in paragraph 78.

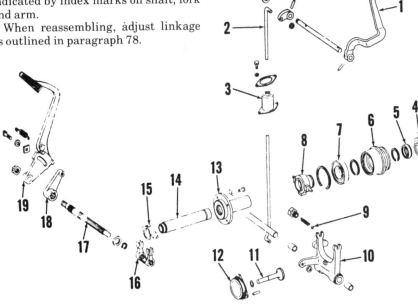

Fig. 66–Exploded view of clutch operating linkage and associated parts used on early Syncro Range models.

1. PTO clutch lever	6. Collar	11. PTO brake shaft	15. Operating collar
2. Link	7. Release bearing	12. PTO brake shoe	16. Fork
3. Boot	8. Sleeve	13. Support	17. Shaft
4. Release bearing	9. Detent	14. Operating tube	18. Arm
5. Collar	10. Fork		19. Pedal

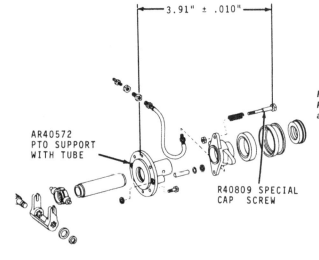

AR40572
PTO SUPPORT
WITH TUBE

3.91" ± .010"

R40809 SPECIAL
CAP SCREW

Fig. 66A–Exploded view of PTO clutch operating parts and installed length of return spring on late models.

SYNCRO-RANGE TRANSMISSION

(For Power Shift Models, Refer to Paragraph 98)

The "Syncro-Range" transmission is a mechanically engaged gear transmission consisting of three transmission shafts and a single, mechanically connected, remote mounted control lever as shown in Fig. 68. The four basic gear speeds are selected by coupling one of the shaft idler gears to the splined main drive bevel pinion shaft, and can only be accomplished by disengaging the engine clutch and bringing the tractor to a stop. The high, low and reverse speed ranges within the four basic speeds are selected by shifting the couplers on the constant

mesh gears on transmission drive shaft. Because of the design of the couplers, shifting can be accomplished by disengaging the engine clutch and moving the control lever, without bringing the tractor to a halt.

NOTE: The rotating speeds of the transmission drive shaft and its idler gears are automatically equalized by the synchronizing clutches, allowing the coupler to engage without clashing. All other phases of shifting are under the direct control of the operator. The fact that clashing of gears is eliminated by the synchronizing clutches does not relieve him of the responsibility of using care and judgment in re-engaging the clutch after shifting is completed.

The idler gears and bearings on the main shaft and bevel pinion shaft are pressure lubricated by a separate transmission oil pump.

INSPECTION
Syncro-Range Models

86. To inspect the transmission gears, shafts and shifters, first drain the transmission and hydraulic fluid and remove the operator's platform; then remove the transmission top cover. Examine the gears for worn or broken teeth and the shifter linkage and cam slots for wear.

CONTROL QUADRANT

Paragraph 87 covers disassembly and overhaul of the shifter controls mounted in tractor steering support. Removal, inspection, overhaul and adjustment of shift mechanism inside the transmission housing is included with transmission gears and shafts.

Syncro-Range Models

87. R&R AND OVERHAUL. To overhaul the control quadrant, remove the cowl and raise the dash high enough to clear quadrant. Disconnect shifter rods (3 & 8—Fig. 68) at upper end. Remove snap ring (1) or ⅜ in. cap screw and washers from end of quadrant shaft. Remove control support cover, hold shaft (2) from turning and remove nut (11). The unit may now be completely disassembled.

If lever (6) or pivot (5) are damaged, proceed as follows: Clamp the lower curved portion of lever (6) in a soft jawed vise and slip a 5/32-inch cotter pin into each spring pin to prevent pin from collapsing, and use Vise-Grip pliers to withdraw the spring pins. When reassembling, leave at least 3/16-inch of each spring pin protruding from lever as shown in Fig. 69. If spring pin is driven in too far, pivot bushing will be damaged.

When reassembling, install shaft (2 —Fig. 68) loosely in bracket (10) and

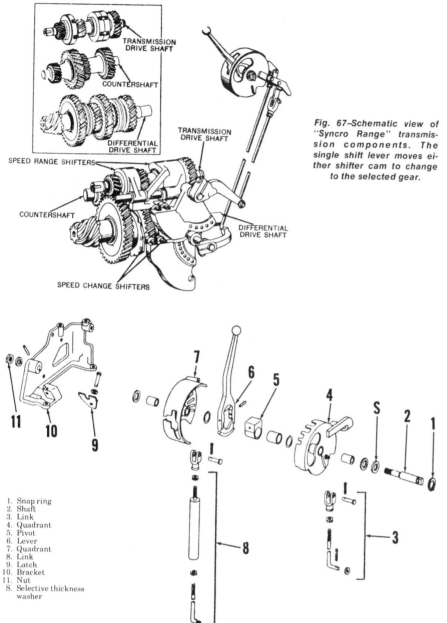

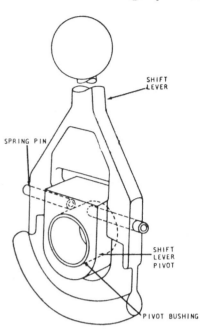

Fig. 67–Schematic view of "Syncro Range" transmission components. The single shift lever moves either shifter cam to change to the selected gear.

Fig. 69–Schematic view of shift lever and lever pivot showing correct installation of spring pins if unit is disassembled. Pins should protrude 3/16-inch as shown. If driven flush, bushing will be damaged and pins cannot be easily removed.

1. Snap ring
2. Shaft
3. Link
4. Quadrant
5. Pivot
6. Lever
7. Quadrant
8. Link
9. Latch
10. Bracket
11. Nut
S. Selective thickness
   washer

Fig. 68–Exploded view of shifter controls. The notch in latch (9) fits around lower rocker of lever (6) to lock the opposite quadrant (4 or 7) when the other is moved.

install washer and nut. Install remaining parts in proper order, leaving off the two outer washers and snap ring (1) or cap screw. Make sure the left hand thrust washer has not slipped off shoulder of shaft (2), then tighten the shaft nut (11). The steel washer (S) is available in thicknesses of 0.018, 0.036 and 0.060. Install washers of sufficient number and thickness to limit the end play of shaft components to 0.015 when snap ring (1) is installed.

TRANSMISSION DISASSEMBLY AND ASSEMBLY

Paragraphs 88 and 89 outline the general procedure for removal and installa-

tion of the main transmission components. Disassembly, inspection and overhaul of the removed assemblies is covered in overhaul section beginning with paragraph 90, which also outlines those adjustment procedures which are not an exclusive part of assembly.

Syncro-Range Models

88. **DISASSEMBLY.** Any disassembly of transmission gears, shafts and controls requires that tractor first be separated (split) between transmission and clutch housing as outlined in paragraph 83. Transmission must be disassembled in the approximate sequence outlined, however, disassembly need not be completed once defective or

damaged parts are removed.

Remove the transmission top cover and rockshaft housing or transmission rear cover. Remove the detent spring caps from right side of housing. Slide upper shift rail (1—Fig. 70) forward out of housing bores. Move shifter cam (6) counter-clockwise to its lowest detent position, remove shaft nut and carefully withdraw shaft (7) and arm (9) to prevent damage to oil seal (8). Rotate shifters (2 & 3) out of cam slots and remove the shifters and cam (6).

Block up one rear wheel and turn bevel ring gear until one of the flat surfaces of differential housing is toward transmission oil pump, then unbolt and remove pump and inlet and outlet hoses.

Working through the front bearing retainer and using a brass drift, drive the transmission drive shaft rearward to force the rear bearing cup part of the way out of housing. Tape or clip the synchronizer clutches together to keep them from separating while shaft is being removed. Remove front bearing retainer and front bearing cup, using care not to damage or lose the shims, then lift out the transmission drive shaft assembly.

Remove the nut from inner end of speed change shifter cam shaft (17—Fig. 70) and remove shaft and arm (20). Withdraw shifter rail (12) forward out of transmission housing and lift out shifter cam (16) and shifter forks (13, 14 & 15).

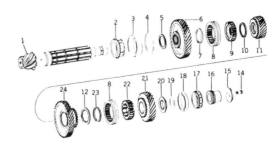

Fig. 71–Exploded view of bevel pinion shaft and associated parts used on Syncro Range models.

1. Pinion shaft
2. Bearing cone
3. Bearing cup
4. Shim
5. Thrust washer
6. Gear
7. Snap ring
8. Collar
9. Shifter gear
10. Snap ring
11. Gear
12. Thrust washer
13. Snap ring
14. Cap screw
15. Washer
16. Bearing race
17. Bearing cone
18. Bearing cup
19. Shim
20. Thrust washer
21. Gear
22. Shifter gear
23. Snap ring
24. Gear

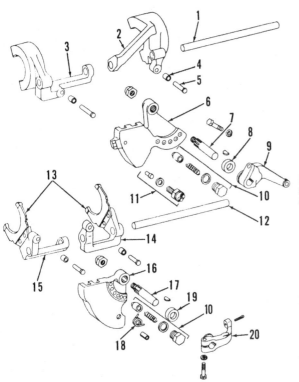

Fig. 70–Exploded view of shift rails, forks and associated parts used on Syncro Range models.

1. Rail
2. High-low fork
3. Reverse fork
4. Cam roller
5. Pin
6. Actuating cam
7. Shaft
8. Oil seal
9. Shift arm
10. Detent
11. Starting safety switch
12. Shift rail
13. Forks
14. Shift fork
15. Shift fork
16. Actuating cam
17. Shaft
18. Park lock spring
19. Oil seal
20. Shift arm

Remove the differential assembly as outlined in paragraph 127. Remove the cap screws retaining the pto idler thrust washer to front of shaft, and lift off the washer and idler gear, being careful not to lose any of the loose needle rollers. Use a jaw-type puller to remove the idler gear inner race (16—Fig. 71) from front of shaft. Bump the pinion shaft (1) rearward as far as possible using a soft drift and heavy hammer. Insert a C-Clamp through the hole provided in front face of housing and clamp front gear (21) to housing front wall. Continue bumping shaft rearward until front bearing cone (17) can be removed, then remove the bearing, shims (19) and washer (20).

Continue to move the parts forward on shaft, unseating snap rings (23, 13, 10 and 7) as they are exposed. With snap rings unseated, continue to move shaft rearward removing gears and associated parts from top opening as shaft is removed.

NOTE: Snap rings are the same diameter but progressively thicker toward rear of shaft, so that no snap ring can drop into a groove forward of the one which it occupies.

Remove the countershaft front bearing retainer (1—Fig. 72) and shim pack (2). Using a hammer and soft drift and working from rear of countershaft, drive front bearing cup (3) from transmission case; then lift out the countershaft.

Overhaul the transmission main components as outlined in paragraphs 90 through 97 and assemble as outlined in paragraph 89.

89. **ASSEMBLY.** Install countershaft and front bearing cup. Install front bearing retainer (1—Fig. 72) using the removed shim pack (2). Install and tighten the retaining cap screws then check countershaft end play using a dial indicator. End play should be 0.001-0.004 and is adjusted with shims (2) which are available in thicknesses of 0.006, 0.010 and 0.018.

If bevel pinion shaft or transmission housing were renewed; or if shaft bearings were not properly adjusted when transmission was disassembled; adjust the shaft and bearings as outlined in paragraph 93.

Special John Deere service tools are almost essential in installing the bevel pinion shaft and gears. The tools required are as follows:

JDT-2 Snap Ring Expanding Cone
JDT-3 Snap Ring Retainers (4)
JDT-9 (1 & 2) Installing Arbor (Modified)

Refer to Fig. 71 for order of assembly. All components must be installed on shaft as shaft is inserted. To expand the four shaft snap rings, use snap ring pliers and the expanding cone. Expand the ring and move it down the cone until ends are separated 1⅜-inches, then install one of the four retainers in open ends. Leave retainers in snap rings until all parts are installed on bevel pinion shaft; then remove the retainers and allow snap rings to seat in their grooves. Sort the snap rings by thickness before assembling the shaft and install on shaft in order, thickest snap ring to the rear.

Attach the installing arbor (JD-9-1) to front of housing and insert dummy shaft (JDT-9-2) counterbored end first, into front of arbor. Install the shaft components on dummy shaft in proper order beginning with front gear (21). When all units are in place on dummy shaft, insert pinion shaft (1) from the rear through the gears, replacing the dummy shaft. After shaft is in place, seat the snap rings in their grooves then install spacer (20), the previously determined thickness of shim pack (19), bearing cone (17) and inner race (16). Installation of bearings will be easier if they are heated to a temperature not to exceed 300° F.

Install speed change shifter forks and cam, and insert shift rail through forks. Install cam shaft and arm, making sure that index marks are aligned as shown in Fig. 74. Tighten the retaining nut to provide a shaft end play of 0.001-0.003.

Fig. 72–Exploded view of transmission countershaft.

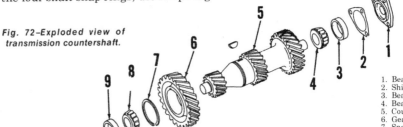

1. Bearing quill
2. Shim
3. Bearing cup
4. Bearing cone
5. Countershaft
6. Gear
7. Snap ring
8. Bearing cone
9. Bearing cup

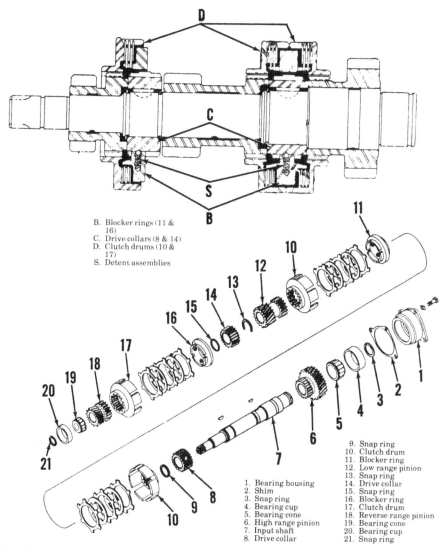

B. Blocker rings (11 & 16)
C. Drive collars (8 & 14)
D. Clutch drums (10 & 17)
S. Detent assemblies

9. Snap ring
10. Clutch drum
11. Blocker ring
12. Low range pinion
13. Snap ring
14. Drive collar
15. Snap ring
16. Blocker ring
17. Clutch drum
18. Reverse range pinion
19. Bearing cone
20. Bearing cup
21. Snap ring

1. Bearing housing
2. Shim
3. Snap ring
4. Bearing cup
5. Bearing cone
6. High range pinion
7. Input shaft
8. Drive collar

Fig. 73–Exploded view of transmission input shaft showing component parts. Synchronizer clutches are cross sectioned in upper view.

To install the transmission drive shaft and shifter mechanism, place shaft in transmission housing and install front bearing cup, shims and retainer. Tighten cap screws and install rear bearing cup and transmission oil pump; then check drive shaft end play using a dial indicator. Adjust end play to 0.001-0.004 by adding or removing shims (2—Fig. 71) as required. Shims (2) are available in thicknesses of 0.006, 0.010 and 0.018.

Place shifter cam (6—Fig. 70) in housing and install shifters (2 & 3) and rail (1). Make sure rollers (4) are in place and engage the slots in cam (6). Install shaft (7) carefully to keep from damaging oil seal. Make sure "V" mark on shaft indexes with corresponding mark on cam 1 (Fig. 74). Install slotted nut and adjust end play to 0.002-0.005.

OVERHAUL

To overhaul the transmission, first disassemble the unit as outlined in para-

graph 88. **Disassemble, overhaul and inspect the components as outlined in the appropriate following paragraphs; then reassemble as in paragraph 89.**

Syncro-Range Models

90. **SHIFTER CAMS AND FORKS.** Refer to Fig. 70 for exploded view. Examine shifting grooves in cams (6 & 16) for wear or other damage. Park lock spring (18) must have enough tension to shift the front shift coupling into engagement. Spring is secured to the cam by a spring pin. Shift forks (13) are riveted to fork carriers (14 & 15) and are renewable as an assembly only.

91. **TRANSMISSION PUMP.** The removed transmission pump may be disassembled by removing the cover and lifting out the gears. Check pump cover and parts for wear, scoring or other damage. The pump has a capacity of 6½ gpm at 2100 engine rpm. Pump delivery can be tested with tractor assembled, by following the

procedures outlined in paragraph 156. Tighten pump cover cap screws to a torque of 21 ft.-lbs. when reassembling.

92. **BEVEL PINION SHAFT.** Except for bearing cups in housing and rear bearing cone on shaft, the bevel pinion shaft is disassembled during removal. Refer to Fig. 71 for an exploded view. Gear (6) contains a bushing which is not available separately. All other gears ride directly on the shaft.

The bevel pinion shaft is available only as a matched set with the bevel ring gear. Refer to paragraph 129 for information on renewal of ring gear. Refer to paragraph 93 for mesh position and bearing adjustment procedure if bevel gears and/or housing are renewed.

93 PINION SHAFT ADJUSTMENT. The cone point (mesh position) of the main drive bevel gear and pinion is adjustable by means of shims (4—Fig. 71) which are available in thicknesses of 0.003 and 0.005. The cone point will only need to be checked if the transmission housing or ring gear and pinion assembly are renewed. To make the adjustment, proceed as follows:

The correct cone point of housing and pinion are factory determined and assembly numbers are etched on left upper gasket flange of housing and rear face of pinion. To determine the thickness of shim pack, add 1.755 to the number etched on pinion shaft, then subtract the total from number stamped on housing. If no number appears on housing, subtract the total from the number 8.600. The result is the correct shim pack thickness. To add or remove shims, use a punch and drive out rear bearing cup (3).

The bevel pinion bearings are adjusted to a pre-load of 0.004-0.006 by means of shims (19). If adjustment or checking is required, proceed as follows before installing the shaft gears.

First make sure that cone point is correctly adjusted as outlined above, then install shaft (1), bearing cone (2), washer (20), the removed shim pack (19) plus one 0.010 shim, and bearing cone (17). Install a 1¾ inch ID by 1½ inch long spacer instead of inner race (16); then install plate (15) and cap screws (14). Tighten the cap screws securely, making sure bearing cone (17) is bottomed against shim pack, then measure shaft end play using a dial indicator. When disassembling to prepare for final assembly, remove shims equal to the observed end play plus 0.005. Assemble the shaft and gears as outlined in paragraph 89.

94. **COUNTERSHAFT.** Refer to Fig. 72. The countershaft is a one-piece unit except for the bearings and high-

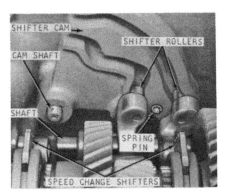

Fig. 74–Partially assembled view of shifter cam showing "V" timing marks on shifter cam and shaft.

speed gear (6). The high-speed gear is keyed to shaft and retained by snap ring (7); and may be removed with a press after removing the snap ring. Countershaft bearings should have 0.001-0.004 end play when shaft is properly installed.

95. TRANSMISSION DRIVE SHAFT. To disassemble the removed transmission drive shaft, proceed as follows:

Remove snap ring (21—Fig. 73) and use a press or bearing puller to remove bearing cone (19). Withdraw range pinion (18), synchronizer drum (17), synchronizer plates and blocker (16).

CAUTION: The four detent balls and springs (S) will be released when blocker is withdrawn. Use care not to lose these parts.

Remove snap ring (15) and use a press or bearing puller to remove drive collar (14).

NOTE: Apply force to collar only. DO NOT attempt to pull low range pinion (12) at same time collar is removed.

Remove snap ring (13) and low range pinion (12). Remove both synchronizer drums (10), the remaining synchronizer plates and blocker (11), being careful not to lose the detent balls and springs.

Remove snap ring (9), then pull collar (8) and high range pinion (6) from shaft. Bearing cone (5) can be pressed from shaft after removing snap ring (3) if renewal is indicated.

96. SYNCHRONIZER CLUTCHES. The purpose of the synchronizer clutches is to equalize the speeds of the transmission drive shaft and the selected range pinion for easy shifting without stopping the tractor. The synchronizer clutches operate as follows:

Range drive collars (C—Fig. 73) are keyed to shaft. Synchronizer clutch drums (D) are splined to pinions and blocker rings (B) are centered in drive collar slots by the detent assemblies (S). Synchronizer clutch discs are connected alternately by drive tangs to clutch drum and blocker ring. When the engine clutch is disengaged and shift control lever is moved to change gear speeds, the first movement moves clutch discs into contact. The difference in rotative speeds of shaft and pinion causes the blocker to try to rotate on drive collar. The drive lugs inside the blocker ring ride up the ramps in drive collar causing the blocker to be temporarily locked in center of drive collar. The clutch discs are thus compressed and speeds of shaft and pinion equalized. When the speeds become equal, the thrust force on blocker is relieved and shifting pressure causes the drive lugs to move back down the ramps in drive collar slot. The synchronizer drum is now permitted to move toward drive collar until splines are engaged and range pinion coupled to shaft.

97. INSPECTION AND ASSEMBLY. Inspect the transmission drive shaft for scoring or wear in areas of range pinion rotation, and make sure oil passages are open and clean. Carefully inspect the blocker rings for damage to the drive lugs and inspect friction faces of blocker rings and synchronizer drums. Check synchronizer discs for wear, using a micrometer. The

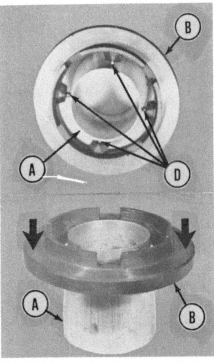

Fig. 75–Using the installing cone, JDT4, to install the detent assemblies in blocker. Use heavy grease to hold the balls and springs (D) in blocker drive lugs, then carefully start the assembly over narrow edge of cone (A). Push blocker (B) to large edge of cone, then use the cone to transfer the blocker to the drive collar located on the shaft.

thickness of new discs is 0.078. Renew discs in sets if thickness of any disc measures less than 0.060. Check drive tangs on discs for thickening due to peening. If thickness of drive tang is twice that of friction surface, renew the discs.

Reassemble the transmission drive shaft by reversing the disassembly procedure. A special installing cone is required to install detent assemblies in blocker rings and to install blocker assemblies on drive collar. See Fig. 75 for view of tool and brief procedural description.

POWER SHIFT TRANSMISSION

Tractors are optionally available with a full power shift transmission which provides 8 forward and 4 reverse speeds by moving a shift lever, without stopping tractor or touching the foot operated inching pedal.

OPERATION
Power Shift Models

98. POWER TRAIN. The power shift transmission is a manually controlled, hydraulically actuated plane-tary transmission consisting of a clutch pack and planetary pack shown schematically in Fig. 76.

Hydraulic control units consist of three clutch packs (C1, C2 & C3) and four disc brakes (B1 through B4). In addition, a multiple disc clutch (PTO) is used in the pto train. All units are hydraulically engaged, and mechanically disengaged when hydraulic pressure to that unit is released. The power train also contains a hand operated single disc transmission disconnect clutch (DC) mounted on the flywheel, a foot operated inching pedal, a mechanical disconnect for towing, and a mechanical park pawl.

Three hydraulic control units are engaged for each of the forward and reverse speeds. In 1st speed, Clutch 1 is engaged and power is transmitted to the front planetary unit by the smaller input sun gear (C1S). Brake 1 is engaged, locking the front ring gear to housing and the planet carrier walks around the ring gear at its slowest speed. Clutch 3 is also engaged, locking the rear planetary unit, and output shaft turns with the planet carrier.

Second speed differs from 1st speed only by disengaging Brake 1 and engaging Brake 2, causing planet carrier and output shaft to rotate together at a slightly faster speed.

Third speed and 4th speed are identical to 1st & 2nd except that Clutch 1 is disengaged and Clutch 2 engaged, and power enters the front planetary unit through the larger input sun gear (C2S).

Fifth speed and 6th speed differ from 3rd and 4th speeds in the rear planetary unit. Clutch 3 is disengaged and Brake 4 engaged, and the output shaft turns faster than the planet carrier through the action of the rear planet pinions and output sun gear.

In 7th and 8th speeds, both Clutch 1 and Clutch 2 are engaged, locking the input planetary unit, and planet carrier turns with input shaft at engine speed. Engaging the three clutch units locks both planetary units, therefore 7th speed is a direct drive, with transmission output shaft turning with, and at the same speed as, the engine. Eighth speed is an overdrive, with transmission output shaft turning faster than engine speed.

Reverse speeds are obtained by engaging Brake 3, which locks the output planetary ring gear to housing, and the output shaft turns in reverse rotation through the action of the two sets of output planetary pinions.

It will be noted that the front planetary unit is an input unit (TI), controlled by the two front clutch units in clutch pack and two front brake units in planetary pack. Two input control units must be engaged to transmit power, and five input speeds are obtained by selectively engaging the input brakes and clutch units.

The rear planetary unit is an output unit (TO) controlled by the two rear brakes and rear clutch. One of the rear control units must be engaged to complete the connection of the power train. Two forward ranges and one reverse output range are provided, depending on which rear control unit is engaged.

The accompanying table lists the control units actuated to complete the power flow in each shift position:

99. CONTROL SYSTEM. The control valve unit consists of manually actuated speed selector and direction selector valves which operate through four hydraulically controlled shift valves to engage the desired clutch and brake units. The valve arrangement prevents the engagement of any two opposing control units which might cause transmission damage or lockup.

Power to operate the transmission system is supplied by an internal gear hydraulic pump mounted on the transmission input shaft, which also supplies the charging fluid for the tractor main hydraulic system. Fluid from the hydraulic pump first passes through a full flow oil filter to the main transmission oil gallery, where the pressure is

	C1	C2	B1	B2	B3	B4	C3
1st Fwd.	C1		B1				C3
2nd Fwd.	C1			B2			C3
3rd Fwd.		C2	B1				C3
4th Fwd.		C2		B2			C3
5th Fwd.		C2	B1			B4	
6th Fwd.		C2		B2		B4	
7th Fwd.	C1	C2					C3
8th Fwd.	C1	C2				B4	
1st Rev.	C1		B1		B3		
2nd Rev.	C1			B2	B3		
3rd Rev.		C2	B1		B3		
4th Rev.		C2		B2	B3		
Boxes at right show clutch and brake units controlled by dual accumulators			Front accumulator orifice (Right-Hand)			Rear accumulator orifice (Left-Hand)	

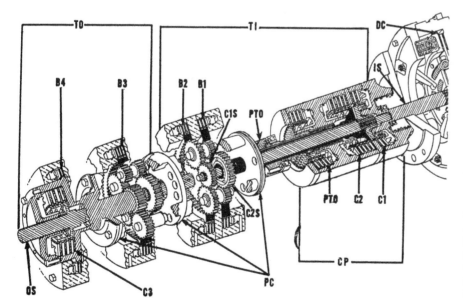

regulated at 150 psi for the transmission control functions. Excess oil passes through the regulating valve to the oil cooler and main hydraulic pump.

Fluid from the transmission main oil gallery is routed through the inching pedal valve to clutches 1 and 2; and through a spring-loaded accumulator to the brake actuating pistons and to Clutch 3.

On SINGLE accumulator system, the shifting engagement rate can be adjusted by opening or closing the accumulator charging orifice to accommodate varying operating conditions and load. Closing the orifice slows the rate of pressure rise and smooths the shifting action under light load. Opening the orifice causes more abrupt shifting under no load, but reduces slipping during shift under heavy load.

On DUAL accumulator system, the accumulator charging orifices are available in three sizes to vary the shifting rate. Dual accumulator system is used on tractors after serial number 022000.

Refer to Fig. 77 for a schematic view of control valves. The direction selector valve (E) and shift valve (4) control the

Fig. 76–Schematic view of Power Shift Transmission showing primary function of units. Disconnect Clutch (DC) is for cold weather starting only, and is not to be used for starting or stopping tractor motion. The Power Take-Off (PTO) clutch and gear are located in, but not a part of, transmission power train.

B1. Brake 1—Low Input
B2. Brake 2—High Input
B3. Brake 3—Reverse Output
B4. Brake 4—High Output
C1. Clutch 1—Low Input

C2. Clutch 2—High Input
C3. Clutch 3—Low Output
CP. Clutch Pack
C1S. C1 Sun Gear
C2S. C2 Sun Gear

DC. Disconnect Clutch (Starting Only)
IS. Input Shaft
OS. Output Shaft
PC. Planet Carrier
PTO. Power Take-Off drive units

TI. Transmission Input (Consisting of clutch pack and front half of planetary pack)
TO. Transmission Output (Consisting of rear half of planetary pack and output shaft)

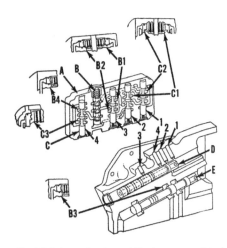

Fig. 77–Schematic view of the two manual and four hydraulic valves which control the Power Shift transmission.

A. Shift valve housing
B. Dump valve
C. Shift valve
D. Speed selector valve
E. Direction selector valve
1. Shift valve 1
2. Shift valve 2
3. Shift valve 3

4. Shift valve 4
B1. Brake 1
B2. Brake 2
B3. Brake 3
B4. Brake 4
C1. Clutch 1
C2. Clutch 2
C3. Clutch 3

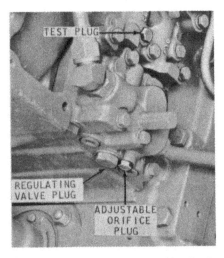

Fig. 81–View of pedal valve housing showing location of pressure test plug, regulating valve plug and adjustable orifice plug.

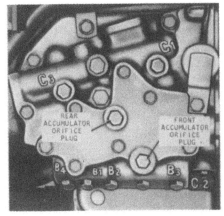

Fig. 82–View of transmission control valve showing front and rear accumulator orifice plugs on tractors after serial no. 022,000.

routing of pressure to the output control units (Clutch 3, Brake 3 and Brake 4). The speed selector valve contains four pressure ports (1, 2, 3 & 4) which control the movement of the hydraulic shift valves (C) by pressurizing the closed end (opposite the return spring) when port is open to pressure. Neutral position is provided by the selector valves or by depressing the inching pedal.

When the direction selector valve is moved to the forward detent position, system pressure is routed to Shift Valve (4). In the low range positions (1st through 4th gears), the speed selector valve charging port (4) is open to pressure, and Clutch 3 is actuated. In the high range positions except 7th gear (which is direct and uses all three

clutch units), charging pressure is cut off to Shift Valve (4), shift valve return spring moves the valve upward and Brake 4 is actuated. When the direction selector valve is moved to reverse detent position, system pressure is cut off from Shift Valve (4) and routed directly to Brake 3, without passing through a shift valve. In the neutral detent position, system pressure is cut off from all three output control units.

Shift valves 1, 2 & 3 direct the system pressure to the input control units (Clutches 1 & 2 and Brakes 1 & 2).

Shift Valve (1) directs pressure to Clutch 2 when hydraulically actuated; and to Clutch 1 when charging port (1) is closed. Refer to Figs. 78 and 79.

Shift Valve (2—Fig. 77) routes pressure to Shift Valve (3) when hydraulically actuated, and permits the simultaneous engagement of Clutches 1 and 2 (7th & 8th speeds) when charging port (2) is closed. Refer to Figs. 78, 79 and 80.

Shift valve (3—Fig. 77) directs pres-

sure to Brake 2 when hydraulically actuated; to Brake 1 when charging port (3) is closed.

ADJUSTMENT
Power Shift Models

100. The multiple disc clutches and brakes require no adjustment. Linkage adjustments do not change materially because of wear, but should be checked and readjusted if necessary (if changes or malfunctions occur), as part of a regular troubleshooting procedure. A change in operating conditions may require readjustment of shift engagement rate as outlined in paragraph 101.

101. **SHIFT RATE.** Single accumulator system. To adjust the shifting engagement rate on tractors before serial number 022001, remove the adjustable orifice plug (Fig. 81) and turn the exposed slotted-head adjustment screw clockwise to slow the shift rate or counter-clockwise for faster shifting. Turn screw ½-turn at a time, then recheck after each adjustment.

Dual accumulator system. Tractors after serial number 022000 use dual accumulators which are located in the transmission control valve housing on right side of tractor, and have removable accumulator orifices. These orifices are accessible by removing either of the accumulator orifice plugs (Fig. 82), and are available in three sizes. The rear (left hand) orifice affects shifting of the units in the output (rear) section of transmission, while the front (right hand) orifice affects shifting of the two brake units in the input (front) section (see Fig. 76).

If shift time is too slow at either accumulator, remove that orifice and install the next larger size. If shift time is too fast, install the next smaller size orifice. Recheck after each orifice change.

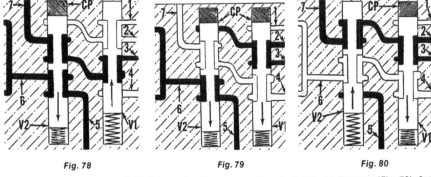

Fig. 78   Fig. 79   Fig. 80

Three views of Interlocking Shift Valves showing valve position for 1st, 2nd & Reverse (Fig. 78); 3rd, 4th, 5th & 6th (Fig. 79); and 7th & 8th (Fig. 80). Black indicates pressurized passages.

V1. Shift Valve 1
V2. Shift Valve 2
CP. Control pressure
1. Return passage

2. Pressure port, Clutch C2
3. Inlet pressure port
4. Return passage
5. Inlet pressure port

6. Pressure port, Shift Valve 3
7. Pressure port, Clutch C1

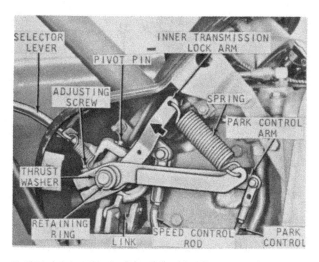

Fig. 83–View of shift linkage with cowl removed, showing points of adjustment.

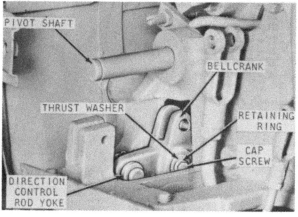

Fig. 84–Closeup view of partially disassembled shift linkage showing direction control bellcrank.

necessary, until yoke pin can just be reinserted.

On late models with one-piece park control arm, move shift lever to "NEUTRAL" position and disconnect park control cable yoke from arm; pull slack out of cable and arm and adjust cable length, if necessary, until yoke pin can just be installed.

To adjust pedal valve linkage, refer to Fig. 86. Disconnect the valve operating rod yoke from valve operating arm. Fully depress pedal, turn valve operating arm fully counterclockwise and thread the yoke on or off rod until yoke pin hole is aligned. Lengthen the rod ½ turn and reconnect.

103. **PRESSURE TEST AND ADJUSTMENT.** Before checking the transmission operating pressure, first check to be sure that transmission oil filter is in good condition and that oil level is at top of "SAFE" mark on dipstick. Place towing disconnect lever in "TOW" position and install a 0-300 psi pressure gage in "CLUTCH" plug hole (Test Plug—Fig. 86). Start and warm the tractor; gage should read 140-160 psi with engine running at 1900 rpm, pedal released and speed control lever in any position. If pressure is not as indicated, remove the regulating valve plug (shown in Fig. 81) and add or remove shims as required.

If pressure cannot be adjusted, other

102. **LINKAGE ADJUSTMENT.**
To adjust the shift linkage, first remove cowl, refer to Fig. 83 and proceed as follows:

Move shift lever to "NEUTRAL" position. Remove the pin securing speed control rod to lever pivot and push down on rod to feel the detented positions. Pull the rod to the uppermost DETENTED position and adjust length of rod if necessary, until yoke pin can just be inserted through yoke and pivot arm.

Disconnect direction control rod yoke (Fig. 84) from bellcrank. Move shift lever in neutral slot midway between forward and reverse slots in dash; then adjust the length of direction control rod if necessary, until yoke pin can be reinserted with valve spool in center detent position.

On early models with two-piece transmission lock arm, move shift lever to "NEUTRAL" position and refer to Fig. 83. Push inner transmission lock arm counter-clockwise as far as possible as indicated by arrow, then measure the clearance between arm and adjusting screw. Turn screw if necessary, until clearance is 0.030. Disconnect park control cable yoke from park control arm and pull up on cable until all slack is removed. Adjust the yoke, if

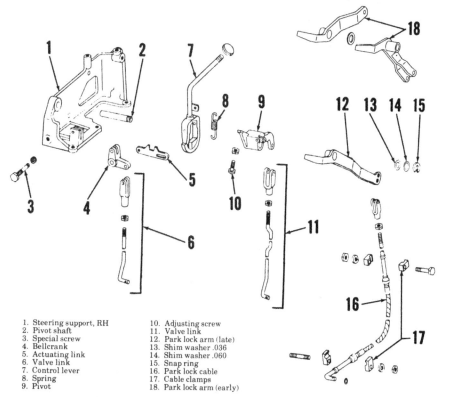

1. Steering support, RH	10. Adjusting screw
2. Pivot shaft	11. Valve link
3. Special screw	12. Park lock arm (late)
4. Bellcrank	13. Shim washer .036
5. Actuating link	14. Shim washer .060
6. Valve link	15. Snap ring
7. Control lever	16. Park lock cable
8. Spring	17. Cable clamps
9. Pivot	18. Park lock arm (early)

Fig. 85–Exploded view of typical shift linkage used on Power Shift Models.

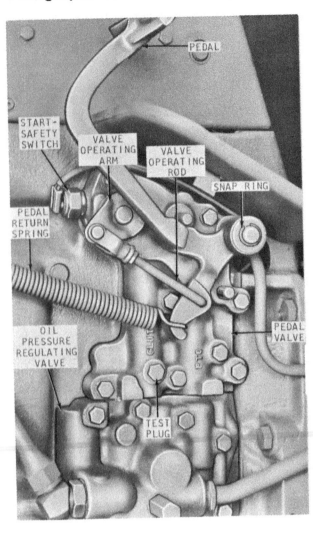

Fig 86–View of pedal valve and associated parts.

control valve housing.

105. BEHAVIOR PATTERNS. Erratic behavior patterns can be used to pinpoint some system malfunctions.

ODD SHIFT PATTERN. If tractor slows down when shifted to a faster speed; speeds up when shifted to a slower speed; or fails to shift when selector lever is moved; a sticking selector valve is indicated. Refer to paragraph 104 and table accompanying paragraph 98. Overhaul the control valve as outlined in paragraph 117.

SLOW SHIFT. First check and adjust the shifting engagement rate as outlined in paragraph 101. Other possible causes are; improper regulating valve adjustment; improper pedal linkage adjustment; plugged fluid filter; malfunctioning regulating valve, pedal valve or oil filter relief valve; broken accumulator spring; sticking accumulator piston; or slipping clutch or brake unit or units.

FAST SHIFT. Shifting engagement rate improperly adjusted; see paragraph 101. Another cause could be high system pressure.

ROUGH PEDAL ENGAGEMENT. If tractor jumps rather than starts smoothly when pedal valve is actuated, a sticking pedal valve or broken pedal valve spring is indicated.

SLIPPAGE UNDER LOAD. If transmission slips, partially stalls or stalls under full load, first check the adjustments of pedal valve as outlined in paragraph 102, then check transmission pressures as outlined in paragraphs 103 and 104. If trouble is not corrected, one of the clutch or brake units; or transmission disconnect clutch, is malfunctioning.

If a clutch or brake unit is suspected of slipping, it will be necessary to determine which of the three units is at fault in that speed. Refer to the table in paragraph 98 to determine which three units are involved in the speed range in question. Then prove one unit at a time by choosing a speed that utilizes that particular unit, and if that speed does not slip, choose another speed that changes only one unit whenever possible. In this way the slipping unit can be isolated. NOTE: 4th to 5th and 5th to 4th change two units at once, as do 6th to 7th and 7th to 6th.

If one or more units are found to be slipping in every gear in which the unit is engaged, remove and overhaul the transmission as outlined in paragraphs 108 through 126.

TRACTOR FAILS TO MOVE. If tractor fails to move when transmission is engaged, first check to see that

possible causes are:

1. Incorrect pedal valve linkage adjustment; correct as outlined in paragraph 102.

2. Malfunctioning regulator valve, pedal valve or oil filter relief valve; overhaul as outlined in paragraph 115 or 116.

If adjustment of operating pressure does not correct the malfunction, leave pressure gage installed and completely check pressure as outlined in paragraph 104.

TROUBLE SHOOTING
Power Shift Models

104. **PRESSURE TEST.** To make a complete check of the transmission hydraulic system pressures, first check and adjust operating pressure as outlined in paragraph 103; then proceed as follows:

With engine speed at 1900 rpm, depress the inching pedal while noting pressure gage reading. Gage pressure should drop to zero with pedal fully depressed.

Release the pedal slowly; gage pressure should rise at a smooth, even rate

until approximately 80 psi is registered with pedal ½ to 1 inch from top; then abruptly to operating pressure with further pedal movement.

Failure to perform as outlined could indicate maladjustment of pedal valve linkage (see paragraph 102) or malfunction of the valve (overhaul as outlined in paragraph 115).

System leakage or malfunctioning shift valves can be determined by installing 0-300 psi gage (or gages) in clutch and brake passage ports in control valve housing and transmission housing. Port plugs are indicated in Fig. 87. Install one or more gages in passage ports and check at 1900 engine rpm by shifting through complete speed range for each test.

The gage should register at approximately system pressure when that control unit is actuated as shown in table accompanying paragraph 98. If the pressure registered in one or more control passage ports is more than 15 psi below system pressure, leakage of that unit is indicated. If pressure is observed on any unit when that unit should not be pressurized, check for sticking shift valves or leakage within

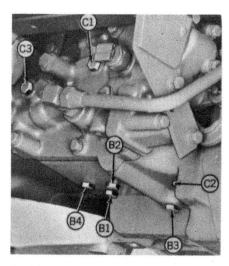

Fig. 87–Port plugs for checking pressure to power shift components are as shown.

transmission disconnect clutch and tow disconnect clutch are fully engaged. NOTE: If transmission disconnect clutch is disengaged or fails to hold due to malfunction, the tractor main hydraulic system will be inoperative after the supply of fluid in oil cooler and pump is exhausted.

CAUTION: Do not attempt to operate tractor with disconnect clutch disengaged. Serious damage could result to main hydraulic pump.

If both disconnect units are engaged, check to see that park pawl operates properly and is correctly adjusted. Park pawl is engaged by cam action and disengaged by a return spring. If spring breaks or becomes unhooked, pawl may remain engaged even though linkage operates satisfactorily. To examine or renew the park pawl return spring, remove transmission housing cover as outlined in paragraph 122.

TRACTOR CREEPS IN NEUTRAL. A slight amount of drag is normal in the clutch and brake units, especially when transmission oil is cold. Excessive creep is usually caused by warped clutch or brake plates; observe the following:

If tractor creeps when inching pedal is depressed and pedal properly adjusted, either Clutch 1 or Clutch 2 is malfunctioning. Check as follows: With engine speed at 1500 rpm and transmission fluid at operating temperature, shift to 2nd speed on a flat, hard surface. Depress the inching pedal; if tractor continues to roll forward at approximately the same speed, Clutch 1 is malfunctioning, if tractor speed increases, Clutch 2 is dragging.

Disconnect speed selector control rod yoke from control arm. With selector lever in "NEUTRAL" position, move the disconnected control rod down one detent notch. If tractor creeps forward

with throttle set at 1500 rpm, Clutch 3 or Brake 4 is dragging; if tractor creeps backward, Brake 3 is malfunctioning.

NOTE: Dragging clutch or brake units, aside from causing creep, will contribute to loss of power, heat and excessive wear. Creep is merely an indication of possibly more serious trouble which needs to be corrected for best performance or to prevent future failure.

REMOVE AND REINSTALL

Power Shift Models

106. **PEDAL AND REGULATING VALVES.** Pedal valve housing and regulating valve housing attach to left side of clutch housing using common gaskets and gasket plate. Housings may be removed separately, but both should be removed to renew the gaskets.

To remove the housings, first remove left battery and battery box. Remove pedal return spring (See Fig. 86) and disconnect rod from valve operating arm. Remove retaining snap ring and withdraw pedal and rod assembly. Disconnect wiring from start-safety switch, and lube pipe and inlet and outlet pipes from housings. Remove access cover and disconnect pto valve operating arm and spring, then unbolt and remove the housings, gasket plate and gaskets. On tractors before serial number 022001 accumulator and spring can be withdrawn from clutch housing bore for service or inspection.

Overhaul the pedal valve housing as outlined in paragraph 115 and regulating valve housing and accumulator as in paragraph 116.

When installing, use light, clean grease to position gaskets and gasket plate, making sure gaskets are installed on proper sides of plate as shown in Fig. 88. Make sure accumulator springs and pistons are in place before installing plate and gaskets. (Early models only). Install regulator valve housing and retaining cap screws, then install pedal valve housing. Tighten retaining screws evenly and securely, and complete the assembly by reversing the disassembly procedure. Adjust as outlined in paragraphs 101, 102 and 103.

107. **CONTROL AND SHIFT VALVES.** To remove the control and shift valve housing, drain transmission and remove right battery and battery box. Disconnect control valve inlet pipe at both ends and remove the pipe. Remove the cotter pins which retain control rods to control arms, remove the retaining cap screws; then remove valve housing, disconnecting the linkage as housing is removed.

Overhaul the removed unit as out-

lined in paragraph 117, and install by reversing the removal procedure. Make sure linkage is connected to control arms as housing is positioned. Tighten retaining cap screws evenly and securely. Adjust as outlined in paragraph 102.

108. **TRACTOR SPLIT.** To obtain access to engine flywheel, engine disconnect clutch and linkage; or power shift transmission main components, it is first necessary to detach (split) clutch housing from engine block. Follow the general procedure outlined in paragraph 80, except for the following.

If the regular John Deere support stand is used, transmission rear oil filter and element must be removed for clearance to install rear stand.

Disconnect main oil supply pipe at regulating valve housing, and wires from start-safety switch, and on 2520 models the wire to transmission oil filter indicator light.

109. **ENGINE DISCONNECT CLUTCH.** The engine disconnect clutch unit can be removed from flywheel after detaching clutch housing from engine as outlined in paragraphs 80 and 108.

NOTE: Starting with 1971 models, the 2520 could be ordered with no linkage to disengage the disconnect clutch on power shift tractors. The clutch disc is used to transmit engine power to the transmission, but may not be disengaged at any time. Overhaul clutch as outlined in paragraph 118.

When reinstalling, use aligning tool JDE-52 or equivalent and tighten cover retaining cap screws to a torque of 35. ft.-lbs. Secure by bending lock tabs. Adjust the three clutch fingers

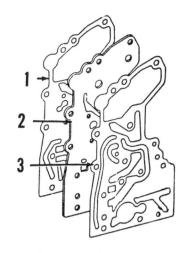

Fig. 88–Make sure gaskets are installed on proper side of plate (2) when installing pedal and regulating valve housings.

1. Outer gasket
2. Gasket plate
3. Inner gasket

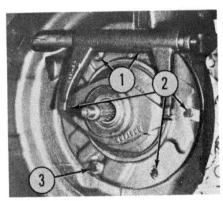

Fig. 89–Adjust the three clutch fingers evenly using tool JDE-51 as shown.

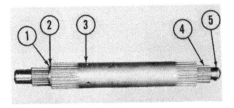

Fig. 91–Three lengths of cap screws are used to secure clutch pack mounting flange to clutch housing.

1. Short cap screws
2. Intermediate screws
3. Long cap screw

evenly, using Tool JDE-51 as shown in Fig. 89.

110. **CLUTCH LINKAGE.** On models so equipped, the engine disconnect clutch linkage is shown exploded in Fig. 90. The clutch is disconnected (for cold starting only) when control lever is pulled rearward until the spring-loaded latch snaps into position in control arm notch. To re-engage the clutch, hold control lever and pull release knob, then slowly allow lever to move forward until fully engaged.

CAUTION: Never run tractor for more than a few seconds with engine clutch locked in disconnect position. The main hydraulic pump cannot obtain fluid unless transmission is operating, and serious damage may result. Disconnect clutch is to be used for COLD STARTING, ONLY.

Release bearing (12) and carrier (11) should be packed with high-temperature lubricant when unit is reassembled.

111. **DISCONNECT CLUTCH ADJUSTMENT.** Upper end of clutch control lever (1—Fig. 90) should have ample movement from the time release bearing contacts clutch fingers until release catch (3) snaps into disconnect position. With clutch housing connected to engine and engine not running, adjust the linkage as follows:

With clutch in operating position, remove the pin which connects control rod yoke (7) to operating arm (5). Pull back on control lever until lockout latch (3) snaps into position. Push down on yoke (7) until release bearing (12) contacts clutch fingers, then thread yoke on control rod (9) until pin holes are aligned in yoke (7) and lever (5). Lengthen rod (9) 6 full turns of yoke (7) and reinsert the yoke pin. Check for complete engagement and disengagement of disconnect clutch before releasing tractor for service.

112. **CLUTCH PACK AND TRANSMISSION PUMP.** The transmission pump and clutch pack can be removed as a unit after detaching clutch housing from engine as outlined in paragraphs 80 and 108; and removing disconnect actuating linkage.

Note the length and location of flange mounting screws as they are removed. Three different lengths are used as shown in Fig. 91.

Connecting shafts (Fig. 92) may be

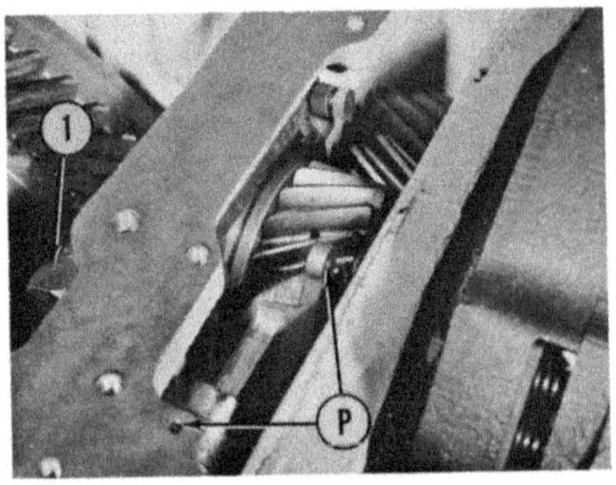

Fig 92–The connecting shaft assembly splines into operating clutch hubs of clutch pack and sun gears of front planetary unit.

1. Snap ring
2. Thrust washer
3. C2 (outer) clutch shaft
4. C1 (inner) clutch shaft
5. Sealing ring

Fig. 93–Top view of transmission housing with covers removed, showing Clutch 3 pressure line (1) and park pawl spring attaching points (P).

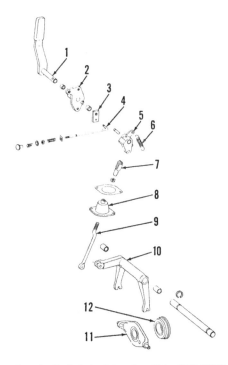

Fig. 90–Exploded view of engine disconnect clutch linkage showing component parts

1. Control lever
2. Support
3. Latch pawl
4. Release rod
5. Lever arm
6. Spring
7. Yoke
8. Boot
9. Rod
10. Fork
11. Carrier
12. Release bearing

Fig. 94–To remove planetary pack after cover is off, remove output shaft, pto drive gear and attaching cap screws (S). Grasp with lifting tool at points indicated and lift from transmission housing

removed with clutch pack; or may be withdrawn after clutch unit is out. Be sure that sealing ring (5) is in good condition and that shafts are in position when unit is reassembled.

Overhaul the removed clutch pack and pump as outlined in paragraphs 119 and 120.

To assist in easier installation of clutch pack, use alignment studs in the two side holes of housing and position gasket on housing using light grease. Make sure oil passages in clutch housing and in gasket are properly aligned. Insert connecting shafts with sealing ring (5) to rear. Install clutch pack with oil passages in mounting flange aligned with those of gasket and clutch housing. Install the two short retaining cap screws in upper holes (1 —Fig. 91); long cap screw in lower hole (3). Install remainder of cap screws (2), tighten all screws evenly and securely and lock in place by bending up corners of tab washers.

113. **CLUTCH HOUSING.** The clutch housing must be removed for access to the pto drive gear train. To remove the clutch housing, first split tractor between engine and clutch housing as outlined in paragraph 80 and 108; and remove clutch pack as in paragraph 112.

Remove rockshaft shields on models so equipped and remove steering support rear panel. Mark and diagram location of hydraulic tubing if necessary; then remove hydraulic tubes and system control linkage. Disconnect park cable and cable housing at rear and light wires at front connections. Remove pto front bearing quill.

Support clutch housing and steering support assembly from a hoist and remove clutch housing flange cap screws. NOTE: The two upper, center screws are accessible through inside front of clutch housing. Pry clutch

housing from its doweled position on transmission case and swing housing assembly away from rear unit.

Use a new gasket and O-rings when reinstalling clutch housing. O-rings may be held in position with grease. Tighten upper center flange cap screws (from inside front of housing) to a torque of 170 ft.-lbs. and remainder of flange screws to a torque of 300 ft.-lbs.

114. **PLANETARY PACK.** To remove the transmission planetary pack, first detach (split) engine from clutch housing as outlined in paragraphs 80 and 108 and remove clutch pack as in paragraph 112. Remove rockshaft housing as outlined in paragraph 166.

Working through clutch housing front opening, unseat the retaining snap ring and pull pto drive gear and bearing forward out of transmission housing. Disconnect pressure tube (1— Fig. 93) from rear bearing quill and

Fig.95–Cutaway view of Power Shift Transmission showing location of parts.

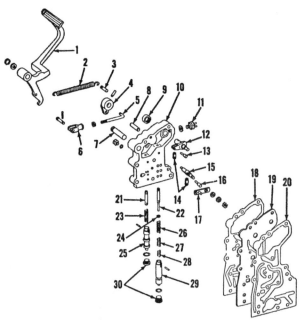

Fig. 96–Exploded view of pedal valve housing and associated parts.

1. Pedal
2. Return spring
3. Anchor pin
4. Operating arm
5. Operating rod
6. Yoke
7. Pivot shaft
8. Stop pin
9. Oil seal
10. Housing
11. Start safety switch
12. Operating shaft
13. Link pin
14. Link
15. PTO valve operating shaft
16. Link pin
17. PTO valve operating arm
18. Outer gasket
19. Gasket plate
20. Inner gasket
21. PTO valve rod
22. Pedal valve rod
23. Spring
24. Spring retainer
25. PTO valve
26. Upper spring
27. Center spring
28. Lower spring
29. Pedal valve
30. Plug

surface of housing. If operating shaft oil seal (9) must be renewed, install seal with lip toward inner side of housing.

Valves (25 & 29) must slide smoothly in their bores and must not be scored or excessively loose. Check the pto and pedal valve springs for distortion and against the values which follow:

PTO Valve Spring
    Free Length, Inches 1 25/32
    Lbs. Test @
        Inches .... 16.5-20.5 @ 1⅜
Pedal Valve Spring (Lower)
    Free Length, Inches .... ½
    Lbs. Test @
        Inches .... 3.3-4.0 @ 11/32
Pedal Valve Spring (Center)
    Free Length, Inches 1 5/16
    Lbs. Test @
        Inches .... 15-18 @ 1 3/32
Pedal Valve Spring (Upper)
    Free Length, Inches 2
    Lbs. Test @
        Inches ...... 11.5-14.0 @ 1 13/32

116. REGULATING VALVE AND ACCUMULATOR. Refer to Fig. 98 for an exploded view of regulator valve housing and associated parts. Refer to Paragraph 106 for removal and installation information.

On tractors before serial number 022001 the single accumulator piston and spring can be withdrawn from clutch housing after regulating valve has been removed. After serial number 022001 the dual accumulators were located in the transmission control valve housing on right side. Check the

remove quill. Using a long brass drift and working through front, center of planetary pack, jar output shaft rearward until rear bearing cup is dislodged from housing bore. Withdraw planetary output shaft rearward out of housing and output gear. Remove the four retaining cap screws (S—Fig. 94) and, using a hoist and ice tongs or similar tool, grasp planetary pack as indicated by arrows and lift the unit

straight upward out of transmission housing.

Overhaul the removed planetary pack as outlined in paragraph 121. Before reinstalling the planetary pack, inspect or renew the four brake-apply-passage O-rings in bottom of transmission housing.

Planetary output shaft should be installed with 0.000-0.002 bearing preload. To check bearing adjustment, make a trial installation of shaft, bearing cup and quill, using one additional shim between bearing quill and housing. Tighten retaining cap screws securely and measure shaft end play using a dial indicator. Remove the shaft assembly. Remove shims from shim pack equal in thickness to the measured end play plus 0.001. Keep remainder of shim pack together with bearing quill for permanent installation after planetary unit is installed.

Lower the planetary unit straight downward when installing, being careful not to dislodge the brake passage O-rings. Tighten the four retaining cap screws evenly and complete the assembly by reversing the disassembly procedure.

OVERHAUL
Power Shift Models

115. PEDAL VALVE. Refer to Fig. 96 for an exploded view of pedal valve housing and associated parts, and to Fig. 97 for a partially disassembled view. Refer to paragraph 106 for removal and installation information.

Pedal shaft (7—Fig. 96) is renewable and is a press fit in housing. Install shaft with inner edge of snap ring groove 1 21/32 inches from machined

Fig. 97–Partially disassembled view of pedal valve housing and valves. To disassemble the pedal valve, drive out the roll pin (P). Refer to Fig. 96 for parts identification.

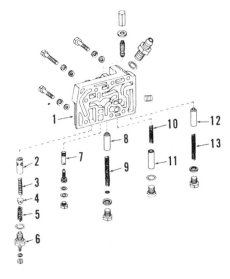

Fig. 98–Exploded view of regulating valve housing.

1. Regulating valve housing
2. Filter relief valve
3. Relief valve spring
4. Spring guide
5. Indicator light spring
6. Indicator light sender
7. Adjustable orifice
8. Pressure regulating valve
9. Regulating valve spring
10. Return pressure spring
11. Return pressure valve
12. Cooler relief valve
13. Cooler relief spring

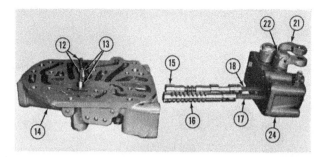

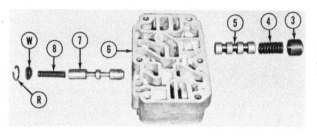

Fig. 99–Partially disassembled view of control valve housing. Refer to Fig. 101 for parts identification.

Fig. 100–Partially disassembled view of shift valve housing. Refer to Fig. 101 for parts identification.

(2520) Oil Filter Relief Valve Spring (3)

Free Length, Inches 2 5/16
Lbs. Test @
Inches . . . . .19-23 @ 1½
(2520 Only) Indicator Light Spring (5)
Free Length, Inches 0.743
Lbs. Test @
Inches . . . . .12-15 @ ½
Regulating Valve Spring (9)
Free Length, Inches 4.29
Lbs. Test @
Inches . . . . .50-61 @ 3 7/16
Return Pressure Relief Spring (10)
Free Length, Inches 2 5/16
Lbs. Test @
Inches . . . . .11-13 @ ¾
Oil Cooler Relief Spring (13)
Free Length, Inches 3.80
Lbs. Test @
Inches . . . . 33.5-40.5 @ 3.22

valves, piston and their bores for sticking or scoring. Valves and piston must move freely in bore without excessive clearance.

On 2510 tractors the valves are all four interchangeable, but the valve springs must be marked or tested to be sure they are properly installed. On 2520 tractors the oil filter relief valve and spring assembly was replaced by a different valve (2) and a sending unit (6) for the transmisison oil filter indicator light. Check the springs against the values which follow, numbers refer to location in Fig. 98:

(2510) Oil Filter Relief Valve Spring (3)

Free Length, Inches 2¾
Lbs. Test @
Inches . . . . .27-33 @ 1 13/16

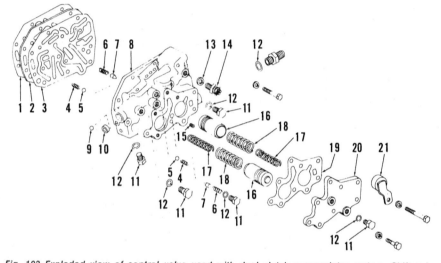

Fig. 102–Exploded view of control valve used with dual clutch accumulator system. Shift valve housing, cover, direction and speed control valves and linkage are the same as single accumulator assembly.

1.	Gasket	11.	Plug
2.	Plate	12.	"O" ring
3.	Gasket	13.	Aluminum washer
4.	Spring (2 used)	14.	Start safety switch
5.	Ball (2 used)	15.	Orifice (2 used)
6.	Spring (2 used)	16.	Accumulator piston
7.	Detent plunger (2 used)	17.	Spring (inner)
8.	Control valve housing	18.	Spring (outer)
9.	Ball	19.	Gasket
10.	Plug	20.	Cover
		21.	Start switch cover

Single Accumulator Piston Return Spring
(Not Shown)
Free Length, Inches 3.96
Lbs. Test @
Inches . . . . .48-57 @ 3.66
186-226 @ 2.78

117. CONTROL VALVE. The shift valve housing is attached to inner face of control valve housing by a cover and six cap screws. Refer to Fig. 99 for a partially disassembled view of control valve housing and associated parts; and to Fig. 100 for a view of shift valve housing. Refer to paragraph 107 for data on removal and installation of control and shift valve unit.

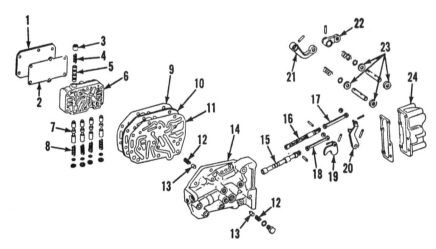

Fig. 101–Exploded view of control and shift valves as used with single accumulator system.

1. Cover
2. Gasket
3. Plug
4. Spring
5. Dump valve
6. Shift valve housing
7. Shift valve
8. Spring
9. Gasket
10. Plate
11. Gasket
12. Spring
13. Detent plunger
14. Control valve housing
15. Direction valve
16. Speed control valve
17. Link
18. Link
19. Operating arm
20. Operating arm
21. Arm
22. Arm
23. Oil seals
24. Cover

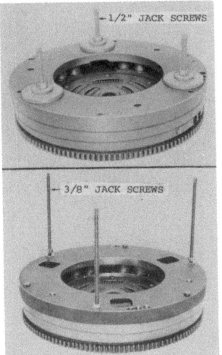

Fig. 103–Methods used to disassemble engine disconnect clutch on models with no release fingers or linkage.

To disassemble the removed unit on single accumulator tractors, proceed as follows: Remove the six cap screws retaining shift valve housing (6—Fig. 101) and lift off the cover, housing, gasket plate (10) and gaskets (9 & 11). Note that the two gaskets are not interchangeable. Mark the removed gaskets "Outer" and "Inner" as they are removed, to aid in correct installation of new gaskets when unit is reassembled. Lift out inner detent springs (12) and plungers (13) to prevent loss. Invert control valve housing (14) and remove outer detent retaining plugs, springs and plungers. Remove the six cap screws retaining cover (24) to housing and withdraw cover, control valves and operating linkage as a unit from control valve housing. Remove plug (3) from shift valve housing and withdraw spring (4) and dump valve spool (5). Remove the four snap rings (R-Fig. 100) and washers (W), then withdraw shift valve springs (8) and spools (7). The four shift valve spools and springs are interchangeable.

Tractors after serial number 022000 used a dual clutch accumulator system, which has been added into the Control Valve housing (see Fig. 102). Disassemble as above, except that the outer cover and gasket retains the two accumulator pistons and the inner and outer springs, which are installed facing opposite to each other. The two replaceable accumulator charging orif-

ices can be removed at this time.

Clean all parts in a suitable solvent and check for scoring or other damage, and for free movement of valve spools in bores. Control valve actuating mechanism need not be disassembled unless renewal of parts is indicated.

Check the springs against the values which follow, numbers refer to location in Fig. 100:

Tractors up to Serial number 022000
Dump Valve Spring (4)
    Free Length, Inches 1.23
    Lbs. Test @ Inches ...... 16-20 @ ¾
Shift Valve Springs (8)
    Free Length, Inches 1.44
    Lbs. Test @ Inches .... 7-8.5 @ 0.81
Tractors after Serial number 022000
Dump Valve Spring (4)
    Free Length, Inches 1 7/16
    Lbs. Test @ Inches .....6-8 @ 13/16
Shift Valve Spring (8)
    1¼ .... Free Length, Inches
    Lbs. Test @ Inches ...... 16-19 @ ¾

Reassemble by reversing the disassembly procedure using Fig. 101 and 102 as a guide. Tighten the six cap screws retaining shift valve housing to control valve to a torque of 20 ft.-lbs.

118. **DISCONNECT CLUTCH.** The disconnect clutch is attached to engine flywheel and can be removed as outlined in paragraph 109. Clutch disc can be lifted out when clutch is removed.

Starting with the 1971 models, the 2520 could be ordered without the linkage for the engine disconnect clutch on Power Shift tractors. This means that the hand lever and linkage, release bearing assembly, and clutch fingers are omitted. The clutch disc is still used to transmit engine power to the transmission. CAUTION: This clutch assembly CANNOT be removed

by simply removing the six cover cap screws as is done with disconnect clutches having the release fingers; on this assembly there is nothing to connect the clutch cover to the pressure plate. Depending on spring condition the spring force can be extremely high.

Therefore, to replace the clutch disc, one of the following methods must be used:

1. Make three ½-inch jack screws and obtain three large washers (see Fig. 103). Install the three jack screws through the square opening in the clutch cover and thread them into the clutch plate as shown. Run the hex nuts down against the large washers. This ties the pressure plate and cover together, just as if the clutch release fingers were installed. The six cap screws may now be removed from the cover and the clutch assembly removed. If the clutch assembly requires disassembly, back off the hex nuts to relieve spring force.

2. If you desire to use ⅜-inch jack screws, remove three of the cover cap screws and install the jack screws in their place as shown in Fig. 103 (lower). This method is not as fast if clutch is not to be serviced, as reassembly will take longer. Reassemble in reverse order of disassembly.

Check the removed disc for loose rivets, broken springs or cracks. Facings should be smooth and free of grease or oil. Hub splines should not show excessive wear. The thickness of a new clutch disc is 0.427-0.447.

Friction surface of flywheel must not be heat-checked or scored, with not more than 0.006 runout. Flywheel may be remachined provided not more than 0.060 is removed from friction surface nor surface lowered to more than 2.909

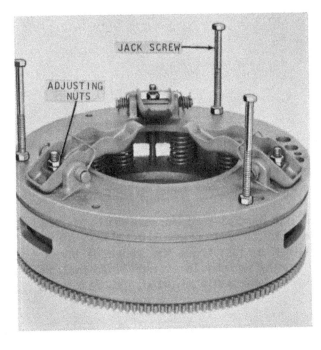

Fig. 104–Use jack screws and flywheel for clutch disassembly. Refer to paragraph 118 for procedure.

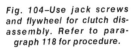

JACK SCREW

ADJUSTING NUTS

Fig. 105–Loosen nuts on jack screws to release spring pressure.

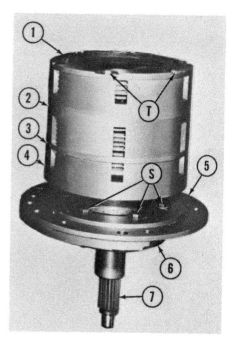

Fig. 107–Assembled view of removed clutch pack showing component parts.

S. Cap screws
T. Through bolts
1. PTO pressure plate
2. PTO—C2 clutch drum
3. C1—C2 pressure plate
4. C1 clutch drum
5. Manifold plate
6. Pump housing
7. Input shaft

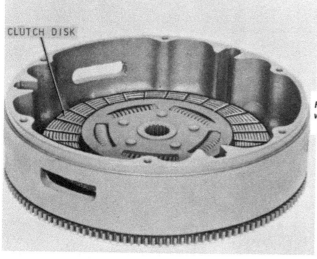

Fig. 106–Install clutch disc with short hub toward pressure plate as shown.

in all clutches. The remaining plate or plates (Clutch C2 uses two) are thinner and identified by a notch in one lug. Clutch C1 uses one disc on early models before Serial Number 008958 and two discs on later models (Or when late model clutch drum is installed on early tractors). Clutch C2 uses three clutch discs, the pto clutch uses two.

Plates and discs for all clutches are installed alternately, either side up, beginning with the thicker (piston) plate next to piston, and ending with an internally splined clutch disc next to pressure plate. Examine pressure plates for scoring or wear and renew as indicated.

All clutch pistons use Belleville spring washers as piston return

below surface of cover mounting flange.

To disassemble the clutch cover and pressure plate, remove the flywheel or use a spare flywheel, and attach clutch cover with disc installed using three jack screws, flat washers and nuts as shown in Fig. 104, 105 & 106. Tighten jack screw nuts against cover and remove the adjusting nuts from release fingers. Back off nuts on jack screws until spring pressure is relieved; remove the jack screws and lift off the cover, springs and pressure plate.

Check pressure plate for scoring of friction surface and excessive wear of drive lugs. Runout must not exceed 0.006. Check clutch springs for rust, pitting or distortion. Clutch springs should have a free length of approximately 3¼ inches and test 105-129 lbs. when compressed to a height of 1¾ inches.

119. CLUTCHES. To overhaul clutch pack, remove assembly as outlined in paragraph 112. To disassemble the removed clutch pack, use a holding

fixture with a 2-inch hole or drill a 2-inch hole near edge of table or bench. Insert input shaft (7—Fig. 107) and release bearing sleeve through hole, with pto clutch pressure plate (1) up. Bend down the locking tab washers and remove through-bolts (T) then lift off pto clutch pressure plate (1), the pto clutch discs and clutch hub. PTO and C2 clutch drum (2), C1 and C2 pressure plate (3) and C1 clutch drum can be separated after jarring slightly with the heel of the hand.

C1 clutch drum can be lifted off input shaft and manifold assembly after unseating and removing the snap ring on rear splines of input shaft.

Clutch plates and discs are interchangeable for all clutches, including Clutch 3 in the planetary pack. The clutch hubs, however, will not interchange. Refer to Fig. 109. The internally splined clutch disc (D—Fig. 108) should measure 0.112-0.118. Two thicknesses of externally lugged clutch plates (P) are used. One (thicker) piston plate is placed against the piston

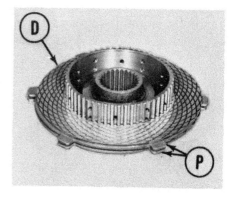

Fig. 108–Clutch discs (D) and plates (P) are interchangeable for all clutches and may be installed either side up.

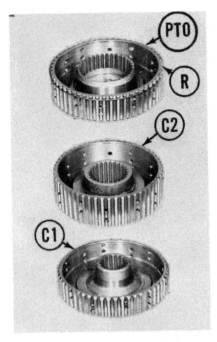

Fig. 109–Clutch hubs are not interchangeable. Note snap ring (R) on PTO clutch hub. PTO and C1 hubs must be inverted when installed.

Fig. 111–Spring compressor for removing clutch pistons.

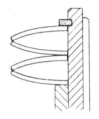 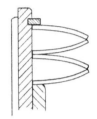

Fig. 112–C3 clutch piston housing partially disassembled. Return spring, spring retainer and snap ring are not shown.

1. Piston   3. Sealing ring
2. Sealing ring   4. Guide dowel

Fig. 113–Cross sectional schematic view showing correct method of installing Belleville washers. Some units used only two washers (1 set) which must be paired.

springs. Two sets (4 washers) are used on all clutches except Clutch C1 in early models Before Serial Number 008958 which used one set (2 washers). Clutch piston must be assembled and disassembled in a press, using a suitable straddle-mounted fixture and compressing spring until the retaining snap ring can be unseated and removed. (See Fig. 111).

NOTE: A fixture can be constructed using about 6 inches of 3½-inch steel pipe. Make sure both ends of pipe are square with centerline. Machine a 1 X 1½ inch notch in one end of pipe for working space to unseat snap ring.

Piston can be worked out of clutch drum by grasping the strengthening ribs with pliers after spring washers have been removed. Refer to Fig. 112. Piston (1) is sealed with an expanding, cast iron ring (2) at outer edge and a neoprene ring (3) on cylinder hub.

Dowel (4) enters a hole in piston to prevent piston rotation. Renew sealing rings (2 and 3) whenever piston is removed, and check piston surfaces of piston and cylinder for scoring or other damage. Reassemble by reversing the disassembly procedure, making sure dowel (4) enters locating hole in piston.

Examine the oil manifold sleeve in C1 clutch drum. Renew drum assembly if sleeve is damaged. Bushing in pto and C2 clutch drum is renewable. Inside diameter of a new bushing is 2.002-2.004. Check new bushing size after installation.

Disassemble, inspect and overhaul the transmission pump, manifold plate and input shaft assembly before reassembling clutch pack, if service is indicated. Follow the procedures outlined in paragraph 120.

Reassemble clutch pack by reversing

the disassembly procedure, making sure clutch hubs are properly positioned and that clutch plates and discs are installed alternately, starting with the thick (piston) lugged plate next to piston and ending with an internally splined clutch disc next to pressure plate. Tighten clutch through bolts evenly to a torque of 28 ft.-lbs. and secure by bending lock plates.

120. **TRANSMISSION PUMP, MANIFOLD PLATE AND INPUT SHAFT.** To overhaul the transmission

Fig. 110–Clutch 1 drum with piston and springs installed. Refer to paragraph 119 for disassembly procedure.

1. Snap ring   3. Piston
2. Belleville washers   4. Clutch drum

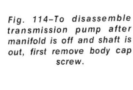

Fig. 114–To disassemble transmission pump after manifold is off and shaft is out, first remove body cap screw.

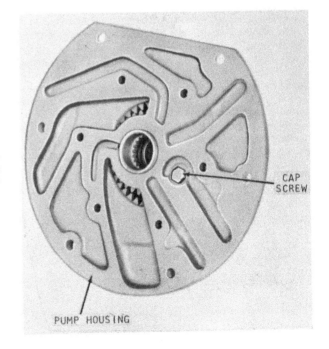

Fig. 115–When installing assembled pump on manifold, make sure steel check balls are in place as shown.

Fig. 118–Planet carrier and gears, assembled view.

C. Cap screw	2. End plate
IP. Input shaft	3. Planet carrier
OP. Output unit	4. Ring gear
1. Retainer	

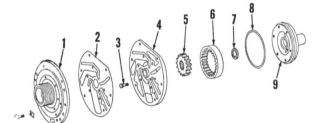

Fig. 116–Exploded view of transmission pump, manifold and associated parts.

1. Manifold
2. Gasket
3. Cap screw
4. Pump housing
5. Drive gear
6. Internal gear
7. Oil seal
8. O-ring
9. Pump body

the disassembly procedure. Tighten cap screw (3) to a torque of 20 ft.-lbs. Place manifold (1) hub down on a bench. Position gasket (2) and drop the two steel check balls into manifold as shown in Fig. 115, before positioning pump assembly on manifold. Tighten the six cap screws retaining pump to manifold to a torque of 20 ft.-lbs. and secure by bending lock plates. Reassemble clutch pack as outlined in paragraph 119.

121. **PLANETARY PACK.** To disassemble the removed planetary pack, place unit on a bench, output end up as shown in Fig. 117. Remove the six cap screws (C) and lift off Clutch 3 piston housing (1). Remove the five through-bolts (T) and disassemble brake piston housing and associated parts until planet carrier assembly can be removed as shown in Fig. 118.

Each planet pinion shaft is retained to its end plate by a steel ball as shown in Fig. 119. The shaft is a slip fit in end plate but prevented from movement by the locking action of the steel ball and by retainer (1—Fig. 118). Each planet pinion contains two rows of loose needle rollers separated by a spacer.

pump, manifold plate and input shaft, first disassemble clutch pack as outlined in paragraph 119. After C1 clutch drum has been removed, bend down locking tabs on cap screws (S—Fig. 107) and remove the screws; then slide input shaft and manifold plate assembly out of pump housing. Remove

and save the two steel check balls (Fig. 115) as pump and manifold units are separated. Remove the one cap screw (Fig. 114) and separate pump housing from body and gear assembly.

Pump gears (5 & 6—Fig. 116) can be lifted out of pump body (9) after housing (4) is removed. Oil seal (7) should be installed in pump body with lip of seal to rear.

Input shaft and bearing assembly can be removed from manifold plate after unseating and removing the retaining snap ring from rear of manifold hub. Press bearing from shaft, if renewal is required, after removing shaft sealing rings and bearing retaining snap ring.

Renew gasket (2), O-ring (8) and cast iron sealing rings on manifold plate (1) and input shaft, whenever unit is disassembled. Examine gears and housings for scoring, wear, cracks or other damage and renew as required. NOTE: On early tractors the pump internal gear (6) has a chamfer on both sides and can be assembled either way. Later models have a wider internal gear with a chamfer on one side only. The chamfer must be assembled toward the pump body (9). Assemble by reversing

Fig. 117–Assembled view of removed planetary pack.

C. Cap screw	3. B3 piston housing
P. Pressure ports	4. B2 piston housing
T. Through bolts	5. Pressure plate
1. C3 piston housing	6. B1 piston housing
2. B4 piston housing	7. Piston return springs

Fig. 119–Planet pinion shafts (S) are retained in end plates by steel balls (B). Refer to paragraph 121 for procedure.

Fig. 120–Output (rear) planetary unit removed from planet carrier. Retainer (R) must be left with unit to hold pinion shafts.

Fig. 122–B1 piston housing with piston removed, showing component parts and details of assembly. Pistons and rings are interchangeable but housings are not.

2. Piston
3. Piston ring
4. Sealing ring

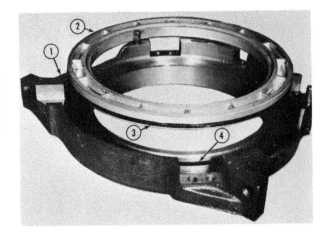

End plates are doweled to planet carrier (3). To prevent loss of any of the small parts on disassembly of the unit, proceed as follows:

Remove the cap screws retaining end plate to input (front) end of planet carrier. Lift off the pressed steel shaft retainer and remove B1 ring gear. Reinstall shaft retainer but not the cap screws. Invert the unit, making sure it is supported so that shaft retainer is held in contact with end plate; then, tap the planet carrier and output planetary assembly loose from its doweled position on input end plate. Disassemble the output planetary unit using the same general procedure, making sure shafts remain in end plate and pinions remain on shafts as shown in Fig. 120 and 121.

Lift off the planet pinions one at a time, keeping bearings in sets as they are removed. Catch and save the steel locking balls as shafts are removed from end plates. All planet pinion bearing rollers are interchangeable, but should be kept (and/or renewed) in sets.

Wash all parts in a suitable solvent. Examine gears for worn, chipped or broken teeth and inside bearing surface of planet pinions for scoring or wear. Examine pinion shafts for nicks, wear or scoring. Fiber thrust washers are used on input end plate, both sides of planet carrier, and output (rear) side of B3 piston housing. Renewable bushings are located in input planetary end plate, Brake 3 piston housing and Clutch 3 piston housing. Inspect bushings for wear or scoring and renew as required. Installed diameters of new bushings are as follows:

Planet carrier end plate . . . 2.130-2.132
Brake 3 piston housing . . 3.755-3.7575
Clutch 3 piston housing:

    Front bushing 1.691-1.693
    Rear bushing 1.753-1.755

Both the internally splined brake discs and externally lugged brake plates measure 0.117-0.123 in thickness when new. The brake plate which is installed next to piston has drilled holes in the four extended lugs to serve as pressure points for the brake return springs. Brakes B1, B2 and B3 are each equipped with two plates and two discs. Brake B4 has one plate and one disc. Check for warped, worn or scored brake discs and plates, and for damage to lugs or splines.

Brake pistons can be removed from piston housings after unit is separated. Refer to Fig. 122. Using two pairs of pliers, grasp piston (2) by two opposing strengthening ribs and work piston from cylinder. Examine piston and cylinder for scoring or other damage, and renew the neoprene sealing rings (3 & 4) whenever piston is removed. Overhaul C3 piston housing and check clutch plates as outlined in paragraph 119.

Assemble planet carrier by reversing the disassembly procedure, being sure to observe the following: Check to see that the steel balls are installed in planet pinion shafts and end plate (Fig. 119) to prevent shafts from turning. Make sure fiber thrust washer is in place on front end plate (E—Fig. 123) before positioning Clutch 2 sun gear (2S). Install double pinions with all "V" timing marks (T) pointing to center as shown to allow all gears to properly mesh at the same time without binding. Position a fiber thrust washer on each side of center web of planet carrier using grease, as unit is assembled. Position B1 ring gear over input planet pinions before permanently installing front pinion shaft retainer. Tighten retaining cap screws securely and fasten by bending retainer tabs, but be certain not to distort shaft retainers.

To assemble the planetary pack, place B1 piston housing (6—Fig. 117) closed end down on a bench, with piston and new rings installed. Install four guide studs in threaded holes and alignment dowels in remaining holes as shown in Fig. 124. Position B1 ring gear and planet carrier assembly as shown. Install end brake plate (with holes) over dowels and studs, then place eight long brake return springs in position over the studs and dowels.

Alternately install two brake discs and one intermediate brake plate. Install B1 and B2 brake pressure plate and B2 ring gear, then install eight more long brake return springs. Starting with an internally splined brake disc, install two brake discs and

Fig. 121–Input (front) planetary unit removed from planet carrier. Retainer (R) holds pinion shafts.

Fig. 123–When assembling input (front) planetary unit, timing "V" marks (T) on pinions must all point to center.

B. Bearing rollers
E. End plate
T. Timing marks

1S. C1 sun gear
2S. C2 sun gear

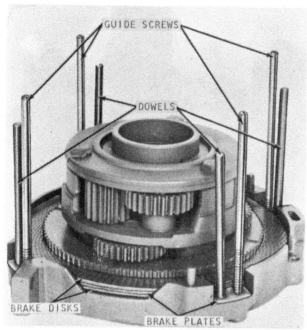

Fig. 124–Use four assembly guide screws in place of through bolts when assembling planetary pack. Refer to paragraph 121.

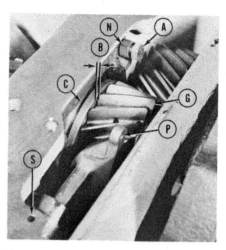

Fig. 126–To adjust the tow disconnect, first loosen locknut (N). Lightly pry output gear (G) rearward and turn adjusting screw (A) until clearance (B) between collar (C) and gear is 0.005-0.010 when lever is in engaged position. (P) and (S) are attaching points for park pawl spring.

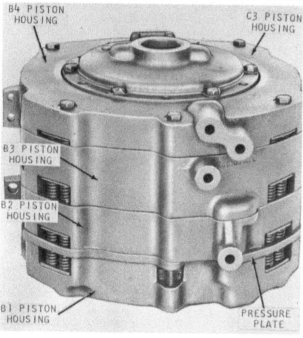

Fig. 125–Assembled view of planetary pack showing parts identified and pressure passage ports properly aligned.

turn spring may be unhooked or broken. The unit can be inspected and spring renewed after removing transmission top cover. Proceed as follows:

Remove operator's platform, steering support rear panel and interfering hydraulic tubes and linkage. Unbolt and remove transmission top cover. Remove damaged or broken spring and install new spring by hooking between points (P & S—Fig. 126). Recheck park pawl adjustment (paragraph 102) after new spring is installed, then reassemble by reversing the disassembly procedure.

Park pawl actuating cam can only be removed after removing planetary pack as outlined in paragraph 114. Refer to Fig. 127. Remove tow disconnect fork and coupling and output reduction gear as outlined in paragraph 123. Remove park pawl shaft retaining snap ring and slide shaft out of housing while park pawl is removed from above. Remove park pawl arm (Fig. 128) and Woodruff key, then remove cam assembly from inside housing. Examine thrust bearing and washers on cam shaft and renew shaft O-ring seal. Check the parts for wear or damage, making sure cam roller turns freely on shaft. Renew questionable parts and reassemble by reversing disassembly procedure. When installing park pawl arm (Fig. 128), slide arm on shaft until 0.005-0.010 end play exists, then tighten clamping screw securely.

123. **TOW DISCONNECT.** To remove the tow-disconnect mechanism, first remove planetary output shaft as outlined in paragraph 114. Unscrew retaining nut (Fig, 129) and retaining nut on disconnect fork shaft (Fig. 128). Remove disconnect fork, collar (C—

one intermediate brake plate, then install end brake plate over dowels and springs. Install B2 piston assembled in piston housing open end down, with oil port opposite mounting lugs on B1 piston housing as shown in Fig. 125.

Follow the same procedure in assembling brake B3, clutch C3 and brake B4, using Fig. 125 as an assembly guide. When properly assembled, all passage ports must be aligned as shown. Remove the aligning studs and install the five through-bolts (T—Fig. 117). Tighten through-bolts to a torque of 35 ft.-lbs. and secure by bending tabs on lock plates.

Before reinstalling planetary pack, apply 50-80 psi air pressure to each of

the passage ports in turn, and listen for air leaks and note action of brake plates. If leaks are noted or if brake return springs do not compress, recheck assembly procedure and correct the trouble before reassembling. Install new O-ring seals in the four brake control oil passages in the transmission case before reassembling the tractor.

REDUCTION GEARS, TOW DISCONNECT AND PARK PAWL
Power Shift Models

122. **TRANSMISSION TOP COVER AND PARK PAWL.** If park pawl remains engaged even though linkage is properly adjusted and operates satisfactorily, the park pawl re-

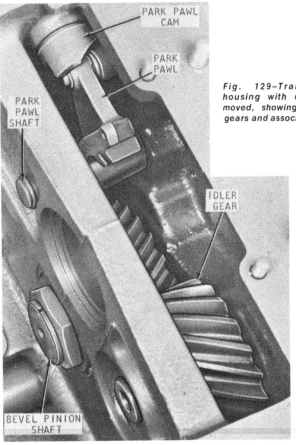

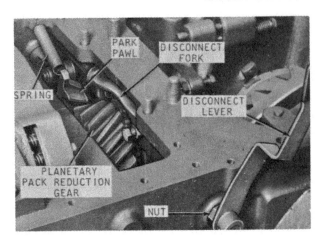

Fig. 129–Transmission housing with covers removed, showing reduction gears and associated parts.

Fig. 127–Top view of transmission housing with planetary pack removed, showing reduction gears and park pawl.

shaft and drift the shaft rearward until front bearing cone (10) and shims (8) can be removed. Withdraw shaft and rear bearing cone from rear while lifting gear (6) and spacers (5 & 7) out through top opening.

The bevel pinion shaft is available only as a matched set with bevel ring gear. Refer to paragraph 129 for information on renewal of ring gear. If rear bearing cup (3) must be renewed, keep cone point adjusting shim pack (4) intact and reinstall the same shim of shims of equal thickness, unless gears and/or housing are renewed. If either

Fig. 126), reduction gear (G) and output shaft front bearing cone. Front bearing cup is retained by a snap ring and can be drifted rearward out of housing after gear has been removed.

Renew sealing washers on fork shafts and install with lever in vertical position with disconnect collar engaged. Adjust as follows, after output shaft is installed and bearings adjusted as outlined in paragraph 114.

Block or hold reduction gear (G— Fig. 126) in rearmost position on shaft; then, loosen locknut (N) and turn adjusting screw (A) if necessary, until

clearance (B) between rear face of gear and front face of collar (C) is 0.005-0.010. Tighten locknut (N) and recheck the adjustment.

124. IDLER GEAR AND SHAFT. Idler gear (Fig. 127) can be removed after removing planetary unit as outlined in paragraph 114 and tow disconnect and reduction gear as in paragraph 123.

Remove idler shaft lubrication pipe and front snap ring (1 and 6—Fig. 131) and drive idler shaft (7) forward out of housing. Spacer (2), and gear (5) can be lifted through top opening as shaft is removed. Use care when removing idler gear to prevent loss of bearings (3) and spacer (4). Renew any parts indicating wear or damage. Assemble in reverse order of disassembly, making sure that roll pin (8) is firmly installed in shaft. Coat bearings and spacer with clean light grease to hold them in position while inserting shaft into housing.

125. BEVEL PINION SHAFT. Bevel pinion shaft and bearings can be removed for service after removing planetary unit as outlined in paragraph 114 and differential assembly as in paragraph 127. Remove idler shaft and gear by following procedures outlined in paragraph 124.

Remove the self-locking nut (11— Fig. 132) from front end of bevel pinion

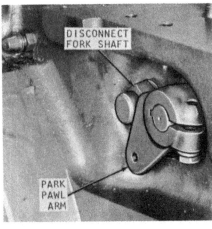

Fig. 128–Right side of transmission housing showing park pawl arm and disconnect fork shaft.

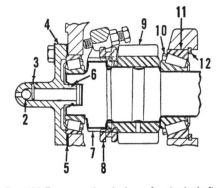

Fig. 130–Cross sectional view of output shaft and associated parts showing Clutch 3 pressure passage seals and shaft bearings.

2. Check ball	7. Output shaft
3. Roll pin	9. Output gear
4. Bearing quill	10. Bearing cone
5. Shim pack	11. Bearing cup
6. Sealing ring	12. Snap ring

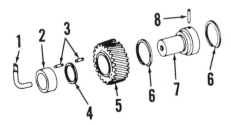

Fig. 131–Exploded view of idler gear and associated parts.

1. Lubrication pipe	5. Gear
2. Spacer	6. Snap ring
3. Bearing	7. Shaft
4. Spacer	8. Roll pin

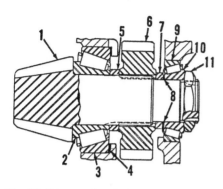

Fig. 132–Cross sectional view of bevel pinion shaft and associated parts.

1. Pinion	7. Spacer
2. Bearing cone	8. Shim pack
3. Bearing cup	9. Bearing cup
4. Shim pack	10. Bearing cone
5. Spacer	11. Shaft nut
6. Gear	

the gears or housing are renewed, check and adjust cone point as outlined in paragraph 126.

When reinstalling bevel pinion shaft and bearings, make a trial installation using the removed shim pack (8) plus one additional 0.010 shim. Tighten shaft nut, then measure shaft end play using a dial indicator. Note the measurement, remove nut and front bearing cone; then remove shims (8) equal in thickness to the measured end play plus 0.005, to obtain the recommended 0.004-0.006 preload of pinion shaft bearings.

NOTE: If main drive bevel gears and/or transmission housing have been renewed, cone point (mesh position) of gears must be checked and adjusted BEFORE adjusting bearing preload.

126. **CONE POINT ADJUSTMENT** The cone point (mesh Position) of the main drive bevel gear and pinion is adjusted by means of shims (4—Fig. 132) which are available in thicknesses

of 0.003, 0.005 and 0.010. The cone point will only need to be checked if the transmission housing or ring gear and pinion assembly are renewed. To make the adjustment, proceed as follows:

The correct cone point of housing and pinion are factory determined and assembly numbers are etched on left upper housing flange and rear face of pinion. To determine shim pack thickness, add 1.755 to the number stamped on rear face of pinion on 2510, or 1.442 on 2520, then subtract the sum from the number etched on transmission housing. If no number appears on Housing, subtract the sum from 8.600. The result is the correct shim pack thickness. Remove and measure the combined thickness of shim pack (4—Fig. 132), then add or remove shims as required. Pinion shaft bearings must be readjusted to 0.004-0.006 preload AFTER cone point has been adjusted.

DIFFERENTIAL AND MAIN DRIVE BEVEL RING GEAR

REMOVE AND REINSTALL
All Models

127. To remove the differential assembly, first drain transmission and hydraulic fluid.

On models with 3-point hitch, remove rockshaft housing as outlined in paragraph 166; knock out the spring pin retaining the extension to top of hydraulic load control arm and remove extension and cam follower. Back out the set screws (A—Fig. 133) retaining caps (B) to each side of drawbar frame and remove the caps. Using a brass drift and working from right side of

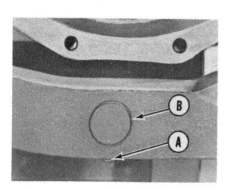

Fig. 133–To remove the drawbar frame, differential assembly or load control shaft seals or bushings, it is first necessary to remove load control shaft. Remove set screws (A) and retaining caps (B), then drive load control shaft out from RIGHT side using a brass drift.

tractor, bump the load control shaft out far enough to clear yoke at lower end of control arm. Pivot the arm forward as far as possible and; when removing differential unit, rotate unit to provide maximum clearance between flat side of differential and load control arm.

On models without 3-point hitch, remove seat and differential case top cover.

On models with differential lock, remove lock control valve, outer link, and three oil pipes to valve.

Fig. 134–Exploded view of bevel ring gear and differential assembly used on models not equipped with differential lock. Ring gear (7) is available only in a matched set with bevel pinion. Shaft (6) is retained in housing by the special cap screw (12) when unit is assembled.

1. Quill
2. Bearing cup
3. Bearing cone
4. Shim
5. Housing
6. Pinion shaft
7. Bevel gear
8. Axle gear
9. Differential pinion
10. Cover
11. Cap screw
12. Special cap screw

On all models, block up tractor and remove both final drive units as outlined in paragraph 134. Remove brake backing plates and brake discs and withdraw both differential output shafts.

On tractors with 'Syncro-Range' transmission, place transmission shift control lever in 'TOW' positon, turn differential assembly until one of the flats in differential housing is uppermost, then remove transmission oil

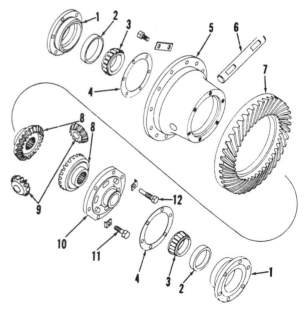

pump with lines attached. If equipped with differential lock, remove the oil pressure inlet pipe.

Place a chain around differential housing as close to bevel ring gear as possible, attach a hoist and lift the differential enough to relieve the weight on carrier bearings. Remove both bearings quills using care not to lose, damage or intermix the shims located under bearing quill flanges. Differential assembly may now be lifted from transmission case.

Overhaul the removed differential as outlined in paragraph 128.

When installing, place an additional 0.010 shim on left bearing quill, tighten retaining cap screws and measure differential end play using a dial indicator. Preload the carrier bearings by removing shims equal in thickness to the measured end play plus 0.002-0.005. Shims are available in thicknesses of 0.003, 0.005 and 0.010.

After the correct carrier bearing preload is obtained, attach a dial indicator, zero indicator button on one bevel ring gear tooth and check the backlash between bevel ring gear and pinion. Proper backlash is 0.008-0.015. Moving one 0.005 shim from one bearing quill to the other will change backlash by about 0.010.

When bearing preload and gear backlash are established, tighten the differential bearing quill capscrews to a torque of 85 ft-lbs. and bend up lock plates.

DIFFERENTIAL OVERHAUL
All Models

128. To overhaul the removed differential assembly, index housing and cover, remove retaining cap screws and lift off cover. Refer to Fig. 134 for exploded view of differential assembly used on models without differential lock. Refer to paragraph 133 for disassembly of differential lock components on models so equipped.

On models without differential lock, one cover retaining cap screw (12—Fig. 134) is extended to lock differential pinion shaft to housing. Models with differential lock use a separate dog-point set screw (22—Fig. 135).

Thrust washers are not used on differential pinion or axle gears. Both differential pinions and the pinion shaft should be renewed as a set if any of the three are damaged. Renew axle side gears or differential housing if worn or scored in thrust areas. Overhaul differential locking mechanism as outlined in paragraph 133 on models so equipped. Refer to paragraph 129 for installation of main drive bevel gear if renewal is indicated.

When reassembling models without differential lock, make sure the special cap screw (12—Fig. 134) is installed in the proper hole to secure differential pinion shaft, tighten all cap screws to a torque of 85 ft-lbs. and bend up lock plates.

BEVEL RING GEAR
All Models

129. The main drive bevel ring gear and pinion are available as a matched set only. To renew the ring gear and pinion, first remove the differential assembly as outlined in paragraph 127.

Remove the cap screws retaining

Fig. 136–Schematic view of hydraulically actuated differential lock.

1. Axle gear	6. Pressure line
2. Clutch plates	7. Pedal
3. Piston	8. Links
4. Pressure line	9. Brake pedals
5. Valve	

main drive bevel gear to differential housing and remove gear using a drift and heavy hammer. The main drive bevel gear is a press fit on housing. Heat ring gear evenly to a temperature of approximately 300°. and positon the gear. Install the retaining cap screws and lock plates and tighten cap screws to a torque of 85 ft-lbs. Renew pinion shaft as outlined in paragraph 92 for models with 'Syncro-Range' transmission or paragraph 125 for models with power shift.

DIFFERENTIAL LOCK

Tractors may be optionally equipped with a hydraulically actuated differential lock which may be engaged to insure full power delivery to both rear wheels when traction is a problem. The differential lock consists of a foot operated control and regulating valve and a multiple disc clutch located in differential housing which locks the left axle gear to differential case.

OPERATION AND ADJUSTMENT
Models So Equipped

130. Refer to Fig. 136. When pedal (7) is depressed, pressurized fluid from the hydraulic system is directed to clutch piston (3), locking axle gear (1) to differential case. Elimination of the differential as a working part causes both differential output shafts and main drive bevel gear to turn together as a unit, transmitting power to both rear wheels equally despite variations in traction.

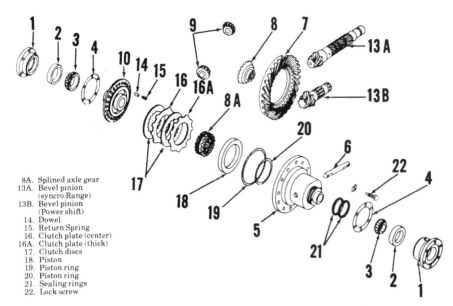

8A. Splined axle gear
13A. Bevel pinion (syncro Range)
13B. Bevel pinion (Power shift)
14. Dowel
15. Return Spring
16. Clutch plate (center)
16A. Clutch plate (thick)
17. Clutch discs
18. Piston
19. Piston ring
20. Piston ring
21. Sealing rings
22. Lock screw

Fig. 135–Exploded view of main drive bevel gears, differential lock and differential assembly. Refer to Fig. 134 for parts identification except for the above.

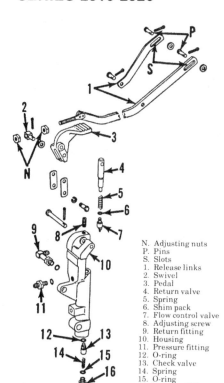

Fig. 137–Exploded view of differential lock control valve and linkage.

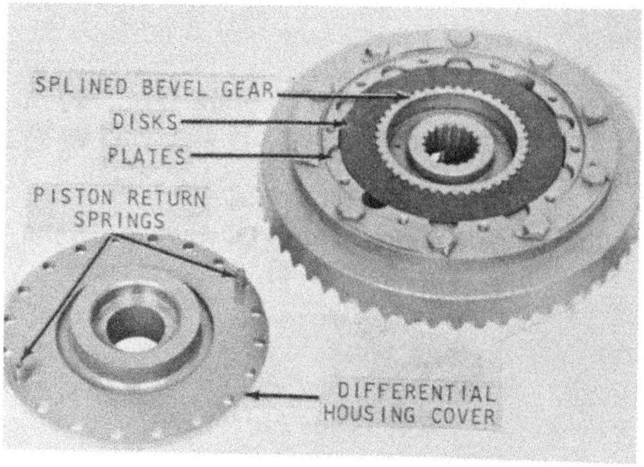

N. Adjusting nuts
P. Pins
S. Slots
1. Release links
2. Swivel
3. Pedal
4. Return valve
5. Spring
6. Shim pack
7. Flow control valve
8. Adjusting screw
9. Return fitting
10. Housing
11. Pressure fitting
12. O-ring
13. Check valve
14. Spring
15. O-ring
16. Pressure fitting

Fig. 139–Differential unit with housing cover removed, showing clutch plates and return springs.

To release the differential lock, slightly depress either brake pedal (9), which releases system pressure by acting through linkage (8).

131. **ADJUSTMENT.** Differential lock valve and linkage should be adjusted for operating pressure and pedal free play, and release linkage adjusted for length. Proceed as follows:

Operating pressure will only need to be checked if valve does not operate properly, or if incorrect pressure is suspected. To check the pressure, install a

Fig. 138–Differential lock valve and associated parts. Outer link connects to brake pedals and releases differential lock when either brake is applied.

0-1000 psi pressure gage in port indicated in Fig. 138. With engine operating at rated speed, depress differential lock actuating pedal. Gage pressure should be 420-480 psi; if it is not, disassemble valve as outlined in paragraph 132 and add or remove adjusting shims (6—Fig. 137) as required.

After valve has been disassembled or if linkage adjustment is required, disconnect release link swivel (2) from pedal (3) and depress pedal until heavy spring pressure is encountered, but not far enough for valve linkage to snap over-center. Release the pedal and turn adjusting screw (8) in or out until it barely clears pedal stop with pedal in normal release position. Recheck carefully before reconnecting swivel, and readjust if necessary, to assure minimum clearance without touching.

Reconnect release swivel (2) and back off both adjusting nuts (N) several turns. With engine running, fully depress pedal (3) until control valve is in locked position. Lightly pull linkage (1) to rear until front ends of slots (S) contact release pins (P) in brake pedals

and all slack is removed; then turn both nuts (N) into contact with swivel without moving linkage or pedal. Tighten both nuts (N) securely. When properly adjusted, the slightest movement of either brake pedal should release the differential lock, but valve will not release unintentionally because of linkage vibration.

OVERHAUL
Models So Equipped

132. **CONTROL VALVE.** To remove the differential lock control valve, bleed down main hydraulic system pressure by actuating brakes with engine not running. Disconnect release linkage and the three oil pipes at the valve. Remove the two retaining cap screws and lift off the valve, noting the location of the two spacer bushings on retaining cap screws.

Remove inlet fitting (16—Fig. 137). Disconnect pedal to valve linkage, remove pedal pivot pin and lift off pedal; then withdraw return valve (4), spring (5), shim pack (6), flow control valve (7), check valve (13) and spring (14) from housing bore.

All parts are renewable individually. Examine parts for wear or scoring and springs for distortion, and renew any parts which are questionable. Keep shim pack (6) intact for use as a starting point when readjusting operating pressure. Renew O-rings whenever valve is disassembled.

Assemble by reversing the disassembly procedure, using Fig. 137 as a guide. Make sure spacer bushings are properly positioned between valve and axle housing, reinstall valve and tighten retaining cap screws to a torque of 130 ft.-lbs. Check and adjust operating pressure and linkage as outlined in paragraph 131.

133. **DIFFERENTIAL CLUTCH.** The multiple disc differential clutch can be overhauled after removing the unit as outlined in paragraph 127. Remove the cover retaining cap screws

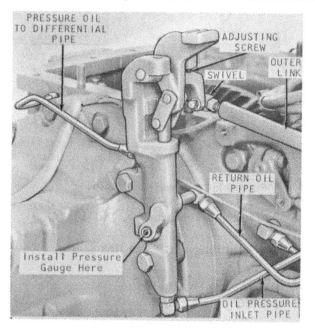

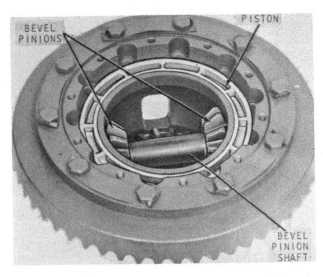

Fig. 140–Differential lock piston can be removed after lifting out clutch plates

Fig. 142–Assembled differential unit, showing pressure passage seals and pressure passage.

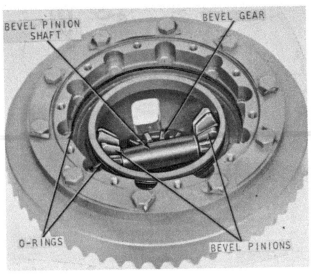

Fig. 141–Differential assembly with piston removed, showing sealing rings and differential gears.

and lift off cover and the two piston return springs as shown in Fig. 139. Clutch discs, plates and splined bevel gear can now be lifted from housing.

Remove piston (Fig. 140) with air pressure or by grasping two opposing strengthening ribs with pliers. Renew sealing O-rings (Fig. 141) whenever

piston is removed. Overhaul differential assembly as outlined in paragraph 128 if required, while differential clutch is disassembled.

Examine sealing surface in bore of right bearing quill and renew quill if sealing area is damaged. Renew the cast iron sealing rings (Fig. 142) on differential housing if broken, scored or badly worn. Check clutch plates and discs and renew if scored, warped, heat discolored or worn to a thickness of 0.100 or less.

Clutch plate (16A—Fig. 135) has ten external lugs and should be installed next to clutch piston. Plate (16) has eight lugs and is installed between the internally splined clutch discs. The two clutch return springs (15) are installed in the blank spaces left by the missing lugs on plate (16). Guide dowels (14) for springs are pressed into cover (10). Retain the springs with light grease during clutch assembly.

REAR AXLE AND FINAL DRIVE

Tractors are available in high clearance (Hi-Crop) models equipped with drop housings containing a final reduction bull gear and pinion; or standard models which have a planetary reduction final drive gear located at inner ends of rear axle housings.

REMOVE AND REINSTALL
All Models

134. To remove either final drive as a unit, first drain the transmission and hydraulic fluid, suitably support rear of tractor and remove rear wheel or wheels.

On standard models, remove fenders and light wiring if so equipped. On Hi-Crop models remove 3-point hitch lift links and draft links or entire drawbar assembly.

On all models, if tractor is equipped with breakaway coupling for hydraulic remote cylinders, remove operator's platform and interfering hydraulic lines.If removing right-hand axle housing on a tractor with a differential lock, disconnect the right battery, differential lock valve, pipes and link.

Support final drive assembly with a hoist, remove attaching cap screws and swing the unit from transmission housing.

On Hi-Crop models remove the six stud nuts securing drop housing to shaft housing, remove the two retainer plugs and thread jack screws into retainer plug holes. Tighten jack screws evenly to force housings apart.

When reinstalling, tighten the retaining cap screws to a torque of 130 ft.-lbs. on standard models or 150 ft.-lbs. on Hi-Crop models. Complete the installation by reversing the removal procedure.

OVERHAUL
All Models Except Hi-Crop

135. To disassemble the removed

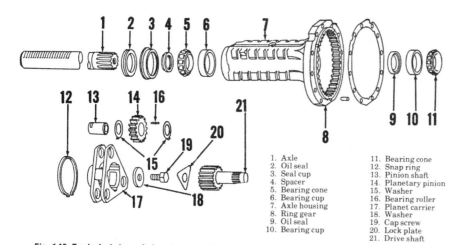

1. Axle
2. Oil seal
3. Seal cup
4. Spacer
5. Bearing cone
6. Bearing cup
7. Axle housing
8. Ring gear
9. Oil seal
10. Bearing cup
11. Bearing cone
12. Snap ring
13. Pinion shaft
14. Planetary pinion
15. Washer
16. Bearing roller
17. Planet carrier
18. Washer
19. Cap screw
20. Lock plate
21. Drive shaft

Fig. 143–Exploded view of planetary type final drive assembly used on all models except Hi-Crop.

final drive unit, remove lock plate (20 —Fig. 143) and cap screw (19) at center of planet pinion carrier and withdraw the planet carrier assembly.

Planet pinion shafts (13) are retained in carrier by snap ring (12). To remove, expand the snap ring and, working around the carrier, tap each shaft out while snap ring is expanded. Withdraw the parts, being careful not to lose the 31 loose bearing rollers in each planet pinion. Examine shaft, bearing rollers and gear bore for wear, scoring or other damage and renew as indicated. NOTE: Bearing rollers should be renewed as a set. New dimensions are as follows:

Planet Pinion ID 2.3835-2.3849
Carrier Pinion Shaft
 Bore ID 1.9458-1.9478

After planet carrier has been removed, axle shaft (1) can be removed from inner bearing and housing by pressing on inner end of axle shaft. Remove bearing cone (5) and spacer (4) if they are damaged or worn. When assembling, heat spacer (4) and bearing (5) to approximately 300° F. and install on shaft making sure they are fully seated. Heat inner bearing cone (11) to 300° F. Have planet carrier assembled and ready to install. Insert bearing on shaft and install and partially tighten the retaining cap screw (19). Leave a barely noticeable end play in axle bearings. Temporarily install lock plate (20) and, using a torque wrench calibrated in inch-pounds, check and record the rolling torque. Remove lock plate (20) and tighten until rolling torque is increased 20-70 inch-pounds from previous reading, then reinstall lock plate (20) and complete the assembly by reversing the disassembly procedure.

Hi-Crop Models

136. OUTER HOUSING AND GEARS. If only the outer housing, gears, shafts, bearings or oil seals are being overhauled, the complete final drive assembly will not need to be removed. Suitably support the tractor and remove wheel and tire unit. Remove draft link or disconnect drawbar from drop housing.

Remove the six stud nuts securing drop housing to shaft housing, remove the two retainer plugs and thread jack screws into retainer plug holes. Tighten jack screws evenly to force housings apart.

To disassemble the removed drop housing, first remove the bull gear cover and wheel axle inner bearing cover, then remove inner bearing nuts with a spanner wrench. Unseat snap ring on inner side of bull gear, then press out rear axle shaft. Axle shaft is equipped with two oil seals, an inner seal which is pressed into housing and a two-piece outer seal in housing and on shaft. When assembling, heat bearing cones to a temperature of 300° F. to facilitate installation. Install axle shaft nut and tighten until drop housing wall is deflected 0.002 in area of shaft bearing, to establish the recommended bearing preload. Tighten the drop housing retaining stud nuts to a torque of 275 ft.-lbs. and fill each final drive housing with 3½ pints of SAE 90 multi-purpose gear lubricant.

137. SHAFT HOUSING AND INNER GEARS. To overhaul the drive shaft housing and associated parts, first remove final drive assembly as outlined in paragraph 134 and outer drop housing as in paragraph 136.

Remove cotter pin and slotted nut from inner end of final drive shaft, then remove shaft using a brass drift. When reinstalling, heat bearing cones to a temperature of 300° F. for easy installation and tighten inner nut to provide end play of 0.004 for shaft bearings. Drive shaft inner oil seal may be renewed when shaft is out.

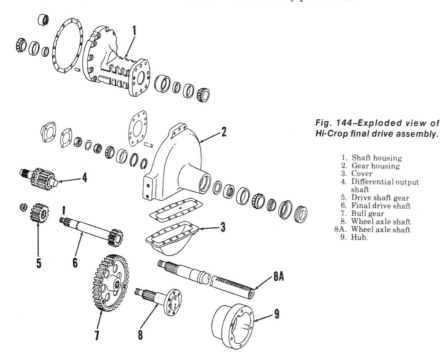

Fig. 144–Exploded view of Hi-Crop final drive assembly.

1. Shaft housing
2. Gear housing
3. Cover
4. Differential output shaft
5. Drive shaft gear
6. Final drive shaft
7. Bull gear
8. Wheel axle shaft
8A. Wheel axle shaft
9. Hub.

BRAKES

OPERATION AND ADJUSTMENT
All Models

138. The hydraulically actuated single disc brakes are located on the differential output shafts and are accessible after removing final drive units as outlined in paragraph 134, and the output shaft and backing plate.

Power is supplied by the system hydraulic pump through foot operated

Fig. 145—Adjust the operating rod for each brake to a length of 2⅞ inches as shown.

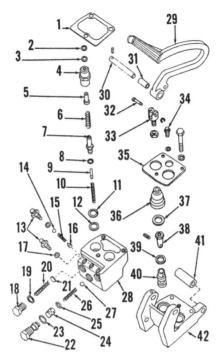

Fig. 147—Exploded view of typical hydraulic power brake operating valve. A master piston (4) applies the brakes if power or accumulator pressure is not available.

1. Gasket	22. Plug
2. Backup ring	23. O-ring
3. O-ring	24. Filter
4. Manual piston	25. O-ring
5. Valve plunger	26. Spring
6. Spring	27. Check valve ball
7. Nipple	28. Housing
8. O-ring	29. Pedal
9. Brake valve	30. Shaft
10. Spring	31. Bushing
11. Backup ring	32. Pin
12. O-ring	33. Yoke
13. Connector	34. Connector
14. Steel ball	35. Boot retainer
15. Spring	36. Boot
16. Check valve stop	37. Washer
17. O-ring	38. Operating rod
18. Plug	39. Seal
19. O-ring	40. Guide
20. Spring	41. Spacer
21. Valve disc	42. Bracket

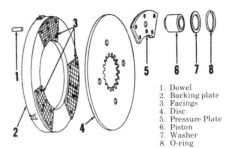

Fig. 148—Exploded view of the wet type hydraulic brake operating parts located on differential output shafts.

1. Dowel
2. Backing plate
3. Facings
4. Disc
5. Pressure Plate
6. Piston
7. Washer
8. O-ring

control valves when engine is running, or manually by means of master cylinders when hydraulic system pump is inoperative. On 2520 models after serial number 022000, a nitrogen filled accumulator is used to store oil under pressure to be supplied to the brake valve immediately if the engine should stop, or for any reason the hydraulic pressure should fall. The accumulator is fastened to the right rear of engine cylinder head.

The only adjustment provided is an adjustable yoke on upper end of each operating valve. This adjustment is used to equalize the height of the brake pedals. To adjust, loosen the locknut and turn operating rod in or out until center of pedal connecting pin is 2⅞-3 inches from surface of boot retainer and both pedals are equal in height. (See Fig. 145).

For service on the hydraulic pump, refer to paragraph 162. To service control valve, actuating cylinders and brake discs, refer to the appropriate following paragraphs.

Fig. 146—To bleed the brakes, loosen the bleed screw (B) and actuate the valve. Bleeding action is internal.

NOTE: Leakage or wear of the brake control valves can usually be determined by performing the leakage test outlined in paragraph 159. Other tests of the main hydraulic system which might affect brake operation are covered in paragraphs 155 through 158.

BLEEDING
All Models

139. When brake system has been disconnected or disassembled, bleed the system as follows:

Start the engine, loosen lock nut on bleed screw (B—Fig. 146) and back out the bleed screw two full turns. Bleed return passage is internal, retighten locknut to prevent external leakage, depress brake pedal and hold depressed for 30 seconds or longer, to flush air from the system. Tighten bleed screw then release the pedal. Bleed opposite brake following the same procedure.

To test the system, stop the engine and depress each brake pedal once. Solid pedal action should be obtained on next application. If pedal action is spongy or pedal travel exceeds 5¾ inches, repeat the bleeding procedure.

CONTROL VALVE
All Models

140. To remove the control valve, disconnect the pressure and discharge lines at valve housing. Remove the attaching cap screws and lift off the control valves and pedals as a unit.

To disassemble the removed unit, remove the two operating rod to pedal connecting pins (32—Fig. 147) and drive the spring pin from pedal shaft (30). Tap out the shaft and remove the two pedals. Remove the cap screws attaching pedal bracket (42) to valve body (28) and lift off the bracket and operating rod assemblies. It will not be necessary to disassemble operating rods unless seal or parts renewal is indicated.

Use Fig. 147 as a guide when disassembling the control valve. Manual brake pistons (4), brake valve plungers (5) and plunger return spring (6) can be lifted out. Use a deep socket to remove inlet valve nipple (7). Clean the parts thoroughly and check for sticking, scored, or excessively worn pistons or valves, or distorted springs.

When reassembling, use new O-rings and gaskets. Use Fig. 147 as an assembly guide. Tighten the inlet valve nipple to a torque of 40 ft-lbs. When control valve has been installed, check for equal pedal height and bleed the system as outlined in paragraph 139.

DISCS AND SHOES
All Models

141. To remove the brake discs or operating cylinder, first remove the final drive unit as outlined in paragraph 134. Remove the backing plate, output shaft and brake disc. The three stationary shoes are riveted to the backing plate. The three actuating shoes are pressed on the operating cyl-

Fig. 149—Transmission housing with final drive removed, showing brake shoes (S) and cylinder (C). The three cylinders are internally bled by bleed screw (B).

POWER TAKE-OFF

OPERATION

All Models

143. All models are available with 1000 rpm front pto and either a 1000 rpm rear pto or 540-1000 rpm dual speed rear pto which deliver the rated pto speeds at 2100 rpm engine speed.

Syncro-Range

144. The power take-off on 2510 and 2520 "Syncro-Range" models before serial number 017000 had a direct mechanical linkage from the pto lever to the pto clutch fingers, and to the pto brake. The pto clutch was applied and held by spring pressure. See paragraph 81 for overhaul.

The 1969 model change (017000 and later) uses the pto lever to actuate a two-way valve (Fig. 152) which routes oil pressure to two pto operating pistons (28—Fig. 63-A) which move a bearing and collar (11-12) against the clutch fingers. The fingers transmit the hydraulic force to the pto clutch and hold it applied until the pto lever is released to the neutral position, at which time the pto collar return springs (9) push the collar back from the clutch fingers and allow the clutch to release. As the pto lever is moved rearward against spring pressure to the brake position, the two way valve moves upward directing system pressure to the pto brake piston (32 Fig. 63-A). When the pto lever is released, a brake return spring returns the lever

Fig. 151—PTO shift lever and associated parts on 2520.

and valve to the neutral position. If the neutral point is not properly adjusted, pressure can be routed to either the apply collar or the brake piston, which can damage the clutch, or the differential drive gear shaft front bearing, by forcing it to carry a thrust load while running with the brake partially applied. See Fig. 151 and 152 for parts breakdown.

Power Shift

145. The pto is driven by a multiple disc, hydraulically operated clutch within the power shift transmission. Pressure is supplied by transmission oil pump. Refer to Fig. 153 for exploded views of pto gear train and associated parts. Refer to paragraph 115 for pto shift valve service, and to paragraph 119 for overhaul data on pto clutch.

inders which can be withdrawn from transmission housing after disc is removed. Facings are available in sets of three and should only be renewed as a set.

Operating pistons are 1.4995-1.5005 in diameter before serial number 022001, and 1.8745-1.8755 on later units, and have a diametral clearance of 0.0025-0.0065 in cylinder bores. Refer to Fig. 148 for an exploded view of brake parts.

BRAKE ACCUMULATOR

142. **R&R AND OVERHAUL.** To remove the accumulator, first relieve pressure by opening the bleed screw and depressing brake pedal. Remove tubing and brackets. **CAUTION: Overhaul of brake accumulator should not be attempted unless charging equipment is available. Gas side of accumulator piston is charged to 500 psi with NITROGEN gas.**

Bleed gas before attempting to disassemble by removing plug (25—Fig. 150) and depressing valve (20). With pressure removed, push cylinder end (23) into cylinder, then unseat and remove snap ring (8). Check all parts for wear or damage and assemble by reversing the disassembly procedure. Recharge the cylinder using approved charging equipment and DRY NITROGEN ONLY, to a pressure of 500 psi. DO NOT use air or any combustible gas as oxidation and condensation are harmful to the oil piston seal and the accumulator.

Reinstall unit on tractor, run engine at 2100 rpm to pressurize the hydraulic system, then bleed brakes. Stop the engine and immediately depress both pedals firmly. If a solid pedal is not felt, bleed brakes again and recheck.

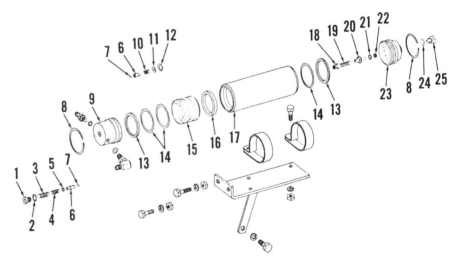

Fig. 150—Exploded view of brake accumulator and associated parts and mounting brackets.

1. Plug	8. Retainer ring	14. "O" ring	20. Valve
2. "O" ring	9. Accumulator cap	15. Piston	21. Packing
3. Spring	10. Spring	16. Packing	22. Washer
4. Spring	11. Washer	17. Cylinder	23. Cylinder end
5. Washer (2)	12. Retainer ring	18. Spring guide	24. "O" ring
6. Guide	13. Backup ring	19. Spring	25. Plug
7. Ball			

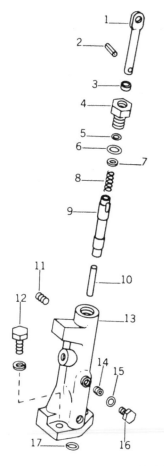

Fig. 152–Exploded view of PTO clutch valve.

1. Shaft	10. Pin
2. Pin	11. Plug
3. Seal	12. Cap screw
4. Plug	13. Valve housing
5. "O" ring	14. Set screw (3)
6. "O" ring	15. "O" ring (2)
7. Washer	16. Plug
8. Spring	17. "O" ring (3)
9. Valve	

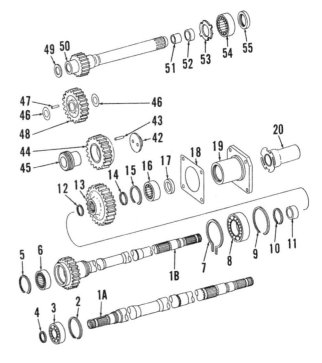

Fig. 153–Exploded view of pto shafts, gears and associated parts used on Syncro Range models and located in transmission housing.

1A. Shaft (single speed)
1B. Shaft (dual speed)
2. Snap ring
3. Bearing
4. Snap ring
5. Snap ring
6. Bearing
7. Snap ring
8. Bearing
9. Snap ring
10. Snap ring
11. Bushing
12. Snap ring
13. Gear
14. Snap ring
15. Snap ring
16. Bearing
17. Oil seal
18. Gasket
19. Bearing quill
20. Guard
42. Plate
43. Needle roller
44. Idler gear
45. Inner race
46. Washer
47. Needle roller
48. Gear
49. Thrust washer
50. Drive (clutch) shaft
51. Bushing
52. Oil seal
53. Washer
54. Bearing
55. Oil seal

Linkage Adjustment

146. On Syncro-Range tractors before serial number 017000, the pto shift linkage used over-center clutch fingers that were self adjusting provided the clutch disc was serviceable. The lever needed only to move forward far enough to allow the fingers to snap over center, (Fig. 63) move far enough rearward to apply the brake, and move slightly forward for neutral.

On 2520 tractors from serial number 017000 on, the pto clutch used system hydraulic pressure to apply and hold the clutch in the engaged position. The lever adjustment is very important to insure a positive neutral position.

The most accurate method for checking the pto for correct adjustment is to use a 0 to 3000 psi pressure gage in the test ports. This procedure is as follows:

1. Remove the brake pressure test plug on the lower left-hand side of the clutch housing and install the gage.

NOTE: This is located below the words pto brake.

2. Move the pto lever rearward toward the brake position, and mark on the dash where the rear of the pto lever is located when the gage starts recording pressure.

3. Continue moving the lever rearward until it stops. The pressure reading should be the same as system pressure (2200 to 2300 psi). If the pressure reading is different than system pressure, the valve is incorrectly adjusted.

4. Remove gage and reinstall plug. Remove the test plug located on the inner side of the pto valve near the word "clutch," and install the gage.

5. Move the pto lever forward toward the engaged position and mark the rear location of the lever when the gage starts recording pressure.

6. Continue moving the lever forward to the overcenter lock position and check the engaged pressure. If the pressure is less than system pressure, readjust the valve linkage.

7. Move pto lever to disengaged position. The lever edge should be halfway between the two marks. If it is not, it will be necessary to readjust the brake-return spring.

When the pto valve and lever adjustments are correct, the pto lever neutral position will be halfway between start of pressure to the pto operating piston and the pto brake. This neutral is held by the brake-return spring balancing against the light spring inside the signal spring. When the pto lever is moved forward far enough to move the pto valve operating rod ⅛ inch, the link should be just contacting the outer signal spring.

147. To adjust the linkage on power shift models, push the pto lever forward into the fully engaged position. With the valve operating arm (17—Fig. 96) in its lowest position, adjust yoke on clutch operating rod (Fig. 151) until pin will just fit through holes in yoke and arm.

Models So Equipped

148. **PTO CLUTCH VALVE.** To overhaul the pto clutch valve, remove left lower control support cover, disconnect oil line and operating rod to valve, and remove two mounting capscrews. Use Fig. 152 as a guide and disassemble valve. Check for worn or scored parts and renew any in questionable condition. Reassemble in reverse order of disassembly.

R&R AND OVERHAUL
All Models

149. **PTO FRONT BEARING.** To renew the pto front bearing (16—Fig. 153 or 154) or oil seal (17), first drain the transmission and hydraulic system fluid. Remove front pto guard (20) then unbolt and withdraw quill (19). Bearing (16) is retained in quill by snap ring (15), and two punch holes are provided in quill to aid in bearing removal. Working through the holes, drift bearing rearward out of quill and install with a piloted arbor and a press. Install oil seal (17) with lip facing

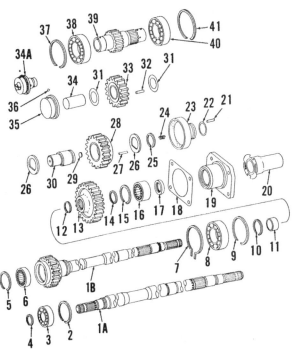

21. Dowel pin
22. O-ring
23. Brake shoe
24. Spring
25. Snap ring
26. Thrust washer
27. Needle roller
28. Idler gear
29. Expansion plug
30. Idler shaft
31. Washer
32. Needle roller
33. Idler gear
34. Idler shaft (early)
34A. Idler shaft (late)
35. Quill
36. Ball
37. Snap ring
39. Drive shaft
40. Bearing
41. Snap ring

toward bearing.

150. PTO DRIVE GEARS AND CLUTCH SHAFT. To renew the pto drive gears or transmission pto shaft, first drain transmission and hydraulic system fluid; then detach (split) transmission from clutch housing as outlined in paragraph 83 (Syncro-Range Models) or 113 (Power Shift Models).

On "Syncro-Range" models, remove pto clutch shaft (50—Fig. 153) as outlined in paragraph 84, if renewal is required. Pto countershaft idler gear (48) and bearings can be slipped from shaft after tractor is split. Hold front and rear washers in contact with gear to retain the loose needle rollers as gear is withdrawn. Idler gear (44) mounts on front end of the transmission main drive bevel pinion shaft and is retained by plate (42) and two cap screws. Inner race (45) is a press fit on pinion shaft and removal procedure is outlined in paragraph 88. Pto drive gear (13) is retained to front of transmission pto shaft (1A or 1B) by snap ring (14), and can be withdrawn after snap ring is removed.

On power shift models, refer to Fig. 154. To disassemble the gears after clutch housing is removed, withdraw pto drive shaft and gear (39) and bearings (38 & 40) as a unit. Withdraw idler gear (33), thrust washers (31) and loose needle rollers. Remove and save spring (24), unseat and remove snap ring (25), then withdraw thrust washer (26), brake idler gear (28) and the contained loose needle rollers (27). Remove snap ring (14) and pto drive gear (13).

On all models, remove transmission

pto shaft (1A or 1B) if necessary, as outlined in paragraph 151.

Renew any gears which are chipped, worn or otherwise damaged. Examine bearing surfaces of gear hubs and idler shafts for wear or scoring. Renew loose needle rollers in complete sets for any one bearing.

Assemble by reversing the disassembly procedure, using clean multipurpose grease to retain loose needle rollers during assembly.

151. PTO OUTPUT SHAFT AND REDUCTION GEARS. To remove the output shaft (and reduction gears on dual speed models), detach trans-

mission from clutch housing as outlined in paragraph 83 or 113. Unseat snap ring (14—Fig. 153 or 154) and withdraw pto drive gear (13). On dual speed models, remove snap ring (67—Fig. 155) and withdraw the output shaft (66 or 68) which is in operating position. Unbolt and remove rear bearing quill (63), then lift out the loose 540 rpm drive gear (56).

Compress snap ring (7—Fig. 153 or 154) and unseat from its groove; then tap transmission pto shaft (1B) rearward. (Make sure the snap ring is properly compressed and free from groove in case before driving on pto shaft.) When bearing (8) is free from housing bore, withdraw the shaft. NOTE: Flat side of differential housing must be turned down to provide clearance for removal of pto shaft. Withdraw idler gear shaft (70—Fig. 155) rearward while lifting out the gear (76) as shown in Fig. 156. Be careful not to lose the loose needle rollers (74—Fig. 155).

On single speed pto models, the transmission pto shaft (1A—Fig. 153 or 154) and rear bearing quill (63—Fig. 155) will be removed as a unit, after removing pto drive gear from front of shaft and the cap screws retaining quill to rear of housing.

Bushing (11—Fig. 153 or 154) in transmission housing has an installed ID of 1.626-1.629. Renew bushing using a piloted driver, if worn or scored. Examine gears and bearings for wear, scoring or other damage and renew as required.

When assembling, use clean grease

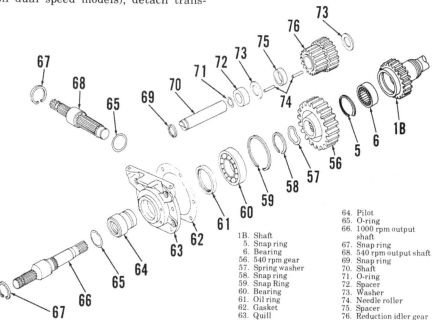

64. Pilot
65. O-ring
66. 1000 rpm output shaft
67. Snap ring
68. 540 rpm output shaft
69. Snap ring
70. Shaft
71. O-ring
72. Spacer
73. Washer
74. Needle roller
75. Spacer
76. Reduction idler gear

1B. Shaft
5. Snap ring
6. Bearing
56. 540 rpm gear
57. Spring washer
58. Snap ring
59. Snap Ring
60. Bearing
61. Oil ring
62. Gasket
63. Quill

Fig. 155—Exploded view of PTO output shaft and reduction gears used on dual speed models. On single speed models, housing (63) carries rear end of shaft (1A–Fig. 153 or 154).

Fig. 156—When removing reduction idler gear and shaft, be careful not to lose the loose needle rollers.

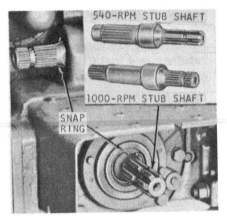

Fig. 157—Rear view of tractor showing the two pto output shafts in operating and storage positions. Shafts can be interchanged by removing snap rings.

shim gaskets (24) at housing flange and output shaft bearings are adjusted by the castellated nut retaining pulley (1). When assembling the pulley unit, proceed as follows:

Install output shaft (14) in housing (3), using a pipe spacer of appropriate length in place of pulley (1). Install the castellated nut and tighten nut to remove all end play without binding. Install shim pack 0.058 thick at (16), then install the assembled input shaft and housing (5) to housing (3), omitting the flange shims (24). Install two flange cap screws 180° apart and tighten finger tight; then measure gap between housings (3 & 5) to determine thickness of shim pack required. Lift off the housing and install steel shims (24) equal to the measured gap. Shims (24) are available in 0.005 (paper) and 0.003, 0.005 and 0.010 (steel). Reinstall the housing and secure with three cap screws tightened snugly. Insert the 1000 rpm pto stub shaft and measure gear backlash at stub shaft splines. Backlash should be 0.003-0.005; if it is not, add or remove shims as necessary to shim pack (16). Shims (16) are available in thicknesses of 0.003, 0.005 and 0.010. Add or remove the same number of shims (24) at shaft flange, then add an extra 0.003 shim to establish the recommended 0.000-0.003 input shaft

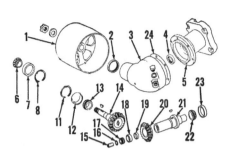

Fig. 158—Exploded view of belt pulley attachment.

1. Pulley	14. Pulley shaft
2. Oil seal	15. Shaft
3. Housing	16. Shims
4. Oil seal	17. Bearing cone
5. Flange housing	18. Bearing cup
6. Bearing cone	19. Cupped plug
7. Bearing cup	20. Gear
8. Snap ring	21. Input shaft
10. Snap ring	22. Bearing cone
11. Snap ring	23. Bearing cup
12. Bearing cup	24. Shims
13. Bearing cone	

end play. (Make sure that a 0.000 reading is zero end play and not actually a preload.) Remove output shaft spacer and install pulley, then tighten pulley nut to obtain a rolling torque of 15-25 inch pounds. Back off the nut if necessary to nearest castellation and secure with cotter pin. Fill pulley gearcase with 2½ pints of John Deere Type 303 Special Purpose Oil.

to position bearing rollers in idler gear, and assemble by reversing the disassembly procedure.

BELT PULLEY

OVERHAUL

All Models So Equipped

152. To disassemble the belt pulley attachment, remove the pulley, drain the gear housing, then unbolt and remove the mounting flange housing (5— Fig. 158). Withdraw the drive shaft (21) and gear assembly. Remove the pulley shaft oil seal (2) and outer bearing cone (6), then remove the pulley shaft from gear housing.

Gear backlash is adjusted by adding or removing shims (16) between bearing cone (17) and housing (3). Input shaft bearings are adjusted by

HYDRAULIC SYSTEM

The hydraulic lift system working fluid is supplied by the tractor main hydraulic system which also provides fluid for the power steering and power brakes. Working fluid for the hydraulic units is available at all control valves at a constant pressure of 2200-2300 psi.

MAIN HYDRAULIC SYSTEM
All Models

153. **OPERATION.** The main hydraulic pump is mounted underneath the tractor radiator and coupled to front of engine crankshaft. This variable displacement, radial piston pump can be either a four piston, or eight piston type, and provides only the fluid necessary to maintain system pressure. When there are no demands on the system, pistons are held away from the pump camshaft by fluid pressure and no flow is present. When pressure is lowered in the supply system by moving a control valve or by leakage, the stroke control valve in the pump meters fluid from the pump camshaft

reservoir, permitting the pistons to operate and supply the flow necessary to maintain system pressure. A maximum of 11½ gallons per minute is available at full stroke on the four piston pump (all 2510 models) and 20 gallons per minute on the eight piston pump, which was optional on 2520 models before serial number 022000, and standard equipment on later units.

The transmission pump provides pressure lubrication for the transmission gears and shafts and on Power Shift models, supplies operating fluid for transmission and power take-off. On all models, excess fluid from transmission pump passes through the full flow system filter to the inlet side of the main hydraulic system pump. If no fluid is demanded by the main pump, the fluid passes into the oil cooler (Power Shift Models) or oil reservoir (early Syncro Range Models) then back to reservoir in transmission housing. Transmission pump capacities at 2100 engine rpm are as follows:

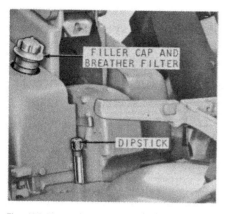

Fig. 159–View of operator's platform showing location of transmission filler cap and dipstick.

2510
Syncro-Range Models ..... 6½ gpm
Power Shift Models ....12 gpm

2520
Syncro-Range .... 9 gpm
Power Shift ....12 gpm

The oil cooler or oil reservoir is mounted in front of the tractor reservoir and provides one gallon of reserve hydraulic fluid to the main pump for peak load demands. The oil cooler contains cooling fins to control fluid temperature. The 2510 has an oil cooler on Power Shift models only. Oil cooler is on all models of the 2520 which have the eight piston pump.

154. **RESERVOIR AND FILTERS.** The hydraulic system reservoir is the transmission housing and the same fluid provides lubrication for the transmission gears, differential and final drive units. The manufacturer recommends that only John Deere Type 303 Special Purpose Oil be used in the system. Reservoir capacity is 8 U.S. Gallons for Syncro Range Models or 11 U.S. Gallons for Power Shift Models. To check the fluid level, stop tractor on level ground and check to make sure that fluid level is in "SAFE" range on dipstick (Fig. 159).

The oil filter element (Power Shift

Fig. 160–Transmission fluid filter is located on left side of transmission housing as shown.

Fig. 161–Power shift models have two filters.

Models have two filters) is located on left side of transmission housing as shown in Fig. 160 and Fig. 161. The filter cartridges may be renewed without draining fluid reservoir by removing filter cover and extracting element.

All filters are provided with a bypass valve which opens to allow oil to bypass the filter when oil is cold or filter plugged. On Syncro Range models, the bypass valve is located in a housing block to the rear of filter cover. To service the valve, remove valve plug below supply line elbow.

On Power Shift Models, the relief valve for front (Inlet) filter is located in Power Shift Regulating Valve housing as shown in Fig. 162. The rear (Systems Return) filter relief valve is located in a housing underneath the filter cover and cartridge. To renew or check the valve, it is first necessary to drain the system and remove rear filter; then unbolt and remove the filter housing.

The transmission housing on 2510 Syncro Range Models contains a check valve at the point where pressure line from transmission pump enters housing. The purpose of check valve is to prevent fluid from oil cooler (or reservoir) from draining back into transmission housing when transmission pump is not operating. To renew the check valve or its spring, first drain the fluid and remove rockshaft housing (or transmission rear cover) as outlined in paragraph 166. Remove the transmission pump outlet (left) line, then thread the outlet pipe bushing in housing with a thread tap. Install a cap screw in tapped hole to serve as a puller and

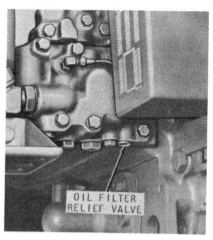

Fig. 162–Shown is relief valve location for front filter on Power Shift models.

remove the bushing. Check valve and spring can then be withdrawn. Thoroughly clean the metal chips from housing and bore, and use a new bushing when reinstalling the valve. On 2520 models, the hydraulic check valve is contained in the oil filter relief valve housing, located just behind oil filter (25—Fig. 164). To inspect or renew check valve, have a container capable of holding all reservoir oil, remove valve from housing and inspect valve and spring for damage and distortion. Renew if necessary and reinstall in housing. Replace any lost oil.

Power Shift Models are equipped with a Manual Bypass valve (Fig. 165) located in Power Shift Regulating Valve Housing. The bypass valve, when open, applies less restriction to the return oil from single acting cylinders, but allows a greater quantity of oil to bypass the oil cooler. Open the valve if necessary when using a front loader or similar implement, but make sure valve is closed for normal tractor operation. Bypass valve can be screwed in or out after removing the hex cap.

155. **SYSTEM TESTS.** Efficient operation of the tractor hydraulic units requires that each component operates properly. A logical procedure for testing the system is therefore needed. The indicated system tests include transmission pump flow test, system pressure tests and leakage tests as outlined in the following three paragraphs. Unless the indicated repairs of

Fig. 163–Relief valve housing for rear filter in power shift models is located behind filter cartridge and filter must be disassembled for renewal or checking.

hydraulic units is obvious because of breakage, these tests should be performed before repairing the individual units.

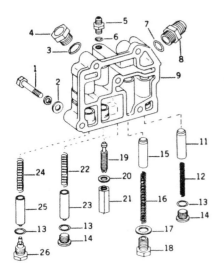

Fig. 164—Exploded view of oil filter relief valve housing used on 2520 models.

9. Housing	19. Bypass screw
11. Filter bypass valve	20. Washer
12. Spring	21. Nut
13. 'O' ring	22. Spring
14. Plug	23. Check valve
15. Relief valve	24. Spring
16. Spring	25. Oil pressure valve
17. Washer	26. Plug
18. Plug	

Fig. 166-Flow meter attaching points for Power Shift transmission pump test.

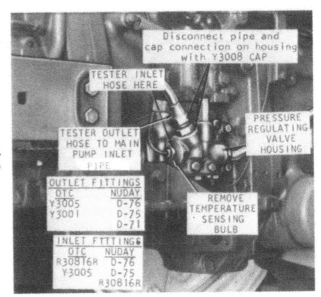

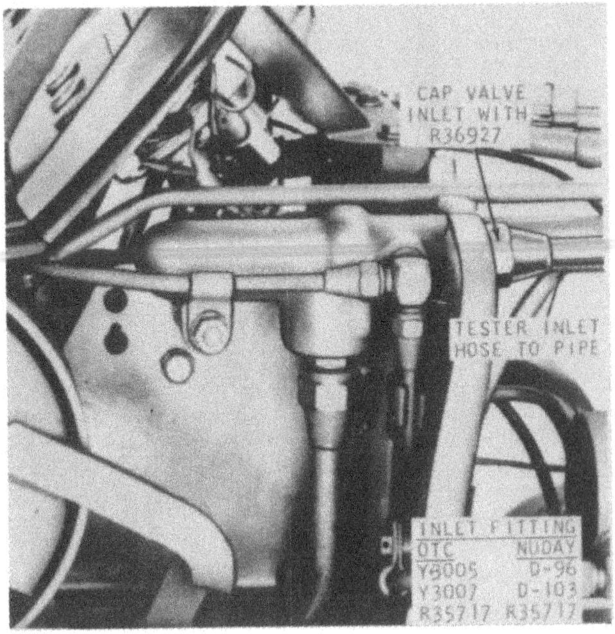

Fig. 167—Inlet connection of flow meter for testing main pump. Refer also to Fig. 168, 169. Fig. 167 Tester inlet hose connects to pipe from main pump as shown for main pump test on all models.

Fig. 165—Power shift models are equipped with a manual bypass valve.

Fig. 165A.-Flow meter installed for transmission pump test on late Syncro-Range tractor.

156. TRANSMISSION PUMP FLOW TEST. A quick test of transmission pump operation can be performed by removing the fluid filter (front filter on Power Shift Models) and turning engine over with starter. A generous flow of fluid will be pumped into filter housing if pump is operating satisfactorily.

An alternate method which more thoroughly tests pump condition is as follows:

Connect flow meter in series as shown in Fig. 165A for Syncro-Range, and for Power Shift use similar test points on pressure regulating valve housing. Before connecting outlet hose on tester to the supply line to main pump, loosen the cap screws on side frame to allow supply line to be connected without bending it. Retighten cap screws before starting engine.

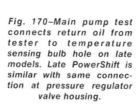

Fig. 170–Main pump test connects return oil from tester to temperature sensing bulb hole on late models. Late PowerShift is similar with same connection at pressure regulator valve housing.

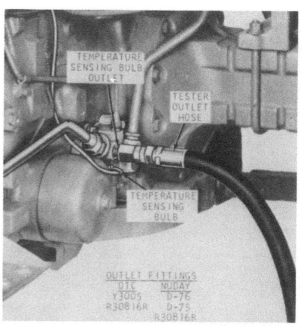

Fig. 168–Outlet hose connection of flow meter when used on early Syncro Range tractor. Refer to Fig. 167 for inlet hose connection.

Open the pressure control valve on meter, start engine and set to 2100 rpm. Check capacity on the test meter and compare flow to specifications in paragraph 153. If capacity is adequate, slowly close pressure control valve on meter. Flow should remain the same until 50-70 psi is reached, when flow will start to decrease. If flow decreases at below 50 psi or not until after 100 psi on 2520 models, service the surge relief valve (15—Fig. 164). If pump output does not meet specifications, check for low oil level, dirty filters, kinks in oil lines, water or foreign material in oil, or leaks. If none of these problems are evident, overhaul transmission pump as outlined in paragraph 91 for Syncro-Range models or paragraph 120 for Power Shift tractors.

157. **MAIN PUMP PRESSURE AND FLOW.** To check the main pump pressure and flow, refer to Figs. 166, 167, 168, 169, and 170. Remove the cowl and hood as shown in Fig. 167 and attach flow meter inlet line to priority valve supply line. On early tractors equipped with rockshaft, lower the rockshaft arms and disconnect rockshaft return pipe as shown in Fig. 168 or 169. Attach flow meter outlet line to return elbow as shown. If tractor is not equipped with rockshaft, remove port plug from housing and install a return elbow.

NOTE: The above return line hookups are valid on pressure tests only. If flow test is desired, it will be necessary to install a tee fitting in main pump supply line to accept the return line from test meter to prevent an incorrect reading on early models. The hookup in Fig. 169 shows return line oil routed so return oil will be directed to main pump supply line on late models.

On late type Syncro-Range models, attach flow meter return line hose after removing transmission temperature sensing bulb as shown in Fig. 170. On late Power Shift models the return line hose can be attached in the same manner by removing the temperature sensing bulb from the regulator valve housing.

Make sure that all hydraulic control levers are in neutral and start and run engine at 2100 rpm. Close the flow meter control valve until there is no flow and check gage pressure which should be 2200-2300 psi. Adjust the standby pressure if necessary, by turning the stroke control adjusting screw (Fig. 171).

Open the flow meter control valve until gage pressure reads 2000 psi, then check system flow which should be approximately 12 gpm for the four piston pump or 20 gpm for the eight piston pump. (See NOTE.) If pressure and flow are not as specified, remove and overhaul the main hydraulic pump as outlined in paragraph 162.

158. **PRIORITY VALVE (PRESSURE CONTROL VALVE) CHECK.** Leave flow meter attached as outlined in paragraph 157 and close flow meter

Fig. 169–Flow meter outlet hose connected to early Power Shift tractor. Refer to Fig. 167 and paragraph 157.

Fig. 171–Main hydraulic pump viewed from underneath tractor, showing location of stroke control valve adjusting screw.

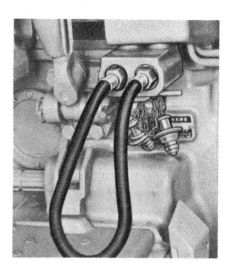

Fig. 172–Jumper hose installed in breakaway coupler for testing priority valve as outlined in paragraph 158.

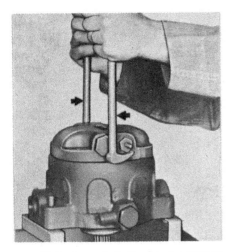

Fig. 173–Stroke control valve housing must be worked from 4-piston pump using pry bars as shown.

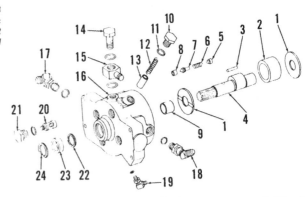

Fig. 174–Exploded view of 4 piston type main hydraulic pump camshaft, camshaft housing and associated parts.

1. Thrust washers
2. Race
3. Needle roller
4. Camshaft
5. Valve guide
6. Valve spring
7. Discharge valve
8. Valve seat
9. Bushing
10. Piston plug
11. O-ring
12. Spring
13. Piston
14. Special screw
15. Banjo fitting
16. Pump housing
17. Inlet fitting
18. Outlet fitting
19. Lube fitting
20. Inlet valve
21. Valve plug
22. Quad ring
23. Oil seal
24. Snap ring

control valve. Attach a jumper hose in breakaway coupler as shown in Fig. 172.

Start engine and operate at 800 rpm (slow idle), then move control lever to pressurize breakaway coupler. Gage will register priority valve setting which should be 1650-1700 psi. If pressure is not within recommended range, adjust the priority valve as outlined in paragraph 164.

159. LEAKAGE TEST. To check for leakage at any of the system valves, move all valves to neutral and run engine for a few minutes at a speed of 2100 rpm. Check all of the hydraulic unit return pipes individually for heating. If the temperature of any return pipe is appreciably higher than the rest of the lines, the valve is probably leaking. Disconnect the return line and measure the flow for a period of one minute. Leakage should not exceed 25cc; if it does, overhaul the system valves as outlined in the appropriate sections of this manual.

160. **LINES AND FITTINGS.** Flared, seamless steel tubing is used for all hydraulic system components. Fittings have SAE Straight Tubing threads with O-ring seals. Do not attempt to substitute pipe thread fittings for components or test equipment.

161. **MAIN HYDRAULIC PUMP.** When external leaks or failure to build or maintain pressure indicates a faulty pump, the main hydraulic pump must be removed for service as follows:

On tricycle tractors remove lower frame plate. On axle models, remove front axle, pivot plate and adapter plate as a unit. On all models disconnect the main pump supply line at rear end and allow fluid from oil cooler or front reservoir to drain into a clean container. Disconnect main supply line, oil cooler line and pressure line

from pump and loosen the clamp screws securing pump drive adapter to pump shaft. Remove the screws securing pump to engine support bracket and move pump forward until drive is disconnected. Pump can now be lifted from tractor for service as required.

Install the pump by reversing the removal procedure. Tighten the mounting cap screws to a torque of 85 ft.-lbs. and coupler clamp screws to 20 ft.-lbs.

162. OVERHAUL. (4 piston and 8 piston types). To disassemble the pump, remove the four cap screws retaining the stroke control valve housing to front of pump and on 4-piston pump remove the housing using two pry bars as shown in Fig. 173. On 8 piston pump before disassembling, attach a dial indicator and measure pump shaft end play which should be 0.002 to 0.006. Adjusting shims (18—Fig. 175) are available in thicknesses of 0.006 and 0.010. Do not lose shims when disassembling pump.

Although the pistons and springs are available as individual parts, it is good shop practice once they have been installed and used, to reinstall them in their original locations. Use a compartmented pan or other means to keep them identified when pump is disassembled.

On 4-piston pump remove discharge valve guides (5—Fig. 174), springs (6) and valves (7). Remove the four piston plugs (10), springs (12) and pistons (13).

Inlet valve assemblies (20) are pressed into housing. To remove the inlet valves, remove plugs (21) and use a small pin punch, working through discharge valve seat (8). Discharge valve seat can be removed with a larger punch after inlet valve is out. Do not remove discharge valve seats unless renewal is indicated.

On 8-piston pump, follow the disassembly sequence for the 4-piston type (refer to Fig. 175) and bump the pump shaft on spline end with a soft faced hammer. Shaft can be removed as an

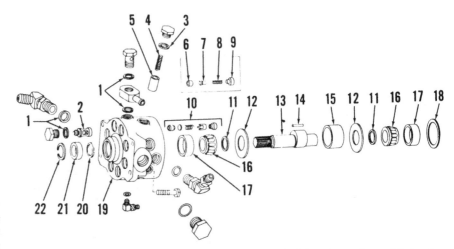

Fig. 175–Exploded view of 8-piston main hydraulic pump camshaft, housing, and associated parts.

1. 'O' ring
2. Inlet valve
3. 'O' ring
4. Spring
5. Pump piston
6. Discharge seat
7. Discharge valve
8. Spring
9. Guide
10. Outlet valve
11. Spacer (2)
12. Thrust washer (2)
13. Pump shaft
14. Roller bearing (36)
15. Race
16. Bearing cone (2)
17. Bearing cup (2)
18. Shim
19. Housing
20. Packing
21. Seal
22. Snap ring

Fig. 176–Remove inlet valve assembly working through discharge valve seat as shown.

assembly with bearings. Remove snap ring (22) and pry out shaft seal (21).

On 4-piston pumps the camshaft bushings (9—Fig. 174 & 10—Fig. 178) are identical pre-sized split bushings. To renew the bushing, curl out one end with a flattened punch as shown in Fig. 180 and remove with pliers. Install with a press, using a 1 inch piloted mandrel to prevent distortion.

Inspect seating surfaces of inlet valve assembly (20—Fig. 174) to make sure they are in good condition. Measure the lift of each of the inlet valves, and check for looseness of stem in guide. There should be no apparent play of valve stem and total lift should be between 0.060 and 0.082. If looseness is apparent, lift is greater than 0.082 or spring is broken or damaged, renew the valve assembly.

When installing the seal in pump housing, drive the seal in only far enough to install snap ring. If seal is

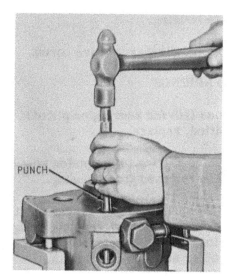

Fig. 177–Remove discharge valve seat working through inlet valve bore.

installed too deep, the oil outlet hole will be covered which would allow pressure buildup and seal failure.

Assemble by reversing the disassembly procedure. Immerse all of the parts, including O-rings and seals, in clean hydraulic fluid when assembling. Tighten the piston plugs (10—Fig. 174) to a torque of 100 ft.-lbs.

Check the stroke control housing as outlined in paragraph 163. Install the housing and tighten the retaining cap screws to a torque of 85 ft.-lbs. After pump is installed, adjust pump pressures as outlined in paragraph 157.

163. STROKE CONTROL VALVE HOUSING. The valves located in the stroke control valve housing control pump output as follows for the 4-piston pump:

The closed hydraulic system has no discharge line except through the operating valves. Peak pressure is thus maintained for instant use. Pumping action is halted when line pressure reaches a given point by pressurizing the camshaft reservoir of pump housing, thereby holding pump pistons outward in their bores.

The cutoff point of pump is controlled by pressure of spring (14—Fig. 178) and can be adjusted by screw (15). When pressure reaches the standby setting, valve (12) opens and meters the required amount of fluid at reduced pressure into crankcase section of pump. Crankcase outlet valve (5) is held closed by hydraulic pressure and blocks the outlet passage. When pressure drops as a result of system demands, crankcase outlet valve (5) is opened by spring (6), dumping the pressurized crankcase fluid into pump inlet and pumping action resumes.

NOTE: Early 4-piston pumps use a ball-type crankcase outlet valve (8) and actuating plunger (17). A modification kit is available through parts stock to update existing pumps and the early parts are no longer serviced. The 8-piston stroke control valve operates much the same as the 4-piston type, but uses slightly different valves to do the job. Refer to Fig. 179 for identification of parts. Inset (20) shows the pump shut-off screw assembly which can be purchased and installed if so desired. This assembly allows the pump to be shut off for cold weather starting to ease the load on the battery and starter.

Disassemble the stroke control valve housing and examine the parts for wear, scoring or other damage. Clean varnish deposits with laquer thinner or equivalent, if present. Make sure that all passages are open and clean. Use new seal when assembling, and lubri-

cate all parts with hydraulic fluid. Install stroke control valve housing as outlined in paragraph 162.

164. **PRIORITY VALVE.** The Pressure Control (Priority) Valve is mounted under engine cowl on right side and is designed to give hydraulic pressure priority to power steering, brake units, and the pto clutch valve on 2520 Syncro-Range models. The valve is shown schematically in Fig. 181 for the 2510, and exploded in Fig. 182 for the 2520. Pressure setting of 1800-1850 psi is adjusted by adding or removing shim washers (5).

165. R&R AND OVERHAUL. To remove the priority valve, first remove cowl and hood, and on the 2510 the instrument panel. Disconnect all hydraulic lines leading to priority valve, then unbolt and remove valve from

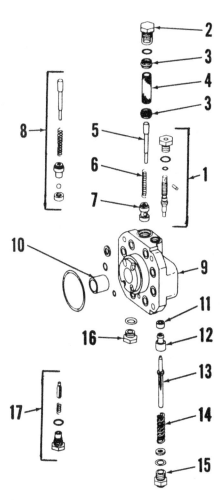

Fig. 178–Exploded view of 4 piston stroke control valve housing showing component parts. Outlet valve parts (8 & 17) were used on early models and are no longer serviced. A kit is available to update early pumps so equipped.

1. Pump shut-off valve
2. Special plug
3. Packing
4. Filter
5. Outlet valve
6. Spring
7. Valve sleeve
8. Early valve
9. Housing
10. Bushing
11. Valve seat
12. Stroke control valve
13. Valve guide
14. Valve spring
15. Adjusting screw
16. Plug
17. Early plug & plunger

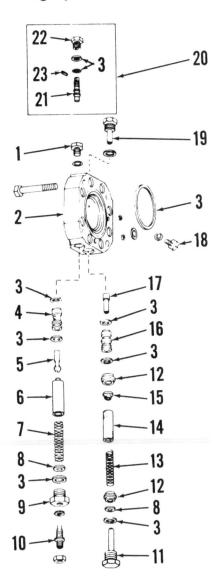

Fig. 179–Exploded view of 8-piston stroke control valve housing and component parts.

1. Plug	13. Spring
2. Housing	14. Filter
3. "O" ring	15. Guide
4. Sleeve	16. Sleeve
5. Stroke control valve	17. Outlet valve
6. Guide	18. Plug
7. Spring	19. Plug
8. Washer	20. Shut off screw
9. Bushing	assembly
10. Adjusting screw	21. Shut off screw
11. Plug	22. Plug
12. Packing	23. Roll Pin

Fig. 180–Curl the camshaft bushings on 4-piston model for removal. Bushings are split.

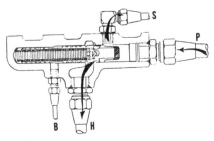

Fig. 181–Cross sectional view of 2510 priority valve. Fluid from main hydraulic pump enters at (P). Steering line (S) is always open to pump pressure; hydraulic systems line (H) is closed off when system pressure drops below 1800-1850 psi. Bleed line (B) drains spring end of valve bore.

remove rockshaft housing covers, driver's seat and platform. Disconnect the three-point lift links on tractors so equipped. Disconnect and remove all interfering wiring and hydraulic lines. Disconnect the rockshaft control valve cable on early models, and control rods on late models. Remove the attaching bolts and lift the housing from tractor with a hoist.

When installing the unit, make sure the cam follower roller in valve housing is positioned to the rear of the follower arm in transmission housing, and lower the unit carefully so as not to damage or bend the draft control mechanism. Complete the assembly by reversing the removal procedure and tighten the attaching cap screws to a torque of 85 ft.-lbs.

167. **CONTROL VALVE HOUSING.** To remove the control valve housing, remove the right rockshaft cover, disconnect control cable or control rods and hydraulic fluid lines. On models so equipped, remove the right breakaway couplings. On 2520 models, be sure the rockshaft valve cover stays with valve. Remove the cap screws retaining control valve housing to rockshaft housing and lift off the complete control valve unit.

168. On 2510 models, remove housing cover (2—Fig. 183) and withdraw springs (6 & 8), balls (9), valves (10) and metering shafts (11). Disassemble the operating mechanism by unscrewing operating shaft quill (15—Fig. 184) then removing operating link

right side of steering support. Refer to Fig. 181 or 182.

Check the valve (3) and valve bores for scoring and orifice (4) to be sure it is open and clean. Spring (6) should have a free length of $4\frac{5}{8}$ inches and test 45-55 lbs. when compressed to a height of $3\frac{1}{2}$ inches. Adjust pressure if necessary by adding or removing shim washers (5).

ROCKSHAFT HOUSING & COMPONENTS
All Models So Equipped

166. **REMOVE AND REINSTALL.** To remove the rockshaft housing, first

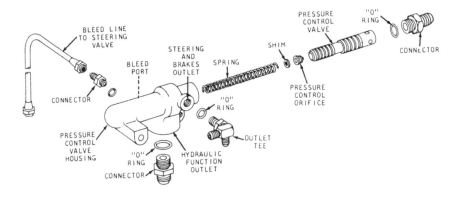

Fig. 182–Exploded view of pressure control (priority) valve used on 2520 models.

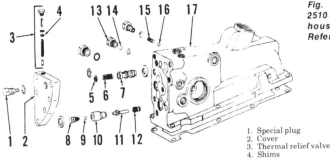

Fig. 183–Exploded view of 2510 rockshaft control valve housing showing valves. Refer to Fig. 184 for control mechanism.

1. Special plug	5. Washer
2. Cover	6. Spring
3. Thermal relief valve	7. Valve
4. Shims	8. Spring
	9. Ball
	10. Valve
	11. Metering shaft
	12. Guide
	13. Plug
	14. Shim
	15. Spring
	16. Ball
	17. Housing

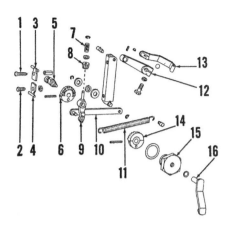

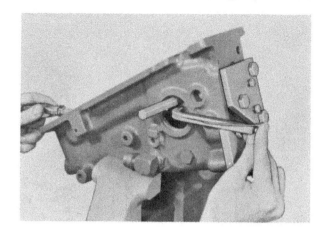

Fig. 186–Adjusting the return valve in 2510 control valve housing.

Fig. 184–2510 Control valve operating mechanism contained in valve housing (17–Fig. 183).

1. Adjusting screw	9. Control arm
2. Adjusting screw	10. Link
3. Operating hinge	11. Spring
4. Operating hinge	12. Load arm
5. Operating shaft	13. Selector lever
6. Operating gear	14. Differential gear
7. Spring	15. Quill
8. Pinion	16. Control lever

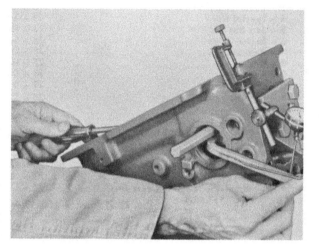

Fig. 187–Adjusting the pressure valve in 2510 control valve housing.

spring (11) and link (10). Differential gears (6 and 14) are pinned to their shafts. End play of operating hinges (3 and 4) is controlled by shims (14—Fig. 183). Specified end play is 0.001-0.005 and shims are available in thicknesses of 0.005, 0.010 and 0.030.

169. When assembling the control valve housing on 2510 models, it is necessary to adjust the operating hinges to a positive but minimum clearance by means of the adjusting screws (1 and 2—Fig. 184). To accurately make the adjustment with a minimum of effort requires the use of special adjusting tools JDH-8, JDH-9, JDH-10, JDH-11 and a dial indicator as shown in Figs. 185, 186 and 187. To make the adjustment, turn housing upside down in a vise and loosen adjusting screws (1 and 2—Fig. 184). Install adjusting plate JDH-8 and adjusting cover JDH-10 as shown in Fig. 185. Springs (8—Fig. 183) are removed before adjusting cover JDH-10 is in-

stalled. Back out the adjusting screws (1 & 2—Fig. 185) in cover to be sure clearance exists and tighten the cover retaining cap screws securely, then tighten the screws (1 & 2) in cover to a torque of approximately 10-12 inch-pounds to be sure valves are tight on their seats. Use screw JDH-9 to secure indicator JDH-11 to valve operating shaft sector gear. Use the tension wire as shown in Fig. 186 to hold adjusting tool away from stop on adjusting plate JDH-8.

With the housing upside down, the lower valve is the return valve and must be adjusted first. With adjusting tool held away from stop (Fig. 186) turn lower adjusting screw (1—Fig. 188) clockwise until indicator moves down and just contacts adjusting plate JDH-8. Tighten the jam nut on adjusting screw (1).

Mount a dial indicator as shown in Fig. 187, with indicator button contacting adjusting tool JDH-11 and remove adjusting plate JDH-8. Turn the inlet (pressure) valve adjusting screw (2—Fig. 188) until total movement at end of tool JDH-11 is 0.009-0.012. Be sure to use only the tension wire when measuring valve movement. Because of the length of the indicator arm, the valve clearance is barely noticeable at valve hinges and measurement without the aid of the tools is almost impossible.

After valves are adjusted, remove the tools and reinstall housing covers and valve springs. Install and time operating link arm, valve operating lever and sector gear. The linkage is correctly timed when pin in operating link arm and cast pointers on housing and operating lever are aligned as shown at (A). Tighten lever quill (15—Fig. 184) securely, install cam follower

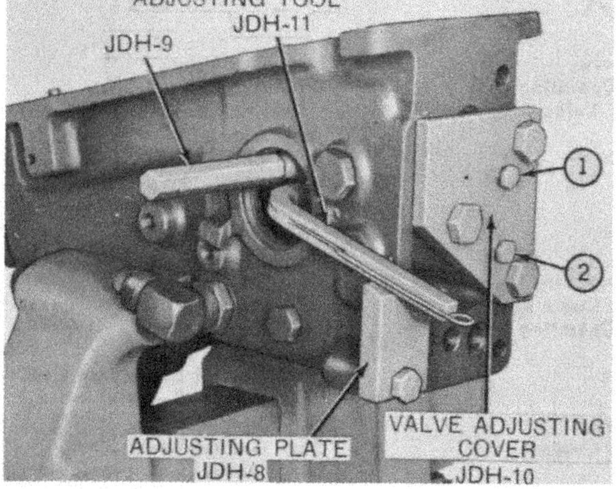

Fig. 185–View of 2510 rockshaft control valve housing with special adjusting tools installed. Refer to Figs. 185 & 186 and paragraph 169.

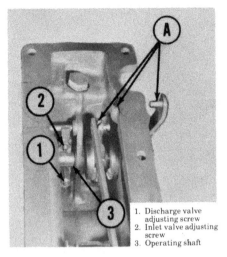

1. Discharge valve adjusting screw
2. Inlet valve adjusting screw
3. Operating shaft

Fig. 188–After 2510 valve is adjusted, make sure that the three indicated points (A) on control arm, housing and control lever are aligned when lever is installed.

link and spring, then install valve housing.

After assembly is completed, move selector lever on rockshaft housing to the lower (D) position and attach control cable. Start engine and check to see that rockshaft moves to full raised and full lowered position when dash control lever is moved the full length of

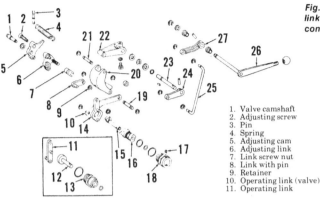

Fig. 190–Rockshaft control linkage and load selector control for early 2520. (with negative stop screw)

1. Valve camshaft
2. Adjusting screw
3. Pin
4. Spring
5. Adjusting cam
6. Adjusting link
7. Link screw nut
8. Link with pin
9. Retainer
10. Operating link (valve)
11. Operating link
12. Operating shaft
13. Lever shaft quill
14. Operating link
15. Operating shaft
16. Bushing
17. Set screw
18. Lever shaft quill
19. Pin
20. Load selector link
21. Pin
22. Load selector arm
23. Load control shaft
24. Lower operating arm
25. Control rod
26. Control lever
27. Upper operating arm

quadrant. If rockshaft raises too slowly, lower the rockshaft, disconnect the pressure pipe from rockshaft valve to rockshaft cylinder. Install inlet hose of a flow meter to the fitting where pipe was removed from valve housing, and place outlet hose in fill tube on valve. With selector lever in (L) load position, engine running 1200 rpm, close the flow meter valve to obtain 2000 psi. Flow should be 5 gpm. If necessary, add shims (5—Fig. 183) to control valve to raise pressure.

170. To disassemble the control valve on 2520 models, remove housing cover (2—Fig. 189), springs and balls (18),

control valves (17), metering shafts (16) and guides (15). Remove shims (3), spring and flow control valve (4), check ball (5), plug (19), and thermal relief valve assembly (10 to 14).

(Before Serial No. 022825) Remove lower operating arm (24—Fig. 190), load selector control shaft (23), and link (20). Remove valve camshaft (1), and avoid losing washer as camshaft is removed. Disconnect spring (4) from control valve adjusting cam (5), remove retainer ring (9) holding linkage to control valve adjusting link (6) and remove linkage.

(Serial No. 022825 and Later) Remove load selector arm (22—Fig. 190-A), control shaft (21) and load control arm (20). Remove valve cam shaft (1), and avoid losing washer as cam shaft is removed. Disconnect spring (4) from valve adjusting cam (5), remove retainer rings holding remainder of linkage and remove linkage.

Check all linkage, springs, valves and housing for wear, scoring, or other damage and replace any parts in question. Valves can be lapped to seats, if necessary, using fine lapping compound. Inspect thermal relief valve assembly and spring, which should have 8 to 10 pounds pressure at a compressed length of 15/32-inch.

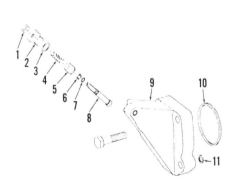

Fig. 189–Exploded view of 2520 rockshaft control valve housing and associated parts.

1. Flow adjusting plug
2. Cover
3. Shim
4. Flow control valve
5. Check ball
6. Plug with pin
7. Housing
8. Backup rings
9. Inlet (2)
10. Thermal relief plug
11. Shim
12. Spring guide
13. Check ball
14. Thermal relief valve seat
15. Guide (2)
16. Metering shaft (2)
17. Control valve (2)
18. Check ball (2)
19. Plug

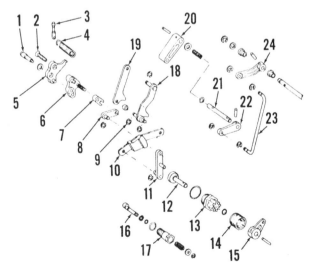

Fig. 189-A–Exploded view of 2520 rockshaft piston cover.

1. Nut
2. Bushing
3. "O" ring
4. Spring
5. Throttle valve
6. Backup ring
7. "O" ring
8. Valve shaft
9. Piston cover
10. Packing
11. Packing

Fig. 190-A–Rockshaft control linkage and load selector control for 2520 (with reverse signal lockout system).

1. Valve camshaft
2. Adjusting screw
3. Pin
4. Spring
5. Valve adjusting cam
6. Adjusting link
7. Link screw nut
8. Link with pin
9. Retainer ring
10. Link with cam
11. Operating link
12. Operating shaft
13. Quill
14. Spring
15. Lower operating arm
16. Negative signal eccentric
17. Bushing
18. Outer selector link
19. Inner selector link
20. Arm
21. Shaft
22. Load selector arm (lower)
23. Load selector rod
24. Load selector arm (upper)

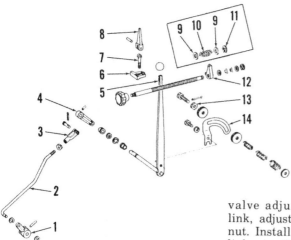

Fig. 191—Rockshaft controls for 2520.

1. Lower operating arm
2. Control rod
3. Yoke
4. Upper operating arm
5. Control lever
6. Height stop
7. Special screw
8. Cam
9. Washer (2)
10. Spring
11. Retainer
12. Lever stop
13. Jam nut
14. Friction plate

When reassembling, install thermal relief valve, spring and shims as required. Refer to Fig. 192 for assembly sequence of early 2520 control linkage. Install valve operating link, link with pin and spring (A). Assemble control valve adjusting cam and adjusting link, adjusting screw and link screw nut. Install in housing and attach to link pin (B). With control valve adjusting cam in position, install valve camshaft and washer, and hook spring to control valve adjusting cam tang (C). Install operating lever shaft quill (18—Fig. 190) and lower operating arm (1—Fig. 191) in housing. Be sure pin on operating link (11—Fig. 190) is in hole of valve operating shaft. Assemble valve operating link, load selector control link, and load selector control arm. Install this assembly in housing and attach valve operating link to operating link (D). Install lower operating arm, shaft, washers and Woodruff key, and tighten cap screw in load selector control arm (20—Fig. 190). Install valve assembly in reverse order of disassembly, and make sure that the roller of the load selector control link goes on the backside of cam follower.

When reassembling rockshaft valve linkage on 2520 tractors after serial number 022825, install in reverse order of disassembly. Refer to view A, B, and C only of Fig. 192, and to Fig. 192-A for assembly sequence. Be sure to assemble numbers (5, 6, and 7—Fig. 190-A) with special screw (2) and install in housing as an assembly. Place the inside hook of spring (14) on the highest lug of quill (13), and install lower arm (15) so that hole for linkage is pointed up. Install valve assembly on rockshaft housing and make sure that the roller of the inner and outer selector links (18 and 19) is on the backside of cam follower.

171. CONTROL VALVE ADJUSTMENT (2520). Place the load selector lever in "D" position for depth adjustment. Remove the plug (19—Fig. 189) to expose valve adjustment screw, and have linkage rod disconnected from valve control lever arm. Adjust control valves by turning the adjusting screw counterclockwise until it bottoms; then clockwise ½-turn. Start the engine and raise rockshaft to the halfway point, then shut off engine. If the rockshaft begins to fall, the valves are too tight and must be loosened by turning the adjusting screw counterclockwise. To check the adjustment, clamp a pair of vise grip pliers to the control lever arm (Fig. 193). At a distance of 10 inches from center of lever shaft, a movement of 3/16 to 3/8-inch should cause the rockshaft to change direction with engine running. CAUTION: Stay clear of

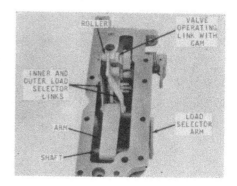

Fig. 192-A—View of properly assembled rockshaft valve after serial number 022825.

Fig. 192—The 2520 control valve housing assembly sequence. (View D is used with negative stop screw model only)

Fig. 193–The 2520 rockshaft control valve housing showing adjustment procedure.

Fig. 195–View of load control arm adjustment opening with plug installed.

lift arms and hitch while making this test.

After adjustment is complete, test the rockshaft operation. If raising speed is too slow, check the flow to rockshaft as follows: Lower the rockshaft and install inlet hose of a flow meter in the rockshaft piston cover after removing the throttle valve assembly from cover. Remove the temperature sensing bulb in oil filter relief valve housing (Fig. 170) on Syncro-Range models, or in the regulator valve housing on Power Shift models (Fig. 166), and install the outlet hose. With the load selector lever in (L) load position, run the engine at 2100 rpm, move the rockshaft lever rearward and close control valve on flow meter until a pressure of 2000 psi is reached. If flow is not 5 gpm, shims (3—Fig. 189) can be added to the flow control valve spring to increase volume flow. Removing shims will decrease flow.

172. 2520 CONTROL LINKAGE ADJUSTMENT (WITH NEGATIVE STOP SCREW). The 2520 tractors be-

fore serial number 022825 were equipped with a negative stop screw which was located in the rear of the draft link support, halfway between draft link attaching pins.

To adjust the negative stop screw, lower the rockshaft, move drawbar sideways to allow access to the stop screw, straighten the tab on locking plate and loosen the lock nut. Turn the adjusting screw out until it contacts the transmission case solidly, then back off the screw approximately ⅛ turn. Retighten the lock nut and bend lock tab against the nut.

Start the engine with selector lever in depth (D) position, raise the rockshaft to full height using valve control lever arm as in Fig. 193, and hold lever arm in that position. Next move the rockshaft console lever rearward to the end of slot in console, adjust the control rod so that it will just go in hole in control lever arm, then shorten control rod at the yoke one full turn and connect it to the control lever arm. Now when the console lever is pulled to approximately 1/16-inch from rear of

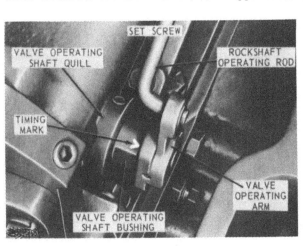

Fig. 194–The 2520 rockshaft linkage on tractors equipped with reverse signal lockout system.

console slot, the rockshaft should raise completely.

173. 2520 CONTROL LINKAGE ADJUSTMENT (WITH REVERSE SIGNAL LOCKOUT). The 2520 tractors starting with serial number 022825 began using a reverse signal lockout system, which uses different parts in the rockshaft valve and eliminates the need for the negative stop screw. This system requires a different adjustment procedure which is as follows: Remove the load control arm adjusting opening plug (Fig. 195). Turn the load control arm extension adjusting screw fully clockwise. Loosen the set screw for the valve operating shaft quill (Fig. 194), position the "V" timing mark on valve operating shaft bushing at 12 o'clock and tighten set screw. With the engine running and load selector lever in (D) depth position, the rockshaft control rod should be adjusted so that rockshaft raises completely when the console lever is moved to the rear until front edge of lever is ¼-inch ahead of the zero on the console guide.

Move the load selector lever to (L) load position and the front edge of rockshaft console lever to zero. Loosen the set screw at valve operating shaft quill (Fig. 194), rotate the valve operating shaft bushing clockwise until rockshaft starts to raise, hold bushing at that position and retighten set screw. Lower the rockshaft with the console lever, then move lever rearward until the front edge of lever is ¼-inch ahead of zero. Leave the load selector lever in (L) load position and turn the load control arm adjusting screw (Fig. 195) counterclockwise until the rockshaft begins to raise. Recheck operation of rockshaft. If adjustment is correct the rockshaft will lower when the console lever's front edge is no farther than the 1½ inch mark on console guide.

174. ROCKSHAFT HOUSING OVERHAUL. To overhaul the removed rockshaft housing, first remove the control valve housing and piston cover. Piston can now be forced from cylinder by rotating the rockshaft. Do not rotate rockshaft so far down as to allow the crank arm to damage the edge of piston cylinder.

Remove the set screw retaining the servo cam to rockshaft and the retaining capscrews in ends of shaft. Remove lift arms and withdraw shaft from left side of rockshaft housing.

Oil seals and shaft bushings can be renewed at this time. Bushings are pre-sized and should be installed with a suitable driver. Be sure to align oil holes in bushings and housing. Shaft to bushing clearance is 0.002-0.008 for new parts. Check cylinder for scoring and renew the housing if cylinder is not serviceable. When reassembling the unit, be sure to align index marks on rockshaft and crank arm. Install cam over rockshaft and place as many washers as can be freely assembled between the cam and right side of housing. Finish the reassembly and tighten capscrews retaining piston cover to a torque of 170 ft.-lbs.

LOAD CONTROL ARM AND SHAFT
All Models So Equipped

175. OPERATION. When the hydraulic lift selector lever is moved to the "L" (Load) position, the operating depth of the three-point hitch is controlled by the draft of the attached implement acting in conjunction with the position of the control lever.

The amount of draft is transmitted by the lower links to the drawbar frame (support), then to the control valve by the load control arm and shaft shown in Fig. 196. The spring steel shaft (13) is anchored in each side of the transmission housing and the drawbar frame is affixed to outer ends of shaft. Positive or negative draft causes the center of the load control shaft to deflect a predetermined amount according to the load encountered. The center arc of the flexing shaft (13) moves the straddle-mounted lower end of load control arm (7) around pivot shaft (9), thus moving the control mechanism to compensate for changes in draft.

The rockshaft follower arm (6) is attached to the load control arm, enabling one adjustment to synchronize the valves for any type of hydraulic control. Adjustment is made by means of adjusting screw (2) as outlined in paragraph 177.

176. REMOVE AND REINSTALL. Removal of the load control arm on tractors equipped with power takeoff can only be accomplished in conjunction with removing the differential assembly and the procedure is outlined in paragraph 127.

On tractors not equipped with pto, removal of the load control arm can be accomplished as follows:

Drain transmission and hydraulic fluid and remove rockshaft control valve housing as outlined in paragraph 167. Remove cam follower (6—Fig. 196) from upper end of load control arm. Remove transmission rear cover and the two cap screws retaining control arm shaft brackets (8), then withdraw the control arm assembly out rear of housing. Install by reversing the above procedure.

If only the load control shaft is to be renewed, proceed as follows: Drain the transmission and hydraulic fluid and remove the set screws which hold the load control shaft retainers at each side of drawbar frame (support). With a brass drift and working from right side of tractor, bump the shaft to the left and out. The drawbar frame will be free to drop at the rear end when shaft is removed.

Bushings in transmission housing and drawbar frame, (support) and the housing oil seals may be renewed at this time. The inside diameter of the bushings is tapered to provide a small bearing area for the flexing shaft. Install the bushings in transmission housing with small ID to inside of housing; and bushings in drawbar frame with small ID to outside, away from transmission housing. The large inside diameters of bushings should be together when frame is installed. Lubricate the shaft and install carefully to prevent damage to the oil seals. Shaft must be installed with the cut-out portions in ends of shaft toward the bottom.

Check the load control shaft, shaft bushings and load control arm against the dimensions which follow:

2510
Load Control Shaft OD . . .0.998-1.000
Load Control Shaft Bushing,
Small ID 1.003-1.005
Load Control Arm Yoke,
Width 1.005-1.015
2520
Load Control Shaft OD . . .1.123-1.125
Load Control Shaft Bushing,
Small ID (Case) 1.128-1.130

177. ADJUSTMENT. To adjust the control mechanism on 2510, remove the plug (Fig. 195). With engine running and selector lever in "D" (Depth) position, move the three-point hitch control lever up the quadrant until leading edge of lever is aligned with the "O" mark on quadrant scale. Adjust rockshaft control cable, if necessary, until rockshaft moves to the full raised

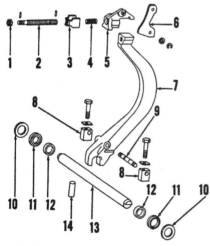

Fig. 196–Load control mechanism used on 2510 models. Shaft (13) flexes under load to move the control valve through arm (7).

1. Nut	8. Bracket
2. Adjusting screw	9. Shaft
3. Lock plate	10. Washer
4. Spring	11. Seal
5. Extension	12. Bushing
6. Follower	13. Shaft
7. Arm	14. Stop pin

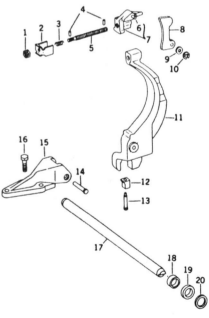

Fig. 197–Load control mechanism used on 2520 models. Operation is much the same as on 2510 models (Fig. 195).

1. Nut	11. Arm
2. Lock plate	12. Follower (arm)
3. Spring	13. Pin
4. Pins	14. Pin
5. Adjusting screw	15. Support
6. Pin	16. Screw
7. Extension	17. Shaft
8. Follower (servo)	18. Bushing (2)
9. Washer	19. Seal (2)
10. Retainer	20. Washer (2)

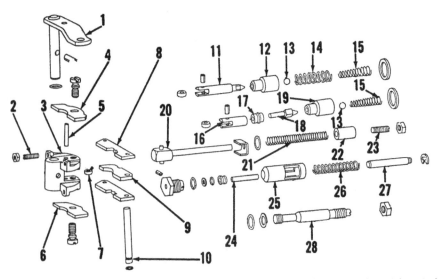

Fig. 198–Exploded view of control mechanism and valves used in 2510 selective (remote) control valve.

1. Arm	8. Detent	15. Spring	22. Plug
2. Adjusting screw	9. Cam	16. Follower	23. Adjusting screw
3. Rocker	10. Shaft	17. Guide	24. Pin
4. Cam	11. Follower	18. Metering shaft	25. Flow control valve
5. Pin	12. Valve	19. Valve	26. Spring
6. Cam	13. Valve ball	20. Guide	27. Guide
7. Roller	14. Spring	21. Spring	28. Metering valve

position. Move quadrant lever to rear position and check to make sure that rockshaft moves to the full lowered position.

Move the selector lever to "L" (Load) position and move quadrant lever until leading edge is aligned with the "1½" mark on quadrant scale.

Move selector lever to "L-D" (Load and Depth) position and make sure that rockshaft moves through the full range of travel. (For 2520 control adjustment refer to paragraphs 172 or 173).

SELECTIVE (REMOTE) CONTROL VALVES
2510 Models So Equipped

178. **OPERATION.** Tractors are optionally equipped with one, two or three selective (remote) control valves for the operation of remote cylinders. The valves are mounted on left front of firewall and controlled by levers on the instrument panel.

As with all other units of the hydraulic system, pressure is always present at the valves but no flow exists until the valve is moved. Refer to Fig. 198 for an exploded view of the 2510 valve mechanism. Each valve assembly is equipped with two ball check pressure and discharge valves which are unseated by plungers when valve is actuated. Flow control valve (25) limits the incoming fluid flow for feathering action when valve lever is moved only slightly, and also seats in housing for a double pressure check when valve is in neutral position. Maximum valve flow

is limited by the adjustable metering valve (28). When valve lever is moved to full raising or lowering, position is maintained by notches in locking detents (8). When the attached remote cylinder reaches the end of its stroke, the valve being used is centered by the flow control valve acting through plunger (24) on centering cam (9). The pressure of detent spring (21) is adjustable from front of valve housing by means of adjusting screw (23).

179. **REMOVE AND REINSTALL.** To remove the selective remote control valve or valves, first remove the cowl and hood, disconnect remote cylinder lines and fluid pressure and return lines. Cap all connections to prevent dirt entry. Disconnect the operating rods at the valves, then unbolt and remove the valve assemblies. Install by reversing the removal procedure.

180. **OVERHAUL.** Service, in most instances, consists of adjustment as outlined in paragraph 181, or in renewing damaged or worn parts. Before removing the valve, conduct a leakage test as outlined in paragraph 159. To disassemble the valve, remove end caps, then withdraw the springs and operating valves. Thoroughly clean the parts and examine the check balls (13 —Fig. 198) and seats (12 and 19) for wear or other damage. Make sure the flow control valve (25) slides freely in its bore and examine the valve and seat. To disassemble the operating mechanism, drive out the roll pin which retains operating shaft (1) to rocker (3) and withdraw the parts. Remove the two special screws which retain operating cams (4 & 6) to housing, then withdraw detent and centering cam shaft and cams.

Use all new O-rings and seals when assembling, and renew any other parts which are questionable. Adjust the valve as outlined in the following paragraphs.

181. **ADJUSTMENT.** Two valve adjustments are made with the valve assembled and in operating condition. These are the metering valve adjustments described in paragraph 182. To synchronize the operating valves, the assembly must be removed as outlined in paragraph 179 and synchronized as in paragraph 184.

182. **METERING VALVE.** To adjust the metering valve, install a 2½ x 8 inch double acting cylinder, operate

Fig. 199–Adjusting the 2510 remote cylinder metering valve. Detent adjustment is also shown.

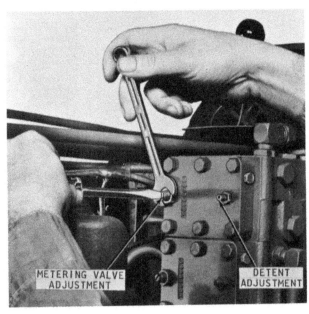

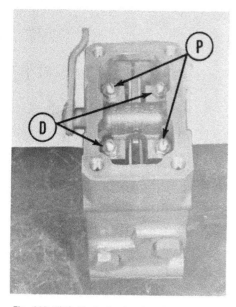

Fig. 200–2510 Control valve housing with rear cover removed. The two diagonally opposite screws (D) control discharge valves; the two screws (P) pressure valves.

Fig. 202–Adjusting plate and dial indicator installed for adjusting 2510 selective control valve. Refer to paragraph 184 for procedure.

the valve a few times to bleed air from the system, then fully retract the cylinder. With engine running at 2100 rpm, move control lever fully to the extend position and check the time required to fully extend the cylinder. The time should be 2-2½ seconds. To adjust, loosen the locknut on metering valve and turn valve as shown in Fig. 199. Tighten the locknut when adjustment is correct.

183. DETENT SPRING. The detent spring and plunger should hold the control lever in operating position until the attached cylinder reaches the end of its stroke, then release and allow

lever to return to neutral. To adjust, attach a double acting cylinder to the valve and check for improper performance. Refer to Fig. 199 for location of adjusting screw. If adjustment is required, loosen the locknut and turn adjusting screw approximately ½ turn and recheck. If condition cannot be corrected by adjustment, disassemble and overhaul the valve as outlined in paragraph 180.

184. VALVE SYNCHRONIZATION. To synchronize the operating valves, remove valve unit as outlined in paragraph 179 and remove both end covers. Loosen the jam nuts on adjusting screws (D & P—Fig. 200). To synchronize the valve, the four operating valve balls must be firmly held on their seats. The manufacturer provides an adjusting cover (JDH-15) which should be modified as shown in Fig. 201 if not previously modified. The ⅝ inch hole drilled in cover will be used for centering the valve lever as later indicated. Install adjusting cover as shown in Fig. 202. Tighten the centering cam pin lock screw (5) and the valve locking screws (1, 2, 3 & 4) using approximately 12 inch pounds torque to be sure valves are solidly seated. Attach a dial indicator as shown, with indicator button contacting valve lever 3 inches from centerline of lever shaft.

NOTE: Dial indicator should have an operating range of 0.125 inch or more. A gage having a range of 0.100 can be used by repositioning the gage while measurements are being taken.

Press down on detent spring plug as shown in Fig. 203 and find center detent position of valve lever; then zero the dial indicator on lever arm at about mid-range of indicator travel. Back out the four adjusting screws on the end nearest to the lever arm (D & P—Fig. 200) one full turn, then adjust the two pressure valve adjusting screws (P) to limit lever travel to 0.060 in each direction when measured with the dial indicator. Tighten the pressure adjusting screw jam nuts, then recheck to be sure adjustment is correct. With pressure valve adjusting screws properly adjusted and locked, adjust the discharge valve adjusting screws (D) to limit valve lever travel to 0.020 in each direction. Tighten the jam nuts then recheck the adjustment.

Remove the adjusting tools, then reassemble and install the valve. Adjust metering valve as outlined in paragraph 182 and detent spring as in paragraph 183, after valve is installed.

2520 Model

185. OPERATION. Tractors are optionally equipped with one, two or

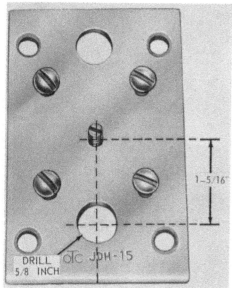

Fig. 201–JDH-15 adjusting plate showing alteration required to find centered position of control lever on 2510 valve.

Fig. 203–Press down on detent spring plug as shown to find neutral detent position of lever on 2510 valve.

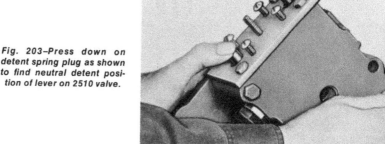

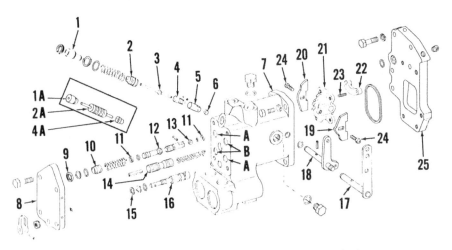

Fig. 204–Exploded view of 2520 selective (remote) control valve.

1. Guide	7. Housing	13. Roller (4)	19. Lower cam
2. Piston	8. Cover	14. Valve	20. Upper cam
3. Pin	9. Packing (6)	15. Thrust washer	21. Rocker
4. Guide	10. Guide (4)	16. Metering valve	22. Keeper (2)
5. Follower	11. Backup ring (6)	17. Arm	23. Set screw (2)
6. Roller	12. Valve	18. Arm	24. Special screw
			25. Cover

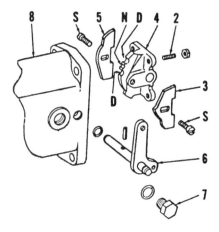

Fig. 207–Cam with pointed lobe goes to numbered side of housing as shown. Tapered ends of both cams is to the top.

three selective (remote) control valves for operation of remote cylinders. Mounting positions of valves are on each side of rockshaft housing.

As with all other units of the hydraulic system, pressure is always present at the valves but no flow exists until the valve is moved. Refer to Fig. 204 for an exploded view of valve mechanism. Each breakaway coupler is equipped with two return valves (12 which position at 'A') and pressure valves (which position at 'B') so arranged that one of each is opened when control lever is moved off center. Detent piston (2 or 2-A) is actuated by pressure differential across metering valve (16) and released by pressure equalization when flow stops at end of piston stroke. Flow control valve (14) maintains an even flow with varying pressure loads.

REMOVE AND REINSTALL. Remove right or left hydraulic cover near rockshaft housing to expose re-

mote valve to be removed. Disconnect linkage rod from valve and remove valve from tractor. Reinstall in reverse order of removal.

186. OVERHAUL. Refer to Fig. 204 for an exploded view of the selective control valve. Clamp the unit in a vise with breakaway coupler tilted upward, and unbolt and remove cover (8) and associated parts carefully as shown in Fig. 205. Identify parts as required for later assembly, then remove valves springs and guides.

Rotate valve body in vise so that rocker assembly is up as shown in Fig. 206. Rocker arm can be disassembled by driving out spring pin.

Assemble by reversing the disassembly procedure. If actuating cam was disassembled, refer to Fig. 207 and 208. Cam with pointed lobe is installed on numbered side of housing as shown, and tapered ends of both cams are to the top.

Adjusting the valve requires use of

special adjusting cover (JDH-15C) and a dial indicator. Remove the two adjusting plugs (7—Fig. 208) and loosen the two cam locking screws (S). Back out the four adjusting screws (2) at least two turns.

Install rocker assembly, pressure and return valves, detent follower, piston, guides and retaining snap ring. Be sure detent follower roller properly rides on rocker. Also make sure that operating valve rollers are turned to ride properly on ramps of cams. Back out all adjusting screws on special plate and install the plate as shown in Fig. 209. Carefully FINGER TIGHTEN the four screws contacting operating valves until valves are seated; then while holding operating lever in center position FINGER TIGHTEN the detent locking screw until detent roller is seated in neutral detent. (N—Fig. 208). Move the operating lever back and forth slightly until follower is felt going into the neutral notch. With operating lever in

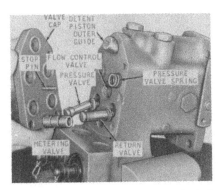

Fig. 205–Clamp housing in a vise as shown to remove rear cap.

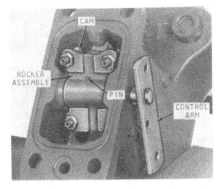

Fig. 206–Front view of housing showing rocker assembly, adjusting screws and associated parts.

Fig. 208–Exploded view of control rocker. Also see Fig. 204.

D. Operating detents	4. Rocker
N. Neutral detents	5. Upper cam
S. Cam clamp screws	6. Arm
2. Set screw	7. Adjusting plug (2)
3. Lower cam	8. Housing

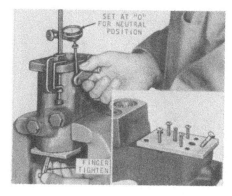

Fig. 209—A special adjusting cover (JDH-15C) is helpful to position rocker for valve adjustment.

Fig. 211—Use a dial indicator to measure rocker arm movement as shown. Refer to paragraph 186.

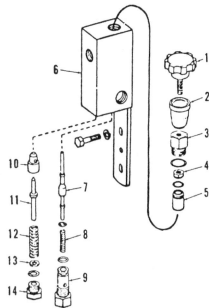

Fig. 212—Exploded view of power weight transfer valve.

1. Valve screw	8. Spring
2. Boot	9. Plug
3. Bushing	10. Relief valve seat
4. Nut	11. Relief valve
5. Shaft	12. Spring
6. Housing	13. Shim
7. Control valve	14. Plug

neutral position, refer to Fig. 210. Turn in the two diagonally opposite Pressure Valve Cam Adjusting Screws until screws, cams and follower rollers are in contact and back out ¼ turn as shown. Turn in the two diagonally opposite Return Valve Cam Adjusting Screws until screws, cams and follower rollers are in contact and back out ⅛ turn. Tighten jam nut on adjusting screws securely. Move the two cams (3 & 5—Fig. 208) into contact with adjusting screws and tighten screws (S) securely.

To double-check the adjustment, smooth a dial indicator reference point 2-inches from lever shaft centerline using a file; and install a dial indicator as shown in Fig. 211. Zero dial indicator while locked in neutral detent, then back out the detent locking screw on adjusting cover. Back out the two adjusting cover screws which contact operating valves on lever side and measure rocker movement which should be 0.013 toward return valve (R) or 0.040 toward pressure valve (P) as shown. (Valves contacting cam 5—Fig. 208 opposite lever side are being checked.) Tighten the two adjusting cover screws on lever side and loosen the other two screws, then check adjustment of valves on lever side. Readjust as necessary for correct rocker movement.

187. **BREAKAWAY COUPLER.** Drive a punch into the expansion plugs (15—Fig. 211-A) and pry out of

housing. Remove retainer rings and springs. Operating levers can then be removed. Drive receptacle assembly from housing. Check steel balls, springs, and all parts for wear and replace as necessary. Replace "O" rings and backup washers. Reassemble in reverse order of disassembly.

POWER WEIGHT TRANSFER VALVE
2520 Models So Equipped

188. **OPERATION.** The power-weight transfer hitch consists of a pressure regulating valve and related parts to transfer a specified amount of front end weight and implement weight to tractor rear tires for improved traction. A conventional double acting remote cylinder is used but only the retracting (rod end) side is pressurized. Refer to Fig. 212 for an exploded view of valve unit and to Fig. 213 for hose routing.

The valve is primarily a switch valve which diverts fluid from the rockshaft cylinder to the control cylinder and Rockshaft Load Control is used for the control lever.

189. **OVERHAUL.** Refer to Fig. 212 for an exploded view of valve unit. Control and relief valves can be re-

moved after removing port plugs (9 & 14). One seat for control valve (7) is on upper surface of plug (9); the other seating surface is in bore of body (6). Control valve moves upward by pressure of spring (8) and inward flow of oil when knob (1) is backed out for rockshaft operation; and control valve seals against upper seat to close off passage to remote cylinder. Turning control knob (1) clockwise mechanically moves control valve into contact with seat on plug (9), closing return passage to rockshaft cylinder and opening passage to remote cylinder. Relief valve spring (12) should test 180-220 lbs. when compressed to a height of 1⅝ inches. Shims (13) may be added if necessary to increase release pressure of relief valve. Renew any parts which are worn, broken or damaged.

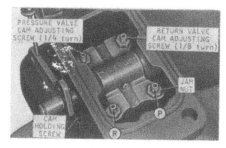

Fig. 210—Pressure and return valves are diagonally opposite in housing as shown.

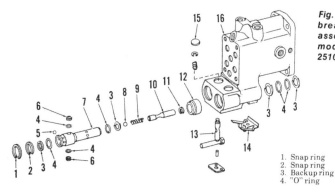

Fig. 211-A—Exploded view of breakaway coupler and associated parts on 2520 model. Internal parts of 2510 coupler are the same as 2520.

	5. Ball
	6. Backup ring
	7. Receptacle
	8. Ball
	9. Spring
	10. Plug
	11. Snap ring
	12. Sleeve
1. Snap ring	13. Operating lever
2. Snap ring	14. Dust plug plate
3. Backup ring	15. Expansion plug
4. "O" ring	16. Housing

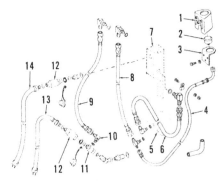

Fig. 213–Hose routing diagram for power/weight transfer valve.

1. Mounting cover	8. Pressure hose
2. Pressure gage	9. Exhaust hose
3. Bracket	10. Elbow
4. Gage hose	11. Tee fitting
5. Rockshaft cyl. cover	12. Coupler
6. Return hose	13. Exhaust hose
7. Valve	14. Transfer cyl. hose

Fig. 214–Exploded view of the high pressure remote cylinder used on all models so equipped.

1.	Cap
2.	Gasket
3.	Spring
4.	Stop valve
5.	Bleed valve
6.	Ball
7.	Spring
8.	Stop rod
9.	Washer
10.	Spring
11.	Cylinder
12.	Spring
13.	"V" packing
14.	Piston
15.	Backup ring
16.	O-ring
17.	Lever
18.	Stop screw
19.	Piston rod
20.	Stop
21.	Oil seal
22.	Backup ring
23.	Arm
24.	Guide

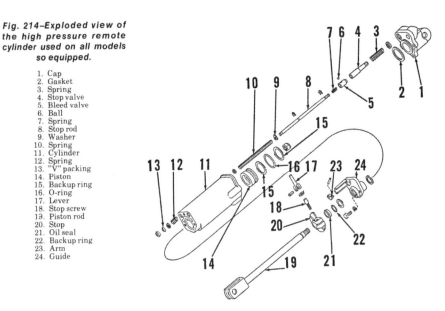

REMOTE CYLINDER
All Models

190. Refer to Fig. 214 for exploded view of the double acting, hydraulic stop remote cylinder. To disassemble, remove end cap (1), stop rod spring (3) and valves (4 and 5), using care not to lose ball (6). Fully retract the cylinder and remove nut from piston end of piston rod (19). To remove the stop rod and springs, drive the groove pin from stop rod arm (23). Assemble rod end of cylinder, rod packing and piston rings, and have piston fully inserted in cyl- inder before installing rod nut. Tighten nut securely. Be sure the piston rod stop (20) is located so that stop lever (17) is opposite the stop rod arm (23). Tighten cap screws securing piston rod guide (24) to a torque of 35 ft.-lbs. and cap screws retaining piston cap (1) to 85 ft.-lbs.

JOHN DEERE

Model ■ 2040

Previously contained in I&T Shop Manual No. JD-41

SHOP MANUAL

JOHN DEERE

SERIES 2040

Tractor serial number is located on right side of transmission case. Engine serial number is stamped on a plate at lower right front engine cylinder block.

INDEX (By Starting Paragraph)

CONDENSED SERVICE DATA

GENERAL

Engine Make	Own
Engine Model	3164D
Number of Cylinders	3
Bore—Inches	4.02
Stroke—Inches	4.33
Displacement—Cu. In.	164
Compression Ratio	16.2:1
Cylinder Sleeves	Wet
Forward Speeds/Reverse Speeds	8/4

TUNE-UP

Firing Order 1-2-3
Valve Tappet Gap, Cold and Hot:
  Exhaust—Inch 0.018
  Intake—Inch 0.014
Valve Face Angle—Degrees 43.5
Valve Seat Angle—Degrees 45
Governed Speeds—Engine Rpm:
  Low Idle 650
  High Idle 2650
  Loaded 2500
Horsepower at Pto Shaft* 40.86

Battery:
  Volts 12
  Ground Polarity ....Neg.
  Capacity Amp/hr. 55
*According to Nebraska Test.

SIZES—CAPACITIES—CLEARANCES

Camshaft Bearing Clearance:
  All—Inch 0.004-0.006
Camshaft Journal Diameters:
  All—Inch 2.200-2.201
Camshaft End Play—Inch ..0.002-0.009
Cooling System—Quarts 10½
Crankcase Oil—Quarts† 7.4
†Including oil filter.
Crankshaft Crankpin:
  Diameter—Inch 2.748-2.749
  Bearing Clearance—Inch .0.0012-0.004
Crankshaft End Play—Inch .0.002-0.008
Crankshaft Main Journal:
  Diameter—Inch 3.123-3.124
  Bearing Clearance—Inch .0.0012-0.004
Transmission, Differential and
  Hydraulic Lift:
  Gallons—Dry/Service 7.9/7.4

FRONT SYSTEM

AXLE AND SUPPORT

1. **AXLE CENTER MEMBER.** Center axle unit (5—Fig. 1) attaches to front support by pivot bolt (6) and rear pivot pin (3—Fig. 2). End clearance of pivot is controlled by shims (3—Fig. 1). Five 0.015-inch shims are installed at factory assembly and recommended maximum end play is 0.015 inch.

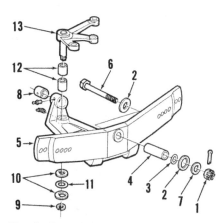

Fig. 1—Exploded view of front axle and associated parts. Slight differences may be noted on some models.

1. Nut	
2. Washer	
3. Shim	8. Bushing
4. Bushing	9. Snap ring
5. Axle	10. Washer
6. Pivot bolt	11. Shim
7. Washer	12. Bushing or bearing
	13. Steering bellcrank

Removing shims reduces the end play. Thrust washers (2) at front and rear of pivot bushing (4) are identical and may be interchanged to compensate for wear when unit is removed. Rear pivot pin (3—Fig. 2) is pressed into front support (1).

The steering bellcrank (13—Fig. 1) is supported in two bushings (12) on some axles, while some other axles are equipped with two needle bearings. Clearance between the steering bellcrank (13) and bushings should be 0.001-0.006 inch on models equipped with bushings. On models with needle bearings at (12), press bearing into bore with lettered side out until flush with inner edge of bore chamfer. On all models, recommended bellcrank end play of 0.004 inch is controlled by varying the number of 0.010-inch thick shims (11). Two shims are normally used; however, installation of additional shims may be necessary to limit end play.

Bolts that clamp axle extensions (7—Fig. 3) to axle main member (5—Fig. 1) should be tightened to 300 ft.-lbs. torque.

2. **SPINDLES AND BUSHINGS.** Refer to Fig. 3 for exploded view. Steering arm (9) is keyed to spindle and retained by a clamp screw. Spindle end play should be maintained at not

more than 0.030 by repositioning steering arm on shaft. Tighten steering arm clamp screw to approximately 85 ft.-lbs. Bushings and associated parts are pre-sized.

TIE RODS AND TOE-IN

3. The recommended toe-in is 1/8 to ¼ inch. Remove cap screw in outer clamp and loosen clamp screw in tie rod end (12—Fig. 3); then turn tie rod tube (11) as required. Both tie rods should be adjusted equally.

On manual steering models, stops on both spindles (1) should contact stops on axle extensions (7) at the same time.

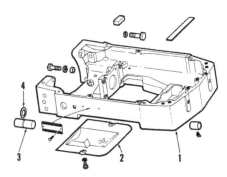

Fig. 2—View of front support casting and associated parts.

1. Front support	3. Rear pivot pin
2. Pan	4. Snap ring

3

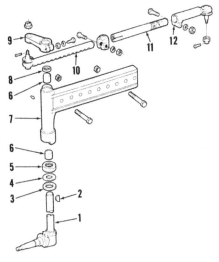

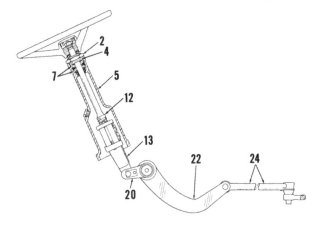

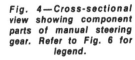

Fig. 4—Cross-sectional view showing component parts of manual steering gear. Refer to Fig. 6 for legend.

Fig. 3—Axle extension, tie rod and associated parts typical of all models.

1. Spindle	7. Axle extension
2. Woodruff key	8. Upper seal
3. Lower seal	9. Steering arm
4. Washer	10. Tie-rod end
5. Thrust bearing	11. Tube
6. Bushing	12. Tie-rod end

On power steering models, none of the spindle stops must contact at extreme turning position. Readjust tie rods if necessary until the correct condition is attained. Tighten tie rod end clamping screws to a torque of 30 ft.-lbs. when adjustment is correct. The width adjusting clamp bolt on adjustable axle models should be tightened to 85 ft.-lbs. torque after changing front tread width.

MANUAL STEERING GEAR

Steering gear is a recirculating ball nut type and the housing is mounted on top side of clutch housing. See Fig. 4 for a cross-sectional view of the manual steering gear unit.

LUBRICATION

All Manual Steering Models

4. Recommended steering gear lubricant is Hy-GARD transmission and Hydraulic Oil or equivalent oil. Fluid should be maintained at level of fill plug located on rear of steering column housing as shown in Fig. 5. Drain plug is the lowest screw that attaches steering shaft cover (14—Fig. 6) to transmission housing.

ADJUSTMENT

All Manual Steering Models

5. Excessive steering column shaft

end and side play can be eliminated by loosening locknut (2—Fig. 6) and turning adjuster (4) until end play is eliminated and barrel bearings (7) slightly preloaded. Tighten locknut (2) to 75 ft.-lbs. torque after adjustment is complete. Refer to paragraph 3 for adjusting front wheel toe-in.

Fig. 5—Steering gear lubricant should be maintained at level of filler plug as shown.

FILLER PLUG

OIL LEVEL

Fig. 6—Exploded view of manual steering gear.

1. Wheel	
2. Locknut	
3. Oil seal	
4. Bearing adjuster	
5. Housing	
6. Gasket	
7. Bearing	
8. Thrust washer	
9. Retainer	
10. Oil seal	
11. Retainer	
12. Shaft & ball nut	
13. Yoke	
14. Cover	
15. Bushing	
16. Gasket	
17. Pin	
18. Cap screw	
19. Clip	
20. Cross shaft	
21. Seal	
22. Steering arm	
23. Cap screw	
24. Drag link	

OVERHAUL

All Manual Steering Models

6. **REMOVE AND REINSTALL.** Either the steering wheel shaft unit or rockshaft assembly can be removed independently of the other if required for service. First drain steering gear as outlined in paragraph 4; then refer to the appropriate following paragraphs.

7. **STEERING WHEEL SHAFT HOUSING.** If tractor is equipped with foot throttle, disconnect pedal linkage and lift foot pedal up out of way. On all models, unbolt and remove steering shaft cover (14—Fig. 6). Refer to inset, Fig. 7 and turn steering wheel until yoke pin is centered in opening as shown; then remove cap screw (18—Fig. 6), clip (19) and yoke pin (17).

NOTE: Yoke pin has tapped hole which will accept one of the cover cap screws for pulling leverage.

Remove steering wheel using a suitable puller. Remove the cap screws securing steering shaft housing to

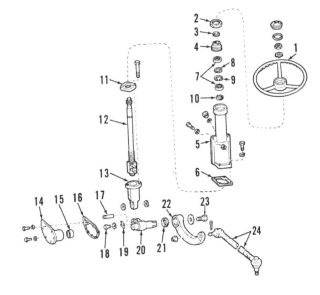

clutch housing and dash, and lift unit from tractor.

Install by reversing the disassembly procedure. Make sure "R" mark on yoke (13) is toward cover opening when pin is installed. Tighten lockplate screw (18) to 84 inch-pounds, steering wheel nut to 50 ft.-lbs. torque. Cap screws retaining housing (5) and cover (14) to transmission housing should be tightened to 36 ft.-lbs. torque. Refill steering gear housing as outlined in paragraph 4.

8. STEERING CROSS SHAFT. The steering cross shaft (20—Fig. 6) can be removed without removing steering wheel shaft, or with wheel shaft and housing removed.

If steering wheel shaft housing is not removed, drain the unit, remove cover (14), cap screw (18) and yoke pin (17) as outlined in paragraph 7; and turn steering wheel until yoke (13) is withdrawn from shaft arm.

Refer to Fig. 7 and remove button plug from left side of clutch housing and remove arm attaching cap screw (23—Fig. 6). Make sure cross shaft arm is still aligned with right cover opening as shown in inset—Fig. 7 and bump steering shaft to right out of clutch housing and arm.

Install by reversing removal procedure. Tighten arm attaching cap screw (23—Fig. 6) fully, strike arm with a hammer, then retighten to 190 ft.-lbs. torque. Complete the assembly by reversing disassembly procedure and fill steering gear as outlined in paragraph 4.

9. OVERHAUL. To disassemble the removed steering shaft housing, loosen locknut (2—Fig. 6) and remove bearing adjuster (4). Pull steering shaft upward until bearings (7) are exposed and remove upper bearing, retainer (9) and thrust washer halves (8). Lift off lower bearing (7) then withdraw shaft downward out of housing and oil seal (10). Yoke (13) can be removed from shaft ball nut by removing the two retaining cap screws. Do not disassemble ball nut as unit is available only as an assembly.

Closed side of yoke (13) should be on same side as ball guides of ball nut (12). Tighten screws securing retainer (11) to yoke (13) to 85 ft.-lbs. torque. Install seal (10) in bore at top of housing (5) with lip toward inside (down). Pack both bearings (7) with grease. Turn steering shaft counter-clockwise until it reaches ball nut stop, then carefully insert steering shaft through seal in housing. Be sure that ball guides of ball nut are toward front of housing as installed on tractor. Install one of the bearings (7) over shaft with outer race

down and inner race (cone) up toward end of shaft. Install both halves of thrust washer (8) and secure with snap ring (9). Push steering wheel shaft into housing until outer race of lower bearing is seated in housing. Install inner race (cone) of upper bearing, then install outer race. Install seal (3) in bearing adjuster (4) and make sure that all parts are well lubricated with grease. Install and tighten adjuster (4) until all end play is eliminated and bearings are slightly preloaded. Tighten locknut (2) to 75 ft.-lbs. torque after adjustment is complete.

Bushings (15) in cover (14) and clutch housing should be installed 0.030 below face of bushing bore. Normal clearance between shaft (20) and bushings is 0.001-0.006 inch. Lip of seal (21) should be toward inside and seal should be flush with bore chamfer. Reinstall steering cross shaft if removed, as outlined in paragraph 8 and steering wheel shaft housing as in paragraph 7. Refill steering gear as in paragraph 4.

POWER STEERING SYSTEM

The power steering system of 2040 models is an open center valve type and a separate gear type pressure pump is mounted ahead of the engine crankshaft with the tractor hydraulic lift system pump.

TROUBLESHOOTING

10. Any problems that may develop in the power steering system usually

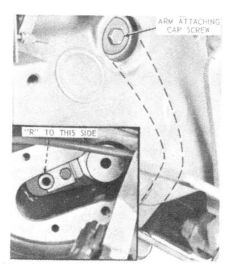

Fig. 7—Removing steering cross shaft. Inset shows yoke arm pin, which should be installed "R" mark visible.

appear as sluggish steering, power steering in one direction only, leakage, or excessive noise in the power steering pump.

Sluggish steering can usually be attributed to:
a. Defective steering valve seat "O" rings which usually produce slow steering in one direction only.
b. Power steering pump not maintaining flow or pressure.
c. Insufficient oil supply to the power steering pump. Sometimes caused by clogged filter, damaged transmission pump or clutch disengaged (disengaging transmission pump).
d. Faulty steering piston "O" rings.

Power steering in one direction only can usually be attributed to:
a. Defective steering valve seat "O" rings.
b. Piston retaining snap ring dislodged (left turn only).

Excessive power steering pump noise can usually be attributed to:
a. Faulty pump, pump drive or other mechanical damage.
b. Air leak on inlet side of pump.
c. Oil foaming.

Excessive oil leakage around upper end of steering wheel shaft, past lip seal can usually be attributed to:
a. Blocked or restricted oil return passage.
b. Improper installation of upper race of lower thrust bearing (15L—Fig. 9). The return passage will be restricted if the large chamfered side of race is not up against lower valve body (23).

SYSTEM CHECKS

A series of checks can be made to determine the overall condition of the power steering system, or to isolate a malfunctioning component. To make these tests, proceed as follows:

12. **OPERATION TEST.** Shut engine off, jack up front end of tractor until wheels are off ground, then turn steering wheel manually in both directions through full limits of travel. Sticking or erratic operation is probably caused by mechanical binding. Check all linkage.

With weight of tractor on front wheels and wheels on smooth hard surface, start engine and turn steering wheel from lock to lock in both directions. Be sure that clutch pedal is released so that transmission oil pump will operate. Steering effort and response should be the same in both directions. Slow steering or excessive effort in one direction may indicate faulty operation of valve body which controls oil for that steering direction. Steering trouble in both directions may be caused by faulty operation of both valve bodies, but may also be caused by clogged filter, damaged transmission pump, damaged power steering pump, improper power steering relief pressure or damaged steering piston seals. Too much front end weight can also result in slow response and increased effort required to turn steering wheel.

Check for steering drift with tractor rolling along straight path, while standing still and with weight removed by jacking up front of tractor. Drift is usually caused by steering valve "O" ring leakage in circuit which controls direction affected. Steering valve leakage can also be caused by improper adjustment of valve body shims (22—Fig. 9 and Fig. 10).

Chatter or noise from the steering valve housing can be caused by defective valve body "O" rings.

13. POWER STEERING PUMP AND RELIEF VALVE CHECK. Some indi-

cations of malfunctioning power steering pump can be caused by plugged transmission oil filter, damaged transmission oil pump, low oil level or other fault which decreases oil supply to the power steering pump.

An accurate test requires use of a hydraulic test unit which incorporates a flow meter, pressure gage, temperature gage and adjustable restriction. Disconnect oil supply line from power steering housing (S—Fig. 11) and attach hydraulic test unit input hose. Outlet hose from test unit should be connected to return tube (R).

NOTE: Be sure to cap fittings when lines are removed to limit oil loss and prevent entrance of dirt. Be sure shut-off valve of test unit is open, then start engine and allow oil to reach 120 degrees F. before testing.

Adjust engine speed to 2000 rpm, close test unit valve long enough to observe relief valve pressure, then open valve.

NOTE: Close test unit valve only as long as necessary for test. Stopping flow too long can result in damage. Relief pressure should be 2100-2130 psi. If relief pressure is incorrect, loosen locknut (1—Fig. 13) and turn adjuster (2) as required. Turning adjuster clockwise will increase relief pressure.

Adjust engine speed to 2500 rpm and begin closing test unit shut-off valve until oil flow just begins to decrease. Pressure should be relief pressure (2100-2130 psi) and flow should be at least 4.75 GPM. If flow is too low, first

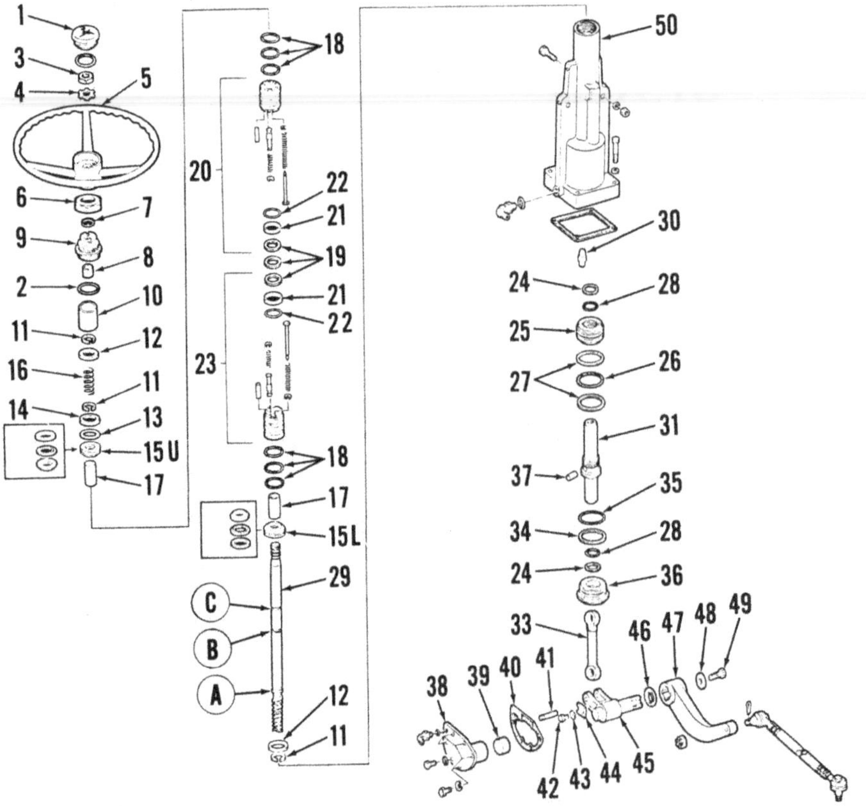

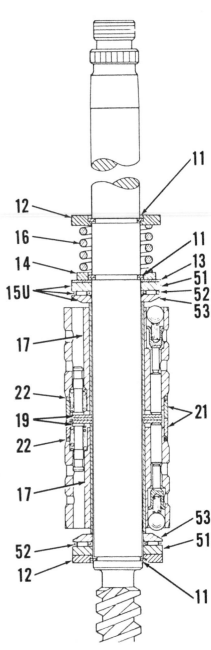

Fig. 9—View of power steering control valve, housing linkage and associated parts.

1. Emblem	15L. Lower thrust bearing	26. "O" ring	39. Bushing
2. "O" ring	15U. Upper thrust bearing	27. Back-up rings	40. Gasket
3. Nut	16. Pressure spring	28. "O" rings	41. Pin
4. Lockwasher	17. Operating sleeve	29. Steering wheel shaft	42. Cap screw
6. Jam nut	18. "O" ring	30. Sleeve	43. Toothed washer
7. Oil seal	19. Special washer	31. Piston rod	44. Lock plate
8. Bushing	20. Upper valve assembly	33. Connecting rod	45. Steering shaft
9. Adjuster	21. Spacers	34. Back-up ring	46. Oil seal
10. Sleeve	22. Shims	35. "O" ring	47. Steering shaft arm
11. Snap ring	23. Lower valve assy.	36. Piston rod guide	48. Special washer
12. Washer	24. Back-up rings	37. Pin	49. Cap screw
13. Washer	25. Piston	38. Steering shaft cover	50. Housing
14. Shims			

Fig. 10—Cross section of steering wheel shaft and control valve. Refer to Fig. 9 for legend except for the following.

51. Flat race
52. Thrust needle bearing   53. Chamfered race

check transmission oil filter, then check transmission.

Power steering relief pressure can also be checked after attaching a suitable pressure gage to power steering pump outlet flange as shown in Fig. 12. Check pressure with engine running at 2000 rpm and steering wheel turned fully against lock.

PUMP

16. The power steering pump should supply at least 4.75 GPM of oil with engine operating at 2500 rpm. The power steering pump and the hydraulic pump are both supplied with oil by the transmission oil pump. Anything that affects operation of the transmission oil pump will also affect the other pumps. Refer to appropriate paragraphs in the Hydraulic Lift System sections for test and servicing procedures.

STEERING VALVE

18. **REMOVE AND REINSTALL.** To remove the steering valve assembly, first remove cap screws from steering shaft cover (38—Fig. 9) and drain oil from steering shaft compartment. If tractor is equipped with a foot accelerator, disconnect rod from foot pedal and swing pedal out of the way. Disconnect return line from steering shaft cover and pressure line from steering valve housing. Remove steering shaft cover. If necessary, center the steering shaft yoke in cover hole, then remove cap screw (42) lock washer (43) and pin retainer (44). Thread a 3/8-inch cap screw into end of pin (41) and remove pin.

Remove steering wheel emblem, straighten tabs of lock washer and remove steering wheel retaining nut (3). Attach a puller and remove steering wheel.

NOTE: Always use a puller to remove steering wheel. Do not drive on upper end of steering shaft.

Remove the metal cover from around steering valve housing, then unbolt steering valve assembly from dash and clutch housing and lift unit from tractor.

Reinstall by reversing removal procedure, start engine and cycle system several times to purge any air which might be present.

19. **OVERHAUL.** To disassemble steering valve, use Fig. 9 and Fig. 10 as a guide and proceed as follows: Place unit in a vise with steering wheel end up. Remove locknut (6), then use a spanner wrench to remove adjuster (9) and seal (7) assembly. Place steering wheel on steering shaft, turn steering wheel counter-clockwise until stop is reached, then remove steering wheel and pull steering shaft upward as far as possible which will expose steering valve parts.

Slide sleeve (10) from shaft (29), then remove upper snap ring (11). Slide upper thrust washer (12) off and remove lower snap ring (11). Remaining valve parts can now be removed from steering shaft.

NOTE: Parts of the valve assemblies (20 and 23) are individually assembled and adjusted by installation of shims (22) at time of original assembly. Be careful when disassembling to not lose or interchange similar parts for the two valves.

With valve parts removed from steering shaft, push steering shaft, piston and piston rod guide assembly out bottom of housing. Remove piston rod guide (36) and piston (25) from piston rod, then press out connecting rod pin (37) and remove connecting rod (33) from piston rod.

NOTE: At this time, it is advisable to determine the condition of the steering shaft and worm (29) and the steering piston rod (31).

Clean and inspect all parts for wear, scoring, burring or other damage. If

a new bushing (8) is installed in adjuster (9), use a piloted driver and install bushing so top end is flush with chamfer in top side of adjuster. Seal (7) is installed in adjuster with lip facing inward. Be especially sure that all oil grooves and passages of valve seats and housing are open and clean.

Use all new "O" rings and back-up rings during reassembly. Lubricate all internal parts as they are installed and reassemble steering valve as follows: If disassembled, insert smaller end of connecting rod (33) in piston rod (31) and insert pin (37). Pin should be flush with outside face of piston rod. Screw piston (25) onto piston rod (31) with dowel holes up. Use special tool (JDH-41-3 or equivalent) to tighten piston to 250 ft.-lbs. torque. Install "O" ring (26) in groove of piston and one back-up ring (27) on each side of "O" ring.

Install one snap ring (11) in lower groove (A) of steering shaft, and position special washer (12) above snap ring so that snap ring is in cut-out section of special washer. Install the thrust bearing (15L) with large chamfered surface up toward lower valve body (23). Install lower operating sleeve (17) and lower valve assembly (23) with special washers (19), spacers (21) and shims (22). Shims should be added between each spacer (21) and control valve (20 or 23) so that spacer protrudes 0.2185-0.2224 inch away from control valve body. Install operating sleeve (17) and upper valve body (20). Install upper thrust bearing (15U) with large chamfered side down toward valve assembly (20). Install washer (13), slide snap ring (11) over shaft against washer (13) into groove (B). Position washer (14) around snap ring and against washer (13), then install spring (16). Install special washer (12) with cut-away up toward end of shaft, then install snap ring (11) over shaft and seat into groove (C) by compressing spring.

Install the six "O" rings (18), one "O" ring (28) and one back-up ring (24) in grooves of housing (50). Install one "O"

Fig. 11—View showing supply line (S) and return line (R) to and from power steering assembly.

Fig. 12—Power steering relief pressure can be checked as described in text by attaching pressure gage to port (P) as shown above.

Fig. 13—Power steering relief pressure is changed by turning adjuster (2) after loosening locknut (1).

ring (28) and one back-up ring (24) in bore of guide (36). Install "O" ring (35) and back-up ring (34) in groove of guide (36). Lubricate all parts, then insert assembled piston, piston rod and connecting rod through guide (36). Guide, rods and piston should be installed in housing (50).

Carefully install the assembled steering wheel shaft and control valves into housing, turn shaft into piston rod, then slide spacer (10) over steering shaft. Assemble bushing (8), seal (7) and "O" ring (2) to adjuster (9).

Lubricate adjuster and position over steering shaft being careful not to damage seal. Tighten adjuster (9) to 50 ft.-lbs. and lock position by tightening locknut to 30 ft.-lbs. torque. Tighten steering wheel retaining nut (3) to 50 ft.-lbs.

ENGINE AND COMPONENTS

R&R ENGINE WITH CLUTCH

28. To remove engine and clutch as a unit, first drain cooling system, remove both batteries, remove front end weights (if so equipped) and drain engine oil (if engine is to be disassembled). Remove side grille screens, hood, tool box and frame side rails. Remove air intake pipe from between intake manifold and air cleaner. Disconnect upper and lower coolant hoses from radiator, wire from fuel gage sending unit in tank and fuel return line from top of fuel tank. Disconnect wire from air cleaner restriction warning switch and move wire out of the way. Close fuel shut-off valve at bottom of fuel tank, then remove the fuel transfer pump and the fuel line between tank and fuel transfer pump. Remove clamps from hydraulic lines along lower right side of engine, detach hydraulic suction line from clutch housing and separate pressure line at union located near joint of engine rear and clutch housing front. On tractors with power steering, disconnect power steering return line from hydraulic suction line fitting and separate power steering pressure line at union located near joint between rear of engine and front of clutch housing. On all models, be sure to cover ends of all lines and open ports to prevent the entrance of dirt. Disconnect drag link at front end (from bellcrank). Support tractor, attach hoist to front support and axle assembly, then unbolt and separate front support and axle assembly from engine.

CAUTION: Use care to keep hydraulic pump drive shaft in alignment with pump couplings until drive shaft is completely free.

29. Disconnect wires from alternator and from starting motor. Disconnect wire from oil pressure sender and speed-hour-meter cable from rear of block behind camshaft. Disconnect ether starting aid pipe from intake manifold of models so equipped. Disconnect shut-off cable and speed control rod from fuel injection pump. Remove temperature gage sensing bulb from cylinder head and position out of the way. Remove screws which attach dash and cowl to flywheel housing and support engine with a hoist. Unbolt engine from clutch housing and pull engine forward until clutch clears clutch shaft. Be sure to keep engine horizontal until completely separated.

Reassemble tractor by reversing the disassembly procedure. On tractors with continuous running pto, the pto shaft mates with the pto clutch disc before transmission shaft mates with main clutch disc. If difficulty is encountered while joining engine to clutch housing, turn crankshaft by hand until both shafts are indexed with both clutch discs and flywheel housing is snug against clutch housing before tightening the retaining cap screws.

Tighten the engine to clutch housing cap screws to 170 ft.-lbs. torque. Tighten the four cap screws which enter from front and attach front support to engine to 170 ft.-lbs. torque and the two cap screws which enter from rear to 130 ft.-lbs. torque.

CYLINDER HEAD

30. To remove cylinder head, first drain cooling system and remove hood. Disconnect battery ground straps. Remove air cleaner tube. Remove exhaust manifold from cylinder head and if tractor is equipped with underslung exhaust, the manifold can be left attached to the exhaust pipe. Discon-nect injector leak-off line from fuel tank, injectors and injection pump and remove complete leak-off line. Disconnect pressure lines from injectors, remove hold-down clamps and spacers and withdraw injectors. Disconnect the cold starting aid (ether) line from the inlet manifold elbow and remove elbow. Unbolt the fuel filters from the cylinder head. Plug all fuel openings. Remove vent tube from rocker arm cover, then remove cover and the rocker arm assembly. Identify and remove push rods and the valve stem caps. Disconnect fan baffle and water outlet elbow from cylinder head, then unbolt and remove cylinder head.

NOTE: Do not turn crankshaft with cylinder head removed unless cylinder sleeves are secured with screws and washers.

When reinstalling cylinder head, use a thin coat of No. 3 Permatex on both sides of head gasket and tighten head bolts to 110 ft.-lbs. (150 N·m) torque in sequence shown in Fig. 23. Be sure oil holes in rear rocker arm shaft bracket and cylinder head are open and clean as this passage provides lubrication for the rocker arm assembly. Head bolts should be retorqued after engine has run about one hour at 2500 rpm under half-load. Loosen the head bolts about 1/6-turn before retightening them to 110 ft.-lbs. (150 N·m) torque. Valve tappet clearance (cold) should be 0.014 inch (0.35 mm) for intake and 0.018 inch (0.45 mm) for exhaust.

31. All valves originally seat against surface machined directly in cylinder head. Exhaust valve seat inserts are available for service if head of valve is recessed more than 0.118 inch (3 mm) below flush with gasket surface of cylinder head. Valve seat angle should be 45 degrees and valve face angle should be 43.5 degrees for all valves. Seat width should be 0.057-0.073 inch for intake valves and exhaust valves which seat directly on cylinder head. Exhaust valve seat width should be 0.051-0.057 inch if equipped with service valve seat insert. Service intake

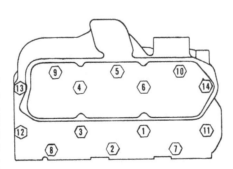

Fig. 23—Tighten cylinder head cap screws to a torque of 110 ft.-lbs., using sequence shown.

and exhaust valves are available with 0.003, 0.015 and 0.030 inch oversize stems. All valves are equipped with hardened steel caps between ends of valve stems and rocker arms.

TAPPET GAP ADJUSTMENT

32. Valve tappet gap for all valves can be set with flywheel being placed in only two positions. Valve tappet gap (cold) is 0.014 inch for inlet valves; 0.018 inch for exhaust.

To set valve tappet gap, turn crankshaft by hand until No. 1 cylinder is at top dead center and TDC timing screw will enter hole in flywheel as shown in Fig. 24. Check the valves to determine whether front cylinder is on compression or exhaust stroke.

NOTE: The rear valve (No. 3 exhaust) is open (rocker arm is tight) when No. 1 piston is at TDC on compression stroke.

Refer to the appropriate diagram (Figs. 25 and 26) and adjust the indicated valves; then turn crankshaft one complete turn until timing screw will again enter TDC hole in flywheel and adjust remainder of valves.

VALVE GUIDES

33. Valve guides are integral with cylinder head and have an inside diameter new of 0.374-0.375 inch which provides 0.002-0.004 inch operating clearance for valves. Maximum allowable valve stem clearance in guide is 0.006 inch and when this value is exceeded, ream valve guide as required to fit next oversize valve stem. Valves are available with stem of 0.003, 0.015 and 0.030 inch oversize.

VALVE SPRINGS

34. Inlet and exhaust valve springs are interchangeable. Springs that are distorted, discolored, rusted, or do not meet the following specifications, should be renewed.

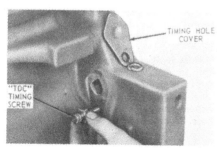

Fig. 24—Reverse timing screw as shown, to find "TDC" timing hole in flywheel. Refer to text.

Free Length (approx.) ....2-1/8 in.
Test Lbs. at 1.81 in. 54-62
Test Lbs. at 1.36 in. 133-153

ROCKER ARMS AND SHAFT

36. Rocker arms are interchangeable and bushings are not available. Inside diameter of shaft bore in rocker arm is 0.790-0.792 inch. Outside diameter of rocker arm shaft is 0.787-0.788 inch. Normal operating clearance between rocker arm and shaft is 0.002-0.005

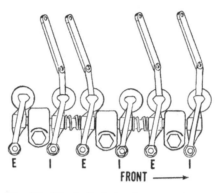

Fig. 25—The indicated valves can be adjusted when No. 1 piston is at TDC on compression stroke. Turn crankshaft one complete turn and adjust remaining valves; refer to Fig. 26.

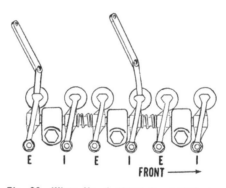

Fig. 26—When No. 1 piston is at TDC on exhaust stroke, the two indicated valves can be gapped. Refer also to Fig. 25.

Fig. 29—Oil pressure relief valve is located as shown. Unit can be adjusted by using shims under forward end of spring. Also see Fig. 46.

inch. Renew rocker arm and/or shaft if clearance is excessive.

Valve stem contacting surface of rocker arm may be refaced but original radius must be maintained.

When reinstalling rocker arm assembly, be sure oil holes and passages are open and clean. Pay particular attention to the rear mounting bracket as lubrication is fed to rocker arm shaft through this passage. Oil hole in rocker arm shaft must face downward when installed on cylinder head.

CAM FOLLOWERS

37. The cylinder type cam followers (tappets) can be removed from below after camshaft has been removed. If

Fig. 30—View of engine timing gear train assembled.

C. Crankshaft gear
G. Camshaft gear
L. Lower idler gear
O. Oil pump gear
P. Injection pump gear
U. Upper idler gear

Fig. 31—With engine at TDC and timing tool (TT) positioned on shafts centerline as shown, camshaft gear timing mark (TM) will be directly under edge of timing tool.

necessary, they can also be removed from above if the cylinder head, rocker arm shaft and push rods are removed. Cam followers are available in standard size only and operate directly in machined bores in cylinder block.

VALVE TIMING

38. Valves are correctly timed when timing mark on camshaft gear is aligned with timing tool (JD254) when tool is aligned with crankshaft and camshaft centerlines as shown in Fig. 31.

TIMING GEAR COVER

39. To remove timing gear cover, first remove the front axle and front support assembly as outlined in paragraph 28.

With front support assembly removed, remove fan, fan belt, alternator and water pump. Remove crankshaft pulley retaining cap screw, attach puller and remove pulley. Remove the oil pressure regulating plug, spring and valve (see Fig. 29). Drain and remove oil pan, then unbolt and remove the timing gear cover. See Fig. 30.

With timing gear cover removed, the crankshaft front oil seal can be renewed. To renew oil seal, coat outside diameter of seal with sealing compound and with seal lip toward inside, support timing gear cover around seal area and press seal into bore until it bottoms.

NOTE: Do not attempt to install the front oil seal in timing gear cover without providing support around seal area. Cover is a light cast aluminum alloy and could be warped or cracked rather easily.

Fig. 33—Before removing camshaft check end play as shown.

CAMSHAFT

40. To remove camshaft, timing gear cover must be removed as outlined in paragraph 39. Remove vent tube, rocker arm cover, rocker arm assembly and push rods. Shut off fuel and unbolt fuel transfer pump from side of cylinder block. Remove injection nozzles, refer to paragraph 30 and remove the cylinder head. Disconnect the speed-hour-meter cable from rear of cylinder block. Use special magnetic tool, wires with 90 degree bend in end or similar procedure to hold cam followers up away from camshaft.

Before removing camshaft, mount a dial indicator as shown in Fig. 33 and measure camshaft end play. Recommended camshaft end play is 0.002-0.009 inch with wear limit of 0.015 inch. Turn camshaft so that thrust plate retaining cap screws are accessible through holes in camshaft gear, then remove the two retaining screws. Pull camshaft and thrust plate from cylinder block.

The camshaft operates in three unbushed bores in cylinder block, the tach-hour-meter drive shaft is pressed into rear of camshaft and the camshaft timing gear is pressed onto front of shaft and aligned with Woodruff key. Check camshaft and associated parts against values which follow:

Camshaft Journal Diameter—
  Desired . . . . 2.200-2.201 in.
  Wear limit . . . . 2.199 in.
Camshaft Bearing Bore
  Diameter . . . . 2.204-2.205 in.
Camshaft to Bearing
  Diametral Clearance—
    Desired . . . . 0.004-0.006 in.
    Wear limit . . . . 0.007 in.
Camshaft Thrust Plate Thickness—
  Desired . . . . 0.156-0.158 in.
  Wear limit . . . . 0.130 in.

The tach-hour-meter drive shaft can be pulled from rear of camshaft by threading exposed end of drive shaft, then using spacers and nut to pull drive shaft from camshaft.

Support camshaft gear and press camshaft from gear if renewal of either is required. When installing camshaft gear, be sure that timing mark is toward front and support camshaft under the front bearing journal.

The easiest method of aligning timing marks is to remove the upper idler gear, turn crankshaft until front piston is at TDC, align camshaft and injection pump drive gear timing marks as shown in Fig. 31 and Fig. 35, then carefully install upper idler gear without moving crankshaft, camshaft or injection pump drive gears.

Tighten camshaft thrust plate cap screws to 35 ft.-lbs. torque. Cap screw for upper idler gear should be tightened to 65 ft.-lbs. torque. Screws attaching timing gear cover to engine should be tightened to 35 ft.-lbs. torque.

IDLER GEARS

42. The upper and lower idler gears are bushed and operate on stationary shafts which are attached to the engine front plate with cap screws. Idler gear end play is controlled by thrust washers. Both idler gears are driven by the crankshaft and the upper idler gear drives the camshaft gear and the diesel injection pump drive gear. The lower idler gear drives the oil pump drive gear (see Fig. 31).

To remove idler gears, remove oil pan and timing gear cover. Remove cap screw and pull gear and thrust washers from shaft. Idler gear shaft can now be removed.

Clean and inspect idler gears and shafts and refer to the following specifications.
Shaft O.D. 1.749-1.750
Bushing I.D. 1.751-1.753
Operating clearance 0.001-0.004
Max. allowable clearance 0.006
End play 0.001-0.007
Max. allowable end play 0.015

Reinstall by reversing removal procedure and be sure camshaft, diesel injection pump and crankshaft are timed as indicated in Figs. 31 and 35. Tighten cap screw retaining upper idler shaft to 65 ft.-lbs. torque and cap screw retaining lower idler shaft to 95 ft.-lbs. torque.

TIMING GEARS

43. **CAMSHAFT GEAR.** The camshaft gear (G—Fig. 30) is keyed and pressed on the camshaft. The fit of gear on camshaft is such that removal of the camshaft, as outlined in para-

Fig. 35—Injection pump gear (P) is correctly timed if timing mark (TM) is directly below timing tool (TT) when tool is placed between centerlines of crankshaft and injection pump shafts.

graph 40 is recommended. Camshaft is correctly timed when centerline of camshaft, timing mark on camshaft gear and centerline of crankshaft are aligned with crankshaft positioned so that front cylinder is at Top Dead Center. Refer to Fig. 31.

44. CRANKSHAFT GEAR. Renewal of crankshaft gear requires removal of crankshaft as outlined in paragraph 56. Gear is keyed and pressed on crankshaft. Support crankshaft under first throw when installing new gear. Installation of gear may be eased by heating gear in oil.

Crankshaft gear has no timing marks but keyway in crankshaft will be straight up when front piston is at TDC.

45. INJECTION PUMP GEAR AND SHAFT. The gear can be removed after removing the timing gear cover and the three retaining screws. Gear timing to pump shaft is assured by the alignment dowel in drive hub and hole in gear. Gear shaft is an integral part of pump and gear must be removed for pump removal. Tighten gear retaining cap screws to a torque of 18 ft.-lbs.

The injection pump drive gear is interchangeable for three and four cylinder models and two timing marks appear on the gear, each identified by a stamped "3" or "4". Use the timing mark which is marked "3" when timing the gears as shown in Fig. 35.

47. TIMING GEAR BACKLASH. Excessive timing gear backlash may be corrected by renewing the gears concerned, or in some instances by renewing idler gear bushing and/or shaft. Refer to the following for recommended backlash maximum limit:

Crankshaft gear to upper idler 0.016 in.
Camshaft gear to upper idler . .0.020 in.
Injection pump gear to
   upper idler . . . . .0.020 in.
Crankshaft gear to lower idler .0.020 in.
Oil pump gear to lower idler . .0.016 in.

ROD AND PISTON UNITS

48. Piston and connecting rod assemblies are removed from above after removing cylinder head and oil pan. Secure cylinder liners (sleeves) in cylinder to prevent liners from moving as crankshaft is turned.

All pistons have the word "FRONT" stamped on head of piston and connecting rods also have the word "FRONT" stamped (embossed) in the web of connecting rod. Connecting rods and pistons are not originally numbered, but should be stamped with correct

cylinder number before removal. When installing rod and piston units, lubricate rod screws and tighten connecting rod screws to 60-70 ft.-lbs. torque.

PISTONS, RINGS AND SLEEVES

49. Pistons are cam-ground, forged aluminum-alloy and are fitted with three rings located above the piston pin. All pistons have the word "FRONT" stamped on the piston head and are available in standard size only.

Top piston ring is of keystone design and a wear gage (JDE-62) should be used for checking piston groove wear. Normal side clearance for middle (2nd) piston ring is 0.0015-0.003 inch with maximum wear limit of 0.008 inch. Installation instructions for piston rings are included in ring kits.

The renewable wet type cylinder sleeves are available in standard size only. Sleeve flange at upper edge is sealed by the cylinder head gasket. Sleeves are sealed at lower edge by packing shown in Fig. 38. Sleeves normally require loosening using a sleeve puller, after which they can be withdrawn by hand. Out-of-round or taper should not exceed 0.005 inch. If sleeve is to be reused, it should be deglazed using a normal cross-hatch pattern.

When reinstalling sleeves, first make sure sleeve and block bore are absolutely clean and dry. Carefully remove any rust or scale from seating surfaces and packing grooves, and from water jacket in areas where loose scale might interfere with sleeve or packing installation. If sleeves are being reused, buff rust and scale from outside of sleeve.

Install sleeve without the seals and measure standout. Check sleeve standout at several locations around sleeve. Also check to be sure that sleeve will slip fully into bore without force. If sleeve cannot be pushed down by hand, recheck for scale or burrs. If sleeve

stand-out is less than 0.001 inch, install one special shim (part number R46906) between liner and cylinder block, then recheck stand-out. If stand-out is more than 0.004 inch, check for scale or burrs; then, if necessary, select another sleeve. After matching sleeves to all the bores, mark the sleeves then refer to the appropriate following paragraph for packing and sleeve installation.

50. RECTANGULAR PACKING AND "O" RINGS. Refer to Fig. 38. Apply liquid soap (such as part number AR 54749) to the rectangular section ring (1) and install over lower end of cylinder liner (sleeve). Slide the rectangular section ring up against shoulder on sleeve, make sure that ring is not twisted and that longer slides are parallel with side of sleeve as shown. Apply the liquid soap to round section "O" rings (2) and install in grooves in cylinder block. Be sure that "O" rings are completely seated in grooves so that installing the sleeve will not damage the "O" rings. Observe the previously affixed mark indicating correct cylinder location, then install sleeves carefully into correct cylinder block bore. Work sleeve gently into position by hand until it is finally necessary to tap sleeve into position using a hardwood block and hammer.

NOTE: Be careful not to damage the packing rings. Check the cylinder sleeve stand-out (Fig. 38) with packing installed. The difference between this measured stand-out and similar measurement taken earlier for same sleeve in same bore without packing will be the compression of the packing. If the compression is less than 0.018 inch,

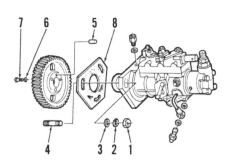

Fig. 36—The injection pump drive gear is attached to pump drive shaft with three screws (7) and lockwashers (6). Gear is timed to pump by dowel (5). Pump is attached to front plate with three studs (4), washers (3), lockwashers (2) and nuts (1). Refer to Fig. 71 for location of studs (4).

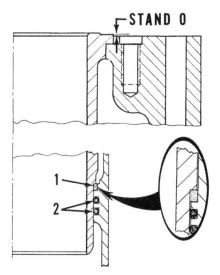

Fig. 38—Cross section of cylinder sleeve showing rectangular section packing (1) and round "O" rings (2). Refer to text for correct installation of cylinder sleeves.

the rectangular section packing ring will not seal properly. Remove sleeve from cylinder block, check packing to be sure that installation has not cut the packing ring. If shoulders on sleeve and in cylinder block do not provide proper compression of the rectangular packing ring, install different sleeve and recheck. If a different sleeve will not provide enough compression of packing sleeve, suggested repair is to install new cylinder block.

53. **SPECIFICATIONS.** Specifications of pistons and sleeves are as follows:
Sleeve Bore . . . . 4.0150-4.0160 in.
Piston Skirt Diameter—
  Bottom of skirt at right angles
  to piston pin . . . . 4.010-4.011 in.
Piston Skirt To Cylinder
  Clearance—
    Desired . . . . 0.004-0.006 in.
    Wear limit . . . . 0.010 in.

PISTON PINS AND BUSHINGS

54. The full floating piston pins are retained in pistons by snap rings. A pin bushing is fitted in upper end of connecting rod and bushing must be reamed after installation to provide a thumb press fit for the piston pin. Piston pin to piston recommended clearance is 0.0001-0.0012 inch with wear limit of 0.0016 inch.
Piston pin diameter is 1.3748-1.3752 inch.

CONNECTING RODS AND BEARINGS

55. The steel-backed, aluminum lined bearings can be renewed without removing rod and piston unit by removing oil pan and rod caps. The connecting rod big end parting line is diagonally cut and rod cap is offset away from camshaft side as shown in Fig. 41. A tongue and groove cap joint positively locates the cap. Rod marking "FRONT" should be forward and locating tangs for bearing inserts should be together when cap is installed.
Connecting rod bearings are available in undersizes of 0.002, 0.010, 0.020 and 0.030 inch as well as standard. Refer to the following specifications:
Crankpin diameter . . . . . . 2.748-2.749 in.
Diametral clearance . . . 0.0012-0.0040 in.
Rod cap torque . . . . 60-70 ft.-lbs.

CRANKSHAFT AND BEARINGS

56. The crankshaft is supported in four main bearings. All main bearing inserts may be renewed after removing oil pan, oil pump and main bearing caps. All main bearing caps except rear

are alike but should not be interchanged. Mark the front main bearing cap "1", the second cap "2" and the third cap "3" for identification prior to removal so they can be reinstalled in their original position. Rear main bearing is flanged and controls crankshaft end play which should be 0.002-0.008 inch. Install main bearing caps with previously affixed identification marks aligned and tighten the retaining screws to 85 ft.-lbs. torque.
To remove the crankshaft, it is necessary to remove engine from tractor. With engine removed, remove oil pan, oil pump, cylinder head and the connecting rod and piston units. Remove clutch, flywheel and flywheel housing. Remove timing gear cover, camshaft, injection pump drive gear and both idler gears, then unbolt and remove engine front plate from cylinder block. Be sure main bearing caps are identified for reinstallation, then remove all bearing caps and lift crankshaft from cylinder block.
Check crankshaft and bearings for wear, scoring or out-of-round condition using the following specifications.
Main journal diameter . . . 3.123-3.124 in.
Crankpin diameter . . . . . . 2.748-2.749 in.
Crankshaft end play—
  Desired . . . . 0.002-0.008 in.
  Maximum limit . . . . 0.015 in.
Journal taper per 1 inch length—
  Maximum limit . . . . 0.001 in.

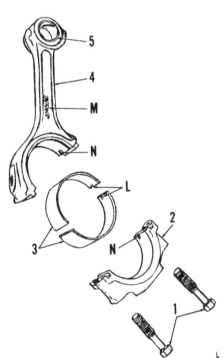

Fig. 41—Refer to text for correct assembly of piston and connecting rod.

1. Cap screws	5. Pin bushing
2. Cap	L. Locating tangs
3. Inserts	M. "FRONT" marking
4. Rod	N. Locating notches

Journal out-of-round—
  Maximum limit . . . . 0.003 in.
Main bearing diametral clearance—
  Desired . . . . 0.0012-0.004 in.
  Maximum limit . . . . 0.006 in.
Connecting rod bearing diametral
  clearance—
  Desired . . . . 0.0012-0.004 in.
  Maximum limit . . . . 0.0062 in.

If crankshaft does not meet specifications, either renew it or grind to the correct undersize. Main bearings are available in standard size and undersizes of 0.002, 0.010, 0.020 and 0.030 inch.

NOTE: The plug located in aft end of crankshaft is a thrust plug for the transmission input shaft. If plug is damaged in any way, drill it out and press a new plug in until it bottoms.

CRANKSHAFT REAR OIL SEAL

58. The lip type crankshaft oil seal is contained in flywheel housing and a wear ring is pressed on mounting flange of crankshaft.
To renew the seal, first remove clutch, flywheel and flywheel housing. If wear ring on crankshaft is damaged, spread the ring using a dull chisel on sealing surface, and withdraw the ring. Install new ring with rounded edge to rear, being careful not to cock the ring. Ring should start by hand and can be seated using a suitable driver such as JD-251. Install seal in flywheel housing working from rear with seal lip to front. When properly installed, seal should be flush with rear of housing bore.

FLYWHEEL

59. To remove flywheel, first remove clutch as outlined in paragraph 111 and 115, then unbolt and remove flywheel from its doweled position on crankshaft.
To install a new flywheel ring gear, heat to approximately 500 degrees F. and install with chamfered end of teeth toward front of flywheel.
The clutch pilot bearing can be bumped out of flywheel after snap ring is removed. Install new bearing with shielded side rearward and pack bearing with a high-temperature grease.
When installing flywheel, tighten the retaining cap screws to 85 ft.-lbs. torque.

FLYWHEEL HOUSING

60. The cast iron flywheel housing is secured to rear face of engine block by

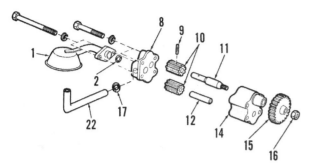

1. Oil pick up
2. "O" ring
8. Cover
9. Groove pin
10. Pump gears
11. Drive shaft
12. Idler shaft
14. Housing
15. Drive gear
16. Nut
17. "O" ring
22. Outlet tube

eight cap screws. Flywheel housing contains the crankshaft rear oil seal and oil pressure sending unit switch. The rear camshaft bore in block is open and the tachometer drive passes through flywheel housing. It is important therefore, that gasket between block and flywheel housing be in good condition and cap screws properly tightened. Tighten all screws evenly to 23 ft.-lbs. then retorque to 35 ft.-lbs.

OIL PUMP AND RELIEF VALVE

61. To remove oil pump, first drain and remove oil pan. Remove the nut retaining oil pump drive gear and pull the gear; then unbolt and remove the oil pump.

With pump removed, use Fig. 43 as a guide and proceed as follows: Remove idler gear and drive gear and shaft (11) from pump housing. Check to see that groove-pin (9) is tight in gear and drive shaft. Pin (9) can be renewed if necessary. Check bearing O.D. of drive shaft (11) which should be 0.630-0.631 in. Check I.D. of housing bore for drive shaft (11) which should be 0.632-0.633 inch. Renew pump if bore exceeds 0.636 inch diameter.

Idler gear shaft (12) can be pressed from pump housing if renewal is necessary. Diameter of new idler shaft is 0.485-0.486 inch.

Width of new gears (10) is 1.620-1.622 inch and if gear width is less than 1.618 inch, they should be renewed. Install gears and shafts in pump body as shown in Fig. 44 and measure between ends of gear teeth and pump

body. This clearance should be 0.001-0.004 inch and pump should be renewed if radial clearance exceeds 0.005 inch. Now place a straight edge across pump body as shown in Fig. 45 and measure between straight edge and end of pump gears. This clearance should be 0.001-0.006 inch and pump should be renewed if axial clearance exceeds 0.008 inch.

Reassemble pump by reversing the disassembly procedure and install pump tightening the screws that attach pump to cylinder block front plate to 35 ft.-lbs. torque and the drive gear retaining nut to 35 ft.-lbs. torque. The nut should be locked in place by staking with center punch at three locations around threads.

The oil pressure is controlled by a relief valve located in the forward end of the cylinder block oil gallery. Relief pressure can be adjusted by adding or deducting shims (6—Fig. 46) located between the spring (5) and plug (7). Oil pressure should be 22 psi minimum with engine at 800 rpm after engine and lubricating oil reaches normal operating temperature. Two aluminum washers (13) may be used if oil pressure is too high with all shims (6) removed. The relief valve spring should have free length of approximately 4.7 inches and should exert 13.5-16.5 lbs. force when compressed to 1.68 inches. The plug retaining the relief valve spring to the timing gear cover should be tightened to 80 ft.-lbs. torque.

The relief valve seat (bushing) is pressed into cylinder block as shown in Fig. 47 and is renewable. When installing the new valve seat, use John Deere special tool JD-248, or equiva-

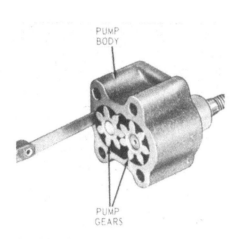

Fig. 44—Clearance between gears and housing should be 0.001-0.004 when measured as shown.

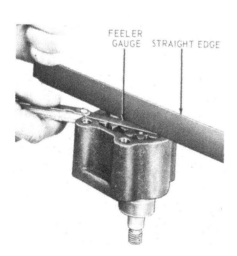

Fig. 45—Clearance between end of gears and cover should be 0.001-0.006 when measured as shown.

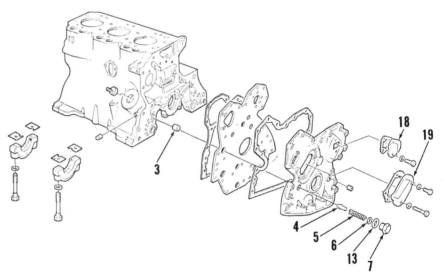

Fig. 46—Exploded view of timing gear cover and front plate showing the engine oil pressure relief valve.

3. Valve seat (bushing)
4. Relief valve
5. Spring
6. Shims
7. Plug
13. Aluminum washers
18. Injection pump cover
19. Timing gear opening cover

lent tool that will contact only the outer diameter of the seat. Press the valve seat into cylinder block until outer recessed edge is flush with bottom of counterbore. Do not press on, or otherwise damage, the raised inner rim of valve seat.

FUEL LIFT PUMP

R&R AND OVERHAUL

64. The fuel lift pump is mounted on right side of cylinder block and is actuated by a lobe on the engine camshaft. The diaphragm type pump moves fuel from the fuel tank through the fuel filters to the fuel injection pump. The pump can be disassembled and cleaned, then reassembled using new diaphragm, spring and gaskets. Parts of pump body, operating levers and check valves are not available. Renewal of complete pump assembly is necessary if pump body, operating levers or check valves are damaged.

DIESEL FUEL SYSTEM

FILTERS AND BLEEDING

65. **FILTERS.** Tractors are equipped with single two-stage fuel filter and sediment bowl units. Renew the filter assembly at least every 1000 hours of operation (oftener under severe conditions or if fuel is contaminated). Remove filter by releasing the retaining spring clip from around filter.

Check the glass sediment bowl for sediment or water, and if necessary, loosen drain plug and operate priming lever of fuel transfer pump to clear deposits from sediment bowl.

66. **BLEEDING.** Whenever fuel system has been run dry, or a line has been disconnected, air must be bled from fuel system as follows: Be sure there is sufficient fuel in tank and that tank outlet valve is open. Loosen bleed screw on top of filter (Fig. 50) and actuate primer lever of fuel transfer pump until a solid, bubble-free stream of fuel emerges from bleed screw, then tighten bleed screw. Loosen pressure line connections at injectors about one turn, open throttle and crank engine until fuel flows from loosened connections, then tighten connections and start engine.

NOTE: If no resistance is felt when operating priming lever of fuel transfer pump and no fuel is pumped, the transfer pump rocker arm is on the high point of pump cam of camshaft. In this case, turn engine to re-position pump cam and release pump rocker arm.

INJECTOR NOZZLES

WARNING: Fuel leaves the injection nozzles with sufficient force to penetrate the skin. When testing, keep your person clear of the nozzle spray.

67. **TESTING AND LOCATING A FAULTY NOZZLE.** If one engine cylinder is misfiring it is reasonable to suspect a faulty injector. Generally, a faulty injector can be located by running the engine at low idle speed and loosening, one at a time, each high pressure line at injector. As in checking spark plugs in a spark ignition engine, the faulty unit is the one that least affects the engine operation when its line is loosened.

Remove the suspected injector as outlined in paragraph 68. If a suitable nozzle tester is available, test injector as outlined in paragraphs 69 through 73 or install a new or rebuilt unit.

68. **REMOVE AND REINSTALL.** To remove an injector, remove hood and wash injector, lines and surrounding area with clean diesel fuel. Use hose clamp pliers to expand clamp and pull leak-off boot from injector. Disconnect high pressure line, then cap all openings. Remove cap screw from nozzle clamp and remove clamp and spacer. Pull injector from cylinder head.

NOTE: Unless the carbon stop seal has failed causing injector to stick, the injectors can be easily removed by hand. If injectors cannot be removed by hand, use John Deere nozzle puller JDE-38 and be sure to pull injector straight out of bore. DO NOT attempt to pry injector from cylinder head or damage to injector could result.

When installing injector, be sure nozzle bore and seal washer seat are clean and free of carbon or other foreign material. Install new seal washer and carbon seal on injector and insert injector into its bore using a slight twisting motion. Install and align locating clamp then install hold-down clamp and spacer and tighten cap screw to 23 ft.-lbs. torque.

Bleed system as outlined in paragraph 66 if necessary.

69. **TESTING.** A complete job of nozzle testing and adjusting requires the use of an approved nozzle tester.

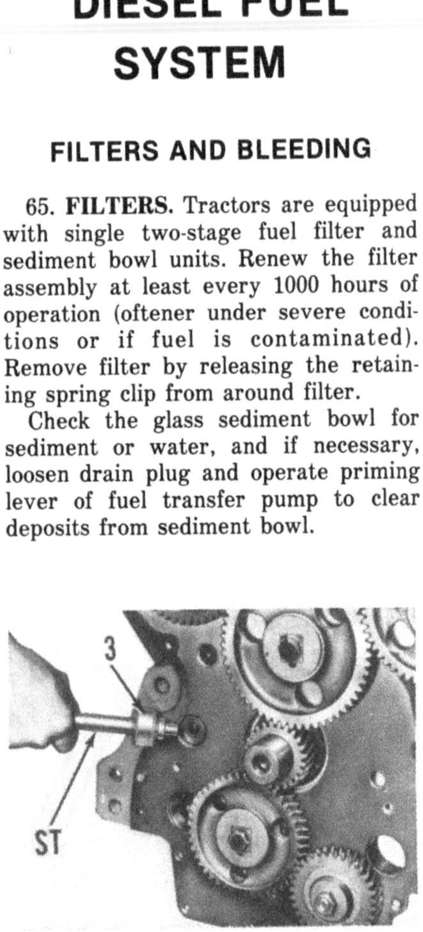

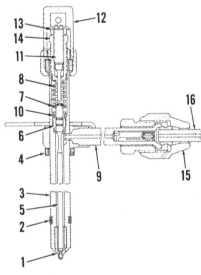

Fig. 54—Cross-sectional view of typical Roosa-Master injector. Nozzle tip (1) and valve guide (6) are parts of finished body and are not serviced separately.

1. Nozzle tip	
2. Carbon seal	10. Seal ring
3. Nozzle body	11. Lift adjusting screw
4. Seal washer	12. Fuel leak off cap
5. Nozzle valve	13. Locknut
6. Valve guide	14. Pressure adjusting
7. Spring seat	screw
8. Pressure spring	15. Compression nut
9. Inlet fitting	16. Pressure line

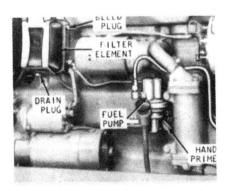

Fig. 47—View showing location of the late engine oil pressure relief valve seat. Seat is renewable; refer to text.

Fig. 50—View of two-stage filter and sediment bowl assembly used on all models.

Only clean, approved testing oil should be used in the tester tank. The nozzle should be tested for spray pattern, opening pressure, seat leakage and back leakage (leak-off). Injector should produce a distinct audible chatter when being tested and cut off quickly at end of injection with a minimum of seat leakage.

NOTE: When checking spray pattern, turn nozzle about 30 degrees from vertical position. Spray is emitted from nozzle tip at an angle to the centerline of nozzle body and unless injector is angled, the spray may not be completely contained by the beaker. Keep your person clear of the nozzle spray.

70. SPRAY PATTERN. Attach injector to tester and operate tester at approximately 60 strokes per minute and observe the spray pattern. A finely atomized spray should emerge at each nozzle hole and a distinct chatter should be heard as tester is operated. If spray is not symmetrical and is streaky, or if injector does not chatter, overhaul injector as outlined in paragraph 74.

71. OPENING PRESSURE. The correct opening pressure is 3000 psi. If opening pressure is not correct but nozzle will pass all other tests, adjust opening pressure as follows: Loosen the pressure adjusting screw locknut (13—Fig. 54), then hold the pressure adjusting screw (14) and back out the valve lift adjusting screw (11) at least one full turn. Actuate tester and adjust nozzle pressure by turning adjusting screw as required. With the correct nozzle opening pressure set, gently turn the valve lift adjusting screw in until it bottoms, then back it out ½-turn to provide the correct valve lift of 0.009 inch.

NOTE: A positive check can be made to see that the lift adjusting screw is bottomed by actuating tester until a pressure of 250 psi above nozzle opening pressure is obtained. Nozzle valve should not open.

Hold pressure adjusting screw and tighten locknut to a torque of 70-75 in.-lbs.

72. SEAT LEAKAGE. To check nozzle seat leakage, proceed as follows: Attach injector on tester in a horizontal position. Raise pressure to approximately 2400 psi, hold for 10 seconds and observe nozzle tip. A slight dampness is permissible but should a drop form in the 10 seconds, renew the injector or overhaul as outlined in paragraph 74.

73. BACK LEAKAGE. Attach injector to tester with tip slightly above horizontal. Raise and maintain pressure at approximately 1500 psi and observe leakage from return (top) end of injector. After first drop falls, the back leakage should be 5 to 8 drops every 30 seconds. If back leakage is excessive, renew injector or overhaul as outlined in paragraph 74.

74. OVERHAUL. First wash the unit in clean diesel fuel and blow off with clean, dry compressed air. Remove carbon stop seal and sealing washer. Clean carbon from spray tip using a brass wire brush. Also, clean carbon or other deposits from carbon seal groove in injector body. DO NOT use wire brush or other abrasive on the Teflon coating on outside of nozzle body between the seals. Teflon coating can be cleaned with a soft cloth and solvent. Coating may discolor from use,

Fig. 57—Using a pin vise and cleaning needle to clean spray tip. Use a 0.008 diameter needle to open holes. Final cleaning should be done with a 0.010 needle on all nozzles except early four-hole type which requires a 0.012 needle.

but discoloration is not harmful.

Clamp the nozzle in a soft jawed vise, loosen locknut (13—Fig. 54) and remove pressure adjusting screw (14), spring (8) and lower spring seat (7).

If nozzle valve (5) will not slide from body when body is inverted, use speical retractor (Fig. 56); or reinstall on nozzle tester with spring and lift adjusting screw removed, and use hydraulic pressure to remove the valve.

Nozzle valve and body are a matched set and should never be intermixed. Keep parts for one injector separate and immerse in clean diesel fuel in a compartmented pan, as unit is disassembled.

Clean all parts thoroughly in clean diesel fuel using a brass wire brush and lint-free wiping towels. Hard carbon or varnish can be loosened with a suitable, non-corrosive solvent.

Clean the spray tip orifices first with an 0.008 cleaning needle held in a pin vise as shown in Fig. 57. On early four-hole nozzles, follow up with a 0.012 needle; on five hole nozzle and late four-hole units, use a 0.010 needle.

Clean the valve seat using a Valve Tip Scraper and light pressure while rotating scraper. Use a Sac Hole Drill to remove carbon from sac hole.

Piston area of valve and guide can be lightly polished by hand, if necessary, using Roosa Master No. 16489 lapping compound. Use the valve retractor to turn valve. Move valve in and out slightly while rotating, but do not apply down pressure while valve tip and seat are in contact.

When assembling the nozzle, back lift adjusting screw (11) several turns out

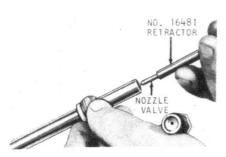

Fig. 56—Use the special retractor as shown, to remove a sticking nozzle valve.

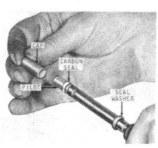

Fig. 58—Use the special pilot or a nozzle storage cap when installing a new carbon seal.

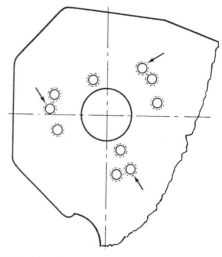

Fig. 71—The studs for mounting the injection pump are shown at (4-Fig. 36). The engine front plate has several mounting holes as shown, but only the holes indicated above by arrows should be used for 2040 tractors.

of pressure adjusting screw (14), and reverse disassembly procedure using Fig. 54 as a guide. Adjust opening pressure and valve lift as outlined in paragraph 71, raising pressure 200 psi above that specified if a new spring is installed.

INJECTION PUMP

The Roto Diesel injection pump is flange mounted on left side of engine front plate and is driven by the upper idler gear of the timing gear train.

Injection pump service demands the use of specialized equipment and special training which is beyond the scope of this manual. This section therefore, will cover only the information required for removal, installation and field adjustment of the injection pump.

87. **REMOVE AND REINSTALL.** To remove the injection pump, first shut off fuel and clean injection pump, lines and surrounding area. Pump can be removed and reinstalled without regard to crankshaft timing position and timing cannot be checked. The only critical requirement of the timing process is correct positioning of the injection pump gear as shown in Fig. 35.

Disconnect or remove fuel inlet, return and pressure lines, throttle rod and stop cable from injection pump.

Drain radiator and remove lower radiator hose, then remove access plate from front of timing gear cover. Remove the three cap screws attaching drive gear to injection pump flange. Support pump, remove the three mounting stud nuts and pull injection pump from timing gear housing. Pump drive gear will be retained by timing gear cover.

When reinstalling pump, turn pump hub until timing slot in hub flange aligns with dowel pin in gear, and reverse removal procedure. Tighten gear mounting cap screws and flange mounting stud nuts to a torque of 18 ft.-lbs. Bleed system as outlined in paragraph 66 and adjust linkage as in paragraph 90.

SPEED AND LINKAGE ADJUSTMENT

90. To adjust diesel engine speeds and control linkage, start engine and run until operating temperature is reached. Disconnect speed control rod

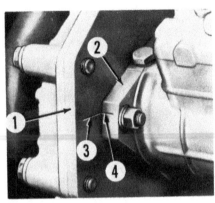

Fig. 72—Installed view of Roto-Diesel injection pump showing timing marks aligned. Pump will be in time with engine if gear timing marks are also in line.

1. Front plate
2. Pump flange
3. Engine timing mark
4. Pump timing mark

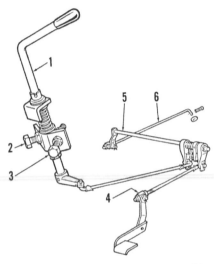

Fig. 74—Schematic of throttle linkage of late models. Refer to Fig. 73 for legend.

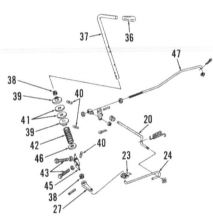

Fig. 76—Exploded view of throttle linkage used on late models. Refer to Fig. 74 for schematic view. Refer to Fig. 75 for legend.

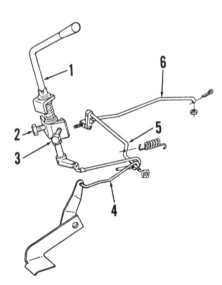

Fig. 73—Schematic view showing throttle linkage adjustments typical of early models. Refer also to Fig. 74.

1. Hand lever
2. Slow idle stop screw
3. Fast idle stop screw
4. Foot throttle yoke
5. Cross shaft
6. Control rod

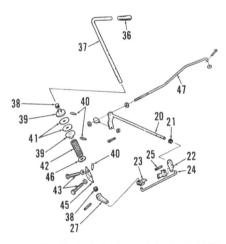

Fig. 75—Exploded view of typical early throttle linkage. Refer to Fig. 73 for schematic view.

20. Control shaft	38. Bushing
21. Bushing	39. Disc
22. Control arm	40. Groove pin
23. Clevis clip	41. Facings
24. Control rod (rear)	42. Spring
25. Spring pin	43. Stop screws
27. Arm	45. Stop
36. Knob	46. Washer
37. Control lever	47. Control rod (front)

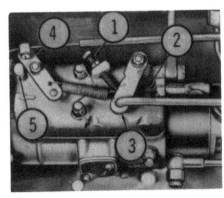

Fig. 79—Installed view of Roto-Diesel pump showing linkage adjustments.

1. Slow idle screw
2. Fast idle screw
3. Pump throttle arm
4. Stop lever
5. Lever stop

from injection pump and move throttle lever (3—Fig. 79) against fast idle adjusting screw (2). Engine speed should be 2650 rpm. If it is not, turn fast idle screw (2) in or out as required. Move pump throttle lever against slow idle adjusting screw (1); engine speed should be 650 rpm. If it is not, adjust slow idle screw. Reconnect throttle control rod (6—Fig. 73 or Fig. 74) to pump and adjust stop screws (2 and 3) as required until pump throttle lever (3—Fig. 79) contacts fast and slow stop screws (1 and 2). If foot throttle is used, adjust foot throttle linkage until pedal pad contacts footrest at the same time pump throttle lever (3) contacts fast idle screw (2).

To adjust shut-off cable, completely push in stop knob and check to be sure stop lever contacts top on injection pump governor cover. If it does not, loosen cable clamp screw and reposition clamp on stop cable.

COOLING SYSTEM

RADIATOR

94. An oil cooler attached to right side of radiator as shown in Fig. 82. Except for disconnecting hoses from the oil cooler, removal procedure for all models is similar.

95. **REMOVE AND REINSTALL.** To remove radiator, first drain cooling system, then remove grille screens and hood. Remove air intake tube. Remove fan shroud from radiator and lay shroud back over fan. Disconnect diesel injector leak-off line from fuel tank, oil inlet line from top of oil cooler and the

outlet line from bottom of oil cooler. Disconnect upper and lower radiator hoses and the upper radiator brace from radiator, then unbolt and remove radiator from tractor.

WATER PUMP

96. **REMOVE AND REINSTALL.** To remove water pump, first remove radiator as outlined in paragraph 95, then remove fan and fan belt. Disconnect by-pass hose from water pump, then unbolt and remove water pump from engine.

Reinstall by reversing removal procedure and adjust fan belt so a 25 lb. pull midway between pulleys will deflect belt 5/8-inch.

97. **OVERHAUL.** To disassemble water pump, use Fig. 83 as a guide and proceed as follows:

Support fan pulley hub in a press and, using a suitable mandrel, press shaft from pulley hub. Suitably support housing on gasket surface and press shaft, bearing, seal and impeller as a unit from housing.

Bearing outer race is a tight press fit in housing bore and is not otherwise secured. It is important therefore, that reasonable precautions be taken during assembly to prevent bearing movement during installation of impeller and pulley hub.

Coat outer edge of seal (6) with sealant and install in housing (4) using a socket or other driver which contacts only the outer flange of seal, then insert long end of bearing (5) through front of housing bore. Use tool No. JD-262 or a similar tool which contacts only outer race of bearing and press bearing into housing until front of bearing is flush with housing bore.

Install insert (7) in cup (8) with "V"

groove of insert toward cup. Parts must be clean and dry. Dip cup and insert in engine oil then press cup and insert into impeller as shown in inset, until cup bottoms in impeller counterbore. Support front end of pump shaft and press impeller (9) on rear of shaft until fins are flush with gasket surface of housing.

Invert the pump assembly and support the unit on rear of shaft which is recessed into impeller hub. **DO NOT** support impeller or housing. With shaft suitably supported, press pulley hub (3) on front of shaft until front face of pulley hub is approximately 5½ inches from gasket surface of housing as shown at (D—Fig. 84). Complete the assembly by reversing the disassembly procedure.

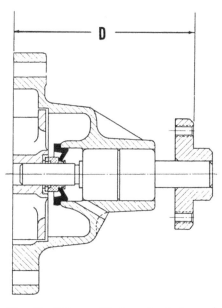

Fig. 84—Distance (D) from front face of pulley hub to gasket surface of body should be approximately 5½ inches when pump is properly assembled.

Fig. 82—Hydraulic oil cooler (OC) is located at side of coolant radiator as shown.

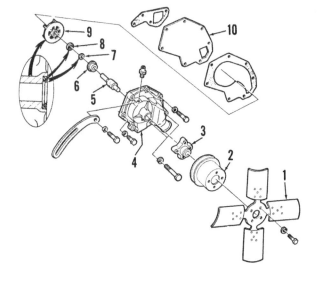

Fig. 83—Exploded view of water pump showing component parts. Inset shows correct installation of parts (7 & 8).

1. Fan blades
2. Pulley
3. Hub
4. Body
5. Shaft & bearing
6. Seal
7. Insert
8. Cup
9. Impeller
10. Cover

ELECTRICAL SYSTEM

ALTERNATOR AND REGULATOR

103. **OPERATION AND TESTING.** A Bosch 12 volt, 28 ampere alternator with separate regulator is used.

To check the charging system, connect a voltmeter to B+ (Output) terminal (1—Fig. 94) on alternator and to a suitable ground. With engine running at 1200 rpm, reading should be 13 volts or above. A lower reading could indicate a discharged battery or faulty alternator or regulator.

With engine not running, disconnect the three-terminal plug (2) from alternator and touch ammeter leads to DF (Green Wire) terminal and B+ terminal. Current draw should be approximately 2 amperes. High readings are caused by shorts or grounds. Low readings may be caused by dirty slip rings or defective brushes.

Connect a jumper wire between B+ and DF terminals and a voltmeter between B+ terminal and ground. Start engine and increase engine speed until voltage rises not to exceed 15.5 volts. If maximum reading is less than 14 volts and battery is fully charged, alternator is defective.

Move jumper wire connection from B+ terminal to D+ terminal. Voltage reading should remain the same. If voltage drops, exciter diodes are defective. If voltage reading was normal with plug disconnected and jumper wire installed, but was low when tested with plug connected, renew the regulator.

104. **OVERHAUL.** Brush holder (1—Fig. 93) can be removed without removing alternator from the tractor. Brush holder should be removed before attempting to separate alternator frame units.

To disassemble the removed alternator, first remove capacitor and brush holder. Immobilize pulley and remove shaft nut, pulley and fan. Mark brush end housing, stator frame and drive end housing for correct reassembly and remove through-bolts. Rotor will remain with drive-end frame and stator will remain with brush-end frame when alternator is disassembled.

Remove the two terminal nuts and three screws securing rectifier (2) to brush end frame and lift out rectifier and stator (3) as a unit. Carefully tag the three stator leads for correct reassembly, then unsolder the leads

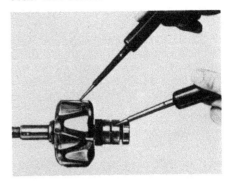

Fig. 94—Installed view of Bosch alternator.

1. Output terminal
2. Terminal plug
3. D+ terminal

Fig. 95—A reading of 4.0-4.4 ohms should exist between slip rings (1 & 2) when tested with an ohmmeter.

from rectifier diodes using an electric soldering iron and minimum heat. Be careful not to get solder on diode plates or overheat the diodes.

Check brush contact surface of slip rings for burning, scoring or varnish coating. Surfaces must be true to within 0.002 inch. Contact surface may be trued by chucking in a lathe. Polish the contact surface after truing using 400 grit polishing cloth, until scratches and machine marks are removed. Check continuity of rotor windings using an ohmmeter as shown in Fig. 96. Ohmmeter readings should be 4.0-4.4 between the two slip rings and infinity between either slip ring and rotor pole or shaft.

Stator is "Y" wound, the three individual windings being joined in the middle. Test the windings using an ohmmeter as shown in Fig. 97. Ohmmeter reading should be 0.4-0.44 between any two leads and infinity between any lead and stator frame.

Alternator brushes and shaft bearings are designed for 2000 hours service life. New brushes protrude 0.4 inch beyond brush holder when unit is removed; for maximum service reliability, renew **BOTH** the brushes and shaft bearings when brushes are worn to within 0.2 inch of holder. Solder copper leads to allow 0.4 inch protrusion using rosin core solder.

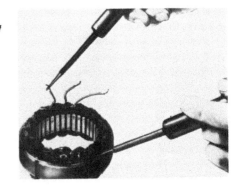

Fig. 96—No continuity should exist between either slip ring and any part of rotor frame.

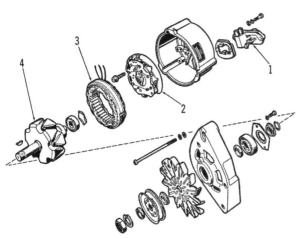

Fig. 93—Exploded view of Bosch alternator.

1. Brush holder
2. Rectifier
3. Stator
4. Rotor

Fig. 97—No continuity should exist between any stator lead and stator frame.

The rectifier is furnished as a complete assembly and diodes are not serviced separately. The rectifier unit contains three positive diodes, three negative diodes and three exciter diodes which energize rotor coils before engine is started. If any of the diodes fail, rectifier must be renewed.

To test the positive diodes, touch positive ohmmeter probe to positive heat sink as shown in Fig. 98, and touch negative test probe to each diode lead in turn. Ohmmeter should read at or near infinity for each test. Reverse the leads and repeat the series; ohmmeter should read at or near zero for the series.

Test negative diodes as shown in Fig. 99. Place negative test probe on negative heat sink and touch each diode lead in turn with positive test probe. Ohmmeter should read at or near infinity for the series. Reverse test leads and repeat the test; ohmmeter should rear at or near zero for the series.

Test exciter diodes by using the D+ terminal as the base as shown in Fig. 100. Ohmmeter should read at or near infinity with positive test probe on terminal screw and at or near zero with negative test probe touching screw.

When assembling alternator, tighten through-bolts to approximately 35-40 inch-pounds and pulley nut to 25-30 ft.-lbs.

STARTING MOTOR

105. A Bosch starting motor is used. Test specifications, including solenoid, are as follows:

R. Bosch 0001 395 016
Brush spring tension ....5.7-6 lbs.
Brush length-Min. ....0.63 in.

No load test:
Volts 11.5
Min. amps 60
Max. amps 90
Min. rpm 4800
Max. rpm 6800

CIRCUIT DESCRIPTION

106. All models are equipped with a 12 volt electrical system with the negative battery post grounded. Either one or two 12 volt (connected parallel) may be used and should provide at least 370 amps cold cranking capacity.

Refer to Fig. 101 for wiring diagram. Refer to the following for explanation of numbers:

1. Head lights
2. Combined rear work light and tail light
3. Warning lamps (tractors without canopy)
4. Socket (for hand lamp)
5. Connector, L.H.
6. Connector, R.H.
7. Connector
8. Flasher relay
9. Light switch
10. Key switch
11. Circuit breaker
12. Fuel gage
13. Engine oil pressure indicator light
14. Alternator indicator light
15. Air cleaner restriction indicator light
16. Batteries
17. Start safety switch
18. Engine oil pressure warning switch
19. Regulator
20. Starting motor
21. Alternator
22. Capacitor
23. Fuel gage sending unit
24. Air cleaner restriction warning switch
25. Turn signal switch (if so equipped)
26. Horn
27. Horn button
28. Resistor (0.7 Ohm) for low beam headlights
29. Front warning lamps (Tractors with canopy only)
30. Rear warning lamps (Tractors with canopy only)
31. Ground straps—Battery "-" to Ground
32. Black—Battery "+" to Starting motor (Terminal 30)
33. Red—Starting motor (Terminal 30) to Alternator (Terminal B+)
34. White—Start safety switch to Starting motor (Terminal 50)
35. Blue—Alternator (Terminal D+) to Regulator (Terminal D+)
36. Brown—Alternator (Terminal D-) to Regulator (Terminal D-)
37. Green—Alternator (Terminal DF) to Regulator (Terminal DF)
38. Blue—Regulator (Terminal D+) to Alternator indicator light
40. Red—Alternator indicator light to Air cleaner restriction indicator light
41. Black—Air cleaner restriction warning switch to Air cleaner restriction indicator light
42. Red—L.H. battery (Terminal +) to Circuit breaker
43. Black—Fuel gage sending unit to Fuel gage (Terminal G)
44. Red—Fuel gage (Terminal +) to Engine oil pressure indicator light
45. Red—Engine oil pressure indicator light to Alternator indicator light
46. Red—Alternator indicator light to Key switch (Terminal ACC)
47. Black—Key switch (Terminal ST) to Start safety switch
48. Green—Engine oil pressure warning switch to Engine oil pressure indicator light
49. Red—Key switch (Terminal BAT) to circuit breaker

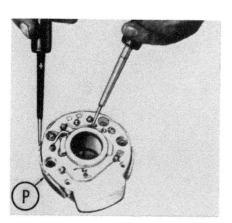

Fig. 98—A near-infinity reading should be obtained when positive probe rests on positive heat sink (P) and negative probe touches diode leads as shown. Reverse the probes and reading should be near zero ohms.

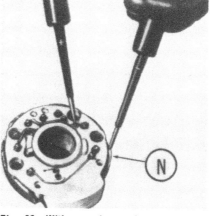

Fig. 99—With negative probe on negative heat sink (N) and positive probe touching diode leads, an infinity reading should be obtained. Reverse the probes and reading should be near zero.

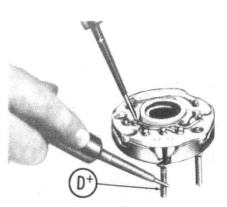

Fig. 100—D+ terminal is used to test exciter diodes, refer to paragraph 104.

50. Red—Circuit breaker to Light switch (Terminal B)
51. Yellow—Light switch (Terminal TL) to L.H. connector
52. Pink—Light switch (Terminal HD) to L.H. connector
53. Red—Turn signal switch (Terminal 2) to Circuit breaker
54. Orange/white—Light switch (Terminal W) to Flasher relay or turn signal switch (Terminal 1)

55. Blue—Light switch (Terminal FL) to L.H. connector
56. Pink—L.H. connector to R.H. connector
57. Orange—R.H. connector to L.H. connector
58. Orange—Flasher relay or turn signal switch (Terminal 4) to L.H. connector
59. Yellow—Light switch (Terminal TL) to Outlet socket

60. Yellow—L.H. connector to Tail light
61. Pink—L.H. connector to L.H. head light
62. Blue—L.H. connector to Rear work light
63. Orange—L.H. connector to wire 66 to L.H. warning lamp
64. Pink—R.H. connector to R.H. head light
65. Orange—R.H. connector to R.H. warning lamp or wire 70

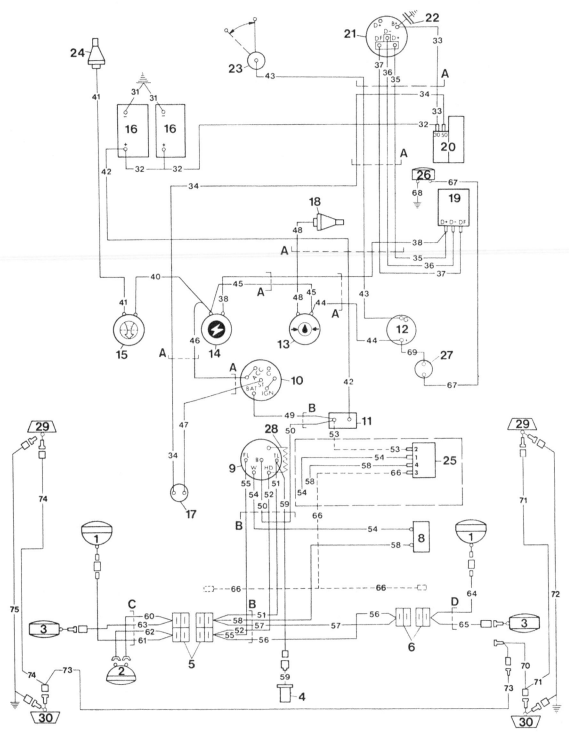

Fig. 101—Wiring diagram for 2040 model tractors. Refer to text for description and legend.

66. Dark green—Turn signal switch (Terminal 3) to Wire 63 (Wire 63 disconnected from connector 5 and attached to wire 66 on models without canopy. Wire 66 attached to wire 73 on models without canopy.)

67. Black/yellow—Horn button to Horn

68. Brown—Horn to ground

69. Red—Horn button to Fuel gage + Terminal

70. Orange—From wire 65 to R.H. rear warning lamp

71. Orange—R.H. rear warning lamp to R.H. front warning lamp

72. Black-R.H. warning lamps to ground

73. Dark green—From wire 65 to L.H. rear warning lamp

74. Dark green—L.H. rear warning lamp to L.H. front warning lamp

75. Black—L.H. warning lamps to ground

ENGINE CLUTCH

ADJUSTMENT

109. **FREE PLAY AND PEDAL POSITION.** Clutch pedal free play should be one inch and should be readjusted when free play decreases to ½-inch. Adjustment is made by disconnecting yoke (Fig. 106) from clutch pedal arm and shortening or lengthening clutch operating rod as required.

Clutch pedal is used in two positions. When tractor has no pto, loosen pedal positioning cap screw (Fig. 107) and move clutch pedal forward until screw contacts rear of slot. Tighten cap screw securely to maintain proper pedal position. On models with continuous

pto and dual clutch, move pedal pad to rear until positioning cap screw contacts front of slot; to allow clutch to fully release.

110. **RELEASE LEVERS (DUAL CLUTCH).** The dual clutch fingers can be adjusted for wear without disassembly or removal of unit from tractor. Proceed as follows:

Disconnect clutch operating rod from clutch pedal arm (Fig. 106) and refer to Fig. 109. Remove access cover from clutch housing and back off clutch operating bolt nuts until operating lever contacts pto clutch plate pins for all three operating levers. Tighten one jam nut until finger begins to pull away from pto clutch plate pin, tighten nut an additional 2½ turns and secure by tightening locknut. Adjust clutch operating rod (Fig. 106) until, with yoke pin inserted, throwout bearing just contacts operating lever. Turn flywheel until each of the other clutch levers are

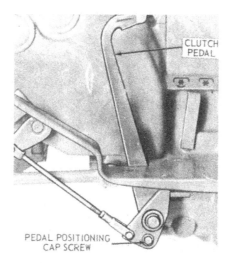

Fig. 107—Pedal position is adjusted by loosening cap screw and shifting pedal in slotted hole.

in position and adjust the lever to lightly contact throwout bearing. Tighten all locknuts securely and adjust clutch pedal free travel as outlined in paragraph 109.

TRACTOR SPLIT

111. To split engine from clutch housing, proceed as follows: Drain cooling system and remove hood. Drain transmission case. Disconnect battery cables, remove battery (or batteries), then remove the two cap screws which retain cowl to flywheel housing. Disconnect wiring from starter solenoid, alternator and oil pressure warning switch, then move wiring harness rearward. Disconnect speed-hour-meter cable from rear of cylinder block and starting aid (ether) pipe from intake manifold. Remove clamps from hydraulic lines along lower right side of engine, detach hydraulic suction line from clutch housing and separate pressure line at union located near joint of engine rear and clutch housing front. On tractors with power steering, disconnect power steering return line from hydraulic suction line fitting and separate power steering pressure line at union located near joint between rear of engine and front of clutch housing. On all models, be sure to cover ends of all lines and open ports to prevent entrance of dirt. Disconnect drag link at front end (from bellcrank). Disconnect wires from fuel gage sending unit in tank, from air cleaner restriction warning switch and from oil pressure sender. Disconnect shut-off cable and speed control rod from fuel injection pump. Remove temperature gage sensing bulb from cylinder head and position out of the way. Support transmission under clutch housing and support engine with a hoist that will permit engine to move forward away from clutch housing. Unbolt engine from clutch housing and move engine

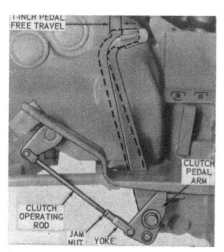

Fig. 106—External clutch adjustment is made by disconnecting rod yoke from pedal arm and adjusting length of operating rod.

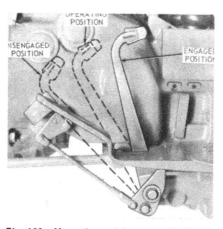

Fig. 108—Move the pedal rearward until cap screw contacts front of slot (Fig. 107) to permit full disengagement of pto clutch.

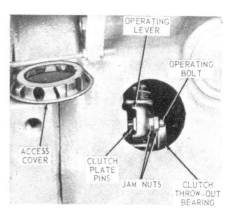

Fig. 109—Wear adjustment can be made as shown. Refer to paragraph 110.

forward until clutch clears clutch shaft. Be sure to keep engine horizontal until completely separated.

When rejoining tractor it may be necessary to bar over engine to facilitate entry of input shafts into clutch discs. Be sure flywheel housing and clutch housing are butted together before tightening retaining cap screws. Tighten retaining cap screws to 170 ft.-lbs. torque.

• R&R AND OVERHAUL

115. Refer to Fig. 119 for an exploded view of dual clutch unit and to Fig. 120 for cross-sectional view. Clutch can be removed after clutch split outlined in paragraph 111. Install new engine clutch disc if total thickness at facing area is less than 0.238 inch. When installing clutch, make sure long hub of transmission clutch disc (2—Fig. 119) is forward. Use a suitable alignment tool and tighten retaining cap screws to a torque of 33-40 ft.-lbs.

To disassemble the clutch cover and associated parts, remove locking nuts (14) and back off adjusting nuts (13) evenly until spring pressure is relieved. Mark the cover (7), rear pressure plate (5) and front pressure plate (3) with paint or other suitable means so balance can be maintained, and separate the units.

Inspect pressure plates (3 and 5) for cracks, scoring or heat discoloration and renew as necessary. Check diaphragm spring (4) for heat discoloration, distortion or other damage and renew if its condition is questionable. Renew transmission clutch disc (2) if facing wear approaches rivet heads, if hub is loose or splines are worn, or if disc is otherwise damaged. PTO disc (6) should be renewed if total thickness at facing area is 0.185-inch or less.

Assemble by reversing the disassembly procedure, observing the previously affixed marks to assure correct balance. The long hub (L—Fig. 121) of the pto clutch disc should be toward rear with the cushion offset (0) toward front as shown. Use a centering tool such as JDE-52-1 to align clutch disc while assembling.

The clutch assembly can be adjusted after attaching to flywheel, before reconnecting clutch housing to engine. Measure distance between pressure face of each clutch lever and rear flange of clutch disc as shown at (A). Loosen locknut (14), turn adjusting nut (13) until distance is 1.466-1.474 inches, then tighten locknuts. Adjust clutch linkage after clutch housing is attached to engine as described in paragraph 109.

CLUTCH SHAFT

117. To remove clutch shaft it is necessary to separate clutch housing from transmission housing. Clutch shaft can be removed from rear of clutch housing.

If tractor has evidence of oil seepage between clutch shaft and powershaft (pto shaft) separate shafts and inspect oil seal and pilot. Press new pilot (cup rearward) into bore of powershaft until it bottoms. Press oil seal in bore of powershaft, with lip rearward, until it contacts pilot.

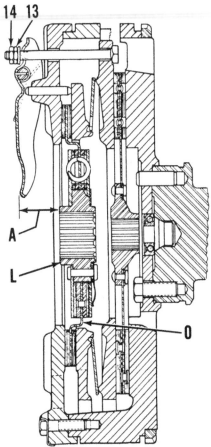

Fig. 121—Cross-section of clutch and flywheel showing long hub (L) and offset (0) of the pto clutch disc. Height (A) is adjusted by turning nuts (13) after loosening locknuts (14).

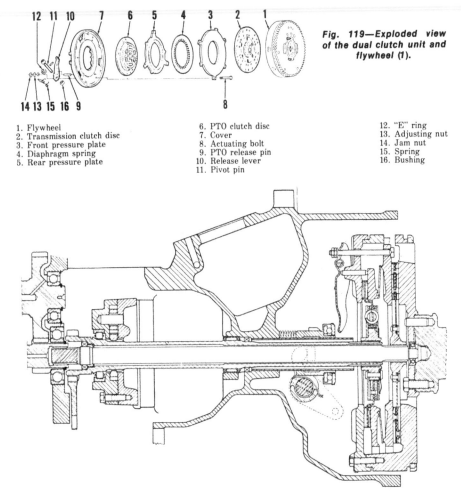

Fig. 119—Exploded view of the dual clutch unit and flywheel (1).

1. Flywheel
2. Transmission clutch disc
3. Front pressure plate
4. Diaphragm spring
5. Rear pressure plate
6. PTO clutch disc
7. Cover
8. Actuating bolt
9. PTO release pin
10. Release lever
11. Pivot pin
12. "E" ring
13. Adjusting nut
14. Jam nut
15. Spring
16. Bushing

Fig. 120—Cross-section of dual clutch assembly, housing and shafts used on 2040 models.

When installing clutch shaft, or powershaft, be sure lugs on shaft align with slots in transmission oil pump drive gear.

The clutch throwout bearing carrier sleeve should be pressed into clutch housing until distance from end of carrier sleeve to engine mounting surface of clutch housing is 2.8 inches.

CLUTCH RELEASE BEARING AND YOKE

118. The clutch release (throw-out) bearing (B—Fig. 123) can be removed after clutch housing is separated (split) from engine. Disconnect return spring and withdraw unit from carrier sleeve.

Throw-out bearing and carrier are available as an assembly and should not be separated.

To remove yoke, disconnect clutch rod from clutch shaft arm, then drive out the two yoke retaining spring pins (RP). Pull clutch shaft out left side of clutch housing and catch yoke as it comes off clutch shaft.

CLUTCH HOUSING

119. Clutch housing normally will not need complete removal for servicing. Clutch control linkage can be serviced when clutch housing is separated from engine. Clutch shaft and pto shafts along with their bearings and oil seals and the transmission oil pump can be serviced after clutch housing is separated from transmission case.

Refer to paragraph 143 for service information on the transmission oil pump. Other service required on clutch housing will be obvious after examination and reference to the following: Clutch shaft bushing is installed with outer end flush with outer edge of bore. Clutch pedal pivot shaft is

renewable and should be installed to protrude 2½ inches from housing. Clutch throwout bearing carrier sleeve is renewable and should be installed so that forward end is 2.8 inches from the machined engine mounting surface of clutch housing. (See Fig. 124). Oil seal for clutch shaft, or pto shaft, located in center of clutch housing can be re-

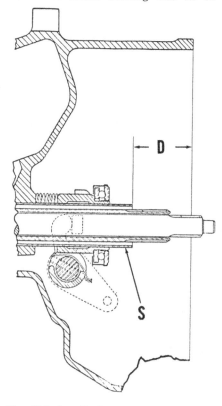

Fig. 124—Install clutch throw-out bearing carrier sleeve (S) into housing so that dimension (D) is 2.8 inches.

newed after removing shafts, transmission oil pump and needle bearing. Seal should be bottomed in bore with lip rearward. Needle bearing is also bottomed in its bore.

TRANSMISSION

Transmissions are constant mesh type using helical cut gears. Two shift levers are used, the left lever selecting high, low and reverse ranges as well as a park position. The right lever controls a four step gear arrangement. Thus with the two shift levers, eight forward speeds (four in high range and four in low range) and four reverse speeds are available. The four reverse speeds approximate in mph the four forward speeds obtained in low range.

TOP (SHIFTER) COVER

135. **REMOVE AND REINSTALL.** To remove the shifter cover, remove the transmission case shield and work it up over shifter lever boots. Detach wire connector from starting safety switch and, if necessary, disconnect rear wiring harness, then unbolt and lift shifter cover from clutch housing.

Any further disassembly required will be obvious after examination of the unit. See Fig. 127.

Reinstall by reversing the removal procedure.

NOTE: Shifter rails and forks are an integral part of the transmission and can be serviced after transmission is

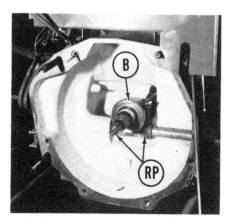

Fig. 123—View showing clutch throw-out bearing (B). Yoke is retained to clutch shaft by roll pins (RP).

Fig. 127—Exploded view of shifter cover, shifter shafts and shifter forks. Shifter shafts are an integral part of transmission.

1. Shifter cover
2. Boot
3. Snap ring
4. Retainer
6. Shift lever
8. Set screw
10. Fork
11. Fork
12. Fork
13. Pin, starter safety switch
14. Fork
15. Plug
16. Spring pin
17. Detent spring
18. Detent & interlock balls (5/16-in.)
19. Shifter shaft, low range & rev.
20. Shifter shaft, 1st, 5th, 2nd, 6th
21. Gasket
22. Shifter shaft, high range
23. Interlock pin
24. Shifter shaft, 3rd, 7th, 4th, 8th
25. Interlock balls (¼-inch)

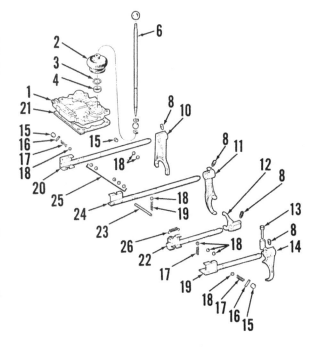

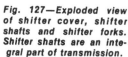

split from clutch housing as outlined in paragraph 136.

TRACTOR SPLIT

136. To split tractor between clutch housing and transmission case, disconnect battery ground straps, drain transmission case, remove shifter cover as outlined in paragraph 135, then remove the two clutch housing to transmission case cap screws located at rear of shifter cover opening under mounting flange. Remove left platform and unhook clutch return spring. Remove right platform, disconnect the two brake pressure lines from brake valve and the main hydraulic pressure line from tee or elbow near transmission oil filter. Remove plate retaining hydraulic pump inlet line and reservoir return line at lower right side of transmission case. Disconnect wiring harness and connector and check to be sure that wires and lines will not interfere with separating. Support transmission case, place a rolling floor jack under front section and block front axle to prevent front end from tipping. Remove remaining clutch housing to transmission cap screws and separate tractor.

NOTE: The cap screw (CS—Fig. 128), located in front of the transmission filter, cannot be completely removed unless filter is removed, however, cap screw can be unscrewed as tractor is split and left in casting hole. DO NOT lose the check valve, located in rear end of the main hydraulic inlet line, which will probably fall out as line comes out of transmission case.

SHIFTER SHAFTS AND FORKS

137. **REMOVE AND REINSTALL.** To remove shifter rails (shafts) and forks it will first be necessary to split tractor as outlined in paragraph 136. Remove seat and disconnect lift links from rockshaft arms. Disconnect return

line (1-Fig. 129), pressure line (2) and pilot line (3) from rockshaft. Detach wiring and lay wires out of the way. Attach a hoist to rockshaft housing, place load selector lever in "L" position, then unbolt and remove rockshaft housing from transmission case.

Remove the starter safety switch pin (13-Fig. 130) from low range shifter fork, and remove self locking set screws (8) from shifter forks.

NOTE: Before removal and installation of shifter shafts and forks, refer to Fig. 130 to determine location of detent and interlock mechanisms as well as for identification of shifter shafts. Also note that the five interlock balls are ¼-inch diameter whereas the eight remaining interlock and detent balls are 5/16-inch diameter.

Move both of the left hand shift shafts (19 & 22) to neutral position, then pull the LNR shift shaft (19) from its bore. Be careful not to lose the three 5/16-inch diameter balls.

NOTE: Do not twist shafts while removing. It is possible for some of the balls to enter holes in shaft making withdrawal of the shaft difficult and possibly damaging parts.

Move the shifter shaft (24) in or out into a gear, which will release the interlock pin (23), then withdraw the PNH shifter shaft (22). Be careful to catch the detent ball and spring as shaft is removed. Lift shift forks from transmission case.

Pull shift shaft (24) from its bore until three of the ¼-inch interlock balls (25) can be removed from hole in shaft, then complete removing shaft. Be careful to catch the rear detent ball,

the two rear 5/16-inch interlock balls and the two remaining ¼-inch diameter interlock balls remaining in forward bore of case.

Pull the remaining shifter shaft (20) from bore and catch the last 5/16-inch diameter detent ball. Lift the shift forks from transmission case.

If interlock pin requires renewal, drive spring pin at right side of transmission case rearward, remove plug (15), then slide interlock pin out right side.

NOTE: Detent bores are not aligned and interlock pin will not enter left detent bore.

Clean and inspect all parts and renew any that are bent, worn or otherwise damaged. Be sure to inspect all balls and renew any which have flat spots that would prevent them from rolling freely.

Reinstall shifter shafts and forks by reversing removal procedure. Start with right hand shifter shaft (20). Move shaft (24) into a gear position before attempting to install shaft (22) so that interlock pin (23) will move to the right. Secure all shifter shaft forks with self locking set screws.

COUNTERSHAFT

138. **R&R AND OVERHAUL.** To remove the countershaft, split tractor as outlined in paragraph 136 and remove shifter shafts and forks as outlined in paragraph 137. Remove the

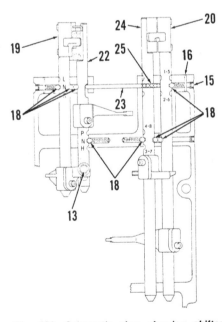

Fig. 130—Schematic view showing shifter shafts and the detent and interlock mechanisms. The rear detents are below shifter shafts instead of sides as indicated. Refer to text for service procedure and Fig. 127 for legend.

Fig. 128—The cap screw (CS) can be unscrewed as tractor is split.

Fig. 129—View of rockshaft housing installed. Refer to text.

pto driven gear from front of transmission. Remove cap screws (C-Fig. 131) from countershaft bearing support. Remove snap ring (52-Fig. 132) from its

groove at rear end of countershaft, then use a screw driver and turn locking washer (48) until splines of washer index with splines of counter-

shaft. Pry bearing support off dowels (12), pull assembly forward and lift gears from transmission case as they come off shaft. See Fig. 133.

Inspect all gears, thrust washers and shift collar for broken teeth, excessive wear or other damage and renew as necessary. If support assembly bearings or shafts, or the snubber brake assemblies (19-Fig. 132) require service, the shafts and bearings can be pressed out after removing retaining snap rings; however, the transmission drive gear (11) must be removed before countershaft can be removed. Snubber brake springs (18) should test 63-77 lbs. when compressed to a length of 1.51 inches. Needle bearing (54) can be removed from its bore after removing snap ring (53).

Fig. 131—Front of transmission case with pto shaft and pto gears removed. Countershaft support (S) must be removed before input shaft or pinion shaft can be removed.

INPUT SHAFT

139. **R&R AND OVERHAUL.** To remove input shaft it is first necessary to remove countershaft as outlined in paragraph 138.

With countershaft removed, the transmission input shaft is removed as follows: Remove transmission oil cup and line (L-Fig. 133). Straighten lock plates and remove input shaft bearing quill (8-Fig. 132) and shims (7) from front of input shaft. Bump input shaft forward until front bearing cup (6) clears its bore, then move input shaft forward, lift rear end of shaft and remove input shaft from transmission case.

With input shaft removed, inspect all gears for chipped teeth or excessive wear. Inspect bearings and renew as necessary. Bump bearing cup (1) forward if removal is required. Inspect needle bearing (4) and renew if necessary.

Install and adjust end play of input shaft as follows: Be sure bearing cup (1) is bottomed in bore and place input shaft in position. Use original shim

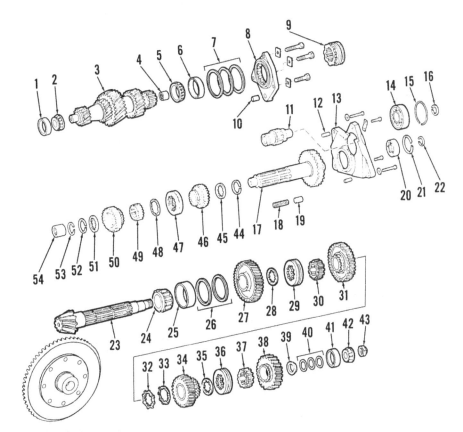

Fig. 132—Exploded view showing transmission shafts and gears. Note that transmission drive gear (11) is mounted in countershaft support (13).

1. Bearing cup	15. Snap ring	29. Shift collar
2. Bearing cone	16. Snap ring	30. Shift collar sleeve
3. Input shaft	17. Countershaft	31. 2nd & 6th gear
4. Needle bearing	18. Brake spring	32. Thrust washer (outer tangs)
5. Bearing cone	19. Brake plug	33. Retaining washer
6. Bearing cup	20. Ball bearing	34. 4th & 8th gear
7. Shims	21. Snap ring	35. Thrust washer (thinnest)
8. Bearing quill	22. Snap ring	36. Shift collar
9. Shifter collar	23. Pinion shaft	37. Shift collar sleeve
10. Dowel	24. Bearing cone	38. 3rd & 7th gear
11. Drive gear	25. Bearing cup	39. Spacer
12. Dowel	26. Shims	40. Shims
13. Support	27. 1st & 5th gear	41. Bearing cup
14. Ball bearing	28. Thrust washer (thickest)	42. Bearing cone
		43. Nut
		44. Snap ring
		45. Thrust washer
		46. Reverse pinion
		47. Shift collar
		48. Thrust washer
		49. Shift collar sleeve
		50. Low range pinion
		51. Thrust washer
		52. Snap ring
		53. Snap ring
		54. Needle bearing

Fig. 133—Be sure to attach oil line (L) when assembling. Bearing quill is shown at (8).

pack (7), or use a new shim pack approximately 0.030-inch thick, install front bearing quill (8) and tighten cap screws to 35 ft.-lbs. torque. Use a dial indicator to check the input shaft end play which should be 0.004-0.006 inch. Refer to Fig. 136. Vary shims as required. Shims are available in thicknesses of 0.003, 0.005 and 0.010 inch. Do not forget front oil line clamp when making final installation. Refer to Fig. 133.

PINION SHAFT

140. R&R AND OVERHAUL. To remove the transmission pinion shaft, remove the input shaft as in paragraph 139 and the differential as in paragraph 144.

With differential removed, remove oil line, nut (43-Fig. 132), bearing (42), shims (40) and spacer (39). Use a screwdriver and turn thrust washers until splines of thrust washers are indexed with splines of countershaft. Pull countershaft rearward and remove parts from transmission case as they come off shaft. Bearing cup (25) and shims (26) can be removed from housing by bumping cup rearward. Be sure to keep shims (26) together as they control the bevel gear mesh position. Bearing cup (41) can be removed from housing by bumping cup forward.

Check all gears and shafts for chipped teeth, damaged splines, excessive wear or other damage and renew as necessary. If pinion shaft is renewed, it will also be necessary to renew the differential ring gear and right hand differential housing as these parts are not available separately. Bearing (24) is installed with large diameter toward gear end of shaft.

NOTE: Mesh (cone point) position of the pinion shaft and main drive bevel

pinion gear is adjusted with shims (26) located between rear bearing cup (25) and housing. If new drive gears or bearings are installed, the mesh position is correctly adjusted by installing two 0.010 inch thick shims (26—Fig. 132). If same pinion shaft and bearing are installed, reinstall the same shims (26).

Install pinion shaft and adjust shaft bearing preload as follows: Use Fig. 132 as a guide and with bearing (24) on pinion shaft, start shaft into rear of housing. With shaft about half-way into housing, place 1st and 5th speed gear (27) on shaft with teeth for shift collar (29) toward front. Place the thickest thrust washer (28) on shaft, then install coupling sleeve (30) and shift collar (29). Move shaft forward slightly and install 2nd and 6th speed gear (31) with teeth for shift collar toward rear. Place thrust washer with outer tangs (32) over shaft, then slide retaining washer (33) over thrust washer (32). Move shaft slightly forward and install 4th and 8th speed gear (34) on shaft with teeth for shift collar toward front.

Fig. 135—View of transmission shafts installed. The input shaft is shown at (3); countershaft and gears at (17); pinion shaft and gears at (23).

Place the thinnest thrust washer (35) on shaft and install shift collar sleeve (37) and shift collar (36). Install 3rd and 7th speed gear (38) on shaft with teeth for shift collar toward rear. Push shaft forward until rear bearing cone (24) seats in bearing cup (25) and use screwdriver to turn thrust washers until splines on thrust washers lock with splines of pinion shaft. Install spacer (39), shims (40), bearing (42) and nut (43), then adjust pinion shaft bearing preload as outlined in paragraph 141.

141. PINION SHAFT BEARING ADJUSTMENT. The pinion shaft bearings must be adjusted to provide a bearing preload of 0.006-inch (5-15 in.-lbs. rolling torque). Adjustment is made by varying the number of shims (40-Fig. 132).

To adjust the pinion shaft bearing preload, proceed as follows: Mount a dial indicator with contact button on front end of pinion shaft and check for end play of shaft (Fig. 137). If shaft has no end play, add shims (40-Fig. 132) to introduce not more than 0.002-inch shaft end play.

NOTE: Do not exceed more than 0.002-inch shaft end play when beginning adjustment as increased end play increases the possibility of inaccuracies due to parts shifting.

If original shims (40) are not being used, install a preliminary 0.035-inch thick shim pack. Shims are available in thicknesses of 0.002, 0.005 and 0.010-inch. Tighten nut (43) to 160 ft.-lbs. torque and measure shaft end play, then remove shims from shim pack (40) equal to the measured shaft end play, PLUS an additional 0.005-inch. This will give the recommended bearing preload of 0.006-inch. Retighten nut to 160 ft.-lbs. torque and stake in position.

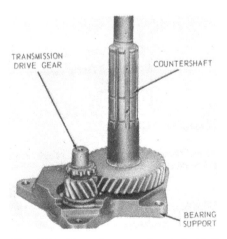

Fig. 134—View of removed countershaft bearing support assembly.

Fig. 136—View showing approved method of checking input shaft end play.

Fig. 137—View showing approved method of checking pinion shaft end play.

TRANSMISSION OIL PUMP

The transmission oil pump is a gear type pump mounted on the rear wall of clutch housing and is driven by the hollow pto shaft. Pump capacity is approximately 9 gpm at 2500 rpm. The transmission pump (2—Fig. 143) draws oil from the transmission case (oil reservoir) through an inlet screen (1) and pushes the oil through the hydraulic system filter (4). Oil which passes through the filter moves directly to the inlet of the main hydraulic pump (7) and

(if so equipped) to the inlet of the power steering pump (8).

Several safeguards including a suction by-pass valve (3), oil filter by-pass valve (5) and a surge relief valve (6) are provided between the transmission oil pump and the main and power steering pumps. The open center system routes unused oil back to the sump or to the pump inlet. Maximum pressure in the system is limited by the relief valve (10) for the power steering system and relief valve (11) for the hydraulic lift system.

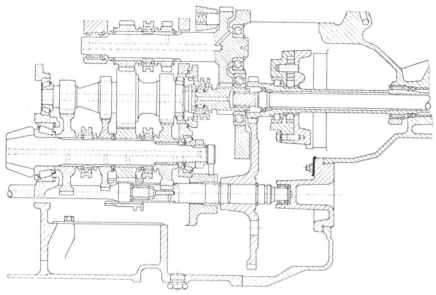

Fig. 138—Cross-section of transmission gears and shaft.

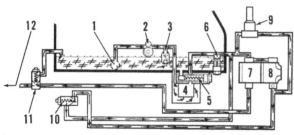

Fig. 143—Schematic of hydraulic system showing the transmission oil pump and related parts.

1. Intake screen
2. Transmission oil pump
3. By-pass valve
4. System filter
5. Filter by-pass valve
6. Surge relief valve
7. Main hydraulic pump
8. Power steering pump
9. Power steering valve & cylinder
10. Power steering relief valve
11. Main hydraulic system relief valve
12. Delivery to main hydraulic system

143. R&R AND OVERHAUL. To remove the transmission oil pump, first split clutch housing from transmission housing as outlined in paragraph 136.

With clutch housing separated from transmission case, pull clutch shaft, or clutch shaft and pto power shaft, from clutch housing. Remove pump inlet and outlet lines from pump, then remove pump from wall of clutch housing and separate pump body from adapter. See Fig. 156.

Clean and inspect all parts for chipping, scoring or excessive wear. If bearing in pump body requires renewal, press new bearing in bore until it bottoms. Pump gears are available as a matched set only. Pump idler shaft is renewable and diameter of new shaft is 0.6240-0.6250 inch. Thickness of new pump gears is 0.507-0.509 inch.

When reassembling pump, coat gears with oil. Tighten pump housing to cover cap screws to 23 ft.-lbs. torque; pump cover to clutch housing cap screws to 35 ft.-lbs. torque. Align slots of clutch shaft, or pto powershaft, with lugs of pump drive gear when installing shafts and be sure seals are on ends of inlet and outlet tubes before rejoining tractor.

DIFFERENTIAL AND FINAL DRIVE

The two pinion differential is equipped with a differential lock. Control hand lever and pedal are located at left side of transmission housing.

Final drives incorporate a planetary gear reduction at inner end of the housing.

DIFFERENTIAL

144. REMOVE AND REINSTALL. To remove differential, drain transmission case then remove final drives as outlined in paragraph 150 and the rock-

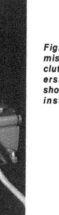

Fig. 144—View of transmission oil pump (P) with clutch shaft and pto powershaft removed. Model shown is not 2040, but installation of pump is similar.

I. Inlet line
O. Outlet line
P. Transmission pump

no defects in the adjustments are noted, the shim packs should be kept intact and reinstalled in their original positions. However, if bevel gears, bearings, bearing quills or transmission case are renewed, the main bevel gears should be checked for mesh (cone point) position, differential carrier bearing preload and gear backlash, and in the foregoing order.

shaft housing as outlined in paragraph 179.

Differential lock must first be removed as follows: Remove clamp screw from lever (1—Fig. 157), and remove the square key (9). Hold the yoke (4) in place, bump shaft (3) rearward and remove Woodruff key (8) when it clears yoke. Continue to bump shaft rearward until plug at rear end of shaft bore comes out, then remove shaft, yoke and collar (7).

Disconnect load control arm spring, slide pivot shaft to the left and lift out load control arm. Remove transmission oil cup and rear oil line. Support the differential, remove both bearing quills (2 and 11—Fig. 158) and keep shims (3) with the correct quill.

The shims (3) located between bearing quills and transmission case control the differential bearing preload and the backlash of main drive bevel gears. Recommended bearing preload is 0.002-0.005 inch and recommended backlash is 0.012 inch. Refer to paragraph 146 for adjustment procedure.

145. OVERHAUL. The bevel ring gear, right hand differential housing and pinion shaft are available as a matched set.

To disassemble unit, remove the eight differential housing bolts and separate housing. Notice bevel pinion shaft is located by a dowel pin.

Procedure for removing bearings (5 and 13—Fig. 158), and bearing cups (4 and 14) is obvious. If any of the pinions (9) or pinion shafts (10) are damaged, all mating parts should also be renewed. If axle (side) gears (8) are damaged or excessively worn, closely examine bores in differential housing as they may also be damaged.

Reassemble by reversing the disassembly procedure and tighten differential housing cap screws to 35 ft.-lbs. Cap screws are self locking.

MAIN DRIVE BEVEL GEARS

146. ADJUSTMENT. If differential is removed for access to other parts and

147. MESH (CONE POINT) POSITION. The fore and aft position of the transmission pinion shaft is controlled by the total thickness of the shim stock (26—Fig. 132) located between pinion shaft rear bearing cup and front wall of differential compartment. When renewing parts the shim pack should consist of two 0.010 in. thick shims.

148. DIFFERENTIAL BEARING ADJUSTMENT. The differential carrier bearings should have a preload of 0.002-0.005 inch and adjustment is made as follows: Install differential and bearing housings with original shim packs, then check differential end play using a dial indicator.

NOTE: When making this adjustment, be positive that clearance exists

Fig. 158—Exploded view of differential and bevel gear assembly. Parts (7R and 12) are included in bevel gear set.

2. Quill, LH
3. Shims
4. Bearing cup
5. Bearing cone
7. Housing, LH
8. Bevel (side) gear
9. Pinion
10. Pinion shaft

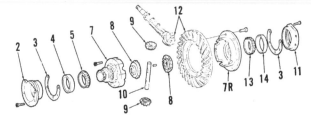

11. Quill, RH
12. Bevel gear set

13. Bearing cone
14. Bearing cup

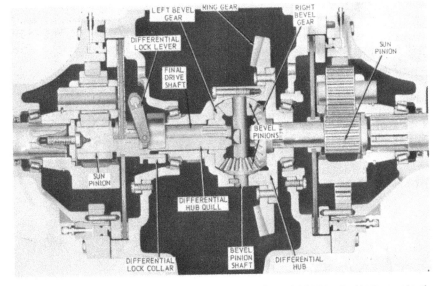

Fig. 157—Exploded view showing parts of differential lock.

1. Lever
2. "O" ring
3. Shaft
4. Yoke
5. Plug
6. Shoe
7. Collar
8. Woodruff key
9. Square key
10. Spring
12. Pedal

Fig. 159—Cross-sectional view showing differential, differential lock and planetary reduction units. Slight differences may be noticed between units shown and those used on 2040 models.

between the main drive bevel ring gear and pinion shaft at all times.

If no differential end play exists, add shims (3—Fig. 158) under right bearing housing (11) to introduce not more than 0.002 inch end play. If more than 0.002 inch end play existed on original check, subtract shims.

Measure end play of differential, then subtract shims equal to the measured end play PLUS an additional 0.003 inch to give the desired 0.002-0.005 inch bearing preload. Shims are available in thicknesses of 0.003, 0.005 and 0.010.

149. BACKLASH ADJUSTMENT. With differential carrier bearing preload adjusted as outlined in paragraph 148, adjust backlash between main drive bevel gear and pinion shaft to 0.012 inch by transferring bearing housing shims (3—Fig. 158) from one side to the other as required. Moving shims from left to right will decrease backlash. Do not remove any shims during backlash adjustment or the previously determined preload adjustment will be changed.

Backlash should be adjusted to as close to 0.012 inch as possible and should be measured by a dial indicator against outer diameter of a ring gear tooth. Ring gear must be installed on differential in such a way that backlash is not less than 0.008 inch or more than 0.016 inch when measured at any other point around circumference of ring gear.

FINAL DRIVE

150. REMOVE AND REINSTALL. To remove final drive, support rear of tractor and remove wheel and tire. Disconnect fender lights and free wiring harness from clamp on final drive housing, then remove fender. If right hand final drive is being removed and tractor has selective control valve, disconnect pressure line, coupler lines and return hose between valve and rockshaft housing and remove control valve. Disconnect the brake line from final drive housing. Attach hoist to final drive, remove attaching cap screws and pull final drive from transmission case.

Reinstall by reversing removal procedure and tighten attaching cap screws to 85 ft.-lbs. torque.

NOTE: If brake disc came off with final drive, install disc so that thickest facing is toward transmission case.

Bleed air from brake system as outlined in paragraph 152 after reinstalling the final drive assembly.

151. OVERHAUL. To overhaul the removed final drive unit, refer to Fig. 161 and proceed as follows:

Remove lock plate (23), cap screw (24) and retainer washer (25), then pull planet carrier assembly (26) from axle. Support outer end of final drive housing (11) so oil seal (2) will clear and press axle out of housing. The axle bearings, bearing cups and oil seals are now available for inspection or renewal. If outer bearing (4) is renewed, heat it to approximately 300° F. and drive it into place while hot.

NOTE: Be sure oil seal is on axle, metal side out, before installing outer bearing. Bearing cups are pressed in bores until they bottom. Seal cup (3) will be pushed out when outer bearing cup is removed. Be sure to reinstall seal cup after bearing cup is installed. If ring gear and/or final drive housing is damaged, renew complete unit (11).

To remove planet pinions (20), expand snap ring (27), lift it from groove of carrier (26) and pull pinion shafts (18). Check carrier, pinions and rollers for pitting, scoring or excessive wear and renew parts as required. If any of the planet pinion rollers are defective renew the complete set.

Reassemble final drive and adjust axle bearings as follows: Coat bores of planet pinions with grease and position rollers (23 in each bore) in pinions. Place a thrust washer on each side of pinions, then place pinions in carrier and insert pinion shafts only far enough to retain rollers and thrust washers. Install snap ring (27) in slots of pinion shafts, then complete insertion of pinion shafts and be sure snap ring seats in groove in carrier. Coat inner seal (13) with grease and install axle in housing. Heat inner bearing (15) to approximately 300° F. and install bearing on inner end of axle. Place carrier assembly on axle, install retaining washer (25) and cap screw (24) and tighten cap screw until bearing is pulled into place and a small amount of axle end play remains. Check the drag torque necessary to turn the axle with the existing axle end play. Maximum torque required to turn axle with loose bearings is 52 in.-lbs. Hold pinion carrier (26), to prevent turning, and tighten screw (24) until 96-148 inch pounds of torque are necessary to turn the axle in bearings. Install lock plate (23) after bearing preload is correctly set. Fill axle outer bearing opening with multi-purpose grease and install oil seal with metal side out.

Use new gasket (12) when reinstalling final drive to transmission case. However, before installing final drive, pull final drive shaft (22) and brake disc and inspect. Brake disc is installed with thickest facing next to transmission case.

Refer to paragraph 159 for information on brake pressure plate and pressure ring.

BRAKES

The brakes are hydraulically actuated and utilize a wet type disc controlled by a brake operating valve

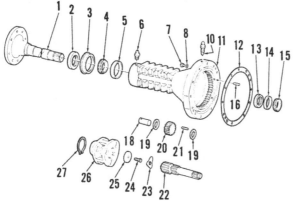

Fig. 161—Exploded view of axle and final drive assembly. Brake bleed screw is shown at (10).

1. Axle, flanged
2. Oil seal
3. Oil seal cup
4. Bearing cone
5. Bearing cup
6. Plug
7. Plug
8. "O" ring
10. Bleed screw
11. Housing
12. Gasket
13. Oil seal
14. Bearing cup
15. Bearing cone
16. Dowel pin
18. Pinion shaft
19. Thrust washer
20. Pinion
21. Rollers (69 used)
22. Final drive shaft
23. Lock plate
24. Cap screw
25. Retaining washer
26. Carrier
27. Snap ring

Fig. 162—View showing location of pedal stop screws. Refer to text for adjustment.

located on right side of clutch housing. See Fig. 162. Brake discs are splined to the final drive shafts and the brake pressure ring is fitted in inner end of final drive housing. Except for a pedal adjustment, no other brake adjustments are required.

BLEED AND ADJUST

152. **BLEEDING.** Brakes must be bled when pedals feel spongy, pedals bottom, or after disconnecting or disassembling any portion of the braking system.

To bleed brakes, start engine and run for at least two minutes at 2000 rpm to insure that brake control valve reservoir is filled.

Attach a bleeder hose (preferably clear plastic) to brake bleed screw (10—Fig. 161) located on top side of final drive housing and place opposite end in filler hole of rockshaft housing. Loosen the bleed screw about ¾ turn, then slowly depress and release brake pedal until oil flowing from bleeder hose is completely free of air bubbles, then depress brake pedal and tighten bleed screw.

NOTE: Do not permit pedal to return sharply with bleeder valve open because air can be drawn into system at the bleed port. Release pedal slowly.

Repeat bleeding operation for opposite side brake.

153. **ADJUSTMENT.** Whenever brake control valve has been disassembled, a brake pedal and equalizing valve adjustment must be made to prevent mechanical interference between brake valve pistons and reservoir check valves.

Before making this adjustment, bleed brakes as outlined as paragraph 152.

154. RIGHT PEDAL. Adjust right hand pedal stop screw (1—Fig. 162) so brake valve piston is fully extended and arm of brake pedal is snug against end of piston without piston being depressed (zero clearance). Apply a force of about 10 lbs. to **LEFT** brake pedal and if left brake pedal settles, turn pedal stop screw for right brake pedal counter-clockwise about ⅓-turn at which time left brake pedal should stop settling. If left brake pedal does not stop settling, a leak in the braking system is indicated and must be isolated and corrected. Refer to paragraph 157.

155. LEFT PEDAL. Adjust left hand pedal stop screw (2—Fig. 162) so brake valve piston is fully extended and arm of brake pedal is snug against end of piston without piston being depressed (zero clearance). Apply a force of about 10 lbs. to **RIGHT** brake pedal and if right pedal settles, turn pedal stop screw for left brake pedal counter-clockwise about ⅓-turn at which time right brake pedal should stop settling. If right brake pedal does not stop settling, a leak in the braking system is indicated and must be isolated and corrected. Refer to paragraph 157.

156. PEDAL HEIGHT. If brake pedal height is not aligned after equalization valves are adjusted as outlined in paragraphs 154 and 155, align pedals by turning stop screw (1 or 2—Fig. 162) on highest pedal about 1/6-turn counter-clockwise.

BRAKE TEST

157. **PEDAL LEAK-DOWN.** With a 60 lb. pressure applied continuously to each pedal for one minute, the pedal leak-down should not exceed one inch. Excessive brake pedal leak-down can be caused by air in the brake system, faulty brake control valve pistons and/or "O" rings, faulty brake pressure ring seals, or faulty brake control valve equalizing valves or reservoir check valves.

Brakes should always be bled as outlined in paragraph 152 before any checking or adjusting of braking system is attempted. Faulty brake control valve pistons or "O" rings will be indicated by external leakage around the brake control valve pistons.

Faulty brake pressure ring seals, or brake control valve, can be determined as follows: Isolate brake from brake control valve by plugging brake line. If leak-down stops, the brake pressure ring seals are defective. If leak-down continues, the brake control valve is faulty and can be checked further by depressing brake pedals individually, then simultaneously. If leak-down occurs in both cases, a defective reservoir check valve is indicated. If leak-down occurs during individual pedal operation but not on simultaneous pedal operation, a faulty equalizer valve is indicated.

Refer to paragraph 158 for brake control valve information to paragraph 159 for brake pressure ring information.

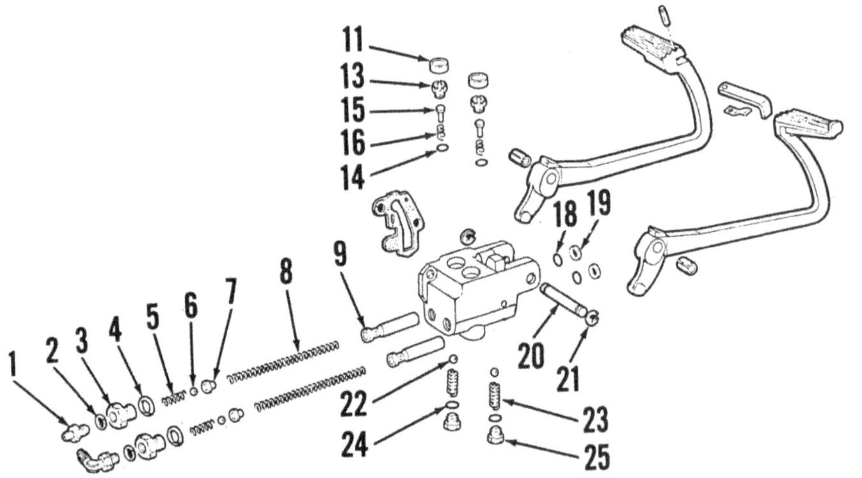

OVERHAUL

158. **BRAKE CONTROL VALVE.** To remove brake control valve, remove right platform and thoroughly clean valve and surrounding area. Disconnect brake lines from rear of control valve, remove the mounting cap screws and remove control valve from clutch housing. Discard gasket located between control valve and clutch housing. Remove "E" ring (21—Fig. 163), pull shaft (20) and remove pedals from control valve. Remove connectors (1), check valve springs (5) and balls (6). Remove seats (3) and ball retainers (7), then push pistons (9) and springs (8) out rear of valve body. Remove cup plugs (11), then using a screwdriver with proper sized bit, remove reservoir check valve assemblies (items 13, 14, 15 & 16). Remove equalizer valve assem-

Fig. 163—Exploded view of brake control valve and pedals. Identical parts are used for left and right brake systems, even though only one of the components may be identified.

1. Connector	7. Retainer	14. "O" ring	20. Pedal shaft
2. "O" ring	8. Spring	15. Check valve	21. Retainer ring
3. Check valve seat	9. Piston	16. Spring	22. Ball
4. "O" ring	11. Cup plug	18. "O" ring	23. Spring
5. Spring	13. Valve seat	19. Oil seal	24. "O" ring
6. Ball			25. Plug

blies (items 22, 23, 24 & 25). "O" rings (18) and oil seals (19) can be removed from piston bores.

Clean and inspect all parts. Piston spring (8) should test 20 lbs. when compressed to a length of 5¾ inches. Renew housing if seats for equalizer balls (22) are damaged. Oil seals (19) are installed with lips toward outside. Pay particular attention to area of reservoir check valve (15) where contact is made with valve piston and renew valve if any doubt exists as to its condition. Brake pedals are fitted with bushings for brake pedal shaft (20) and bushings and/or shaft should be renewed if clearance is excessive.

Lubricate lips of oil seals (19) and all other parts. Use new cup plugs (11) and reassemble by reversing disassembly procedure. Use a new gasket when installing valve on tractor. Bleed brakes as outlined in paragraph 152 and adjust pedals as outlined in paragraphs 154, 155 and 156.

159. BRAKE PRESSURE PLATE, RING AND DISC. To remove brake pressure plate, pressure ring and brake disc, remove final drive housing as outlined in paragraph 150. Pull final

Fig. 164—Final drive shaft and brake disc being removed. Thickest disc facing is next to transmission case.

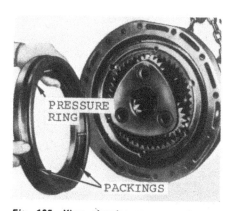

Fig. 165—View showing pressure ring removed from cylinder (groove) in final drive housing. Note that flat chamfered side is toward bottom of cylinder.

drive shaft from differential and remove brake disc from final drive shaft. See Fig. 164. Lift brake pressure plate from dowels in final drive housing. Remove brake pressure ring by prying it out evenly. If pressure ring is difficult to remove, attach a small hydraulic pump to brake line connections, be sure bleed valve is closed, then pump oil behind pressure ring to force it from cylinder (groove). See Fig. 165. Dowels can be removed from final drive housing, if necessary.

Inspect brake disc for worn or damaged facing or damaged splines. If facings require renewal, renew complete disc assembly as facings are not available separately. Inspect pressure plate for scoring, checking, or other damage and renew if necessary. Remove and discard seals from pressure ring and inspect ring for cracks or other damage.

To reassemble brake assembly, proceed as follows: Place brake disc on final drive shaft so thickest facing is next to transmission case and insert final drive shaft into differential and insert final drive shaft into differential. Place new inner and outer seals on pressure ring and lubricate assembly liberally. Start pressure ring into its cylinder (groove) with flat chamfered side first and press into cylinder until it bottoms. Be absolutely sure that neither seal is cut or rolled during installation. Place pressure plate over dowels in final drive housing, hold in place if necessary, then install final drive housing.

Bleed brakes as outlined in paragraph 152.

POWER TAKE-OFF

Tractors may be equipped with 540 rpm continuous running pto. A mechanical shift coupling is provided to

disengage the pto and the dual stage engine clutch can also be used to control the pto.

When service is required on the pto system, the following should be taken into consideration. Work involving rear pto shaft can be done by working from rear of tractor. Work involving the pto driven gears, powershaft clutch shaft, mid-pto shaft or mid-pto shifter assembly will require that the clutch housing be separated from the transmission case. Work involving the rear pto shaft shifter assembly will involve removing the rockshaft housing, high and low range shifter shafts and countershaft assembly in addition to separating the clutch housing from transmission case.

REAR PTO SHAFT

160. R&R AND OVERHAUL. To remove rear pto shaft (12—Fig. 167), drain transmission and remove pto shield and shaft guard, if so equipped. Place rear pto shaft control lever in "OFF" position, then remove cap screws from pto shaft bearing quill (17) and pull shaft and quill assembly from transmission case. Be careful not to pull shift collar from front drive shaft. Bearing (14) can be renewed after removing snap rings (4 and 13). Press new oil seal (15) in quill with lips toward front until it bottoms. For service on remainder of pto assembly, split tractor as outlined in paragraph 136.

Reassemble by reversing removal procedure and mate splines of rear shaft with splines of shift collar as shaft is installed.

DRIVEN GEAR AND FRONT DRIVE SHAFT

162. R&R AND OVERHAUL. To service the pto driven gear and front

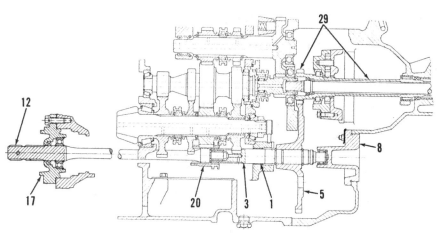

Fig. 166—Schematic drawing showing arrangement of pto shafts, bearings and seals. Refer to Fig. 167 for legend except item (29) which is the pto clutch power shaft.

power shaft, it is first necessary to split tractor as outlined in paragraph 136.

NOTE: The pto clutch power shaft and engine clutch shaft can also be removed at this time by withdrawing them rearward out of clutch housing. Pto drive gear can be pressed from pto clutch power shaft if necessary.

Prior to any disassembly, install rear pto shaft, if removed, and place rear pto shaft shifter lever (23—Fig. 167) in "ON" position. This will slide shift collar onto rear pto shaft (12) and prevent it from dropping to bottom of transmission case. If shift collar comes off, it will be necessary to remove the rockshaft housing to retrieve it.

The drive shaft (3) and gear (5) can be withdrawn. Inspect the drive shaft journals, bushings (1 and 2) and needle bearing (11). Installed diameter of bushing (1) should be 1.376-1.379 inches and journal should be 1.373-1.374 inches. Installed diameter of bushing (2) should be 0.752-0.755 inch and pilot journal at end of rear pto shaft should be 0.749-0.750 inch.

PTO SHIFTER

163. R&R AND OVERHAUL. To remove the rear pto shifter, first remove the transmission countershaft as outlined in paragraph 138, then remove the front drive shaft as described in paragraph 162.

Remove set screw (25—Fig. 167) from fork (21) and slide fork forward on rail until collar is free from pto shaft (12).

Remove roll pin (28), then shift arm (22) and lever (23) can be separated and removed. Removal of remainder of parts will be self evident.

Clean and inspect all parts and renew as necessary. Reinstall shifter mechanism by reversing removal procedure.

HYDRAULIC LIFT SYSTEM

Model 2040 tractors may be equipped with either an open center type hydraulic system or a closed center type hydraulic system. The closed center hydraulic system used on Model 2040 is identical to the closed center system used on Model 2240 tractors. Refer to the Model 2240 "Hydraulic Lift System" section, paragraphs 129-149, located elsewhere in this manual for service procedures covering the closed center system.

The open center type system is covered in the immediately following paragraphs 166 through 188. The open center type hydraulic system utilizes a crankshaft driven gear-type pump that is supplied with fluid from the transmission housing. System priority (routing of fluid) is as follows:

1. Transmission reservoir
2. Transmission oil pump
3. Supply by-pass valve
4. Oil filter and filter relief valve
5. Low pressure relief valve
6. Main hydraulic pump
7. System relief valve
8. Selective control valve
9. Rockshaft control valves
10. Brake valves
11. Transmission and final drive lubrication

Supply by-pass valve opens under negative pressure when transmission pump is inoperative (engine clutch disengaged, etc.), enabling main hydraulic pump to pick up fluid directly from transmission reservoir. The low pressure relief valve regulates inlet pressure of the main pump, normally by-passing a small excess flow from transmission pump when oil is warm and filter is in good condition.

On models with power steering, the power steering fluid is supplied by a second main pump section installed in tandem with

the main hydraulic pump. Initial steering fluid is supplied by the transmission oil pump but return steering fluid is directed to the main pump inlet line, therefore the fluid supply available to the hydraulic system is not decreased by adding the power steering option. NOTE: The only exception to this statement would be complete failure of power steering relief valve, in which case the fluid pumped by the power steering pump would be returned to reservoir and the supply available for hydraulic functions would be reduced.

Transmission and final drive lubrication passage tees into main supply line ahead of brake valve, assuring continuing lubricant flow when brakes are being applied. An orifice in rockshaft dump valve supplies a continuing lubricant flow when rockshaft is actuated, thus the only time lubricant flow is temporarily interrupted is during raising action of the single acting selective control valve.

Refer to the appropriate following paragraphs for testing, removal and overhaul of hydraulic system components.

TESTING

166. PUMP PRESSURE AND FLOW. Main pump system flow is 6.5 gpm at 2500 engine rpm. Main system relief pressure should be 2100-2130 psi. Pressure and flow can be checked by disconnecting rear pressure line (5—Fig. 169) from elbow (4) and teeing a suitable flow meter in the line.

Pressure only can be checked by installing a gage in port (4—Fig. 170) in rockshaft housing as shown, and immobilizing lift arms to test the pressure. Flow is correct if lift arms will move a

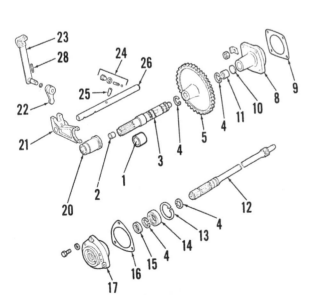

Fig. 167—Exploded view showing gears and shafts of the transmission driven 540 rpm pto.

1. Bushing
2. Bushing
3. Drive (front) shaft
4. Snap ring
5. Driven gear
8. Cover
9. Gasket
10. Thrust washer
11. Needle bearing
12. Power (rear) shaft
13. Snap ring
14. Ball bearing
15. Oil seal
16. Gasket
17. Bearing quill
20. Shift collar
21. Shift fork
22. Shift arm
23. Shift lever
24. Detent
25. Set screw
26. Shift rail
28. Roll pin

Fig. 169—View of pressure relief valve housing and lines.

1. Relief valve
2. Bracket
3. Return tube
4. Inlet elbow
5. Pressure line
6. Mounting screws
7. Outlet pressure line
8. Adjusting screw

load from fully lowered to fully raised position in 2½ seconds with engine running at 2100 rpm and fluid at operating temperature.

To adjust the pressure, loosen jam nut and turn adjusting screw (8—Fig. 169) in or out as required.

ADJUSTMENT

167. ROCKSHAFT LINKAGE. The following external adjustments should be made on rockshaft linkage after first checking fluid level and bringing system to operating temperature. Adjustments should be checked or made, in the order given.

Fig. 170—A suitable gage can be used for checking pressure only. Refer to paragraph 166.

1. Gage
2. Rockshaft housing
3. Connecting hose
4. Pressure port
5. Pressure port

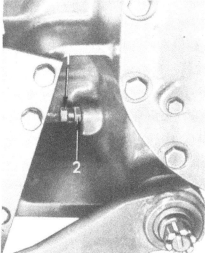

Fig. 171—Negative stop screw (1) should be backed out ¾ turn from the point where it contacts load control arm. Tighten locknut securely when adjustment is correct.

168. NEGATIVE STOP SCREW. Refer to Fig. 171. Hold stop screw (1) and back off locknut (2) at least two turns. With engine not running and no load on lift arms, turn stop screw (1) in until it just contacts load control arm, back screw out ¼-turn and tighten locknut.

NOTE: Contact between stop screw and load control arm can be determined by removing oil filler cap and placing a screwdriver or similar tool against upper end of arm. Refer to Fig. 174.

169. DEPTH CONTROL NEUTRAL RANGE. With engine running at high idle speed, no load on hydraulic system and selector lever (2—Fig. 172) in upper ("D") position, move rockshaft control lever to bottom of quadrant, then slowly upward until lift arms just start to raise. Mark this position on quadrant rim.

Push control lever forward until lift arms start to settle and again mark quadrant. The two marks should be 0.16-0.24 inch apart; if they are not, remove plug (4) and reaching through plug hole turn adjusting screw (1—Fig. 173) clockwise to increase neutral range or counter-clockwise to decrease range.

Fig. 172—View of rockshaft control quadrant showing points of adjustment.

1. Quadrant slot
2. Selector lever
3. Clamp screw
4. Plug

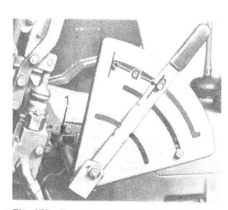

Fig. 173—To adjust load control mechanism, move selector lever (1) to "L" position and move quadrant lever until distance (a) measures 92 mm (3 5/8 inches), then refer to Fig. 174.

170. CONTROL LEVER ADJUSTMENT. With engine running at high idle speed, no load on hydraulic system and selector lever (2—Fig. 172) in upper ("D") position, move control lever downward until lever pin is 0.3 inch from bottom of slot (1) as shown at (a). Loosen clamp nut (3) and move lever arm clockwise until lift arms are fully lowered, then counter-clockwise until lift arms just start to move upward. Tighten clamp nut at this point.

Move control lever slowly rearward until lift arms are fully raised and measure clearance between lever pin and upper end of slot (1). Clearance should be approximately equal to lower clearance (a). Equalize the difference by loosening clamp screw (3) and making minor adjustments as required.

171. LOAD CONTROL ADJUSTMENT. With engine running at high idle speed, no load on hydraulic system and selector lever (1-Fig. 173) in lower ("L") position, move control lever downward until distance (a) measures 3.62 inches.

Remove filler cap as shown in Fig. 174. Loosen locknut and turn adjusting

Fig. 174—With control levers positioned as shown in Fig. 173 turn adjusting screw (1) until lift arms just start to raise. Refer to paragraph 171.

Fig. 175—View of rockshaft housing showing Rate-Of-Drop adjusting screw.

screw (1) counter-clockwise until lift arms start to lower, then clockwise until lift arms just start to rise. Tighten jam nut at this point without moving adjusting screw.

172. RATE OF DROP. The rate-of-drop adjusting screw is located on top of rockshaft housing underneath the seat as shown in Fig. 175. Turn screw clockwise to slow rate of drop, or counter-clockwise to increase the rate.

173. **SELECTIVE CONTROL VALVE.** Valve levers can be equalized (dual valve models) by loosening locknut and turning valve end in connecting ball cap.

To adjust the lowering speed, adjust rate-of-drop screw (Arrow–Fig. 176) located at bottom of each valve housing as shown.

MAIN HYDRAULIC PUMP

The open center hydraulic system is equipped with a crankshaft driven gear-type main hydraulic pump which is mounted ahead of the radiator and driven by the crankshaft through a coupling and drive shaft attached to crankshaft pulley. Refer to Fig. 177 for exploded view of drive unit and to Fig. 178 for installed view.

Refer to Model 2240 "Hydraulic Lift System" section located elsewhere in this manual for service procedures covering the closed center hydraulic system.

174. **REMOVE AND REINSTALL.** To remove the pump, first remove right grille and air cleaner. Remove cotter pin from pump drive shaft and push shaft rearward out of front coupling. Remove retaining clamps and disconnect inlet and pressure lines at pump. Remove cap screws (4–Fig. 178) and lift off pump and mounting plate as an assembly. When installing, tighten the screws attaching drive coupling to pulley and socket head screws attaching pump to mounting bracket to a torque of 35 ft.-lbs.

175. OVERHAUL. Shaft couplings are riveted assemblies and parts are available. The only pump parts available separately are shaft seal and pump sealing rings. If shafts (gears), bearings or housings are damaged, renew the pump. Mark the parts as pump is disassembled, to be sure parts are installed in same position as before removal.

RESERVOIR, FILTER AND VALVES

176. **RESERVOIR, FILTER AND INLET SCREEN.** The transmission housing is reservoir for hydraulic system. Fluid capacity is 9.5 gallons and recommended fluid is John Deere Hy-GARD or equivalent. Oil level dip-stick is located on right side of transmission case and filler plug is at rear of rockshaft housing. When draining and changing filters, only about 7.5 gallons will be removed.

The full flow oil filter is located underneath transmission housing as shown at (1–Fig. 179). Filter element should be renewed every 500 hours, element can be removed without draining transmission. Every 1000 hours, drain and renew transmission fluid, remove and clean intake screen (2) and renew filter element.

177. **FILTER BY-PASS VALVE.** The oil filter by-pass valve is located in transmission housing directly above filter unit as shown in Fig. 180. To service the valve, remove plug (2) and withdraw valve cartridge (1) as an assembly. The piston type valve should open at a pressure differential of 30 psi. Approximately 11 lbs. pressure should be required on end of valve to open by-pass slot and valve should move smoothly. Drive out retaining spring

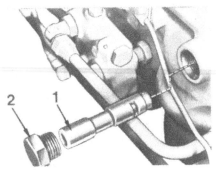

Fig. 180—Filter by-pass valve (1) and plug (2) is located directly above filter unit (1-Fig. 179).

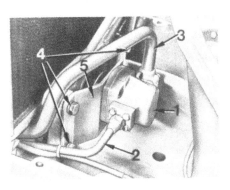

Fig. 178—Installed view of main hydraulic pump.

1. Pump
2. Pressure line
3. Inlet line
4. Mounting cap screws
5. Mounting plate

Fig. 176—Arrow shows Rate-Of-Drop adjusting screw for selective control valves.

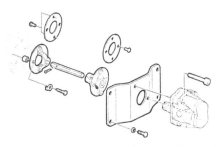

Fig. 177—Exploded view of front mounted main hydraulic pump and drive unit.

Fig. 179—Hydraulic oil filter is located beneath transmission housing as shown at (1). Plug (2) provides access to intake screen.

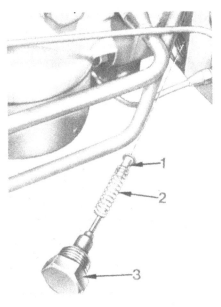

Fig. 181—Low pressure relief valve (1), spring (2) and plug (3) are located immediately forward of system filter as shown.

(roll) pin and remove valve spring and piston if service is indicated.

178. LOW PRESSURE RELIEF VALVE. The low pressure relief valve is located in hydraulic pump inlet passage. Valve is accessible from outside by removing hex plug in clutch housing immediately forward of system filter as shown in Fig. 181.

Valve (1) maintains an inlet pressure of approximately 60 psi to main hydraulic pump and normally by-passes a continuous flow of approximately 1 gpm at rated speed when system is in good condition. Spring (2) should have a free length of approximately 2 inches and a pressure of 12-14 lbs. should be required to compress spring to a height of ¾-inch.

ROCKSHAFT HOUSING AND COMPONENTS

179. REMOVE AND REINSTALL. To remove the rockshaft housing, disconnect battery ground straps, remove transmission shield and disconnect wires from starter safety switch. Disconnect lift links from rockshaft arms and remove seat assembly. Disconnect oil return hose from manifold and rockshaft housing. Disconnect rear quick couplers if so equipped. Remove pilot line between rockshaft housing and dump valve housing. Move selector lever to lower ("L") position and remove retaining cap screws. Attach a suitable hoist and lift rockshaft housing from transmission housing.

Install by reversing the removal procedure, making sure selector lever is in ("L") position and that roller link

mates properly with load control arm cam follower. Tighten retaining cap screws to a torque of 85 ft.-lbs.

180. OVERHAUL. Remove rockshaft arms from rockshaft and selector lever from load control arm. Turn unit until bottom side is accessible. Remove control lever and quadrant. Remove remote cylinder adapter (5—Fig. 182) and unhook spring (3—Fig. 183) from pin on lift cylinder. Remove front, outside cylinder retaining cap screw and remaining cap screws, then lift cylinder from rockshaft housing disengaging selector arm from roller link as cylinder is removed.

NOTE: Do not lose throttle valve ball (3—Fig. 182) which is free to fall out as cylinder is removed.

Remove cam and spacer from rockshaft then pull rockshaft from control arm and housing. Bushing and "O" ring seal will be removed from one side of housing as rockshaft is withdrawn.

When assembling, remove throttle screw (1—Fig. 182) from housing. Install cylinder and the five retaining cap screws finger tight, then tighten

outside (front) cap screw first, to the recommended 35 ft.-lbs. Tighten the other four cap screws to recommended (35 ft.-lbs.) torque after cylinder is properly positioned. Drop throttle valve ball (3) in threaded hole in housing where throttle screw (1) was removed. Reinstall throttle screw and turn down until it firmly contacts ball; tighten lower jam nut (2) until "O" ring is seated in housing groove, turn down an additional ½-turn, point lever (L) to right and tighten upper jam nut firmly down on lever.

181. CYLINDER AND VALVE UNIT. To overhaul cylinder and valve unit, refer to Fig. 182. If piston will not slide from cylinder, remove closed-end plug (9) and push piston out with a convenient tool. Pressure and discharge valves (10 through 15) are identical, but worn-in parts should be kept together. Springs (13) have an approximate free length of 1.33 inches and should test 9 lbs. when compressed to a height of 0.875 inch. Check valve spring (7) has an approximate free length of 2.05 inches and should test 4.5 lbs. when compressed to 0.75 inch.

182. VALVE LINKAGE. Refer to Fig. 183. Linkage should move freely without binding or excessive looseness. Check actuating cam (2) for wear or scoring and other parts for wear, bending, binding or breakage. Use Fig. 183 as a guide for disassembly and reassembly. Adjust upper lift limit as outlined in paragraph 183.

183. ADJUSTMENT. With lift cover fully assembled, loosen set screw (9—Fig. 183) in adjusting clamp (8). Turn neutral range adjusting screw (1) until a minimum positive clearance exists between the two lobes of cam (2) and the two valves (11—Fig. 182). (Slight clearance between cam and one valve when cam touches the other valve.) Insert a 1/8-inch thick shim between rockshaft ram arm and rear housing wall and raise rockshaft arms

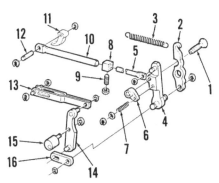

Fig. 183—Exploded view of control valve linkage.

1. Adjusting screw	9. Set screw
2. Valve cam	10. Linkage tube
3. Spring	11. Cam
4. Link	12. Pin
5. Control rod	13. Selector link
6. Adjusting nut	14. Operating link
7. Spring	15. Pivot block
8. Stop clamp	16. Link

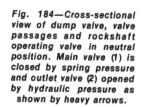

Fig. 182—Exploded view of rockshaft cylinder, valves and associated parts.

1. Throttle screw	
2. Locknut	
3. Steel ball	10. Valve seat
5. Outlet adapter	11. Valve
6. Plug	12. Sleeve
7. Spring	13. Spring
8. Steel ball	14. Plug
9. Plug	15. Snap ring
	L. Drop rate lever

Fig. 184—Cross-sectional view of dump valve, valve passages and rockshaft operating valve in neutral position. Main valve (1) is closed by spring pressure and outlet valve (2) opened by hydraulic pressure as shown by heavy arrows.

C. Circulating fluid	
D. Dump valve	
O. Outlet passage	
P. Pump (pressure) passage	
S. Static fluid	

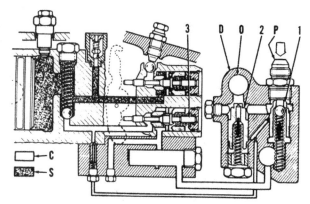

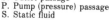

until shim is securely held in place. Move rockshaft control lever if necessary until cam (2—Fig. 183) is in a neutral position (neither valve depressed). Slide clamp (8) against end of tube (10) and tighten set screw (9).

Remove the shim. Move selector lever to lower ("L") position and rockshaft control lever to raised position, then check to be sure cam (2) starts to open upper (discharge) valve before rockshaft ram arm contacts housing wall.

DUMP VALVE

184. OPERATION. Refer to Fig. 184. The dump valve (D) contains two hydraulically actuated, spring loaded valves. When rockshaft control valve is in neutral position, main valve (1) is hydraulically balanced by the orifice passage which equalizes pressure on both sides of valve; and valve and ball are seated by spring pressure. The outlet control valve (2) is hydraulically unbalanced by the slightly higher pressure of circulating oil (C) and is held open against spring pressure, allowing fluid to flow from pump port (P) to outlet port (O) and on to sump.

When rockshaft control lever is moved to "Raise" position, pressure valve (3) is opened as shown in Fig. 185. Pressure is momentarily reduced at spring end of main valve (1) causing valve to move downward opening the passage to spring end of outlet control valve (2). The resultant flow restores hydraulic balance to outlet valve which then closes by spring pressure, shutting off the main hydraulic flow through the opened check ball in main valve (1), through the opened pressure valve (3) to the rockshaft cylinder.

The flow continues until pressure valve (3) is closed; until rockshaft piston reaches the end of its stroke; or until system relief valve pressure is exceeded. When any of these conditions occur, hydraulic balance is restored to main valve (1) which closes by spring pressure unbalancing outlet valve (2) which then opens against spring pressure permitting resumed fluid flow through outlet passage (O) to the sump.

The orifice in closed end of outlet valve (2) provides lubrication flow for transmission gears when outlet passage is closed to main hydraulic flow.

185. OVERHAUL. The dump valve is shown exploded in Fig. 186. Dump valve attaches to lower right side of transmission housing by connector (4), and can be removed after disconnecting pressure, return and pilot lines.

Valve spools (8 and 11) should slide smoothly in their bores without excessive looseness. Springs (7 and 9) are interchangeable, have an approximate free length of 2¼ inches, and should test 11-13 lbs. when compressed to a height of 1 1/8 inches.

Use Figs. 184, 185 and 186 as a guide when reassembling the valve and reinstall by reversing the removal procedure.

LOAD CONTROL (SENSING) SYSTEM

The load sensing mechanism is located in the rear of the transmission case. See Fig. 187 for an exploded view showing component parts.

The load control shaft (21) is mounted in tapered bushings and as load is applied to the shaft ends from the hitch draft links, the shaft flexes forward and actuates the load control arm (12) which

pivots on shaft (10). Movement of the load control arm is transmitted to the rockshaft control valves via the roller link (13—Fig. 183) and control linkage and control valves are opened or closed permitting oil to flow to or from the rockshaft cylinder and piston.

186. R&R AND OVERHAUL. To remove the load sensing mechanism, remove the rockshaft housing assembly as outlined in paragraph 179, the left final drive assembly as outlined in paragraph 150, and the three-point hitch.

Remove cam follower spring (18—Fig. 187), then slide pivot shaft (10) to the left and lift out control arm (12) assembly. Removal of pivot shaft (10) can be completed if necessary by removing snap ring. Remove retainer ring and retainer bushing (3) from right end of load control shaft and bump shaft from transmission case. The negative stop screw (7), located on rear of transmission case behind right final drive, can also be removed if necessary.

Inspect bushings (6) in transmission case and renew if necessary. Drive old bushings out by inserting driver through opposite bushing. New bushings are installed with chamfer toward inside. Check the special pin (20) for damage in area where it is contacted by load control shaft and renew if necessary. Also check contact areas of negative stop screw (7) and load control arm (12). Check load control shaft (21) to be sure it is not bent or otherwise damaged. Wear or damage to any other parts will be obvious.

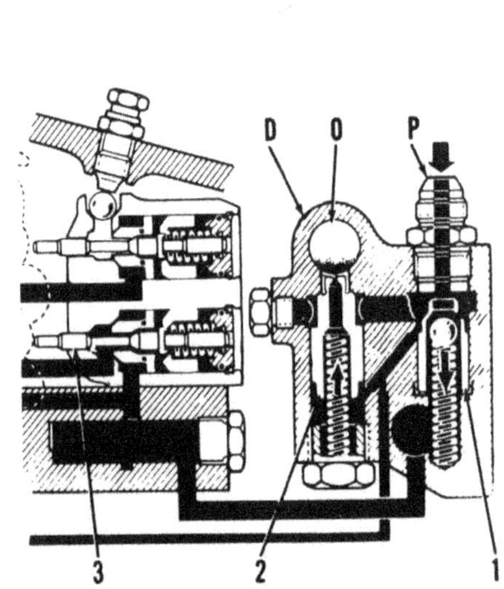

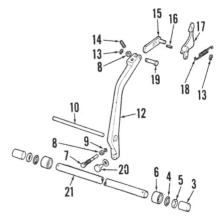

Fig. 187—Load control mechanism is located in rear of transmission case. Flexing of control shaft (21) actuates load control arm (12).

3. Retaining bushing	13. Retaining ring
4. Seal	14. Adjusting screw
5. "O" ring	15. Extension
6. Bushing	16. Spring pin
7. Negative stop screw	17. Cam follower
8. Jam nut	18. Spring
9. "O" ring	19. Pin
10. Pivot shaft	20. Special pin
12. Load control arm	21. Load control shaft

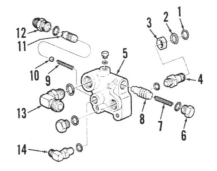

Fig. 186—Exploded view of dump valve showing component parts.

1. "O" ring	8. Outlet valve
2. Special washer	9. Spring
3. Jam nut	10. Check ball
4. Fitting	11. Main valve
5. Valve body	12. Inlet fitting
6. Plug	13. Return fitting
7. Spring	14. Pilot fitting

Fig. 185—Cross-sectional view of dump valve in raising position. Main valve (1) is opened by hydraulic pressure and outlet valve (2) closed by spring pressure as shown by heavy arrows. Refer also to Fig. 184.

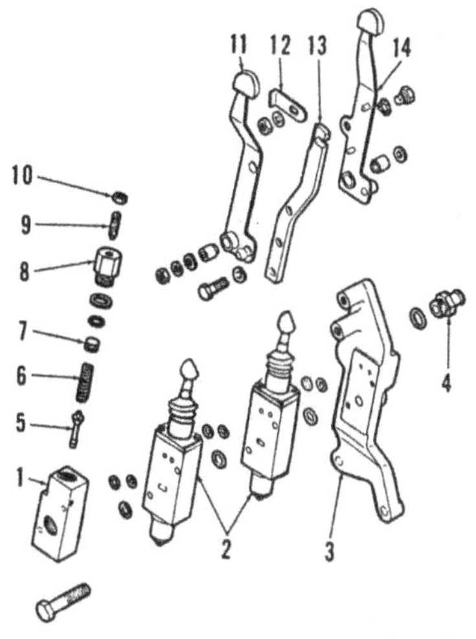

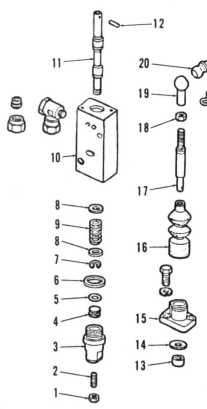

Fig. 188—Exploded view of mounting bracket and pressure relief valve showing selective control valve lever linkage.

1. Valve body
2. Selective control valves
3. Bracket
4. Fitting
5. Valve plunger
6. Spring
7. Spring cap
8. Plug
9. Adjusting screw
10. Locknut
11. Outer lever
12. Locking strap
13. Support
14. Inner lever

Reassemble load sensing assembly by reversing the disassembly procedure and when installing hitch draft links, tighten retaining nuts until end play is removed between link collar and retaining ring, then tighten nut until next

slot aligns with cotter pin hole and install cotter pin. After assembly is completed, adjust the negative stop screw as outlined in paragraph 168.

PRESSURE RELIEF VALVE

187. The pressure relief valve housing (1—Fig. 188) mounts on outside of selective control valves (2) on models so equipped; or directly on mounting bracket (3) on models without remote control.

Relief valve spring (6) should have a free length of approximately 2 inches and should test approximately 117 lbs. when compressed to a height of 1.76 inches. Adjust system pressure as outlined in paragraph 166 after valve is reassembled.

SELECTIVE CONTROL VALVES

188. Selective control valve linkage is shown in Fig. 188 and exploded view of valve is shown in Fig. 189.

Valve spool (11) and body (10) are only available in a valve assembly which includes items (1 through 9). All other components are available individually. Keep valve body and spool together as a matched set if more than one valve is disassembled at a time. Fill hollow of cover (15) with multipurpose grease when reassembling valve unit. Adjust the installed valve if necessary, as outlined in paragraph 173.

Fig. 189—Exploded view of selective control valve showing component parts.

1. Locknut
2. Adjusting screw
3. Retainer
4. Spring cap
5. "O" ring
6. Copper gasket
7. Snap ring
8. Washer
9. Centering spring
10. Body
11. Valve spool
12. Pin
13. Packing
14. Washer
15. Cover
16. Boot
17. Connector
18. Locknut
19. Ball socket
20. Ball

NOTES

JOHN DEERE

Models ■ 2240 ■ 2440 ■ 2630 ■ 2640

Previously contained in I&T Shop Manual No. JD-45

SHOP MANUAL

JOHN DEERE

SERIES 2240-2440-2630-2640

Tractor serial number is located on right side of front support. Engine serial number is stamped on a plate on right side of engine cylinder block.

INDEX (By Starting Paragraph)

INDEX (Cont.)

CONDENSED SERVICE DATA

	2240 Diesel	2440 Diesel	2630 Diesel	2640 Diesel
GENERAL				
Engine Make	Own	Own	Own	Own
Number Cylinders	3	4	4	4
Bore—Inches	4.19	4.02	4.19	4.19
Stroke—Inches	4.33	4.33	5.00	5.00
Displacement—Cubic Inches	179	219	276	276
Compression Ratio	16.2:1	16.3:1	16.3:1	16.3:1
Battery Terminal Grounded	Neg.	Neg.	Neg.	Neg.
Forward Speeds	8	8	8	8
TUNE-UP				
Firing Order	1-2-3	1-3-4-2	1-3-4-2	1-3-4-2
Valve Clearance—				
Inlet (Hot or Cold)	0.014	0.014	0.014	0.014
Exhaust (Hot or Cold)	0.018	0.018	0.018	0.018
Timing Mark Location		Crankshaft Pulley		
Engine Low Idle—Rpm	650	800	800	800
Engine High Idle—Rpm	2650	2650	2650	2650
Working Range	1500-2500	1500-2500	1500-2500	1500-2500
Pto Horsepower at 2500 Rpm	50	60.65	70.37	70.37

CONDENSED SERVICE DATA (Cont.)

	2240 Diesel	2440 Diesel	2630 Diesel	2640 Diesel
SIZES—CAPACITIES—CLEARANCES				
(All Dimensions Are In Inches)				
Crankshaft Journal Diameter	3.123-3.124	3.1230-3.1240	3.1235-3.1245	3.1235-3.1245
Crankpin Diameter	2.748-2.749	2.7480-2.7490	3.0630-3.0640	3.0630-3.0640
Balancer Shaft Journal Diameter	....	1.4995-1.5005	1.4995-1.5005	1.4995-1.5005
Piston Pin Diameter	1.3748-1.3752	1.3748-1.3752	1.3748-1.3752	1.3748-1.3752
Main Bearing Clearance	0.0012-0.0040	0.0016-0.0046	0.0011-0.0041	0.0011-0.0041
Rod Bearing Clearance	0.0012-0.0040	0.0012-0.0042	0.0016-0.0046	0.0016-0.0046
Camshaft Journal Clearance	0.004-0.006	0.004-0.006	0.004-0.006	0.004-0.006
Balancer Shaft Bearing Clearance	....	0.0015-0.0049	0.0015-0.0049	0.0015-0.0049
Crankshaft End Play	0.002-0.008	0.002-0.008	0.002-0.008	0.002-0.008
Camshaft End Play	0.002-0.009	0.002-0.009	0.002-0.009	0.002-0.009
Piston Skirt Clearance	See Para. 32	See Para. 32	See Para. 32	See Para. 32
Cooling System—Qts.	10.4	12	12	12
Crankcase (With Filter)—Qts.	6	6	10	10
Fuel Tank—Gallons	16.5	19½	19½	19½
Trans. & Hydraulic System—Gallons	9.5	10	10	10
TIGHTENING TORQUES—FT.-LBS.				
Cylinder Head	110	95	110	95
Main Bearings	85	85	85	85
Con. Rod Bearings (Oiled)	60-70	60-70	90-100	90-100
Rocker Arm Assembly	35	35	35	35
Flywheel	85	85	85	85

FRONT SYSTEM

AXLE AND SUPPORT

All Models

1. **AXLE CENTER MEMBER.** Center axle unit (5 or 5A—Fig. 1) attaches to front support by pivot bolt (6) and rear pivot pin (7). End clearance of pivot is controlled by shims (3). Five 0.015 inch thick shims are installed at factory assembly and recommended maximum end play is 0.015 inch. Removing shims reduces the end play. Thrust washers (2) at front and rear of pivot bushing (4) are identical and may be interchanged to compensate for wear when unit is removed. Rear pivot pin (7) is pressed into front support (1—Fig. 2).

Steering bellcrank (13—Fig. 1) should have 0.005-0.011 inch clearance in the two bushings (12) pressed into center axle unit. Recommended maximum end play of 0.010 inch is controlled by shims (11) which are 0.010 inch thick. Two shims are normally used; install additional shims when necessary providing snap ring (9) can still be installed.

2. **SPINDLES AND BUSHINGS.** Refer to Fig. 3 for exploded view of typical standard duty parts. The steer-

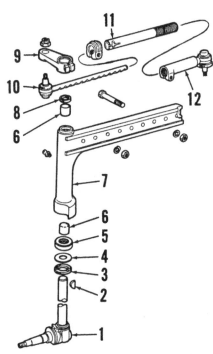

Fig. 3—Axle extension and associated parts of the type used on all models. Heavy duty models are slightly different.

1. Spindle		7. Axle extension
2. Woodruff key		8. Upper seal
3. Lower seal		9. Steering arm
4. Washer		10. Tie-rod end
5. Thrust bearing		11. Tube
6. Bushing		12. Tie-rod end

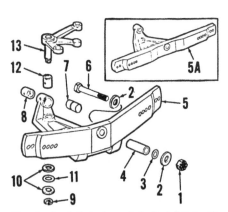

Fig. 1—Exploded view of typical front axle and associated parts. Other (straight and heavy-duty) axles are similar in major details.

1. Nut	8. Bushing
2. Washer	9. Snap ring
3. Shim	10. Washer
4. Bushing	11. Shim
5. Axle	12. Bushing or needle
5A. Axle	bearing
6. Pivot bolt	13. Steering bellcrank
7. Pivot pin	

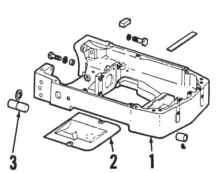

Fig. 2—View of front support casting and associated parts.

1. Front support
2. Pan

3. Rear pivot pin

ing arm (9) is keyed to spindle (1) and is retained by the clamp screw. Spindle end play should be less than 0.030 inch and is adjusted by repositioning the steering arm (9) lower on shaft of spindle (1). Spindle bushings (6) are pre-sized and should require no reaming if carefully installed.

The screw, which clamps steering arm (9) to spindle of standard duty models, should be tightened to 85 ft.-lbs. torque. Tighten slotted nuts of tie rod ends to 55 ft.-lbs. torque, tighten the additional amount necessary to align the slot in nut with hole, then install cotter pin. Screws attaching axle extensions (7) to axle center member (5—Fig. 1) should be tightened to 300 ft.-lbs. torque.

The heavy-duty axle (Fig. 4) has steering arm (9) splined to upper end of spindle (1). End play of spindle should be limited to 0.030 inch by adding washers (W) below steering arm. Tighten the screw, which retains steering arm (9) to spindle (1), to 170 ft.-lbs. torque. Tighten tie rod end nuts to 55 ft.-lbs. torque, plus amount necessary to align slot in nut with hole for cotter pin. Tighten the screws, which attach axle extensions (7) to axle center member, to 300 ft.-lbs. torque.

TIE RODS AND TOE-IN

All Models

3. The recommended toe-in is 1/8 to ¼-inch. Remove cap screw in outer clamp and loosen clamp screw in tie rod end (12—Fig. 3 or Fig. 4), then turn tie rod tube (11) as required. Make sure that tie rods are equal length so that tractor will turn the same in both directions.

None of the spindle stops must contact at extreme turning position. Readjust tie rods if necessary until the cor-

rect condition is attained.

Be sure to tighten the tie rod clamp screws to correct torque after adjustment is complete. Clamp screw torque should be 60 ft.-lbs. for 2630 models. On 2440 and 2640 models, tighten inner (small adjustment) clamp screws to 35 ft.-lbs. torque and the outer (large adjustment) clamp screws to 55 ft.-lbs. torque. On 2240 models, tighten the inner (small adjustment) clamp screws to 30 ft.-lbs. torque and the outer clamp screws to 85 ft.-lbs. torque.

4. **STEERING SHAFT (CROSS SHAFT).** The steering shaft (43—Fig. 6) can be removed without removing steering valve and steering wheel assembly, as follows:

Drain the oil in steering shaft by removing lower screw for cover (36). Remove cover (36), center the yoke of steering shaft in cover opening and remove cap screw (40), lockwasher (41) and pin retainer (42). Thread a cap screw into pin (39) and remove pin. Take out button plug on left side of clutch housing for access to cap screw (47). Refer to Fig. 5 and remove cap screw and steering shaft arm. Inset in Fig. 5 shows cover removed for access to steering shaft yoke, pin retainer and pin. Turn steering wheel full right, make sure steering shaft yoke is aligned with right cover opening and bump steering shaft out right side of clutch housing. Inspect bushing (37—Fig. 6) in cover (36) and bushing in clutch housing. Bushings are renewable.

Install by reversing removal procedure. Screws attaching cover (36) should be tightened to 36 ft.-lbs. torque. After cover (36) is installed, tighten arm attaching screw (47) fully, strike arm (45) with hammer, then tighten cap screw to 170 ft.-lbs. torque.

POWER STEERING SYSTEM

The power steering system is the closed center type and pressurized oil is furnished by the main hydraulic pump via a priority type pressure control valve which is bolted to lower right side of transmission housing. In the event of hydraulic failure, or engine stoppage, steering can still be accomplished with the mechanical advantage built into steering valve assembly.

TROUBLESHOOTING

All Models

5. Problems that develop in the power steering system may appear as sluggish steering, loss of steering, power steering in one direction only, or excessive noise in the power steering unit.

Sluggish steering can usually be attributed to:
   a. Leakage past valve seats usually produces slow steering in one direction only.
   b. Piston sealing ring or "O" ring failure.
   c. Steering valve body (or bodies) leaking.
   d. Clogged filter.

Loss of steering can usually be attributed to:
   a. Insufficient oil supply from transmission oil pump or main pump.
   b. Pressure control valve out of adjustment (other functions receiving priority).
   c. Clogged hydraulic oil filter.
   d. Shut-off screw in main hydraulic pump in "OFF" position.

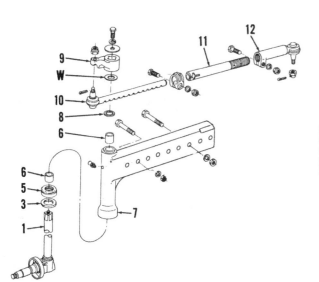

Fig. 4—Exploded view of heavy duty front axle extension and associated parts. Refer to Fig. 3 for legend except (W), which is the adjustment washer.

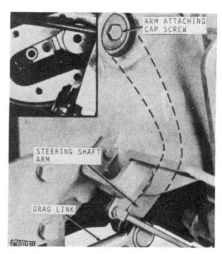

Fig. 5—Remove button plug for access to arm attaching cap screw on left side. Inset shows right side with cover removed.

Chattering system components are usually the result of "O" ring failure in unit.

Since the power steering valve is a completely self-contained unit, its only external requirement for proper operation is adequate hydraulic pressure. The transmission pump serves as a charging pump for the main hydraulic pump, which in turn supplies the steering valve via the pressure control valve. The steering valve receives hydraulic pressure first, in the event the main pump cannot supply enough for all hydraulic functions at the same time.

The pressure control valve output should be tested as outlined in paragraph 8, main hydraulic pump tested as outlined in paragraph 131 and transmission pump tested as outlined in paragraph 101.

Raise the wheels off the ground and with engine stopped, turn steering wheel from one extreme turn position to the other. No binding or hard spots should be encountered. Lower front wheels onto a hard surface, start engine and turn steering wheel from one extreme turn position to the other. Repeat test on a soft surface. Steering effort should be equal in both direc-

tions. Excessive steering effort in one direction only would indicate trouble in steering valve body and/or steering valve for that direction. Excessive effort in both directions might indicate trouble with piston "O" ring (or piston seal ring if one-piece piston/rod assembly used). Internal leakage may be attributed to dirty or worn valve seats in valve bodies (14—Fig. 6), or by a valve body shim pack being adjusted incorrectly. If adjuster oil seal (2) is leaking, the return oil passage may be obstructed. Possible causes of leaks may be incorrect adjustment of lower thrust bearing (17) or damaged gasket (23).

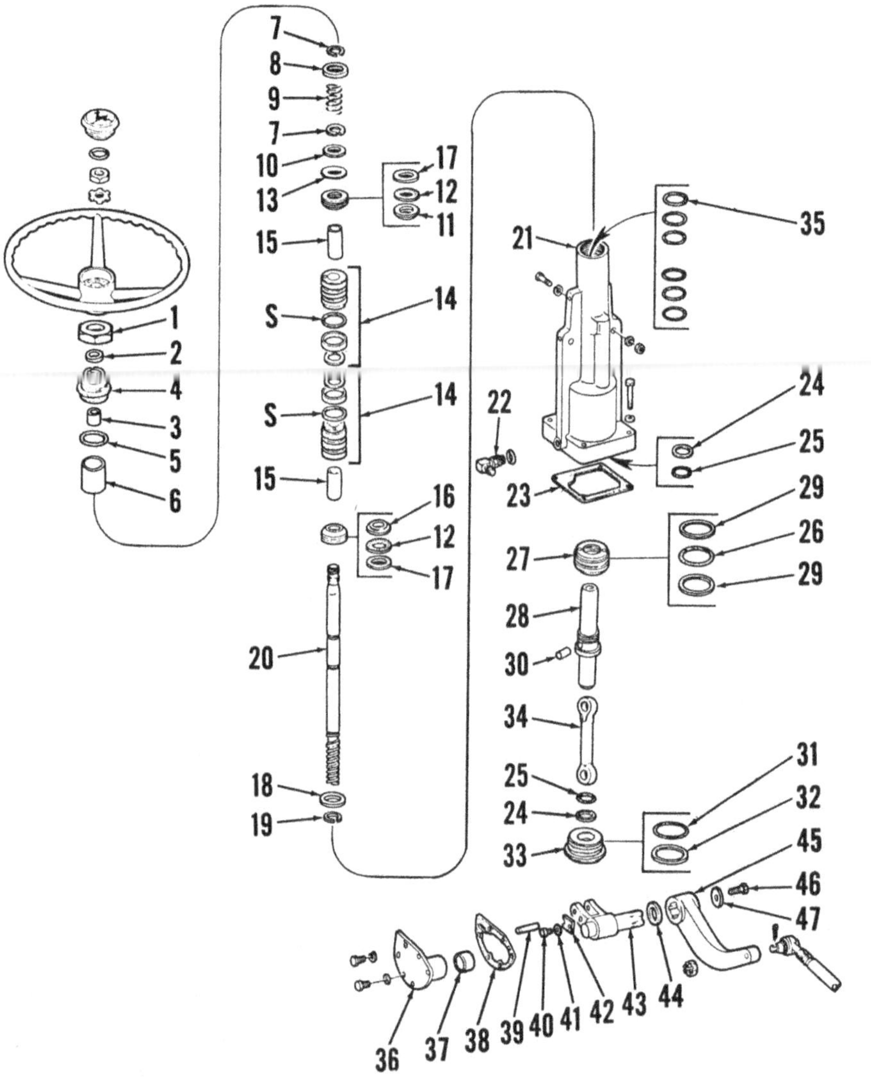

STEERING VALVE

All Models

6. REMOVE AND REINSTALL. Relieve hydraulic oil pressure slowly by disconnecting inlet pressure line, then remove drain plug on right side of clutch housing and drain oil from steering shaft compartment. Remove steering shaft cover (36-Fig. 6). If necessary, center the steering shaft yoke in cover hole, then remove cap screw (40), lockwasher (41) and pin retainer (42). Thread a 3/8-inch cap screw into end of pin (39) and remove pin. Steering valve may now be disconnected from dash and clutch housing and removed as a unit. If unit is to be overhauled, it may be easier to first remove steering wheel before assembly is removed from tractor. Remove steering wheel emblem, straighten tabs of lockwasher and remove steering wheel retaining nut. Attach a puller and remove steering wheel.

NOTE: DO NOT drive on upper end of steering shaft.

Reinstall in reverse order of removal, start engine and cycle system several times to remove any air which may be present.

7. OVERHAUL. With steering valve assembly removed as outlined in paragraph 6, refer to Fig. 6 and proceed as follows:

Remove jam nut (1) and oil seal (2). Using JDH-41-1 spanner wrench, remove adjuster (4) and slide sleeve (6) from housing. Rotate steering wheel shaft (20) clockwise until piston (27) reaches top of cylinder. Hold steering rod (34) from turning and turn shaft (20) until shaft and piston rod (28) come apart. Remove piston assembly and piston rod guide from housing, and carefully remove steering wheel shaft assembly. Using JDH-41-2 spring compressor as shown in Fig. 7, carefully

Fig. 6—Exploded view of power steering valve typical of all models. Items 36 through 47 are contained in top of clutch housing.

S. Special washer	13. Shim	24. Backup ring (2 used)	36. Cover
1. Jam nut	14. Valve body	25. "O" ring (2 used)	37. Bushing
2. Oil seal	15. Sleeve	26. "O" ring	38. Gasket
3. Bushing	16. Thrust washer	27. Piston	39. Pin
4. Adjuster	17. Thrust washer	28. Piston rod	40. Cap screw
5. "O" ring	18. Special washer	29. Backup rings	41. Lockwasher
6. Sleeve	19. Snap ring	30. Pin	42. Pin retainer
7. Snap ring	20. Steering wheel	31. "O" ring	43. Steering shaft
8. Special washer	shaft	32. Backup ring	44. Oil seal
9. Spring	21. Housing	33. Piston rod guide	45. Steering shaft arm
10. Special washer	22. Inlet check valve	34. Steering rod	46. Special washer
11. Thrust washer	23. Gasket	35. "O" rings	47. Cap screw
12. Needle bearing			

compress spring and remove top snap ring (7—Fig. 6). Release spring compressor, slide parts from shaft and remove bottom snap ring. If special spring compressor is not available, a steel pipe of correct diameter and length can be used. Cut an access slot for snap ring removal as shown in Fig. 7 and use the steering wheel retaining nut with a large washer to compress spring. Pay close attention to how parts are installed on shaft.

NOTE: Upper and lower valve body parts (14—Fig. 6) are individually assembled and adjusted by installation of shims (S) at time of original assembly. Be careful when disassembling, to not lose or interchange similar parts for the two valves. If renewal is necessary, renew as a pre-adjusted assembly.

Use a JDH-41-3 spanner wrench to remove piston (27) from piston rod (28), if a two-piece assembly. Pin (30) should be pressed from piston rod for removal of steering rod (34).

Renew adjuster oil seal (2) and bottom seal in bore. Adjuster bushing (3) should be renewed if I.D. is larger than 0.884 inch. Bushing should be installed until flush with bottom of seal bore in adjuster (4). Spring (9) should have 130-170 lbs. tension when compressed to a length of 1.094 inches. Inspect all parts for distortion, wear or rough spots on threads and ring lands. On early Model 2630, measure distance between the two top snap ring grooves. Early model shafts should measure 1.546-1.576 inches and 1.375-1.405 inches on later models. Early models require an additional snap ring washer between washer (8) and top of spring (9). Later shafts (shorter measurement) do not require this washer. Install piston (27) onto piston rod (28) with dowel holes in piston to top (if two-piece unit). Use JDH-41-3 spanner wrench and tighten piston to 250 ft.-lbs. torque. Assemble all parts on steering wheel shaft (20), beginning with snap ring (19). Be sure

large chamfers on thrust washers (11 and 16) go TOWARD valve bodies. Install spring (9) and two snap rings (7) using JDH-41-2 compressor or equivalent. On early 2630 models with longer space between snap ring grooves of shaft (20), install two washers (8) on top of spring. Renew all "O" rings, backup rings and sealing rings. Lubricate all parts generously before assembly. Assemble steering rod (34) into piston rod (28) small end first and press pin (30) until flush with outside surface of piston rod. Slip piston rod guide (33)-onto piston rod and install assembly into housing. Insert assembly into steering wheel shaft and parts into housing and thread shaft into piston rod. Install sleeve (6) and use a suitable sleeve over upper end of steering wheel shaft to protect lips of seal (2) as adjuster (4) is screwed into housing. Tighten adjuster with spanner wrench JDH-41-1 to 75 ft.-lbs. torque on 2040 models, 50 ft.-lbs. torque for all other models. On all models, hold adjuster with spanner wrench, tighten jam nut (1) to 30 ft.-lbs. torque.

Reinstall unit on tractor with front wheels in the straight ahead position. Install steering rod pin (39-Fig. 6) through steering shaft yoke (43). Install pin retainer (42), lockwasher and cap screw (41 and 40). Install cover (36) with new gasket. Install steering wheel with spoke pointing straight down and tighten nut to 50 ft.-lbs. torque. Start engine and cycle system several times to purge any air from steering valve. Stops on axle knuckles should not make contact when steering wheel is turned full right and left. Readjust tie rods if necessary to prevent stop contact.

PRESSURE CONTROL VALVE

All Models

The pressurized oil needed for power steering is furnished by the hydraulic system pump, via a priority type pressure control valve, which is mounted on the lower right side of the transmission case.

8. **TESTING.** If a pressure gage is not available, the pressure control valve operation can be checked as follows: Start engine and run at low idle. Turn steering wheel in both directions and note effort required to turn the steering wheel. Then, while turning the steering wheel in one direction, operate either the rockshaft or a remote cylinder and again note the effort required to turn the steering wheel. There should be no change in the effort. A notable increase in the steering effort indicates faulty operation of the pressure control valve.

If a faulty valve is indicated, test and adjust the pressure control valve as outlined in paragraph 9.

A large quantity of oil leakage past seal (2—Fig. 6) usually indicates a restricted or blocked oil return passage. Blockage can be caused by improper installation of bearing race (16) for the lower thrust bearing. The large chamfer must be toward control valve (14) to prevent restriction of return oil passage. Improper installation of gasket (23) or improper use of sealer on this gasket can also restrict or block the return oil passage.

9. Remove the plug, or the pump shut-off valve assembly, if so equipped, located directly opposite of the main

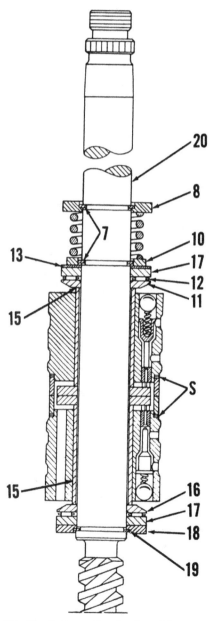

Fig. 7A—Cross-section drawing of the power steering control valve. Refer to Fig. 6 for legend.

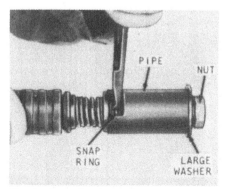

Fig. 7—View showing special tool JDH-41-2 for removing snap ring. Refer to text.

hydraulic pump stroke control valve and install a 3000 psi gage as shown in Fig. 9. Start engine and run at approximately 1900 rpm, then check the lift system operations and the pump stand-by pressure which should be 2220-2280 psi. Readjust the pump stroke control valve (see paragraph 132) to 1500 psi, then attempt to operate either the rockshaft or a remote cylinder. The lift system function should not occur, and if it does, the pressure control valve is faulty and should be repaired as outlined in paragraph 10 before proceeding further.

To continue testing of the pressure control valve, do either of the following: Completely lower rockshaft and place the rockshaft control lever in the raise position; or, retract a remote cylinder and place the selective control valve lever in the extend position. With engine running at 1900 rpm, adjust the main hydraulic pump stroke control (raise pressure) until either of the lift system operations previously mentioned occurs at its normal rate. This is the regulating point of the pressure control valve and should be 1700-1800 psi. If pressure is not as stated, disconnect front oil line, remove fitting (11—Fig. 8) and vary shims (7) as required. Shims are 0.030 inch thick and one shim will change pressure 35-40 psi.

If pressure control valve cannot be adjusted satisfactorily, remove and service valve as outlined in paragraph 10. Readjust the main hydraulic pump stroke control valve to obtain the recommended 2220-2280 psi stand-by pressure, then remove gage and reinstall plug or pump shut-off valve.

10. **R&R AND OVERHAUL.** To remove pressure control valve, drain transmission, then slowly disconnect inlet (front) oil line and if equipped with remote hydraulics, the rear (outlet) oil line. Unbolt and remove unit from transmission housing.

With valve removed, remove fitting (11—Fig. 8), then remove spool (9) (and

stop on large pumps), shims (7) and spring (6) from housing (3). Retain shims for subsequent installation.

Inspect spool and housing for wear, scoring or other damage. New spool diameters are 0.749-0.750 inch at front and 0.725-0.726 inch at rear. Spring has a free length of 4.62 inches and should test 45-55 pounds when compressed to a length of 3½ inches.

When reinstalling, attach oil lines before final tightening of mounting bolts.

ENGINE AND COMPONENTS

R&R ENGINE WITH CLUTCH

All Models

11. To remove engine and clutch as a unit, first drain cooling system and if engine is to be disassembled, drain oil pan. Remove front weights if any are installed. Remove cowling from right and left sides, remove battery door on top of cowl and remove batteries.

Fig. 9—Install 0-3000 psi gage in main pump as shown for pressure control valve test.

Remove the side grille screens, vertical muffler, if so equipped, and hood. Remove tool box and side rails if so equipped. Remove hydraulic line clamps on right side of tractor and separate the hydraulic pump pressure line at connector located at right rear of front support. Remove the line retainer from lower right front of clutch housing.

NOTE: On tractors without Hi-Lo shift and without independent pto, be careful not to lose the check valve located at rear of the hydraulic pump inlet line.

On all models, shut off fuel and disconnect fuel lines from pump. Disconnect drag link and remove from tractor, if so desired. Remove air intake pipe and radiator hoses. Disconnect fuel leak-off pipe and hydraulic pump coupling. Disconnect bleed lines from top of hydraulic reservoir. Place two wooden blocks between front axle and support to prevent tipping sideways. Support tractor, attach hoist to front support and axle assembly and separate front support from engine. Disconnect all wires to wiring harness and lay harness back out of the way. Disconnect tachometer cable from clutch housing and oil line from power steering valve. Disconnect cold weather starting aid line from intake manifold. Remove coolant temperature bulb from water outlet manifold. Disconnect control rod from injection pump and fuel shut-off rod from pump and remove rod. If tractor has an under-slung exhaust, disconnect exhaust pipe from exhaust manifold. Install two JD-244 engine lifting adapters (or equivalent) to cylinder head and attach hoist to a suitable engine sling or to adapters. Place a suitable container under rear of clutch housing to catch residual oil, then remove the cap screws which secure cowl (dash) to flywheel housing. Keep engine horizontal with hoist, unbolt engine from clutch housing and pull engine forward until clutch clears clutch shaft.

Reassemble tractor by reversing the disassembly procedure. On tractors with continuous running pto, the pto shaft mates with the pto clutch disc before transmission shaft mates with main clutch disc. If difficulty is encountered while joining engine to clutch housing, bar over engine until both shafts are indexed with both clutch discs and flywheel housing is snug against clutch housing before tightening the retaining cap screws.

Tighten the engine to clutch housing and the 5/8-inch front support to engine cap screws to 170 ft.-lbs. torque. Tighten the 9/16-inch front support cap screws to 130 ft.-lbs. torque.

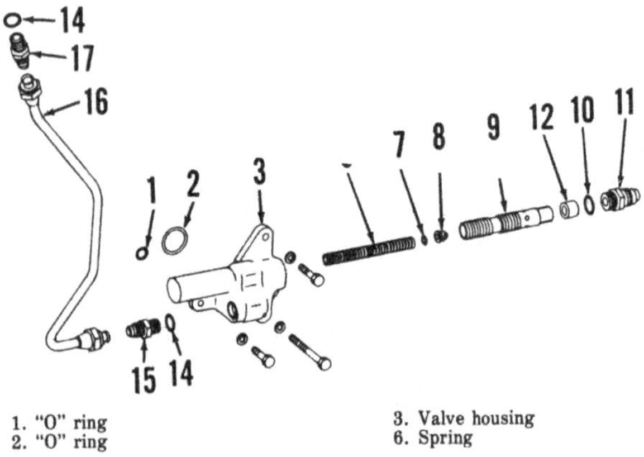

Fig. 8—Exploded view of pressure control valve. Line (16) is pressure line for selective control valve. Orifice (8) is integral with valve (9). On tractors with large pump, a stop (12) is placed between items (9) and (11).

7. Shim (.030)
8. Orifice
9. Control valve
10. "O" ring
11. Connector
12. Stop
14. "O" ring
15. Connector
16. Pressure line
17. Connector

1. "O" ring
2. "O" ring
3. Valve housing
6. Spring

CYLINDER HEAD

All Models

12. To remove cylinder head, first drain cooling system and remove hood. Disconnect battery ground straps. Remove air cleaner tube. Remove exhaust manifold from cylinder head and if tractor is equipped with underslung exhaust, the manifold can be left attached to the exhaust pipe. Disconnect injector leak-off line from fuel tank, injectors and injection pump and remove complete leak-off line. Disconnect pressure lines from injectors, remove hold-down clamps and spacers and withdraw injectors. Disconnect the cold starting aid from the inlet manifold fitting. Unbolt the fuel filter from the cylinder head. Plug all fuel openings. Remove vent tube from rocker arm cover, then remove cover and the rocker arm assembly. Identify and remove push rods and the valve stem caps. Disconnect water outlet elbow from cylinder head, then unbolt and remove cylinder head. Notice that the four headbolts between push rods on 2440 and 2640 models (with block marked R60832) are different than the other 14 cylinder head bolts.

When installing the head, install head gasket dry and tighten head bolts to 110 ft.-lbs. torque on 2240 and 2630 models, and to 95 ft.-lbs. on 2440 and 2640 models, in sequence shown in Fig. 10. Be sure oil holes in rear rocker arm shaft bracket and cylinder head are open and clean as this passage provides lubrication for the rocker arm assembly. Align spring pin in head with pin

hole in rocker arm shaft. Tighten rocker arm shaft to cylinder head cap screws, intake manifold cap screws and the exhaust manifold retaining cap screws all to 35 ft.-lbs. torque. Head bolts should be retorqued after engine has run about one hour at 2500 rpm under half-load. Loosen the head bolts about 1/6-turn before retightening them to above listed torque. Valve tappet gap, hot or cold, is 0.014 inch for intake and 0.018 inch for exhaust on all models.

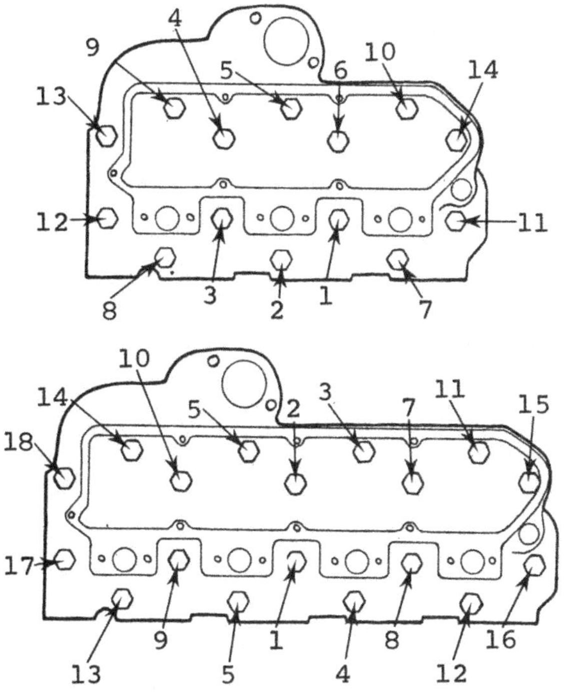

VALVES AND SEATS

All Models

13. Exhaust valves for all models seat against renewable inserts which are pressed into the cylinder head. Service inserts are slightly larger to insure tight fit and bore in cylinder head must be machined to permit installation. Most models are equipped with valve rotators on exhaust valves and all valves are fitted with renewable hardened steel caps on valve stem ends. Refer to the following service specifications:

2240 Models
Valve face angle—degrees 43.5
Valve seat angle—degrees 45
Valve stem diameter—
　standard . . . . .0.371-0.372 in.
Valve recessed in
　cylinder head—maximum . . .0.118 in.
Valve seat width—
　Intake . . . . .0.057-0.073 in.
　Exhaust . . . . .0.051-0.057 in.

2440 Models
Valve face angle—degrees 43.5
Valve seat angle—degrees 45
Valve stem diameter—
　standard . . . . .0.3715-0.3725 in.
Valve recessed in cylinder head—
　Limits
　　Intake . . . . .0.0230-0.0470 in.
　　Exhaust . . . . .0.0380-0.0720 in.
Valve seat width—Intake
　and Exhaust . . . . .0.0781-0.0937 in.

2630 and 2640 models
Valve face angle—degrees 44.5
Valve seat angle—degrees 45
Valve stem diameter—
　standard . . . . .0.3715-0.3725 in.
Valve recessed in cylinder head—
　Limits
　　Intake . . . . .0.0230-0.0470 in.
　　Exhaust . . . . .0.0380-0.0720 in.
Valve seat width—Intake and
　Exhaust . . . . .0.060-0.070 in.

Valves are available with 0.003, 0.015 and 0.030 inch oversize stems. Seats can be narrowed using 20 and 70 degree stones.

TAPPET GAP ADJUSTMENT

All Models

14. Valve tappet gap for all valves can be set with flywheel being placed in only two positions. Valve tappet gap (hot or cold) for all engines is 0.014 for inlet and 0.018 for exhaust.

To set valve tappet gap, turn crankshaft by hand until No. 1 cylinder is at top dead and center and TDC timing screw will enter hole in flywheel as shown in Fig. 11. JD-281 engine rotation tool may also be used to obtain TDC. Check the valves to determine whether front cylinder is on compression or exhaust stroke.

Refer to the appropriate diagram (Fig. 12 or 13) and adjust the indicated valves; then turn crankshaft one complete turn until timing screw will again enter TDC hole in flywheel and adjust remainder of valves.

Fig. 10—Tighten cylinder head cap screws to a torque of 110 ft.-lbs. for 2240 and 2630 models and 95 ft.-lbs. for 2440 and 2640 models. Use the sequence shown at top for three cylinder 2240 models; the lower sequence for all four cylinder models.

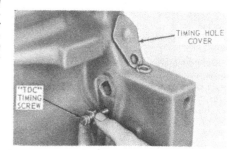

Fig. 11—Reverse timing screw as shown on all models to find "TDC" timing hole in flywheel. Refer to text.

VALVE GUIDES

All Models

15. Valve guides are integral with cylinder head and should have 0.002-0.004 inch operating clearance for valves. Original inside diameter of guide bore is 0.374-0.375 inch for three cylinder 2240 models, 0.3475-0.3755 inch for other (four cylinder) models. Maximum allowable valve stem clearance in guide is 0.006 inch. If clearance is 0.008 inch or less, guides may be knurled and original valves used, if valve stem wear is not excessive. Use knurling tool EXACTLY as recommended by the manufacturers. When clearance exceeds 0.08 inch, ream valve guide as required to fit next oversize valve stem; valves are available with 0.003, 0.015 and 0.030 inch oversize stems.

VALVE SPRINGS

All Models

16. Inlet and exhaust valve springs are interchangeable. Springs that are distorted, discolored, rusted, or do not meet the following specifications, should be renewed.

Free length (approx.) ....2-1/8 in.
Installed height—
    Valve closed ....1-13/16 in.
    Valve open ....1-23/64 in.
Test lbs. at 1-13/16 in............54-62
Test lbs. at 1-23/64 in.........133-153

VALVE ROTATORS

Models So Equipped

17. Positive type valve rotators are factory installed on the exhaust valves of most engines.

Normal servicing of the rotators consists of renewing the units. It is important however, to check operation of the rotators. If rotator is removed, see that it turns freely in one direction only. If rotator is installed, be sure valve rotates a slight amount each time it opens.

ROCKER ARMS AND SHAFT

All Models

18. Rocker arm and shaft assemblies for all models use identical parts.

Rocker arms are interchangeable and bushings are not available. Inside diameter of shaft bore in rocker arm is 0.790-0.792 inch. Outside diameter of rocker arm shaft is 0.787-0.788 inch. Normal operating clearance between rocker arm and shaft is 0.002-0.005 inch. Renew rocker arm and/or shaft if clearance is excessive.

Valve stem contacting surface of rocker arm may be refaced but original radius must be maintained.

When reinstalling rocker arm assembly, be sure oil holes and passages are open and clean. Pay particular attention to the rear mounting bracket as lubrication is fed to rocker arm shaft through this passage. Oil hole in rocker arm shaft must face downward when installed on cylinder head.

CAM FOLLOWERS

All Models

19. The cylinder type cam followers (tappets) can be removed from below after camshaft has been removed. If necessary, they can also be removed from above after cylinder head, rocker arm shaft and push rods are removed. Identify followers so they can be in-

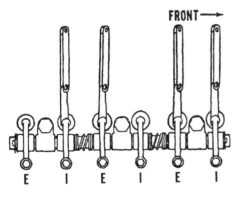

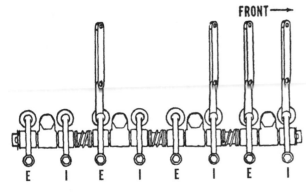

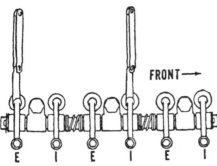

Fig. 12—With No. 1 piston at TDC on compression stroke, gap can be adjusted on the indicated valves. Refer to drawing at top for three cylinder 2240 models, lower drawing for other (four cylinder) models. Turn crankshaft one complete turn and adjust remaining valves as shown in Fig. 13.

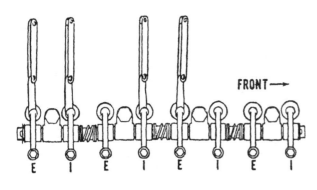

Fig. 13—With the No. 1 piston at TDC on exhaust stroke, the valves indicated can be adjusted. The top drawing is for three cylinder models, the lower drawing is for four cylinder models.

stalled in the same bore, if they are to be reused. Cam followers are available in standard sizes only and operate directly in machined bores in cylinder block.

It is recommended that new cam followers always be installed if a new camshaft is being installed.

VALVE TIMING

All Models

20. Valves are correctly timed when timing mark on camshaft gear is aligned with timing tool (JD254) when tool is aligned with crankshaft and camshaft centerlines as shown in Fig. 16.

TIMING GEAR COVER

All Models

21. To remove timing gear cover, first remove the front axle and front support assembly as outlined in paragraph 11.

With front support assembly removed, remove fan, fan belt, alternator and water pump. Remove crankshaft pulley retaining cap screw, attach puller and remove pulley. Remove the oil pressure regulating plug, spring and valve from 2240 and 2440 models as shown in Fig. 14. Drain and remove oil pan, then unbolt and remove the timing gear cover. See Fig. 15.

With timing gear cover removed, the crankshaft front oil seal can be renewed. To renew oil seal, coat outside diameter of seal with sealing compound and with seal lip toward inside, support timing gear cover around seal area and press seal into bore until it bottoms.

NOTE: Do not attempt to install the front oil seal in timing gear cover without providing support around seal area. Cover could be warped or cracked rather easily.

Tighten timing gear cover cap screws to 35 ft.-lbs. torque.

CAMSHAFT

All Models

22. To remove camshaft, timing gear cover must be removed as outlined in paragraph 21. Remove vent tube, rocker arm cover, rocker arm assembly and push rods. Shut off fuel and unbolt fuel pump from cylinder block. Use a magnetic holding tool set to hold tappets away from camshaft lobes, or remove tappets and identify for reassembly into original bores.

Before removing camshaft, end play should be checked. End play should be 0.002-0.009 inch (new) with a maximum of 0.015 inch on a used engine.

Turn engine until thrust plate retaining cap screws can be reached through holes in camshaft gear, then remove cap screws and pull camshaft and thrust plate from cylinder block. Be careful not to allow camshaft to drag in bores in block.

NOTE: If upper idler gear is not being removed, mark camshaft gear and upper idler gear so camshaft can be reinstalled in its original position. If upper idler gear is being removed, turn engine to TDC and align camshaft gear timing marks as shown in Fig. 16 when installing the upper idler gear.

Support camshaft gear, press camshaft from gear and remove Woodruff key.

If tachometer drive shaft at aft end of camshaft requires renewal, thread exposed end and install a nut, then attach a puller to nut and remove the tachometer drive shaft from camshaft.

Camshaft is carried in three unbushed bores in cylinder block. When checking camshaft journal diameters, also check inside diameter of the cam-shaft journal bores using the following data.

Camshaft journal O.D....2.200-2.201 in.
   Wear limit ....2.199 in.
Camshaft bearing bore
   I.D..................2.204-2.205 in.
Normal Diametral
   Clearance ....0.003-0.005 in.
     Wear limit ....0.007 in.
Camshaft end play ......0.002-0.009 in.
   Maximum allowable ....0.015 in.
Thrust plate thickness ...0.156-0.158 in.
   Wear limit ....0.151 in.

When installing new camshaft gear, be sure timing mark is toward front and support camshaft under front bearing journal. When installing new tachometer drive shaft, be sure drive slot is toward rear and support camshaft under rear journal.

When installing camshaft in cylinder block be sure timing mark is aligned as shown in Fig. 16 and tighten thrust plate cap screws to 35 ft.-lbs. torque. If upper idler gear was removed, be sure outer thrust washer aligns with spring pin in idler shaft and tighten cap screws to 65 ft.-lbs. torque.

BALANCER SHAFTS

Four Cylinder Models

All of the four cylinder engines used in 2440, 2630 and 2640 models are equipped with two balancer shafts which are located below the crankshaft on opposite sides of the crankcase. Each shaft is carried in three renewable bushings which are located in bores in the cylinder block. The right hand balancer shaft is driven by the lower idler gear and the left hand balancer shaft is driven by the oil pump gear. See Fig. 15. Shafts rotate in opposite directions at twice engine speed and are designed to dampen the vibration which is inherent in four cylinder engines.

Fig. 14—Oil pressure relief valve on 2240 and 2440 models is located as shown. Unit can be adjusted by using shims under forward end of spring. Also see Fig. 25.

Fig. 15—View of exposed timing gear train. Balance gears and shafts are not used on three cylinder 2240 models.

B. Balance shaft gear
C. Crankshaft gear
G. Camshaft gear
L. Lower idler gear
O. Oil pump gear
P. Injection pump gear
T. Thrust screw
U. Upper idler gear

Fig. 16—With engine at TDC and timing tool (TT) positioned on shafts centerline as shown, camshaft gear timing mark (TM) will be directly under edge of timing tool.

23. To remove the balancer shafts, first remove the timing gear cover as outlined in paragraph 21. Remove lower idler gear and oil pump gear.

At this time, check end play of both balancer shafts. End play should be 0.002-0.008 inch and if end play exceeds 0.015 inch, renew thrust plates during assembly.

Identify balancer shafts as to right and left. Models 2630 and 2640 have bolt-on balancer weights. On these models, remove and identify weights as right or left and front and rear. On all models, unbolt and remove thrust plates and carefully withdraw balancer shafts from cylinder.

Use the following specifications and check balancer shaft, bushings and thrust plates.

Shaft journal O.D. . . . .1.4995-1.5005 in.
Bushing I.D. . . . . .1.5020-1.5040 in.
Shaft operating
  clearance. . . . . .0.0015-0.0045 in.
Max. allowable
  clearance . . . . .0.006 in.
Shaft end play . . . . .0.002-0.008 in.
Max. allowable
  end play . . . . .0.015 in.
Thrust plate thickness . . .0.117-0.119 in.

Renew any parts which do not meet specifications. The two front balancer shaft bushings for either shaft can be renewed with engine in tractor, however, if either of the two rear bushings require renewal, remove the engine, flywheel and flywheel housing. This will permit staking of the rear bushings.

When installing bushings, use a piloted driver (JD-249 or equivalent) and install bushings from front so that front of bushing is flush with chamfer at front of bore and oil holes are aligned with oil holes in cylinder block. With bushings pressed in place, they

must be staked as follows: Use John Deere tool number JD-255 and place the half-round portion in I.D. of bushing so the staking ball (B—Fig. 17) is in round relief in bushing groove directly opposite to bushing oil hole. Turn the square half of tool so the correct size dowel (D) is toward lead (dowel) hole and position square half of tool on crankcase boss. Check alignment of cap screw holes and if necessary, turn half-round portion of tool end-for-end. Install cap screws, BE SURE staking ball is in the relief, then tighten cap screws evenly until the half-round half of tool butts against bushing I.D. This will indent the bushing into dowel hole and stake bushing in bore.

If new gears are being installed on balancer shafts, be sure timing mark is toward front and support shaft on both sides of front journal with tool JD-247, or equivalent. Press gear on shaft until it is flush with end of shaft. Be sure gear is within 0.001 inch of being flush with end of shaft as this controls shaft end play.

Reinstall balance shafts by reversing removal procedure, however before

Fig. 18—To time the right balance weight on four cylinder models, place timing tool (TT) between centerlines of crankshaft and right balance shaft. The timing mark (TM) on gear (B) will be directly below timing tool when shaft is correctly timed.

installing the lower idler gear, set engine on TDC and align crankshaft centerline, timing mark and balance shaft centerline using John Deere timing tool JD-254 or equivalent, as shown in Figs. 18 and 19. Check clearance between gear and thrust plate with a feeler gage. Clearance should be 0.002-0.008 inch.

IDLER GEARS

All Models

24. All engines are equipped with upper and lower idler gears. Idler gears are bushed and operate on stationary shafts which are attached to the engine front plate with cap screws. Idler gear end play is controlled by thrust washers. Both idler gears are driven by the crankshaft. The upper idler gear drives the camshaft and the diesel injection pump drive gear. The lower idler gear drives the oil pump drive gear. On four cylinder models, the right hand balance shaft is driven by the lower idler gear and the left hand balance shaft is driven by the oil pump drive gear. Refer to Fig. 15.

Check idler gears for excessive end play before removal. To remove idler gears, remove oil pan and timing gear cover, wedge a clean rag between gears, then remove cap screw and pull gear and thrust washers from shaft. Idler gear shaft can now be removed.

Clean and inspect idler gears and shafts and refer to the following specifications.

Shaft O.D. . . . . .1.749-1.750 in.
Bushing I.D. . . . . .1.751-1.753 in.
Operating clearance . . . . .0.001-0.004 in.
Max. allowable clearance. . . . . .0.006 in.
End play . . . . .0.001-0.007 in.
Max. allowable end play . . . . . .0.015 in.

Fig. 17—View of tool (JD-225) used to stake balance shaft bushings on four cylinder models. Half-round (left hand) part of tool contains staking ball and is used in I.D. of bushing. Refer to text.

B. Staking ball
D. Dowels
CS. Cap screws

Fig. 19—To time the left balance weight on four cylinder models, place timing tool (TT) between centerlines of crankshaft and left balance weight. The timing mark (TM) on gear (B) will be directly below timing tool when shaft is correctly timed.

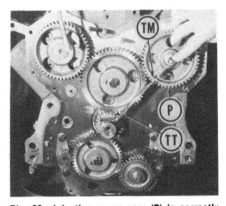

Fig. 20—Injection pump gear (P) is correctly timed if timing mark (TM) is directly below timing tool (TT) when tool is placed between centerlines of crankshaft and injection pump shafts. A four cylinder model is shown, but procedure is same for three cylinder engine.

Make sure that upper idler gear oil hole is not plugged. Bushings should be installed flush with either side of gear. If necessary to renew spring pin in idler shaft, be sure projection from shaft is 0.138-0.177 inch for upper idler gear shaft, 0.197-0.275 inch for lower idler gear shaft, to prevent bottoming in thrust washer.

Reinstall by reversing removal procedure and be sure camshaft, injection pump and both balance shafts (on four cylinder models) are timed as indicated in Figs. 16, 18, 19 and 20. Tighten shaft cap screws to 65 ft.-lbs. torque on upper idler shaft and to 95 ft.-lbs. torque on lower idler shaft.

TIMING GEARS

All Models

25. **CAMSHAFT GEAR.** The camshaft gear (G—Fig. 15) is keyed and pressed on the camshaft. The fit of gear on camshaft is such that removal of the camshaft, as outlined in paragraph 22 is recommended. Camshaft is correctly timed when centerline of camshaft, timing mark on camshaft gear and centerline of crankshaft are aligned and crankshaft is at TDC-1 as shown in Fig. 16.

26. **CRANKSHAFT GEAR.** Renewal of crankshaft gear requires removal of crankshaft as outlined in paragraph 35. Gear is keyed and pressed on crankshaft. Support crankshaft under first throw when installing new gear. Installation of gear may be eased by heating gear in oil, not to exceed 360 deg. F.

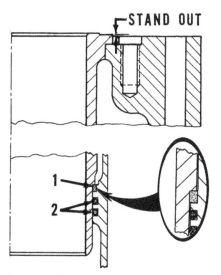

Fig. 21—Cross-section of cylinder sleeve showing rectangular section packing (1) and round "O" rings (2). Refer to text for correct installation of cylinder sleeves.

27. **INJECTION PUMP GEAR AND SHAFT.** The injection pump gear and shaft unit can be withdrawn from timing gear housing and pump after cover is off. Gear is keyed to shaft and retained by washer and nut. Tighten nut to a torque of 35 ft.-lbs.

Two timing marks appear on the gear, each identified by a stamped "3" or "4". Use the "4" timing mark when timing four cylinder engines. Use the "3" timing mark when timing the gears in three cylinder engines. Refer to Fig. 20.

28. **TIMING GEAR BACKLASH.** Excessive timing gear backlash may be corrected by renewing the gears concerned, or in some instances by renewing idler gear bushing and/or shaft. Refer to the following for recommended backlash:

Three Cylinder Models
Crankshaft gear to upper
   idler gear ....0.003-0.012 in.
   Max. limit ....0.016 in.
Camshaft gear to upper
   idler gear ....0.003-0.014 in.
   Max. limit ....0.020 in.
Injection pump drive gear to
   upper idler gear ......0.003-0.014 in.
   Max. limit ....0.020 in.
Crankshaft gear to
   lower idler gear.......0.003-0.014 in.
   Max. limit ....0.020 in.
Oil pump gear to lower
   idler gear ....0.0016-0.0150 in.
   Max. limit ....0.016 in.

Four Cylinder Models
Crankshaft gear to upper
   idler gear ....0.0027-0.0116 in.
Camshaft gear to upper
   idler gear ....0.0028-0.0135 in.
Injection pump drive gear to
   upper idler gear ....0.0028-0.0135 in.
Crankshaft gear to lower
   idler gear ....0.0016-0.0147 in.
Right balance shaft gear to
   lower idler gear.....0.0018-0.0156 in.
Oil pump gear to lower
   idler gear ....0.0016-0.0147 in.
Left balance shaft gear to
   oil pump gear.......0.0020-0.0140 in.

ROD AND PISTON UNITS

All Models

29. Piston and connecting rod assemblies are removed from above after removing cylinder head and oil pan. Secure cylinder liners (sleeves) in cylinder with cap screws and washers to prevent liners from moving as crankshaft is turned.

All pistons have the word "FRONT" stamped on head of piston. Connecting rods also have the word "FRONT"

stamped (embossed) in the web of connecting rod. Replacement rods are not numbered and should be stamped with correct cylinder number. When installing rod and piston units, lubricate rod screws and carefully tighten connecting rod screws to correct torque of 60-70 ft.-lbs. for 2240 and 2440 models, 90-100 ft.-lbs. for 2630 and 2640 models.

PISTONS, RINGS AND SLEEVES

All Models

30. All pistons are cam-ground, forged aluminum-alloy and are fitted with three rings located above the piston pin. All pistons have the word "FRONT" stamped on the piston head and are available in standard size only.

Top piston ring is of keystone design and a wear gage (JDE-62) should be used for checking piston groove wear. Shoulder on wear gage should not touch ring land, or piston is not fit for further use.

Ring side clearance in groove should not exceed 0.005 inch for the second compression ring or the lower, oil control ring for all four cylinder (2440, 2630 and 2640) models. Side clearance for second compression ring in groove of piston for 2240 (three cylinder) models should be 0.0015-0.0030 inch and should not be permitted to exceed 0.008 inch.

On all models, the renewable wet type cylinder sleeves are available in standard size only. Sleeve flange at upper edge is sealed by the cylinder head gasket. Sleeves are sealed at lower edge by packing of the type shown in Fig. 21. Sleeves normally require loosening using a sleeve puller, after which they can be withdrawn by hand. Out-of-round or taper should not exceed 0.005 inch. If sleeve is to be reused, it should be deglazed using a normal cross-hatch pattern.

When reinstalling sleeves, first make sure sleeve and block bore are absolutely clean and dry. Carefully remove any rust or scale from seating surfaces, packing grooves and from water jacket in areas where loose scale might interfere with sleeve or packing installation. If sleeves are being reused, buff rust and scale from outside of sleeve.

Install sleeve without the seals and measure standout. Check sleeve standout at several locations around sleeve. Also check to be sure that sleeve will slip fully into bore without force. If sleeve cannot be pushed down by hand, recheck for scale or burrs. If sleeve standout is less than 0.001 inch, install one special shim (part number R46906) between liner and cylinder block, then recheck standout. If standout is more than 0.004 inch, check for scale or burrs; then, if necessary, select another sleeve. After matching sleeves to all the bores, mark the sleeves then refer to the appropriate following paragraph for packing and sleeve installation.

31. RECTANGULAR PACKING AND "O" RINGS. Refer to Fig. 21. Apply liquid soap (such as part number AR-54749) to the rectangular section ring (1) and install over lower end of cylinder liner (sleeve). Slide the rectangular section ring up against shoulder on sleeve, make sure that ring is not twisted and that longer sides are parallel with side of sleeve as shown. Apply the liquid soap to round section "O" rings (2) and install in grooves in cylinder block. Be sure that "O" rings are completely seated in grooves so that installing the sleeve will not damage the "O" rings. Observe the previously affixed mark indicating correct cylinder location, then install sleeves carefully into correct cylinder block bore. Work sleeve gently into position by hand until it is finally necessary to tap sleeve into position using a hardwood block and hammer.

NOTE: Be careful not to damage the packing rings. Check the cylinder sleeve stand-out (Fig. 21) with packing installed. The difference between this measured stand-out and similar measurement taken earlier for same sleeve in same bore without packing will be the compression of the packing. If the compression is less than 0.005 inch, the rectangular section packing ring will not seal properly. Remove sleeve from cylinder block, check packing to be sure that installation has not cut the packing ring. If shoulders on sleeve and in cylinder block do not provide proper compression of the rectangular packing ring, install different sleeve and recheck. If a different sleeve will not provide enough compression of packing ring, suggested repair is to install new cylinder block.

32. SPECIFICATIONS. Specifications of pistons and sleeves are as follows:

2240 (4.19 Bore) Engines
Sleeve Bore ....4.192-4.194 in.
Piston Skirt Diameter
 (bottom) ....4.185-4.188 in.
Piston Skirt Clearance ...0.004-0.009 in.
 Wear limit ....0.010 in.
2440 (4.02 Bore) Engines
Sleeve Bore ....4.0150-4.0164 in.
Piston Skirt Diameter
 (bottom) ....4.0089-4.0109 in.
Piston Skirt Clearance
 (Selective Fit) ......0.0041-0.0075 in.
2630 And 2640 (4.19 Bore) Engines
Sleeve Bore ....4.1922-4.1936 in.
Piston Skirt Diameter
 (bottom) ....4.1883-4.1904 in.
Piston Skirt Clearance .0.0018-0.0053 in.

PISTON PINS AND BUSHINGS

All Models

33. The full floating piston pins are retained in pistons by snap rings. A pin bushing is fitted in upper end of connecting rod and bushing must be reamed after installation to provide a thumb press fit for the piston pin. Desired clearance between piston pin and bore in piston is 0.0001-0.0009 inch. Piston pin diameter is 1.3748-1.3752 on all models.

CONNECTING RODS AND BEARINGS

All Models

34. The steel-backed, aluminum lined bearings can be renewed without removing rod and piston unit by removing oil pan and rod caps.

Connecting rod big end parting line is diagonally cut and rod cap is offset as shown in Fig. 23. A tongue and groove cap joint positively locates the cap. Rod marking "FRONT" should be forward and locating tangs for bearing inserts should be together when cap is installed.

Connecting rod bearings are available in undersizes of 0.002, 0.010, 0.020 and 0.030 inch, as well as standard. Refer to the following specifications:
Crankpin Diameter, New—
 2240 and 2440 models2.7480-2.7490 in.
 2630 and 2640 models3.0630-3.0640 in.
Diametral Clearance, Desired—
 2240 models ....0.0012-0.0040 in.
 2440 models ....0.0012-0.0042 in.
 2630 models ....0.0016-0.0046 in.
 2640 models ....0.0016-0.0046 in.
Diametral Clearance, Wear Limit—
 2240 model ....0.006 in.
Rod Cap Screw Torque—
 2240 and 2440 models ...60-70 ft.-lbs.
 2630 and 2640 models ..90-100 ft.-lbs.

CRANKSHAFT AND BEARINGS

All Models

35. The crankshaft of three cylinder engines (2240 model) is supported in four main bearings. The crankshaft of four cylinder engines is supported in five main bearings. Main bearing inserts of all engines may be renewed after removing oil pan, oil pump and main bearing caps. All main bearing caps except rear are identical and are numbered so they can be reinstalled in their original position.

NOTE: Main bearing caps are available separately and are fitted to cylinder block using a set gage and shims. If shims are found under main cap as engine is disassembled, that bearing cap has been renewed, and shims must be kept with the same cap. Be sure shims are not mixed or lost and that bearing cap is returned to same location.

Rear main bearing is flanged and controls crankshaft end play which should be 0.002-0.008 inch. Install main bearing caps in their original positions and tighten the retaining screws to 85 ft.-lbs. torque.

To remove the crankshaft, it is necessary to remove engine from tractor. With engine removed, remove oil pan, cylinder head and the connecting rod and piston units. Remove clutch, flywheel and flywheel housing. Remove timing gear cover, camshaft, injection pump drive gear and shaft and both idler gears. Remove oil pump drive gear and oil pump, balance shaft thrust plates, and then remove engine front plate from cylinder block. Be sure main bearing caps are identified for reinstallation, then remove all bearing caps and lift crankshaft from cylinder block.

Check crankshaft and bearings for wear, scoring or out-of-round condition

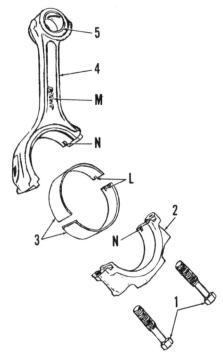

Fig. 23—Connecting rod assembly used in all engines.

1. Cap screws
2. Cap
3. Inserts
4. Rod
5. Pin bushing
L. Locating tangs
M. "FRONT" marking
N. Locating notches

using the following specifications:

Main Journal Diameter, New—
    2240 and 2440 . . . . . . 3.1230-3.1240 in.
    2630 and 2640 . . . . . . 3.1235-3.1245 in.
Crankpin Diameter, New—
    2240 and 2440 . . : . . . . 2.7480-2.7490 in.
    2630 and 2640 . . . . . . 3.0630-3.0640 in.
Crankshaft End Play—
    Recommended . . . . 0.002-0.008 in.
    Wear limit . . . . 0.015 in.
Journal Taper (per inch)—
    Max. allowable . . . . 0.001 in.
Journal Out-Of-Round—
    Max. allowable . . . . 0.003 in.
Main Bearing Diametral Clearance—
    2240 Desired . . . . . . 0.0012-0.0040 in.
      Wear limit . . . . 0.006 in.
    2440 Desired . . . . . . 0.0012-0.0040 in.
      Wear limit . . . . 0.006 in.
    2630 and 2640 . . . . . 0.0011-0.0041 in.
      Wear limit . . . . 0.0055 in.

If crankshaft does not meet specifications, either renew it or grind to the correct undersize. Main bearings are available in standard size and 0.002, 0.010, 0.020 and 0.030 inch undersizes.

CRANKSHAFT REAR OIL SEAL

All Models

36. A lip type crankshaft rear oil seal is contained in flywheel housing and a sealing ring is pressed on mounting flange of crankshaft.

To renew the seal, first remove clutch, flywheel and flywheel housing. If wear ring on crankshaft is damaged, spread the ring using a dull chisel on sealing surface, and withdraw the ring. Install new ring with rounded edge to rear, being careful not to cock the ring. Ring should start by hand and can be seated using a suitable driver such as JD-251. Install seal in flywheel housing working from rear with seal lip to front. When properly installed, seal should be flush with rear of housing bore.

FLYWHEEL

All Models

37. To remove flywheel, first remove clutch as outlined in paragraph 70 through 74, then unbolt and remove flywheel from its doweled position on crankshaft.

To install a new flywheel ring gear, heat to approximately 300 deg. F and install with chamfered end of teeth toward front of flywheel.

The clutch pilot bearing is factory-sealed and cannot be lubed. Turn bearing by sticking a finger into inner race and checking for rough or dry operation. Renew any bearing that is questionable.

When installing flywheel, tighten the retaining cap screws to 85 ft.-lbs. torque.

FLYWHEEL HOUSING

All Models

38. The cast iron flywheel housing is secured to rear face of engine block by eight cap screws. Flywheel housing contains the crankshaft rear oil seal and oil pressure sending unit switch. Flywheel must be removed for access to flywheel housing cap screws. The rear camshaft bore in block is open and the tachometer drive passes through flywheel housing. It is important therefore, that gasket between block and flywheel housing be in good condition and cap screws properly tightened. Tighten all screws evenly to 23 ft.-lbs., then retorque to 35 ft.-lbs. Tighten flywheel cap screws to 85 ft.-lbs. torque.

OIL PUMP

Models 2240-2440

39. To remove oil pump, first drain and remove oil pan. Remove timing gear cover as outlined in paragraph 21. On 2440 models, the left balance shaft is driven by the oil pump gear, so the engine should be set at TDC before removing the oil pump. On all models, loosen the nut retaining oil pump drive gear, tap nut with a hammer to free gear from shaft, and pull the gear; then unbolt and remove the oil pump.

With pump removed, use Fig. 25 as a guide and proceed as follows: Remove idler gear (10) and drive gear and shaft (5) from pump housing. Check to see that groove-pin (4) is tight in gear and drive shaft. Pin (4) can be renewed if necessary. Check bearing O.D. of drive shaft (5) and I.D. of housing. Shaft should have a normal operating clearance of 0.001-0.003 inch in housing bore and if clearance exceeds 0.007 inch, renew drive shaft, pump gear and pin as an assembly.

Idler gear shaft (8) can be pressed from pump housing if renewal is necessary. Diameter of new idler shaft is 0.485-0.486 inch.

Install gears and shafts in pump body as shown in Fig. 26 and measure between ends of gear teeth and pump body. This clearance should be 0.001-

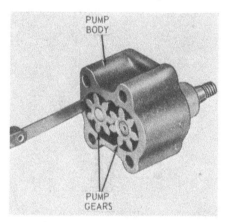

Fig. 26—Radial clearance between pump gears and housing should be measured as shown.

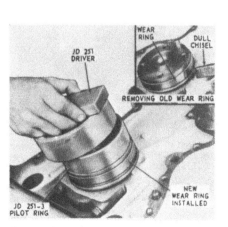

Fig. 24—Installing crankshaft rear oil seal using John Deere tool JD-251. Refer to text.

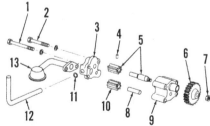

Fig. 25—Exploded view of engine oil pump used on 2240 and 2440 engines. See Fig. 14 for relief valve.

1. Cap screw
2. Cap screw
3. Pump cover
4. Pin
5. Drive shaft & gear
6. Drive gear
7. Nut
8. Idler shaft
9. Housing
10. Idler gear
11. "O" ring
12. Tube
13. Intake screen

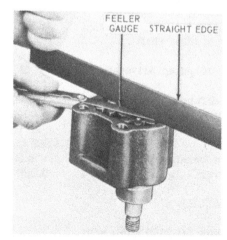

Fig. 27—Axial (end) clearance of gears in pump housing is measured as shown.

0.004 inch. Wear limit between gears and housing is 0.005 inch. Now place a straightedge across pump body as shown in Fig. 27 and measure between straightedge and end of pump gears. This clearance should be 0.001-0.006 inch and the wear limit is 0.008 inch. If either of these clearances are excessive, renew parts as necessary. Tighten pump cover and mounting bolts to 35 ft.-lbs. torque. Tighten the oil pump drive gear retaining nut to 35 ft.-lbs. torque and lock in position by staking.

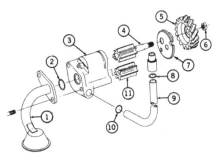

Fig. 28—Exploded view of engine oil pump used on 2630 and 2640 engines. See Fig. 30 for relief valve.

1. Intake tube	6. Nut
2. "O" ring	7. Pump cover
3. Housing	8. "O" ring
4. Drive gear & shaft	9. Discharge tube
5. Pump drive gear	10. "O" ring

Fig. 29—View showing location of the 2240 and 2440 engine oil pressure relief valve seat. Seat is renewable; refer to text. Also see Fig. 14.

Fig. 30—Remove regulating (relief) valve as an assembly on 2630 and 2640 engines.

On 2440 models, it is necessary to time the left side balancer as outlined in paragraph 23 when installing the oil pump.

Models 2630-2640

40. To remove oil pump, first drain and remove oil pan. Remove timing gear cover as outlined in paragraph 21, then set engine at TDC. Loosen the nut retaining oil pump drive gear, tap nut with a hammer to free gear from shaft and pull the gear; then unbolt and remove the oil pump.

With pump removed, use Fig. 28 as a guide and proceed as follows: Remove intake tube and screen (1) and separate pump housing (3) and cover (7). Inspect gears, cover, idler shaft in housing and housing for wear, nicks or grooves. If idler shaft or housing (3) is damaged, both must be renewed. I.D. of gear bore in pump should be 1.5340-1.5350 inches. Clearance between gears and housing, measured as shown in Fig. 26, should be 0.003-0.006 inch. Check cover (7) for score marks and renew if damaged.

When reassembling, renew "O" rings (2, 8 and 10), make sure gears rotate freely in housing and tighten mounting bolts to 35 ft.-lbs. torque. Time the left hand balancer shaft as outlined in paragraph 23. Install oil pump drive gear and tighten nut to 35 ft.-lbs. torque.

RELIEF VALVE

Models 2240-2440

41. Engine oil pressure relief valve is located in timing cover as shown in Fig. 14. With timing cover removed, inspect valve seat bushing in cylinder block for damage. Renew bushing (Fig. 29) if necessary, using tool JD-248 or equivalent tool that will bear only on outer diameter of seat, and press valve

seat into cylinder block until outer recessed edge is flush with bottom of counterbore. Do not damage valve seat surface, which is the raised inner rim of bushing.

Relief spring free length should be 4.7 inches. When compressed to a length of 1.68 inches, spring should have 13.5-16.5 lbs. tension.

Models 2630-2640

42. Engine oil pressure relief valve is located behind the engine oil cooler housing, on right front of engine as shown in Fig. 30. Drain cooling system and remove oil cooler and oil filter assembly. Pull valve from engine as an assembly. Drive spring pin (4—Fig. 31) from outer housing (5) and remove valve and spring. Valve (7) should be smooth and free. Spring should have 20.45-25.05 lbs. tension when compressed to a length of 1.34 inches. Reassemble valve and spring into housing and reinstall spring pin. Install assembly into engine with spring pin end out.

ENGINE OIL COOLER AND BY-PASS VALVE

Models 2630-2640

43. The engine oil cooler-filter assembly is mounted on right front of engine and includes the oil cooler by-pass valve (16—Fig. 31). Cooler assembly must be removed for cleaning when engine oiling system is cleaned, or for access to pressure relief valve. Cooler need not be removed when cleaning cooling system, since cooler will be cleaned along with a cooling system clean-out.

Drain cooling system and remove oil cooler and filter assembly. Remove end caps (9 and 14) and filter bracket (3). Check cooler core (10) for clogged

Fig. 31—Exploded view of engine oil cooler, filter and relief valve used on 2630 and 2640 models.

1. Filter
2. Nipple
3. Bracket
4. Spring pin
5. Valve housing
6. Spring
7. Relief valve
8. Coolant return hose
9. Cap
10. Cooler core
11. Cooler housing
12. "O" ring
13. "O" ring
14. Cap
15. Spring
16. Cooler by-pass valve
17. Seat
18. Coolant inlet hose

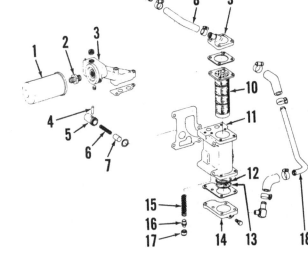

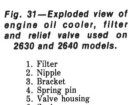

tubes or damaged fins. Clean passages with a stiff brush and compressed air. By-pass valve seat (17) can be removed with a blind-hole puller. Valve (16) and spring (15) should show no damage. 2630 model spring should have 7.8-9.6 lbs. tension when compressed to a length of 1.460 inches. 2640 model spring should have 15.1-20.1 lbs. tension at a length of 1.62 inches. Drive valve seat (17) into housing until it bottoms. Tighten mounting cap screws to 35 ft.-lbs. torque.

AIR INTAKE SYSTEM

All Models

44. Intake air enters through an air cleaner, which may contain a single element or a dual element.

The single, or outer (primary) air cleaner element should be cleaned and reused. A new single or outer air cleaner element should be installed after cleaning old element six times, if element is damaged or at least once each year.

The inner, safety (or secondary) element should be renewed, once each year, when clogged, if damaged or every third change of the primary (outer) filter element, whichever occurs first. Do not remove the secondary air filter element unless a new element is to be installed. Do not attempt to clean the secondary air cleaner element.

Tractor may be equipped with a restriction indicator, which will show a red signal whenever restriction reaches an excessive level. If tractor is not equipped with a restriction indicator, clean element every 200 hours or whenever excessive smoke level or a power loss is evident.

FUEL LIFT PUMP

R&R AND OVERHAUL

All Models

45. Refer to Fig. 33 for view of the

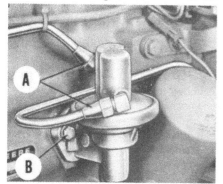

Fig. 33—Installed view of fuel lift pump typical of type used on all models.

A. Connectors   B. Mounting cap screw

Airtex lift pump. The pump should maintain 3.5-4.5 psi pressure at slow idle speed. Individual repair parts are not available, so pump must be renewed as a complete assembly.

Close fuel shut-off valve at tank and disconnect fuel lines at pump. On 2630 models, remove oil dipstick and tube from engine before removing pump.

When pump is reinstalled on engine, bleed fuel system as outlined in paragraph 47.

DIESEL FUEL SYSTEM

FILTERS AND BLEEDING

All Models

46. **FILTERS.** Tractors are equipped with single two-stage fuel filter and sediment bowl units. Renew the filter assembly if necessary, by releasing the retaining spring clip from around filter.

Check the transparent sediment bowl for sediment or water, and if necessary, loosen drain plug and operate priming lever of fuel transfer pump to clear deposits from sediment bowl.

47. **BLEEDING.** Whenever fuel system has been run dry, or a line has been disconnected, air must be bled from fuel system as follows: Be sure there is sufficient fuel in tank and that tank outlet valve is open. Loosen bleed screw on top of filter (Fig. 34) and actuate primer lever of fuel transfer pump until a solid, bubble-free stream of fuel emerges from bleed screw, then tighten bleed screw. Loosen pressure line connections at injectors about one turn, open throttle and crank engine until fuel flows from loosened connections, then tighten connections and start engine.

NOTE: If no resistance is felt when operating priming lever of fuel transfer pump and no fuel is pumped, the transfer pump rocker arm is on the high point of pump cam of camshaft. In this case, turn engine to reposition pump cam and release pump rocker arm.

INJECTOR NOZZLES

All Models

WARNING: Fuel leaves the injection nozzles with sufficient force to penetrate the skin. When testing, keep your person clear of the nozzle spray.

48. **TESTING AND LOCATING A FAULTY NOZZLE.** If one engine cylin-

der is misfiring it is reasonable to suspect a faulty injector. Generally, a faulty injector can be located by running the engine at low idle speed and loosening, one at a time, each high pressure line at injector. As in checking spark plugs in a spark ignition engine, the faulty unit is the one that least affects the engine operation when its line is loosened.

Remove the suspected injector as outlined in paragraph 49. If a suitable nozzle tester is available, test injector as outlined in paragraphs 50 through 54 or install a new or rebuilt unit.

49. **REMOVE AND REINSTALL.** To remove an injector, remove hood and wash injector, lines and surrounding areas with clean diesel fuel. If an injector near the alternator is being removed, disconnect battery ground cable to prevent a short circuit through tools. Pull leak-off hose from injector. Disconnect high pressure line, then cap all openings. Remove cap screw from nozzle clamp and remove clamp and spacer. Pull injector from cylinder head.

NOTE: Unless the carbon stop seal has failed causing injector to stick, the injectors can be easily removed by hand. If injectors cannot be removed by hand, use John Deere nozzle puller JDE-38 and be sure to pull injector straight out of bore. DO NOT attempt to pry injector from cylinder head or damage to injector could result.

When installing injector, be sure nozzle bore and seal washer seat are clean and free of carbon or other foreign material. Install new seal washer and carbon seal on injector (see Fig. 38) and insert injector into its bore using a slight twisting motion. Install and align locating clamp then install hold-down clamp and spacer and tighten cap screw to 20 ft.-lbs. torque. Bleed fuel system as outlined in paragraph

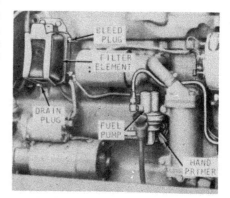

Fig. 34—View of two-stage filter and sediment bowl assembly typical of type used on all models.

47. Tighten pressure line connector to 35 ft.-lbs. torque.

50. **TESTING.** A complete job of nozzle testing and adjusting requires the use of an approved nozzle tester. Only clean, approved testing oil should be used in the tester tank. The nozzle should be tested for spray pattern, opening pressure, seat leakage and back leakage (leak-off). Injector should produce a distinct audible chatter when being tested and cut off quickly at end of injection with a minimum of seat leakage.

NOTE: When checking spray pattern, turn nozzle about 30 degrees from vertical position. Spray is emitted from nozzle tip at an angle to the centerline of nozzle body and unless injector is angled, the spray may not be completely contained by the beaker. Keep your person clear of the nozzle spray.

51. **SPRAY PATTERN.** Attach injector to tester and operate tester at approximately 60 strokes per minute and observe the spray pattern. A finely atomized spray should emerge at each nozzle hole and a distinct chatter should be heard as tester is operated. If spray is not symmetrical and is streaky, or if injector does not chatter, overhaul injector as outlined in paragraph 55.

52. **OPENING PRESSURE.** The correct opening pressure is 3000 psi for used injectors and 3200 psi for new ones, or when new spring is installed.

If opening pressure is not correct but nozzle will pass all other tests, adjust opening pressure as follows: Loosen the pressure adjusting screw locknut (4—Fig. 35), then hold the pressure adjusting screw (6) and back out the valve lift adjusting screw (5) at least one full turn. Actuate tester and adjust nozzle pressure by turning adjusting screw as required. With the correct nozzle opening pressure set, gently turn the valve lift adjusting screw in until it bottoms, then back it out ½ turn, to establish a nominal valve lift of 0.090 inch.

Hold pressure adjusting screw and tighten locknut to 70-75 in.-lbs. torque.

NOTE: A positive check can be made to see that the lift adjusting screw is bottomed by actuating tester until a pressure of 250 psi above nozzle opening pressure is obtained. Nozzle valve should not open.

53. **SEAT LEAKAGE.** To check nozzle seat leakage, proceed as follows: Attach injector on tester in a horizontal position. Raise pressure to approximately 2600 psi, hold for 10 seconds and observe nozzle tip. A slight dampness is permissible but should a drop

form in the 10 seconds, renew the injector or overhaul as outlined in paragraph 55.

54. **BACK LEAKAGE.** Attach injector to tester with tip slightly above horizontal. Raise and maintain pressure at approximately 1500 psi and observe leakage from return (top) end of injector. After first drop falls, the back leakage should be 3 to 10 drops every 30 seconds. If back leakage is excessive, renew injector or overhaul as outlined in paragraph 55.

55. **OVERHAUL.** First clean outside of injector thoroughly. Place nozzle in a holding fixture and clamp the fixture in a vise. NEVER tighten vise jaws on nozzle body without the fixture. Refer to Fig. 35. Loosen locknut (4) and back out pressure adjusting screw (6) containing lift adjusting screw (5). Slip nozzle body from fixture, invert the body and allow spring seat (9) and spring (8) to fall from nozzle body into your hand. Catch nozzle valve (10) by its stem as it slides from body. If nozzle valve will not slide from body, use the special retractor (16481) as shown in Fig. 36; or reinstall on nozzle tester with spring and lift adjusting screw removed, and use hydraulic pressure to remove the valve.

NOTE: DO NOT use wire brush or other abrasive on the Teflon coating on outside of nozzle body between the seals. Teflon coating can be cleaned with a soft cloth and solvent. Coating may discolor from use, but discoloration is not harmful.

Nozzle valve and body are a matched set and should never be intermixed. Keep parts for each injector separate and immerse in clean diesel fuel in compartmented pan as injector is disassembled.

Clean all parts thoroughly in clean diesel fuel using a brass wire brush. Hard carbon or varnish can be loosened

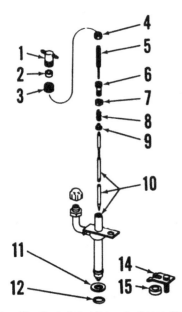

Fig. 35—Exploded view of Roosa-Master injector used on all models.

1. Leak-off cap	8. Spring
2. Grommet	9. Spring seat
3. Nut	10. Valve & body assy.
4. Lift screw nut	11. Upper washer
5. Lift screw	12. Nozzle seal
6. Pressure screw	14. Clamp
7. Nut	15. Spacer

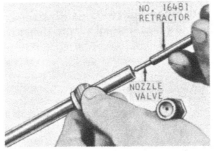

Fig. 36—Use the special retractor as shown, to remove a sticking nozzle valve.

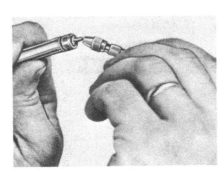

Fig. 37—Using a pin vise and cleaning needle to clean spray tip. Use a 0.008 diameter needle to open holes. Final cleaning should be done with a needle 0.001 inch smaller than orifice.

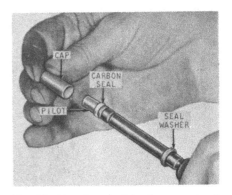

Fig. 38—Use the special pilot or a nozzle storage cap when installing a new carbon seal.

with a suitable non-corrosive solvent.

Clean the spray tip orifices using a pin vise (Fig. 37) and the appropriate size cleaning needle. Model 2440 nozzles have 0.011-inch diameter orifices, 2240, 2630 and 2640 model nozzles have 0.012 orifices. Finish cleaning operation with a needle 0.001 inch smaller than orifice.

Clean the valve seat using a Valve Tip Scraper and light pressure. Use a Sac Hole Drill to remove carbon from inside of tip.

Piston area of valve can be lightly polished by hand if necessary, using Roosa-Master no. 16489 lapping compound. Use the valve retractor (Fig. 36) to turn valve. Move valve in and out slightly while turning but do not apply down pressure while valve tip is in contact with seat.

Valve and seat are ground to a slight interference angle. Seating areas may be cleaned up if necessary using a small amount of 16489 lapping compound, very light pressure and no more than three to five turns of valve on seat. Thoroughly flush all compound from valve body after polishing.

When assembling, back out lift adjusting screw (5), and reverse the disassembly procedure using Fig. 35 as a guide. Adjust opening pressure and valve lift as outlined in paragraph 52 after valve is assembled.

INJECTION PUMP

Roosa-Master JDB injection pumps are used on 2440, 2630 and 2640 models; C.A.V. Roto-Diesel injection pumps are used on 2240 models. Be sure to refer to the appropriate following paragraphs for service.

Injection pump service demands the use of specialized equipment and special training which is beyond the scope of this manual. This section therefore, will cover only the information required for removal, installation and field adjustment of the injection pump.

Models 2440-2630-2640

56. REMOVE AND REINSTALL. To remove the injection pump, first shut off fuel and clean injection pump, lines and surrounding area. Turn engine until No. 1 piston is at TDC on compression stroke, using timing pin in flywheel.

NOTE: Pump can be removed and reinstalled without regard to crankshaft timing position, however, TDC-1 position is necessary if timing is to be checked. If timing is not to be changed, scribe timing marks on pump flange and engine front plate which can be realigned when pump is reinstalled.

Disconnect or remove fuel inlet, return and pressure lines, throttle rod and solenoid wire from injection pump. Remove mounting stud nuts and carefully slide pump straight to rear until clear of pump shaft and seals.

The pump shaft contains two soft plastic seals which are installed back to back as shown in Fig. 39. A special tool (Roosa-Master 13369) is required to install seals as shown. Only the rear seal can be installed without removing shaft from drive gear. If both seals must be renewed, drain cooling system and remove lower radiator hose. Remove access plate from front of timing gear cover, back gear nut out until flush with end of shaft, carefully bump shaft from drive gear, then withdraw shaft after removing the nut. Gear will remain in engagement with idler gear if timing gear cover is not removed.

To install new seals, first examine seal grooves in shaft carefully and remove any roughness or burrs. Coat seal liberally with Lubriplate and install from each end of shaft using the special installing tool. If shaft is removed and both seals renewed, reinstall shaft in pump before installing pump on engine. Reference mark (dot)

on shaft tang and pump slot must align as shown in inset, Fig. 40.

The special Seal Installation Tool (Roosa-Master 13371 or equivalent) must be used when installing pump (or shaft in pump). Also use extreme care. If resistance is felt, remove the pump and re-examine rear seal. If lip has been turned back, renew the seal.

If pump is not to be timed, realign the previously installed scribe marks and tighten stud nuts, then complete the assembly by reversing the removal procedure.

To time the pump, first be sure that crankshaft is at TDC-1. Remove timing cover from side of pump housing and turn the pump until governor weight timing line and cam timing line are in register as shown in Fig. 41. Tighten retaining cap screws to a torque of 35 ft.-lbs.

Bleed fuel system as outlined in paragraph 47 and if necessary, adjust throttle linkage as in paragraph 59.

57. TIMING. To check the timing without removing injection pump, turn crankshaft until No. 1 piston is coming up on compression stroke, then remove engine timing pin and cover as shown in inset, Fig. 41. Insert timing pin, long end first, into threaded hole in housing and continue turning crankshaft until pin slides into timing hole in flywheel.

Shut off fuel and remove timing cover from side of injection pump. With crankshaft at TDC-1, the timing scribe line on governor weight retainer should align with cam timing line as shown. If it does not, loosen pump mounting stud nuts and rotate pump in slotted holes until scribe lines are aligned. Hold pump in this position and retighten mounting stud nuts.

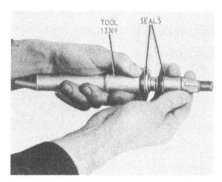

Fig. 39—A special tool is required to properly install shaft seals as shown.

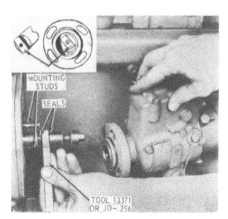

Fig. 40—On Roosa-Master injection pumps, align reference marks (inset) and use the special seal tool when installing pump.

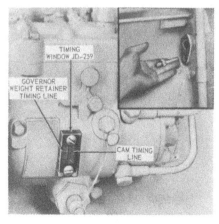

Fig. 41—With engine at TDC-1 when checked with timing pin (inset), governor weight retainer timing line and cam timing line must register as shown. Timing window need only be installed to check intermediate advance as shown in Fig. 42.

57A. ADVANCE TIMING. The injection pump is provided with automatic speed advance which is factory set and will not normally need to be checked or reset. Minor adjustments can, however, be made without removal or disassembly of the pump. To check the advance mechanism, proceed as follows:

Shut off fuel, remove pump timing hole cover and install timing window as shown in Fig. 42. Turn on fuel and bleed fuel system, then start and run engine at high idle speed.

Total advance should be as follows:
2440 . . . . .7°
2630 . . . . .7½°
2640 . . . . .6°

If total advance is not correct at high idle speed, renew or overhaul pump. If maximum advance was correct, reduce engine speed to 1100 rpm. On all models, advance should be four degrees. If necessary, loosen locknut and adjust trimmer screw as shown in Fig. 42. Tighten locknut and reinstall seal cap and wire, then remove timing window and reinstall timing hole cover.

Models 2240

58. REMOVE AND REINSTALL. To remove the injection pump, first shut off fuel and clean injection pump, lines and surrounding area. Pump can be removed and reinstalled without regard to crankshaft timing position and timing cannot be checked. The only critical requirement of the timing process is correct positioning of the injection pump gear as shown in Fig. 20.

Disconnect or remove fuel inlet, return and pressure lines, throttle rod and stop cable from injection pump. Drain radiator and remove lower radiator hose, then remove access plate from front of timing gear cover. Remove the three cap screws attaching drive gear to injection pump flange. Support pump, remove the three mount-

ing stud nuts and pull injection pump from timing gear housing. Pump drive gear will be retained by timing gear cover.

When reinstalling pump, turn pump hub until timing slot in hub flange aligns with dowel pin in gear, and reverse removal procedure. Tighten gear mounting cap screws and flange mounting stud nuts to a torque of 18 ft.-lbs. Bleed system as outlined in paragraph 47 and adjust linkage as in paragraph 59A.

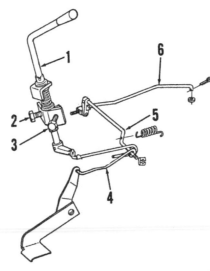

Fig. 43—Schematic view showing throttle linkage adjustments, typical of early type. Foot throttle model shown.

1. Hand lever
2. Slow idle stop screw
3. Fast idle stop screw
4. Foot throttle yoke
5. Cross shaft
6. Control rod

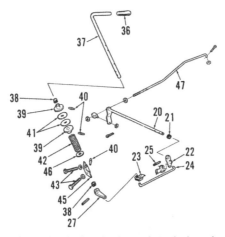

Fig. 43A—Exploded view of typical early throttle linkage. Refer to Fig. 43 for schematic view.

20. Control shaft	38. Bushing
21. Bushing	39. Disc
22. Control arm	40. Groove pin
23. Clevis clip	41. Facings
24. Control rod (rear)	42. Spring
25. Spring pin	43. Stop screws
27. Arm	45. Stop
36. Knob	46. Washer
37. Control lever	47. Control rod (front)

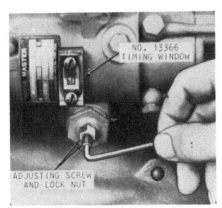

Fig. 42—Adjusting intermediate advance on Roosa-Master injection pumps.

SPEED AND LINKAGE ADJUSTMENT

Models 2440-2630-2640

59. To adjust engine high and low idle speeds and the control linkage, proceed as follows: Start engine and bring to operating temperature. Disconnect control rod (6—Fig. 43) from injection pump, move throttle arm to high idle position (against override spring) and check the engine high idle speed which should be 2650 rpm for tractors without a foot throttle, or 2800 rpm for tractors with a foot throttle. If engine high idle is not as stated, break seal on pump high idle adjusting screw, turn screw as required and reseal. Now slowly move pump throttle lever to low idle position and check engine speed which should be 800 rpm. If engine low idle is not as stated, adjust pump low idle screw as required. If minor adjustments to high and low idle adjusting screws do not obtain correct speeds, pump must be renewed or overhauled.

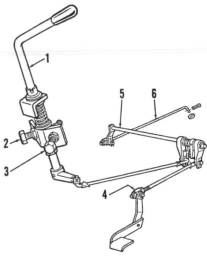

Fig. 43B—Schematic of throttle linkage of late models. Refer to Fig. 43 for legend.

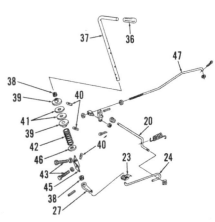

Fig. 43C—Exploded view of throttle linkage used on late models. Refer to Fig. 43B for schematic view. Refer to Fig. 43A for legend.

With engine speeds adjusted at injection pump, reconnect control rod to pump throttle lever. Raise cowl top door, remove right side cowl panel and adjust linkage as follows: Locate the stop screws (2 and 3—Fig. 43) on lower end of throttle hand lever. Move throttle hand lever to low idle position, observe low idle speed and make sure idle stop screw (2) is against dash. If low idle speed is not 800 rpm, loosen locknut, run stop screw in one full turn and retighten locknut. This will make sure linkage moves pump arm to slow idle before set screw contacts dash.

If tractor does not have a foot throttle, turn high idle screw (3) in several turns, then move hand throttle lever to high idle position. Continue to pull hand lever down until top of injection pump throttle lever has about 1/16-inch over-travel. Keep hand lever in this position, back out the high idle screw until it contacts stop and tighten jam nut.

If tractor is equipped with a foot throttle, adjust linkage as follows: Depress foot throttle against platform and adjust foot pedal rod length until upper end of injection pump throttle lever has about 1/16-inch over-travel. Tighten the clevis jam nut. Place hand throttle lever in high idle position and adjust idle screw (3) to obtain the recommended high idle speed of 2650 rpm, then tighten jam nut.

Models 2240

59A. To adjust diesel engine speeds and control linkage, start engine and run until operating temperature is reached. Disconnect speed control rod from injection pump and move throttle lever (3—Fig. 44A) against fast idle adjusting screw (2). Engine speed should be 2650 rpm. If it is not, turn fast idle screw (2) in or out as required. Move pump throttle lever against slow idle adjusting screw (1); engine speed should be 650 rpm. If it is not, adjust slow idle screw. Reconnect throttle control rod (6—Fig. 43 or Fig. 43B) to pump and adjust stop screws (2 and 3) as required until pump throttle lever (3—Fig. 44A) contacts fast and slow stop screws (1 and 2). If foot throttle is used, adjust foot throttle linkage until pedal pad contacts footrest at the same time pump throttle lever (3) contacts fast idle screw (2).

To adjust shut-off cable, completely push in stop knob and check to be sure stop lever contacts top on injection pump governor cover. If it does not, loosen cable clamp screw and reposition clamp on stop cable.

COOLING SYSTEM

RADIATOR

All Models

60. All models have an oil cooler attached to right side of radiator as shown in Fig. 45. Removal procedure for all models is similar.

61. **REMOVE AND REINSTALL.** To remove radiator, first drain cooling system, then remove grille screens and hood. Remove air intake tube. Remove fan shroud from radiator and lay shroud back over fan. Disconnect injector leak-off line from fuel tank. Disconnect radiator from oil cooler. Disconnect upper and lower radiator hoses and the upper radiator brace from radiator, then unbolt and remove radiator from tractor.

WATER PUMP

All Models

62. **REMOVE AND REINSTALL.** To remove water pump, first remove radiator as outlined in paragraph 61, then remove fan and fan belt. Disconnect by-pass hose from water pump, then unbolt and remove water pump from engine.

Reinstall by reversing removal procedure and adjust fan belt so a 20 lb. force mid-way between pulleys will deflect belt ¾-inch.

63. **OVERHAUL.** To disassemble

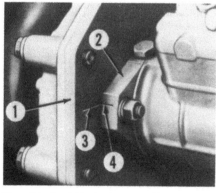

Fig. 44—Installed view of Roto Diesel injection pump showing timing marks aligned. Pump will be in time with engine if gear timing marks are also in line.

1. Front plate	3. Engine timing mark
2. Pump flange	4. Pump timing mark

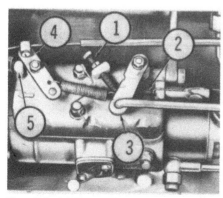

Fig. 44A—Installed view of Roto Diesel pump showing linkage adjustments.

1. Slow idle screw
2. Fast idle screw
3. Pump throttle arm
4. Stop lever
5. Lever stop

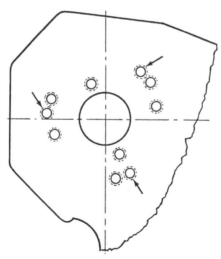

Fig. 44B—The studs for mounting the Roto Diesel injection pump should be installed in only the holes indicated above by arrows on 2240 model tractors.

Fig. 45—Hydraulic oil cooler (OC) is located at side of coolant radiator as shown.

water pump, use Fig. 46 as a guide and proceed as follows:

Support fan pulley hub in a press and, using a suitable mandrel, press shaft from pulley hub. Suitably support housing on gasket surface and press shaft, bearing seal and impeller as a unit from housing.

Bearing outer race is a tight press fit in housing bore and is not otherwise secured. It is important therefore, that reasonable precautions be taken during assembly to prevent bearing movement during installation of impeller and pulley hub.

Coat outer edge of seal (6) with sealant and install in housing (4) using a socket or other driver which contacts only the outer flange of seal. Large I.D. of seal must be installed toward impeller. Next, insert long end of bearing (5) through front of housing bore. Use tool no. JD-262 or a similar tool which contacts only outer race of bearing and press bearing into housing until front of bearing is flush with housing bore.

Install insert (7) in cup (8) with "V" groove of insert toward cup. Parts must be clean and dry. Dip cup and insert in engine oil then press cup and insert into impeller as shown in inset, until cup bottoms in impeller counterbore. Support front end of pump shaft

and press impeller (9) on rear of shaft until fins are flush to 0.010 inch below gasket surface of housing.

Invert the pump assembly and support the unit on rear of shaft which is recessed into impeller hub. DO NOT support impeller or housing. With shaft suitably supported, press pulley hub (3) on front of shaft until front face of pulley hub is flush with end of shaft. Complete the assembly by reversing the disassembly procedure.

ELECTRICAL SYSTEM

ALTERNATOR AND REGULATOR

Motorola Alternator

64. Refer to Fig. 47 for an exploded view of Motorola alternator unit used on 2440, 2630 and 2640 models.

The isolation diode and brush holder can be renewed without removal or disassembly of alternator; all other alternator service requires disassembly.

The primary purpose of the isolation diode is to permit use of charging indicator lamp. Failure of the isolation

diode is usually indicated by the indicator light, which glows with engine stopped and key switch off if diode is shorted, or with engine running if diode is open.

Failure of a rectifying diode may be indicated by a humming noise when engine is running, if diode is shorted; or by a steady flicker of charge indicator light at slow idle speed if diode is open. Either fault will reduce alternator output.

To check the charging system, refer to Fig. 47A and proceed as follows:

(1). With key switch and all accessories off and engine not running, connect a low reading voltmeter to terminals D-F. Reading should be 0.1 volt or less. A higher reading would indicate a short in isolation diode, key switch or wiring.

(2). Turn key switch on but do not start engine. Recheck voltmeter reading which should be 1-3 volts. A higher or lower reading may indicate a defective alternator, regulator or wiring.

(3). Start and run engine at approximately 1500 rpm and, with all accessories off, again check voltmeter reading which should be 15 volts. A lower reading could indicate a discharged battery or defective alternator.

(4). Move voltmeter lead from auxiliary terminal (D) to output terminal

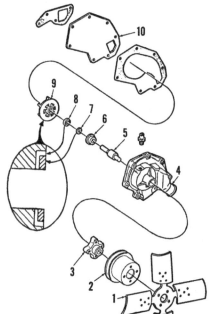

Fig. 46—Exploded view of typical water pump showing component parts.

1. Fan blades	6. Seal
2. Pulley	7. Insert
3. Hub	8. Cup
4. Body	9. Impeller
5. Shaft & bearing	10. Cover

Fig. 47—Exploded view of MOTOROLA alternator unit showing component parts.

1. Nut
2. Pulley
3. Fan
4. Spacer
5. Drive end frame
6. Bearing
7. Snap ring
8. Rotor
9. Bearing
10. Stator
11. Negative heat sink
12. Positive heat sink
13. Insulators
14. Retainer
15. Slip ring end frame
16. Brush holder
17. Cover
18. Isolation diode

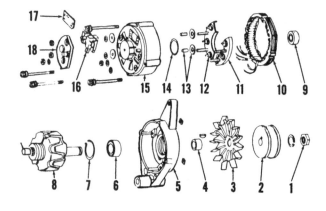

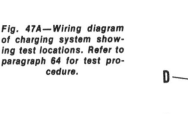

Fig. 47A—Wiring diagram of charging system showing test locations. Refer to paragraph 64 for test procedure.

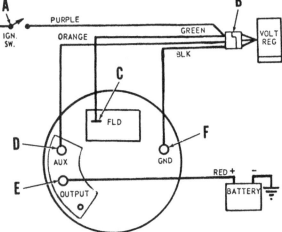

(E) and recheck voltage. Reading should drop one volt from reading in previous test (3), reflecting the resistance designed into isolation diode (18—Fig. 47). If battery voltage (12 volts) is obtained, isolation diode is open and must be renewed.

(5). If a reading lower than the specified 15 volts was obtained when checked as outlined in test 3, stop engine and disconnect regulator plug (B—Fig. 47A). Connect a jumper wire between output terminal (E) and field terminal (C) on alternator brush holder. Connect a suitable voltmeter to terminals (D-F) on alternator. Start engine and slowly increase engine speed while watching voltmeter. If a reading of 15 volts can now be obtained at 1500 engine rpm or less, renew the regulator. If a reading of 15 volts cannot be

obtained, renew or overhaul the alternator.

CAUTION: DO NOT allow voltage to rise above 16.5 volts when making this test. DO NOT run engine faster than 1500 rpm with regulator disconnected.

64A. **OVERHAUL.** The isolation diode and brush holder can be removed without removing alternator from tractor unit, remove through-bolts and attempting to separate the frame units.

To disassemble the removed alternator unit, remove through-bolts and separate slip ring end frame (15—Fig. 47) from drive end frame (5). Rotor (8) will remain with drive end frame and stator (10) with slip ring end frame. Be careful not to damage stator windings when prying units apart.

Examine slip ring surfaces of rotor for scoring or wear and field windings for overheating or other damage. Check bearing surfaces of rotor shaft for visible wear or scoring. Check rotor for grounded, shorted or open circuits using an ohmmeter as follows:

Refer to Fig. 47B and touch the ohmmeter probes to points (1-2) and (1-3); a reading near zero will indicate a ground. Touch ohmmeter probes to the two slip rings (2-3); reading should be 5.5 ohms. A higher reading will indicate an open field circuit, a lower reading a short.

Runout should not exceed 0.002. Slip ring surfaces can be trued if runout is excessive or if surfaces are scored. Finish with 400 grit or finer silicon carbide paper until scratches or machine marks are removed.

Stator is "Y" connected (Fig. 47C) and center connection need not be unsoldered to test for continuity. Each field winding uses two coils as shown. Continuity should exist between any two of the stator leads but not between any lead and stator frame. Because of the low resistance, shorted windings within a coil cannot be satisfactorily checked. Three positive diodes are located in slip ring heat sink (12—Fig. 47) and three negative diodes in grounded-heat sink (11). Diodes should test at or near infinity in one direction when checked with an ohmmeter, and at or near zero when meter leads are reversed. Renew any diode with approximately equal meter readings in both directions. Diodes must be removed and installed using an arbor press or vise and a suitable tool which contacts only outer edge of diode. Do not attempt to drive a faulty diode out of heat sink, as shock may cause damage to other good diodes. If all diodes are being renewed, make certain the positive diodes (marked with red printing) are installed in positive heat sink (12) and negative diodes (marked with black printing) are installed in negative heat sink (11). Use a pair of needle nose pliers as a heat sink when soldering diode leads (see Fig. 47D). Use only rosin core solder and an iron instead of a torch. Excess heat can damage a good diode while it is being installed.

Exposed length of brushes in removed brush holder should be ¼-inch or more. Brushes are available only in an assembly with the holder. Check for continuity between field terminal (A—Fig. 47E) and insulated brush (C); and between brush holder (B) and grounded

Fig. 47B—Removed rotor assembly showing test points to be used when checking for grounds, shorts and opens.

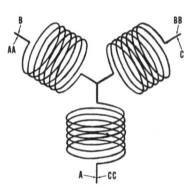

Fig. 47C—Schematic view of typical "Y" connected stator. Center connection need not be unsoldered to check for continuity.

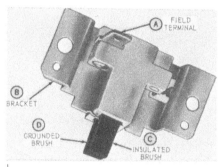

Fig. 47E—Check brush holder for continuity between A-C and B-D.

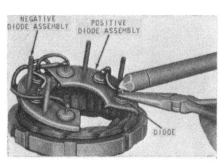

Fig. 47D—Use needle nose pliers as a heat sink, and an iron only, when soldering diode connections.

Fig. 49—Exploded view of Bosch alternator.

1. Brush holder
2. Rectifier
3. Stator
4. Rotor

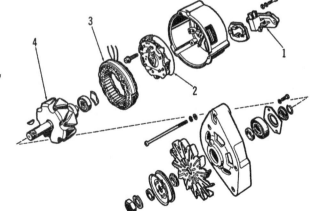
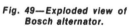

brush (D). Wiggle the brush and lead while checking, to test for poor connections or an intermittent ground.

NOTE: A battery powered test light can be used instead of an ohmmeter for all electrical tests except shorts in rotor windings. When checking diodes however, test light must not be more than 12 volts.

Bosch Alternator

65. OPERATION AND TESTING. A Bosch 12 volt, 28 ampere alternator with separate regulator is used on 2240 models.

To check the charging system, connect a voltmeter to B + (output) terminal (1—Fig. 49A) on alternator and to a suitable ground. With engine running at 1200 rpm, reading should be 13 volts or above. A lower reading could indicate a discharged battery or faulty alternator or regulator.

With engine not running, disconnect the three-terminal plug (2) from alternator and touch ammeter leads to DF (Green Wire) terminal and B + terminal. Current draw should be approximately 2 amperes. High readings are caused by shorts or grounds. Low readings may be caused by dirty slip rings or defective brushes.

Connect a jumper wire between B + and DF terminals and a voltmeter between B + terminal and ground. Start engine and increase engine speed until voltage rises not to exceed 15.5 volts. If maximum reading is less than 14 volts and battery is fully charged, alternator is defective.

Move jumper wire connection from B + terminal to D + terminal. Voltage reading should remain the same. If voltage drops, exciter diodes are defective. If voltage reading was normal with plug disconnected and jumper wire installed, but was low when tested with plug connected, renew the regulator.

65A. OVERHAUL. Brush holder (1—Fig. 49) can be removed without removing alternator from the tractor. Brush holder should be removed before attempting to separate alternator frame units.

To disassemble the removed alternator, first remove capacitor and brush holder. Immobilize pulley and remove shaft nut, pulley and fan. Mark brush end housing, stator frame and drive end housing for correct reassembly and remove through-bolts. Rotor will remain with drive-end frame and stator will remain with brush-end frame when alternator is disassembled.

Remove the two terminal nuts and three screws securing rectifier (2) to brush end frame and lift out rectifier

and stator (3) as a unit. Carefully tag the three stator leads for correct reassembly, then unsolder the leads from rectifier diodes using an electric soldering iron and minimum heat. Be careful not to get solder on diode plates or overheat the diodes.

Check brush contact surface of slip rings for burning, scoring or varnish coating. Surfaces must be true to within 0.002 inch. Contact surface may be trued by chucking in a lathe. Polish the contact surface after truing using 400 grit polishing cloth, until scratches and machine marks are removed. Check continuity of rotor windings using an ohmmeter as shown in Fig. 49C. Ohmmeter readings should be 4.0-4.4 between the two slip rings and infinity between either slip ring and rotor pole or shaft.

Stator is "Y" wound, the three individual windings being joined in the middle. Test the windings using an ohmmeter as shown in Fig. 49D. Ohmmeter reading should be 0.4-0.44 be-

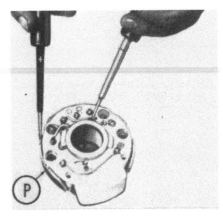

Fig. 49E—A near-infinity reading should be obtained when positive probe rests on positive heat sink (P) and negative probe touches diode leads as shown. Reverse the probes and reading should be near zero ohms.

Fig. 49A—Installed view of Bosch alternator.

1. Output terminal
2. Terminal plug

3. D+ terminal

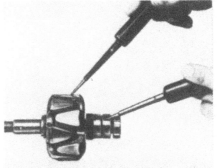

Fig. 49C—No continuity should exist between either slip ring and any part of rotor frame.

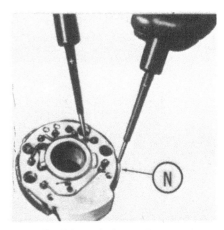

Fig. 49F—With negative probe on negative heat sink (N) and positive probe touching diode leads, an infinity reading should be obtained. Reverse the probes and reading should be near zero.

Fig. 49B—A reading of 4.0-4.4 ohms should exist between slip rings (1 and 2) when tested with an ohmmeter.

Fig. 49D—No continuity should exist between any stator lead and stator frame.

tween any two leads and infinity between any lead and stator frame.

Alternator brushes and shaft bearings are designed for 2000 hours service life. New brushes protrude 0.4 inch beyond brush holder when unit is removed; for maximum service reliability, renew **BOTH** the brushes and shaft bearings when brushes are worn to within 0.2 inch of holder. Solder copper leads to allow 0.4 inch protrusion using rosin core solder.

The rectifier is furnished as a complete assembly and diodes are not serviced separately. The rectifier unit contains three positive diodes, three negative diodes and three exciter diodes which energize rotor coils before engine is started. If any of the diodes fail, rectifier must be renewed.

To test the positive diodes, touch positive ohmmeter probe to positive heat sink as shown in Fig. 49E, and touch negative test probe to each diode lead in turn. Ohmmeter should read at or near infinity for each test. Reverse the leads and repeat the series; ohmmeter should read at or near zero for the series.

Test negative diodes as shown in Fig. 49F. Place negative test probe on negative heat sink and touch each diode lead in turn with positive test probe. Ohmmeter should read at or near infinity for the series. Reverse test leads and repeat the test; ohmmeter should read at or near zero for the series.

Test exciter diodes by using the D+ terminal as the base as shown in Fig. 49G. Ohmmeter should read at or near infinity with positive test probe on terminal screw and at or near zero with negative test probe touching screw.

When assembling alternator, tighten through-bolts to approximately 35-40 in.-lbs. and pulley nut to 25-30 ft.-lbs.

No load test:
Volts 11.5
Min. amps 60

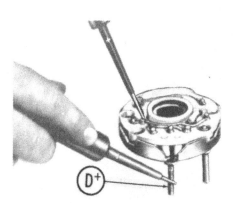

Fig. 49G—D+ terminal is used to test exciter diodes, refer to paragraph 65A.

Max. amps 90
Min. rpm 4800
Max. rpm 6800

STARTING MOTOR

All Models

66. Delco-Remy and Robert Bosch starting motors are used. Test specifications, including solenoid, are as follows:

D-R 1107863 and 1107871
Brush spring tension . . . . .35 oz.
    Brush length min. . . . . .3/8 in.
No load test:
    Volts 9.0
    Min. amps 40
    Max. amps 140
    Min. rpm 8000
    Max. rpm 13,000
    Pinion clearance . . . . . .0.010-0.140 in.

D-R 1109251
Brush spring tension . . . . .80 oz.
    Brush length—min. . . . . .3/8 in.
No load test:
    Volts 9.0
    Min. amps 20
    Max. amps 120
    Min. rpm 9000
    Max. rpm 14,000
    Pinion clearance . . . . . .0.010-0.140 in.

Robert Bosch
Brush spring tension . . . . .5.7-6 lbs.
    Brush length—min. . . . . .063 in.
Armature end play . . . . . .0.004-0.012 in.
Commutator—
    Max. out-of-round . . . . .0.0012 in.
    Min. diameter . . . . .1.555 in.
No load test:
    Volts .11.5
    Min. amps 60
    Max. amps 90
    Min. rpm 4800
    Max. rpm 6800

CIRCUIT DESCRIPTION

All Models

67. All models are equipped with a 12-volt electrical system with the negative battery post grounded. Tractor may have a single battery, or two batteries connected in parallel, which must have equal ratings.

Fig. 51 shows wiring connections to key switch. Fuel gage sender unit has a full tank resistance of 30 ohms and empty tank resistance of 0 ohms, with an even rate of increase as float lever is raised. A good ground must exist for both the sender and receiver units. Oil pressure indicator lamp switch should close at 5.5-10.5 psi.

Main circuit breaker is 20 amperes on all models and the electrical remote control circuit breaker is 25 amperes.

Refer to Fig. 52 for wiring diagram

for 2240 models and the following for description. Other models are slightly different.

1. Head lights
2. Combined rear work light and tail light
3. Warning lamps (tractors without canopy)
4. Socket (for hand lamp)
5. Connector, L.H.
6. Connector, R.H.
7. Connector
8. Flasher relay
9. Light switch
10. Key switch
11. Circuit breaker
12. Fuel gage
13. Engine oil pressure indicator light
14. Alternator indicator light
15. Air cleaner restriction indicator light
16. Batteries
17. Start safety switch
18. Engine oil pressure warning switch
19. Regulator
20. Starting motor
21. Alternator
22. Capacitor
23. Fuel gage sending unit
24. Air cleaner restriction warning switch
25. Turn signal switch (if so equipped)
26. Horn
27. Horn button
28. Resistor (0.7 Ohm) for low beam headlights
29. Front warning lamps (Tractors with canopy only)
30. Rear warning lamps (Tractors with canopy only)
31. Ground straps—Battery "-" to Ground
32. Black—Battery "+" to Starting motor (Terminal 30)

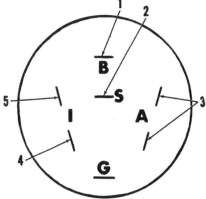

Fig. 51—Key switch viewed from terminal side, showing wiring connections. Terminals I are for ignition which is not used but wires are included in service harness. Tape free ends of wires and band to harness.

1. Red
2. Black
3. Purple or red
4. Resistance wire
5. Gray

33. Red—Starting motor (Terminal 30) to Alternator (Terminal B+)
34. White—Start safety switch to Starting motor (Terminal 50)
35. Blue—Alternator (Terminal D+) to Regulator (Terminal D+)
36. Brown—Alternator (Terminal D-) to Regulator (Terminal D-)
37. Green—Alternator (Terminal DF) to Regulator (Terminal DF)
38. Regulator (Terminal D+) to Alternator indicator light

40. Red—Alternator indicator light to Air cleaner restriction indicator light
41. Black—Air cleaner restriction warning switch to Air cleaner restriction indicator light
42. Red—L.H. battery (Terminal +) to Circuit breaker
43. Black—Fuel gage sending unit to Fuel gage (Terminal G)
44. Red—Fuel gage (Terminal +) to Engine oil pressure indicator light

45. Red—Engine oil pressure indicator light to Alternator indicator light
46. Red—Alternator indicator light to Key switch (Terminal ACC)
47. Black—Key switch (Terminal ST) to Start safety switch
48. Green—Engine oil pressure warning switch to Engine oil pressure indicator light
49. Red—Key switch (Terminal BAT) to circuit breaker

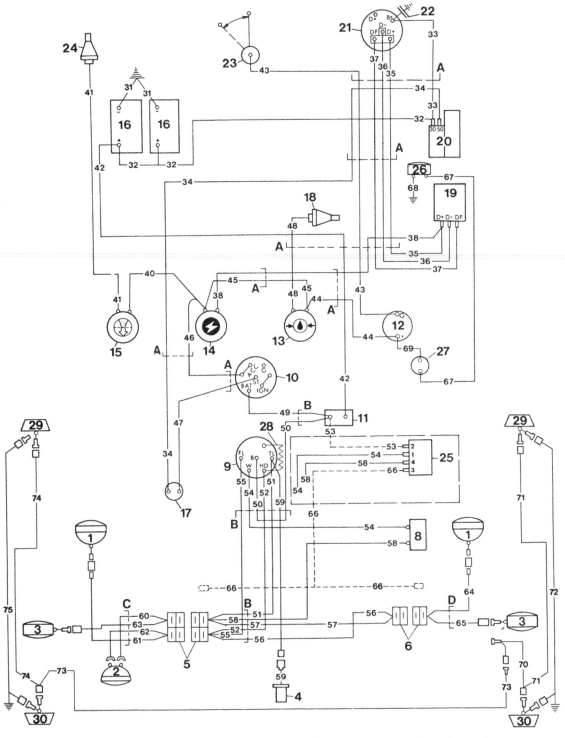

Fig. 52—Wiring diagram for 2240 model tractors. Refer to text for description and legend.

50. Red—Circuit breaker to Light switch (Terminal B)
51. Yellow—Light switch (Terminal TL) to L.H. connector
52. Pink—Light switch (Terminal HD) to L.H. connector
53. Red—Turn signal switch (Terminal 2) to Circuit breaker
54. Orange/white—Light switch (terminal W) to Flasher relay or turn signal switch (Terminal 1)
55. Blue—Light switch (Terminal FL) to L.H. connector
56. Pink—L.H. connector to R.H. connector
57. Orange—R.H. connector to L.H. connector
58. Orange—Flasher relay or turn signal switch (Terminal 4) to L.H. connector
59. Yellow—Light switch (Terminal TL) to Outlet socket
60. Yellow—L.H. connector to Tail light
61. Pink—L.H. connector to L.H. headlight
62. Blue—L.H. connector to Rear work light
63. Orange—L.H. connector to wire 66 to L.H. warning lamp
64. Pink—R.H. connector to R.H. headlight
65. Orange—R.H. connector to R.H. warning lamp or wire 70
66. Dark green—Turn signal switch (Terminal 3) to Wire 63 (Wire 63 disconnected from connector 5 and attached to wire 66 on models without canopy. Wire 66 attached to wire 73 on models without canopy.)
67. Black/yellow—Horn button to Horn
68. Brown—Horn to ground
69. Red—Horn button to Fuel gage + Terminal
70. Orange—From wire 65 to R.H. rear warning lamp
71. Orange—R.H. rear warning lamp to R.H. front warning lamp
72. Black—R.H. warning lamps to ground

73. Dark green—From wire 65 to L.H. rear warning lamp
74. Dark green—L.H. rear warning lamp to L.H. front warning lamp
75. Black—L.H. warning lamps to ground

ENGINE CLUTCH

ADJUSTMENT

Models With Reverser

68. PEDAL FREE PLAY. When clutch pedal is pushed part way down, clutch control valve lowers reverser system pressure, stopping tractor motion. When pedal is fully depressed, flywheel clutch is released stopping power take-off and reverser drive shaft.

To adjust the free play, proceed as follows: With engine stopped, depress clutch pedal through first stage, until

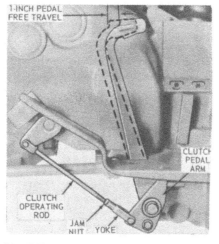

Fig. 54A—External clutch adjustment is made by disconnecting rod yoke from pedal arm and adjusting length of operating rod.

throwout bearing contacts clutch release levers, then measure distance from pedal pad to flywheel housing flange as shown in Fig. 54. Distance should be 5¼ inches for all models. Adjust by disconnecting rod yoke and shortening or lengthening operating rod as required. Any time distance exceeds 5¾ inches, adjustment should be performed.

With tractor in a forward gear and moving, clutch pedal should be pushed in ¾-1¼ inch before any drop occurs in clutch apply pressure. As pedal is depressed further, tractor motion should stop. If the ¾-1¼ free play is absent, check pressure as outlined in paragraph 89.

Other Models

68A. PEDAL FREE PLAY. Clutch pedal free play on all models with independent pto should be one inch and should be readjusted when free play decreases to ½ inch. Adjustment is made by disconnecting yoke (Fig. 54A) from clutch pedal arm and shortening or lengthening clutch operating rod as required.

On models with continuous pto and dual clutch, loosen pedal positioning cap screw (Fig. 54B) and move clutch pedal rearward until screw contacts front of slot. Tighten cap screw securely to maintain proper pedal position.

If so desired, the continuous running pto clutch can be locked out by loosening pedal positioning cap screw, depressing clutch pedal to place cap screw to rear of slot and retightening cap screw.

68B. RELEASE LEVERS (DUAL CLUTCH). The dual clutch fingers can be adjusted for wear without disassembly or removal of unit from tractor. Proceed as follows:

Disconnect clutch operating rod from clutch pedal arm (Fig. 54A) and refer

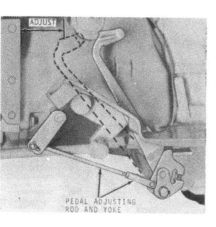

Fig. 54—Adjust clutch pedal free play to 5¼ inches on all reverser models by disconnecting yoke from pedal arm and changing length of rod.

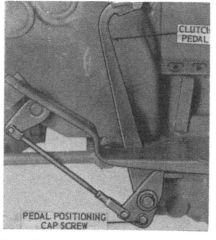

Fig. 54B—Pedal position is adjusted by loosening cap screw and shifting pedal in slotted hole.

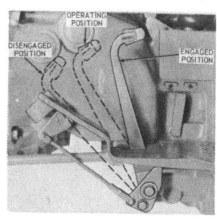

Fig. 54C—On dual clutch models, move pedal rearward until cap screw contacts front of slot to permit full disengagement of pto clutch.

to Fig. 55. Remove access cover from clutch housing and back off clutch operating bolt nuts until operating lever contacts pto clutch plate pins for all three operating levers. Tighten one jam nut until finger begins to pull away from pto clutch plate pin, tighten nut an additional 2½ turns and secure by tightening locknut. Adjust clutch operating rod (Fig. 54A) until, with yoke pin inserted, throwout bearing just contacts operating lever. Turn flywheel until each of the other clutch levers are in position and adjust the lever to lightly contact throwout bearing. Tighten all locknuts securely and adjust clutch pedal free travel as outlined in paragraph 68A.

TRACTOR SPLIT

All Models

69. To split engine from clutch housing, proceed as follows: Drain cooling system and remove muffler and hood. Remove front end weights, if used. Disconnect battery cables, remove battery (or batteries), then remove the two cap screws which retain cowl to flywheel housing. Remove side frame rails. Disconnect wiring from starter solenoid, alternator, oil pressure sender and fuel tank indicator wire, then move wiring harness rearward. Disconnect return lines from hydraulic reservoir and loosen line clip at right side of cylinder head. Disconnect drag link at either end or remove if desired. Disconnect main hydraulic pump pressure line at coupling near front of tractor. Remove line retainer from lower right front side of transmission case and loosen inlet and return lines from housing. Do not lose check valve located in main pump inlet line. Disconnect temperature sending bulb from water outlet elbow and cold weather starting aid line from inlet manifold. Disconnect tachometer cable at clutch housing. Disconnect throttle control rod

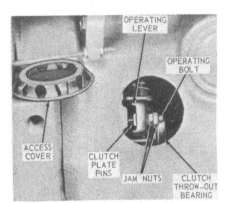

Fig. 55—Wear adjustment of dual clutch levers can be made as shown. Refer to paragraph 68B.

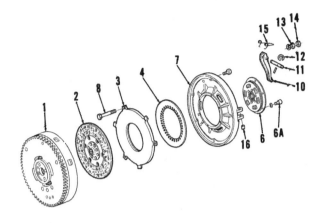

Fig. 56—Exploded view of single stage diaphragm clutch used on 2240 models with independent pto.

1. Flywheel
2. Transmission clutch disc
3. Pressure plate
4. Diaphragm spring
6. Pto drive hub
6A. Allen head screw
7. Cover
8. Actuating bolt
10. Release lever
11. Pivot pin
12. "E" ring
13. Adjusting nut
14. Locknut
15. Spring
16. Bushing

from injection pump. Place floor jack under transmission and attach hoist or suitable support to front section of tractor. Drive wooden blocks between front axle and support to prevent tipping sideways, place a container under rear of clutch housing to catch oil, then unbolt engine from clutch housing and split (separate) tractor.

When rejoining tractor it may be necessary to bar over engine to facilitate entry of input shafts into clutch discs. Be sure flywheel housing and clutch housing are butted together before tightening retaining cap screws. Tighten retaining cap screws to 170 ft.-lbs. torque.

R&R AND OVERHAUL

Single Stage Diaphragm (2240 With Independent Pto)

70. Clutch can be unbolted from flywheel after splitting as outlined in paragraph 69. To disassemble the removed clutch, first remove locking nuts (14—Fig. 56) and back off adjusting nuts (13) evenly until spring pressure is relieved. Mark cover (7) and pressure plate (3) with paint or other suitable means so balance can be maintained, then separate the units.

Inspect pressure plate (3) for cracks, scoring or heat discoloration and renew as necessary. Check diaphragm spring (4) for heat discoloration, distortion or other damage and renew if condition is questionable. Renew clutch disc (2) if facing is worn to less than 0.238 inch thickness, if hub is loose or splines are worn, or if disc is otherwise damaged. Pto drive hub (6) should be renewed if splines are worn, hub is loose or if bent (usually caused by attempting to force tractor together with splines not aligned).

Distance (F—Fig. 56A) from clutch mounting surface to the friction surface should be 2.200-2.210 inches. Thickness of new clutch disc (2) is 0.366-0.390 inch.

Assemble by reversing disassembly procedure, observing the previously affixed marks to assure correct balance. Tighten the screws (6A) which retain the pto hub to 14-20 ft.-lbs. torque. The long hub of clutch disc (2) should be toward engine. Use a centering tool to align clutch disc while assembling. Screws attaching clutch to flywheel should be tightened to 33 ft.-lbs. Adjust the clutch levers so that distance (A—Fig. 56A) is 1.791-1.811 inches and that the distance is exactly

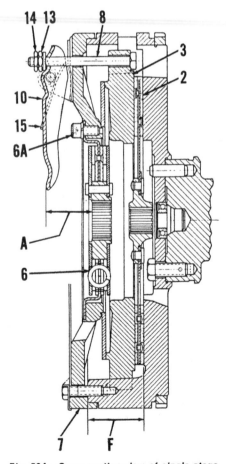

Fig. 56A—Cross-section view of single stage clutch used on 2240 models with independent pto. Refer to text for setting height (A) and for depth dimension (F).

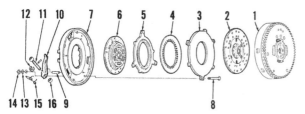

Fig. 57—Exploded view of the dual clutch unit and flywheel (1) used on 2240 models with continuous running pto.

1. Flywheel
2. Transmission clutch disc
3. Front pressure plate
4. Diaphragm spring
5. Rear pressure plate
6. Pto clutch disc
7. Cover
8. Actuating bolt
9. Pto release pin
10. Release lever
11. Pivot pin
12. "E" ring
13. Adjusting nut
14. Jam nut
15. Spring
16. Bushing

the same for all three levers. Adjust by turning nut (13), then lock adjustment by tightening nut (14). Adjust clutch linkage after clutch housing is attached to engine.

Dual Stage Diaphragm Clutch (2240 With Continuous Running Pto)

71. Refer to Fig. 57 for an exploded view of dual clutch unit and to Fig. 57A for cross-sectional view. Clutch can be removed after clutch split outlined in

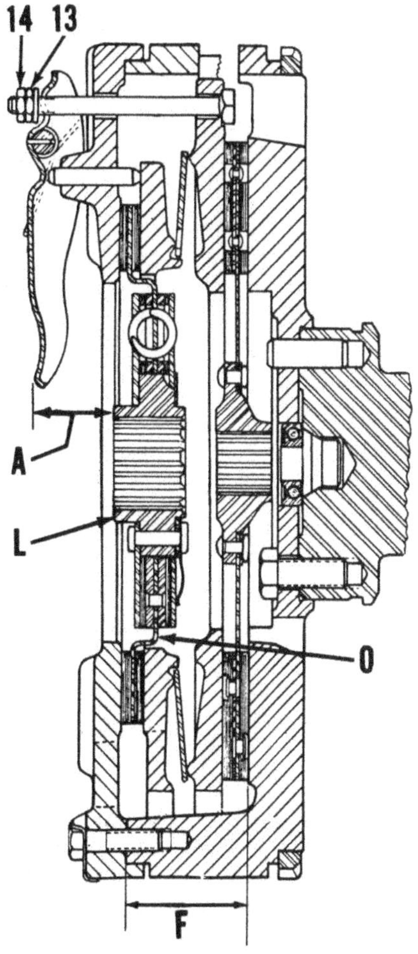

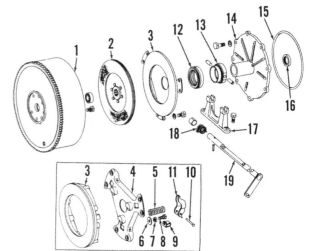

Fig. 57A—Cross-section view of dual clutch used on 2240 models with continuous running pto. Long hub (L) and offset (O) of pto clutch disc should be as shown. Height (A) is adjusted by turning nuts (13) after loosening locknut (14).

paragraph 69. Install new engine clutch disc if total thickness at facing area is less than 0.238 inch. Thickness of new engine clutch disc is 0.366-0.390 inch. When installing clutch, make sure long hub of transmission clutch disc (2—Fig. 57) is forward. Use a suitable alignment tool and tighten retaining cap screws to a torque of 33-40 ft.-lbs.

To disassemble the clutch cover and associated parts, remove locking nuts (14) and back off adjusting nuts (13) evenly until spring pressure is relieved. Mark the cover (7), rear pressure plate (5) and front pressure plate (3) with paint or other suitable means so balance can be maintained, and separate the units.

Distance (F—Fig. 57A) from clutch mounting surface to the friction surface should be 2.200-2.210 inches.

Inspect pressure plates (3 and 5—Fig. 57) for cracks, scoring or heat discoloration and renew as necessary. Check diaphragm spring (4) for heat discoloration, distortion or other damage and renew if its condition is questionable. Renew transmission clutch disc (2) if facing wear approaches rivet heads, if hub is loose or splines are worn, or if disc is otherwise damaged. Pto disc (6) should be renewed if total thickness at facing area is 0.185 inch or less. Thickness of new pto clutch disc is 0.300-0.330 inch.

Assemble by reversing the disassembly procedure, observing the previously affixed marks to assure correct balance. The long hub (L—Fig. 57A) of the pto clutch disc should be toward rear with the cushion offset (0) toward front as shown. Use a centering tool to align clutch disc while assembling.

The clutch assembly can be adjusted after attaching to flywheel, before reconnecting clutch housing to engine. Measure distance between pressure face of each clutch lever and rear flange of clutch disc as shown at (A). Loosen locknut (14), turn adjusting nut (13) until distance is 1.466-1.474 inches, then tighten locknuts. Adjust clutch linkage after clutch housing is attached to engine as described in paragraph 68A.

Auburn Type Single Clutch (2440, 2630, 2640 With Reverser)

72. Refer to Fig. 59 for exploded view of Auburn type single-stage clutch used without pto or with continuous running pto.

Clutch can be removed after clutch split by removing the six retaining cap screws holding clutch cover assembly to flywheel. When installing clutch, make sure long hub faces forward, unless tractor is equipped with a reverser, in which case long hub must face rearward. Use a suitable alignment tool and tighten clutch cover retaining cap screws to a torque of 35 ft.-lbs.

To disassemble the clutch cover assembly, a press and a suitable steel plate to depress all three levers at once will be necessary. If no flywheel of the type used on tractor is available, remove flywheel from engine and use it to hold clutch cover for disassembly. Place flywheel on a press, place clutch disc (2—Fig. 59) and pressure plate assembly (3) on the flywheel. Install

Fig. 59—Exploded view of Auburn type single disc clutch unit used on tractors without pto or with continuous running pto unit.

1. Flywheel
2. Clutch disc
3. Pressure plate
4. Clutch bracket
5. Spring
6. Washer
7. Locknut
8. Adjusting screw
9. Clip
10. Pivot pin
11. Lever
12. Release bearing
13. Carrier w/pins
14. Carrier support
15. Seal ring
16. Oil seal
17. Clutch fork
18. Return spring
19. Shaft

three pressure plate attaching cap screws equally around flywheel and tighten snugly. Use the press to depress clutch release levers (11) as far as possible without bottoming against clutch bracket (4). Loosen locknuts (7) and remove adjusting screws (8). Before releasing levers, make a mark on both the bracket (4) and pressure plate (3) for reassembly so all parts may be assembled exactly as they were, or the factory balance will be lost. If the pressure plate (3) is renewed, the assembly must be rebalanced. Release pressure on levers (11) gradually to prevent springs from flying out. Inspect springs

for rust, distortion, and tension. Spring specifications are as follows:

2630 model
Approx. free length (in.) 2½
Lbs. pressure at 1-11/16 in.,
  Without reverser 158-175
  With reverser 234

2440 and 2640 models
Free length (in.) 2½
Lbs. pressure at 1-13/16 in. 279-308

Check pressure plate for cracks, scoring or heat discoloration. Check cover and release levers for wear at pivot, distortion or other damage. Release levers and pivot pins are available separately. Grind off peened ends of pivot pins for removal.

Assemble the clutch by reversing the disassembly procedure. For tractors without reverser, adjust release levers to a height of 2.012 inches, measured between contact surface of lever and surface of flywheel. Finger height must not vary more than 0.010. Use John Deere Special Tool (JD-227—Fig. 65) modified by grinding a small (0.156 inch deep) step in finger contacting surface,

then use bottom of step to adjust finger height.

For tractors with reverser, adjusting tool JD-7 must be used to adjust release lever height as shown in Fig. 61. A new clutch disc must be installed for accurate measurement, and then a used disc may be installed later if desired. Depress levers several times and recheck setting. Install return clips (9—Fig. 59), when lever adjustments are correct and there is not more than 0.010 between lever heights.

Angle-Link Type Clutch (2440, 2630 And 2640 With Independent Pto)

73. Clutch can be removed after clutch split as outlined in paragraph 69. When installing clutch, make sure long hub faces forward and tighten clutch

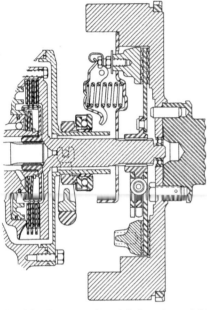

Fig. 60—Cross-section of Auburn type clutch used on 2440, 2630 and 2640 models with reverser.

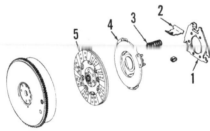

Fig. 62—Exploded view of Angle-Link type clutch used on models with ipto. Model 2440 disc (5) is shown.

1. Spring ring
2. Angle clip
3. Spring
4. Cover assembly
5. Clutch disc

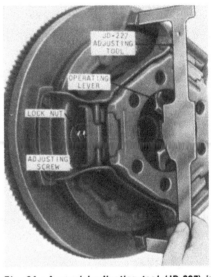

Fig. 64—A special adjusting tool (JD-227) is required to adjust release lever height.

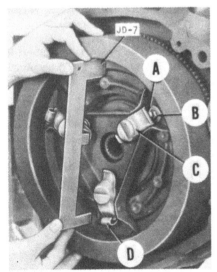

Fig. 61—Adjust release lever height as shown, using new clutch disc only.

A. Adjusting screw
B. Locknut
C. Release lever
D. Return clip

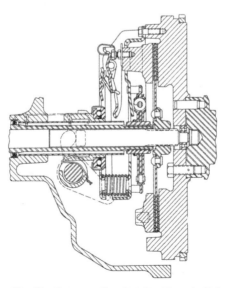

Fig. 63—Cross-section drawing of angle link clutch used on some 2440, 2630 and 2640 models.

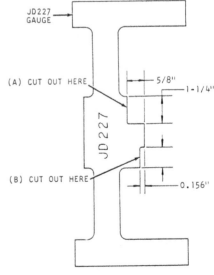

Fig. 65—Modify a JD-227 as shown, to obtain correct release lever adjustment.

cover retaining cap screws to a torque of 35 ft.-lbs.

To disassemble the removed clutch cover, solidly block up under pressure plate on the bed of a press. Use a spacer to distribute the pressure evenly and compress spring ring (1—Fig. 62) until release levers are loose and angle clips (2) can be withdrawn. Remove all three angle clips and release pressure slowly, then lift off spring ring (1) and six springs (3). If pressure plate and cover must be disassembled, mark the two parts with paint so balance can be maintained, remove the three adjusting screws and lift off the cover.

Springs (3) have a free length of 3-5/64 inches, and should test 118-130 lbs. when compressed to a height of 1-11/16 inches. Release levers and pto drive hub/damper assembly are a part of the cover and available only as an assembly.

Assemble by reversing the disassembly procedure. To adjust release levers after assembly, use a new clutch disc and install cover on flywheel as shown in Fig. 64. Using John Deere Clutch Adjusting Tool JD-227, set all three release levers to just touch the bottom of cut-out section of modified adjusting tool center leg. Adjustment must be equal to within 0.010 inch. Reinstall return clips after adjustment is complete. Adjust clutch linkage as outlined in paragraph 68A after tractor is reconnected.

Dual Clutch
(2440, 2630 And 2640 With Continuous Running Pto)

74. Refer to Fig. 66 for an exploded view. Clutch can be removed after tractor split as outlined in paragraph 69. When installing clutch, make sure long hub of transmission clutch disc (2—Fig. 66) is forward. Use a suitable alignment tool and tighten retaining cap screws to a torque of 35 ft.-lbs.

To disassemble the removed clutch cover, place the assembly in a press and apply pressure to cover (6); or attach cover to a spare flywheel using equally spaced threaded rods and nuts. Remove nuts (15 and 16) from operating bolts (10), then loosen the equally spaced nuts evenly until spring pressure is relieved.

Inspect pressure plates (3 and 4) for cracks, scoring or heat discoloration and renew as necessary. Clutch outer springs (8) and color-coded red, should have an approximate free length of 3-1/16 inches, and should test 111 lbs. or more when compressed to a height of 1-7/8 inches. Inner springs (7) are color-coded black. They should have an approximate free length of 3½ inches and should test 50 lbs. or more when compressed to a height of 1¾ inches. Renew springs which are distorted, heat discolored or fail to meet test specifications.

Assemble by reversing the disassem-

bly procedure. Make sure long hub of clutch discs face away from each other. Use JDE-52-1 centering tool to align discs. Tighten adjusting nuts (15—Fig. 66) evenly until release levers (13) just contact pto clutch push pins (11); tighten nuts exactly 2½ turns and secure by tightening jam nuts (16). Check lever adjustment after joining tractor, as outlined in paragraph 68B; and equalize as necessary by making minor adjustments to high and low levers. Adjust clutch linkage as outlined in paragraph 68A.

CLUTCH SHAFT

Models Without Hi-Lo Or Reverser

NOTE: For clutch shaft service on models with Hi-Lo or reverser, see appropriate section.

75. To remove clutch shaft it is necessary to separate clutch housing from transmission housing. Clutch shaft can be removed from rear of clutch housing.

If tractor has continuous-running pto and there is evidence of oil seepage between clutch shaft and powershaft (pto shaft) separate shafts and inspect oil seal and pilot. Press new pilot (cup rearward) into bore of powershaft until it bottoms. Press oil seal in bore of powershaft, with lip rearward, until it contacts pilot.

When installing clutch shaft, or powershaft when tractor is equipped with continuous-running pto, be sure lugs on shaft align with slots in transmission oil pump drive gear.

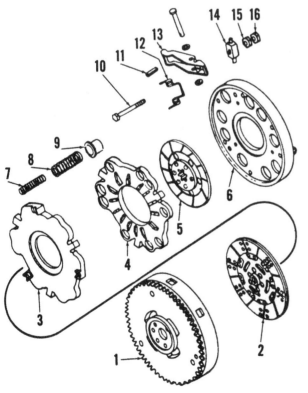

Fig. 66—Exploded view of dual clutch assembly used on some 2440, 2630 and 2640 models. Item 2 on 2630 and 2640 models is button type.

1. Flywheel
2. Transmission clutch disc
3. Front pressure plate
4. Rear pressure plate
5. Pto clutch disc
6. Clutch cover
7. Inner spring
8. Outer spring
9. Spring cup
10. Actuating bolt
11. Pto release pin
12. Spring
13. Release lever
14. Pivot block
15. Adjusting nut
16. Jam nut

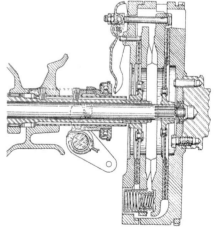

Fig. 67—Cross-section drawing of dual stage clutch used on some 2440, 2630 and 2640 models.

CLUTCH RELEASE BEARING
AND YOKE

Models Without Reverser

76. The clutch release (throwout) bearing (B—Fig. 68) can be removed after clutch housing is separated (split) from engine. Disconnect return spring and withdraw unit from carrier sleeve.

To remove yoke, disconnect clutch rod from clutch shaft arm, then drive out the two yoke retaining spring pins (RP). Pull clutch shaft out left side of clutch housing and catch yoke as it comes off clutch shaft.

When installing a new throwout bearing, align index mark on bearing with notch in bearing carrier.

Models With Reverser

77. The clutch release (throwout) bearing (B—Fig. 68) can be removed after clutch housing is separated (split) from engine. Withdraw unit from carrier support sleeve (14—Fig. 59). It may be necessary to disconnect clutch rod from clutch shaft arm.

To remove yoke, remove the two cap screws in yoke (clutch fork 17). Return spring (18) or shaft (19) may be renewed by driving out roll pin holding return spring. Pivot pins in carrier (13) may be renewed separately.

CLUTCH HOUSING

All Models

78. Clutch housing normally will not need complete removal for servicing. Clutch control linkage can be serviced when clutch housing is separated from engine. Clutch shaft and pto shafts along with their bearings and oil seals and the transmission oil pump can be serviced after clutch housing is separated from transmission case.

Refer to paragraph 101 for service information on the transmission oil pump. Other service required on clutch housing will be obvious after examination and reference to the following: Clutch shaft bushing is installed with outer end flush with outer edge of bore. Clutch pedal pivot shaft is renewable and should be installed to protrude 2½ inches from housing. Clutch throwout bearing carrier sleeve (on all models without reverser) is renewable and should be installed so that forward end is 2-13/16 inches from the machined engine mounting surface of clutch housing as shown at (A—Fig. 69). Oil seal for clutch shaft, or pto shaft, located in center of clutch housing can be renewed after removing shafts, transmission oil pump and needle bearing. Fill new seal with grease, then press seal into housing bore with sealing lip toward rear. Press seal into bore until distance (B) from rear of seal to face of housing bore is 1.3 inches. Needle bearing should be pressed into bore until rear of bearing is flush with bottom of bore chamfer.

HI-LO
SHIFT UNIT

Some tractors may be optionally equipped with a Hi-Lo Shift Unit (inside clutch housing) which consists of a planetary gear set, hydraulically actuated direct drive clutch and hydraulically actuated multiple disc brake unit which locks the planet carrier to clutch housing. Pressure oil to operate the Hi-Lo Shift Unit is supplied by the transmission oil pump which also supplies lubrication for the transmission gears and excess flow supplies fluid for the main hydraulic system pump.

OPERATION

All Models So Equipped

79. Refer to Fig. 70 for schematic view. Input shaft (1) is transmission clutch shaft and also carries hub for Hi-Clutch (2). When Hi-Clutch is engaged, planet carrier (3) is locked to input shaft providing direct drive at crankshaft speed through Hi-Lo shift unit to transmission input shaft (9).

The hub for Lo-Brake unit (5) is splined to oil manifold hub. When Lo-Brake unit is engaged, planet carrier (3) is locked to clutch housing and power is transmitted through planet pinions to sun gear (4) and transmission input shaft at approximately ¾ crankshaft speed.

The spool-type control valve directs pressure oil to either the Hi-Clutch or

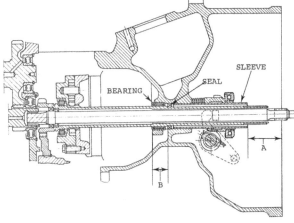

Fig. 69—Refer to text for installation of clutch throwout bearing carrier sleeve, shaft seal and needle bearing.

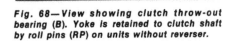

Fig. 68—View showing clutch throw-out bearing (B). Yoke is retained to clutch shaft by roll pins (RP) on units without reverser.

Fig. 70—Schematic view of Hi-Lo Shift Unit showing main components.

1. Input shaft
2. Hi-Clutch
3. Planet carrier
4. Sun gear
5. Lo-Brake
6. Oil pump
7. Pto clutch shaft
8. Pto drive gears
9. Transmission input shaft

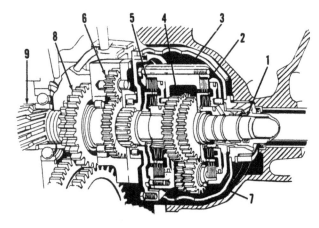

Lo-Brake when dash-mounted control lever is moved. Refer to Fig. 71. The pressure regulating valve maintains approximately 150 psi in control valve on 2630 models and approximately 130 psi on other mdoels. The entire flow except for what leakage exists (through clutches or control valve) passes through the regulating valve to the main hydraulic pump.

TESTING

All Models So Equipped

80. To check the system pressure, remove transmission case shield if so equipped, remove test plug (6—Fig. 72) from right side of shift cover and install a suitable pressure gage. Check the pressure with system at normal operating temperature and engine running at 2100 rpm. On 2630 models, pressure should be 145-155 psi, with or without ipto. On all other models, pressure should be 125-135 psi with or without ipto. Add or remove shims behind spring in regulating valve (8) to adjust pressure.

If regulated pressure cannot be properly adjusted, refer to paragraph 130 for additional hydraulic system checks and to paragraphs 81 and 82 for overhaul procedure.

OVERHAUL

All Models So Equipped

81. **CONTROL AND PRESSURE REGULATING VALVE.** Control and pressure regulating valves are located in shifter cover as shown in Fig. 72. Regulating valve can be adjusted without removing shifter cover, but overhaul of valve unit can only be accomplished after cover is removed; proceed as follows:

Remove the four cap screws retaining transmission shield and remove the

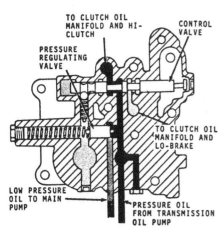

Fig. 71—Cross-sectional view of control valve showing fluid passages.

shield by working it up over shift lever boots. Disconnect Hi-Lo shift lever linkage and, if so equipped, independent pto lever link. Disconnect rear wiring harness. Remove clutch control lever housing (2). Remove the remaining cap screws retaining shifter cover and lift off cover, valves and shift levers.

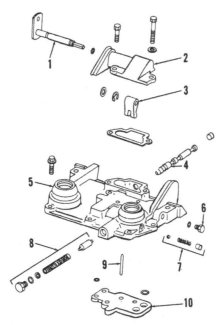

Fig. 72—Exploded view of shift cover and associated parts used on models with Hi-Lo Clutch unit. Refer to Fig. 121 for cover used on models equipped with ipto option.

1. Valve shaft
2. Housing
3. Valve arm
4. Valve spool
5. Shift cover
6. Test plug
7. Detent assembly
8. Regulating valve
9. Detent pin
10. Cover plate

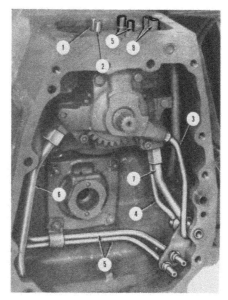

Fig. 73—Rear view of clutch housing showing clutch oil manifold, oil pump and ipto oil lines.

1. Lo-Brake pressure
2. Hi-Clutch pressure
3. Lo-Brake lube
4. Hi-Clutch lube
5. Ipto oil lines
6. Pump inlet
7. Pump outlet
8. Valve supply lines

Pin (9) is inserted behind spring of valve detent (7). Remove plate (10) and withdraw the pin before attempting to withdraw valve spool (4). Detent passage plug (7) must be removed when reassembling, to depress the spring and reinstall pin.

Clean and inspect all parts. Control valve spool and regulating valve piston should slide freely in their bores without binding or excessive looseness. On 2240 models, the control valve detent spring should have an approximate free length of 0.98 inch and should exert 13-16 lbs. force when compressed to 0.7 inch. On 2440, 2630 and 2640 models, the control valve detent spring should have a free length of 0.94 inch and should exert 14-23 lbs. force when compressed to 0.75 inch. The regulating valve spring on 2240 models should have a free length of 3.11 inches and should exert 40-50 lbs. force when compressed to 2.56 inches. The regulating valve spring on 2440, 2630 and 2640 models should have a free length of 3.25 inches and should exert 26.5-32.5 lbs. force when compressed to 2.5 inches.

82. **HI-LO DRIVE UNIT.** To remove the Hi-Lo drive unit, first split tractor between clutch housing and transmission as outlined in paragraph 95. Disconnect and remove hydraulic oil tubes

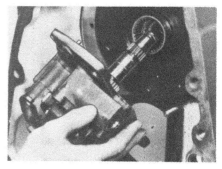

Fig. 74—Clutch oil manifold and transmission oil pump are removed as a unit.

Fig. 75—Support Hi-Lo Shift unit during removal. Note thrust washer located in bore of clutch housing.

from transmission oil pump and clutch housing (Fig. 73). Remove mid-pto if tractor is so equipped.

Unbolt and remove transmission oil pump and manifold assembly as a unit as shown in Fig. 74; then withdraw Hi-Lo Shift Unit as shown in Fig. 75.

To overhaul the removed unit, remove pto drive cover (35—Fig. 76) and withdraw planetary unit and associated parts. Remove the three long cap screws (through-bolts) from alternate holes in brake backing plate (34) and withdraw input shaft (20), Hi-Clutch drum (11) and associated parts from planetary unit. Remove the remaining three cap screws from backing plate (34) and withdraw Lo-Brake plates (33) and hub (32).

Thread a ¼-inch cap screw into planet pinion shaft (25) to act as a puller (Fig. 77) and withdraw the shaft, being careful not to lose locking balls (30—Fig. 76) at end of shaft. Lift out planet pinions (24) and needle bearings. Lift out sun gear (26) and thrust washer (27) after pinions are removed.

To remove pistons and springs (Belleville Washers) from planet carrier and Hi-Clutch drum, use a press and fixture as shown in Fig. 78. Collapse the spring washer pack and unseat snap ring, working through cutout portion of fixture. Refer to Fig. 79 for an exploded view. Install spring washers in pairs as shown in inset.

Hi-Clutch pack (18—Fig. 76) and Lo-Brake pack (33) both use the same

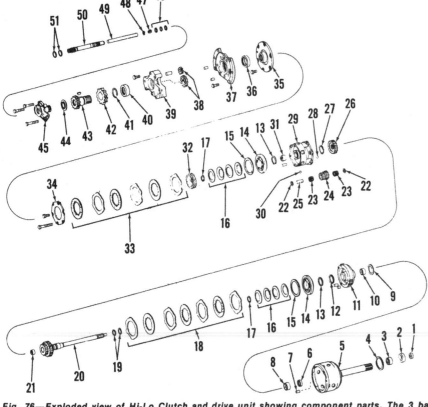

Fig. 78—Use a special tool as shown when compressing clutch spring pack to remove the retaining snap rings.

Fig. 76—Exploded view of Hi-Lo Clutch and drive unit showing component parts. The 3 ball plugs (28) are used to plug lube passages.

1. Ball bearing	14. Piston	27. Thrust washer
2. Oil seal	15. Seal ring	28. Ball plug
3. Bushing	16. Spring washers	29. Planet carrier
4. Thrust washer	17. Snap ring	30. Locking ball
5. Pto clutch shaft	18. Hi-Clutch pack	31. Bushing
6. Oil seal	19. Sealing rings	32. Brake hub
7. Dowel	20. Input shaft	33. Lo-Brake pack
8. Bushing	21. Bushing	34. Backing plate
9. Thrust washer	22. Thrust washer	35. Pto drive cover
10. Bushing	23. Needle bearing	36. Roller bearing
11. Hi-Clutch drum	24. Planet pinion	37. Pump adapter
12. Thrust washer	25. Pinion shaft	38. Pump gears
13. "O" ring	26. Sun gear	

39. Pump body	45. Oil manifold
40. Roller bearing	46. Sealing ring
41. Snap ring	47. Retainer
42. Pto drive gear	48. Sealing ring
43. Pto drive shaft	49. Tube
44. Thrust washer	50. Trans. input shaft
	51. Sealing rings

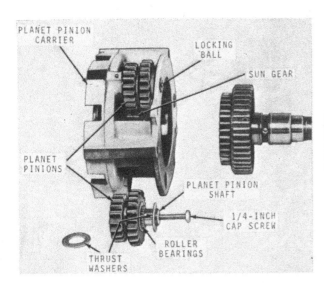

Fig. 77—Partially disassembled view of Hi-Lo Shift unit planet carrier and associated parts. Pinions must be timed when installed as outlined in paragraph 82.

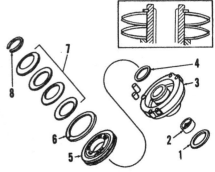

Fig. 79—Exploded view of Hi-Clutch drum and associated parts. Belleville spring washers are installed as shown in inset.

1. Thrust washer	5. Piston
2. Bushing	6. Piston ring
3. Hi-Clutch drum	7. Spring washers
4. "O" ring	8. Snap ring

separator plates and clutch discs, but one more plate is used in Hi-Clutch pack. Externally splined steel separator plates in clutch or brake pack should be renewed if scored, warped, heat discolored or otherwise damaged. Bronze-faced discs should be renewed in sets, even though available individually.

Planet pinions (24) should be renewed in sets. When assembling planetary unit, install thrust washer (27) and output sun gear (26) in carrier with "V" marks on sun gear up. Assemble two needle bearings in each planet pinion (24) with spacer washer at each end of pinion; then install planet pinion with "V" mark on face of pinion aligned with similar mark on sun gear. Push pinion shaft into housing until indentation in shaft is ready to enter carrier, install locking ball (30) and push shaft in until front end is flush with end of housing bore. Make sure "V" mark on all planet pinions properly align with mark on sun gear as pinions are installed.

Lo-Brake unit must be assembled and backing plate (34) installed before Hi-Clutch unit can be installed. Start with a steel separator plate next to piston and end with a bronze disc next to backing plate when assembling Lo-Brake unit. Start and end with a steel separator plate when assembling Hi-Clutch pack. Tighten cap screws and through-bolts to a torque of 23 ft.-lbs.

REVERSER UNIT

Some tractors may be optionally equipped with a Reverser Unit which consists of a planetary gear set, a hydraulically actuated forward drive clutch and hydraulically actuated reverse brake, all located in the clutch housing and operated by a dash mounted directional lever. Pressure oil to operate the reverser unit is supplied by the transmission oil pump which also provides lubrication for the transmis-

sion gears and serves as a supply pump for the main hydraulic system pump.

OPERATION

All Models So Equipped

83. Refer to Fig. 80 for a schematic view of planetary gear unit. Planet carrier drive shaft (2) is splined into hub of forward clutch carrier. When forward clutch is engaged, clutch shaft is locked to carrier and the planetary unit turns together driving tractor in a forward direction.

The reverse brake unit is carried in transmission pump mounting plate and output sun gear (4) is splined into reverse brake hub. When reverse brake is applied, power entering at planet carrier causes the small planet pinion to walk around the fixed secondary sun gear (4) and the large planet pinion drives clutch shaft (1) at 16 percent overdrive ratio in a reverse direction.

The hydraulic control unit contains a shift valve which directs pressure fluid to either the forward clutch or reverse brake piston. No neutral position is provided in the shift valve. Also included is a clutch control valve which interrupts pressure and flow to the shift valve when clutch pedal is depressed, thus stopping forward or reverse motion of the tractor and permitting the gear change unit to be shifted in the normal manner. The design of the mechanical linkage prevents moving shift valve to reverse position when transmission range shifter lever is in "High" position.

Also included in the valve unit is a spring loaded accumulator piston which smooths the engagement during direction changes. An orifice screw is pro-

vided to permit adjustment of the shift rate. The accumulator remains charged when power flow is interrupted by the clutch pedal, thus permitting immediate and accurate control of clutch reengagement rate.

CHECK AND ADJUST

All Models So Equipped

84. Before making any tests, first check to make sure that transmission oil is at the correct level and that transmission is at operating temperature. It is generally advisable to check and adjust the unit in the order given.

85. **REGULATING VALVE.** Remove plug (29—Fig. 81) from bottom of cover (28) and install a 0-300 psi pressure gage. With left hand transmission shift lever in "Park" position and tractor running at 2100 rpm, gage pressure should be 155-165 psi on 2440 models and 165-185 psi on other models. To adjust the pressure, remove plug (1) and add or remove shim washers (7) as required.

86. **CLUTCH CONTROL VALVE.** First make sure regulating valve is correctly adjusted as outlined in paragraph 85, and record the pressure for later reference. Also check and/or adjust clutch pedal free play as outlined in paragraph 68.

Remove plug (13 or 14—Fig. 81) from top of control valve body and install a 0-300 psi pressure gage. Move reverser control lever to pressurize the circuit containing the gage, then check circuit pressure which should equal system pressure as previously recorded. If pressure needs adjusting, adjust as outlined in paragraph 85.

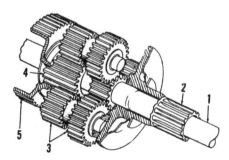

Fig. 80—Three-quarter view of reverser planetary unit showing component parts.

1. Clutch shaft
2. Planet carrier
3. Planet pinions
4. Secondary sun gear
5. Reverse brake hub

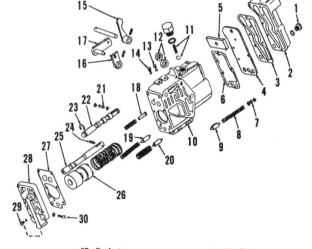

Fig. 81—Exploded view of reverser control valve showing component parts.

1. Plug
2. Cover
3. Gasket
4. Plate
5. Block
6. Gasket
7. Shims
8. Spring
9. Regulating valve
10. Body
11. Detent
12. Oil seals
13. Plug
14. Plug
15. Arm
16. Detent arm
17. Shift shaft
18. Valve stop
19. Cooler by-pass valve
20. Lubrication valve
21. Pin
22. Shift valve
23. Retaining ring
24. Pin
25. Control valve
26. Accumulator piston
27. Gasket
28. Cover
29. Plug
30. Adjusting screw

87. SHIFT ENGAGEMENT RATE. Shift engagement rate when changing direction of travel can be adjusted within specified limits. Nominal shift time is ¾ to 1¼ seconds. Shift engagement rate can be varied by turning adjusting screw (30—Fig. 81) IN to slow shift rate or OUT to speed shift rate. Initial (average) setting is two turns out from closed (bottomed) position. A slow shift rate accompanied by a jerky start could indicate a broken accumulator spring or sticking accumulator piston (26).

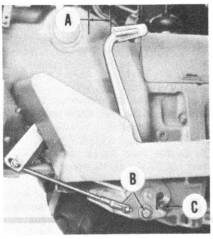

Fig. 82—Reverser clutch pedal linkage showing points of adjustment.

A. ¾-1¼ inch
B. Pedal shaft screw
C. Adjusting screw

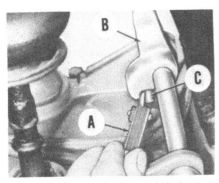

Fig. 83—Adjusting high speed lockout as outlined in paragraph 88.

A. Feeler gage
B. Lockout arm
C. Clearance 0.060

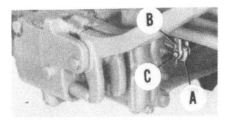

Fig. 84—Adjust link rod yoke to obtain proper clearance for high speed lockout.

A. Shift valve arm
B. Yoke
C. Headed pin

88. HIGH SPEED LOCKOUT. To adjust the high speed lockout, refer to Fig. 83. Move reverser control lever to "FORWARD" position and range shifter lever to "HIGH" position; then measure the clearance between lockout cam on reverser shift shaft and lockout pin as shown. Clearance should be 0.060; if it is not, disconnect link rod (Fig. 84) at lower end and turn adjusting yoke (B) as required until clearance is correct.

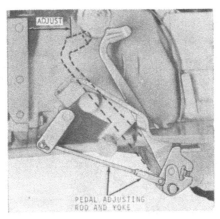

Fig. 85—Adjust clutch pedal free play to 5¼ inches on all reverser models by disconnecting yoke from pedal arm and changing length of rod.

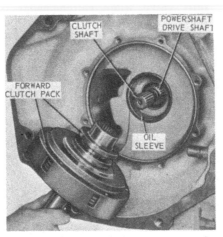

Fig. 86—Forward clutch pack can be removed as shown, after detaching engine from clutch housing.

Fig. 87—Exploded view of forward clutch pack and associated parts.

1. Sleeve
2. Sealing rings
3. Clutch drum
4. Steel ball
5. Bushing
6. Dowel pin
7. Piston ring
8. Piston
9. Piston ring
10. Spring washers
11. Snap ring
13. Drive shaft
14. Clutch discs
15. Separator plates
16. Clutch hub

89. PEDAL FREE TRAVEL. With transmission in PARK, a 0-300 psi pressure gage installed in plug (13 or 14—Fig. 81) and engine running at 2100 rpm, depress pedal slowly until pressure drops. There should be ¾-1¼ inch free travel before pressure starts to drop. If free travel is not correct, refer to Fig. 82 and adjust screw (C) to obtain dimension (A) before pressure drops. Refer to paragraph 85 and adjust pressure, if necessary.

OVERHAUL

All Models So Equipped

90. REVERSER CONTROL VALVE. To remove the reverser control valve, remove right platform and thoroughly clean brake valve, reverser control valve and surrounding area. Drain transmission case. Disconnect brake lines, oil cooler lines and reverser link yoke. Drive out spring pin securing clutch shaft and control valve shaft. Remove the retaining cap screws and lift off brake valve and pedals. Remove the remaining three cap screws and lift off reverser control valve.

Refer to Fig. 81 for exploded view of reverser control valve. Evenly loosen cap screws retaining rear cover (28). Cover is under heavy pressure from accumulator spring when seated. Cooler by-pass valve (19), lubrication valve (20) and pressure regulating valve (9) are interchangeable but springs are not. All parts are available individually and selective fitting is not required. Assemble by reversing the disassembly procedure. Tighten cover retaining cap screws to a torque of 70-75 in.-lbs. Identify valve springs if necessary, by the following test data:

Regulating valve
(9) . . . . .30 lbs. at 2-9/16 in.
Cooler by-pass valve
(19) . . . . .20 lbs. at 2-1/16 in.

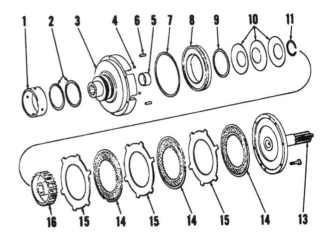

Lubrication valve

(20) ....4½ lbs. at ¾ in.

Clutch valve

(25) ....20 lbs. at 1-1/8 in.

Accumulator piston

(26) ....310 lbs. at 2-5/8 in.

91. FORWARD CLUTCH. Forward clutch unit can be removed after splitting engine from clutch housing as outlined in paragraph 69. Remove clutch fork shaft and throwout bearing carrier and withdraw forward clutch pack as shown in Fig. 86.

Unbolt and remove clutch drive shaft (13—Fig. 87), then remove the three clutch discs (14) and separator plates (15). Lift out clutch hub (16). Refer to paragraph 93 for removal of clutch piston (8) and overhaul of forward clutch.

92. REVERSE BRAKE AND PLANETARY UNIT. To remove the reverse brake and planetary unit, first detach (split) clutch housing from transmission as outlined in paragraph 95. Transmission oil pump, reverse brake assembly and planetary unit can be removed as a unit as shown in Fig. 88; or oil pump may be removed separately by removing pump housing cap screws. To remove either unit, it is first necessary to remove mid-pto shift assembly and drive, if so equipped, as outlined in paragraph 121.

Refer to Fig. 89 for an exploded view of reverse brake and planetary unit. Withdraw planet carrier (26), clutch shaft (23) and associated parts forward out of brake assembly. Withdraw clutch shaft to rear out of planet carrier.

Remove the three cap screws retaining carrier shaft (21) to carrier (26). Push each pinion shaft (35) forward slightly, remove and save locking balls (25), then remove shaft (35) and pinion (33), being careful not to lose the 44

loose needle rollers (32), thrust washers and spacer. Assemble by reversing the removal procedure. Note the stamped numbers on large gear teeth of pinions (33) and corresponding numbers on front of gear on clutch shaft (23). Time planetary unit during assembly by aligning corresponding numbers on clutch shaft and pinion. Overhaul reverse brake as outlined in paragraph 93.

Bushings in clutch housing are renewable. If new bushings are installed, make sure open ends of oil grooves are together as shown in Fig. 90. I.D. of forward clutch bushing is 2.130-2.133 inches and I.D. of reverse brake bushing is 1.755-1.758 inches. Tighten cap screws retaining carrier shaft (21—Fig. 89) to carrier (26) to a torque of 35 ft.-lbs. Tighten cap screws retaining reverse brake housing (9) to clutch housing to a torque of 35 ft.-lbs. and cap screws retaining transmission oil pump to brake housing to 23 ft.-lbs.

93. BRAKE/CLUTCH. A clutch piston disassembly tool can be made by

cutting away a portion of a discarded oil filter cover as shown in Fig. 91; or from a suitable piece of pipe. Depress the spring washer pack in a press as

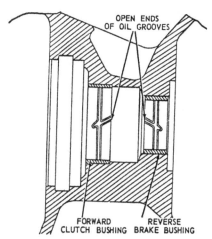

Fig. 90—Cross-sectional view of clutch housing used on reverser models, showing bushings properly installed.

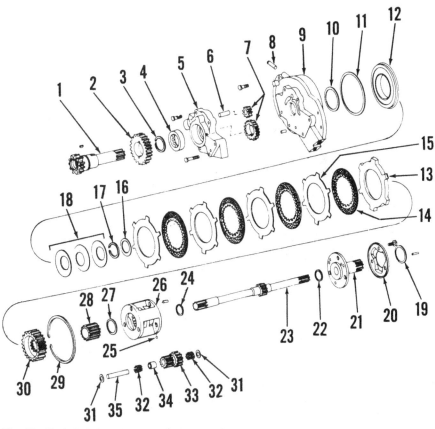

Fig. 89—Exploded view of reverse brake and planetary unit showing component parts.

1. Pto drive shaft	9. Brake housing	18. Spring washers	27. Thrust washer
2. Pto drive gear	10. Sealing ring	19. Thrust washer	28. Secondary sun gear
3. Snap ring	11. Piston ring	20. Baffle	29. Snap ring
4. Bearing	12. Brake piston	21. Carrier shaft	30. Brake hub
5. Transmission pump body	13. Backing plate	22. Thrust washer	31. Thrust washer
6. Dowel pin	14. Brake discs	23. Clutch shaft	32. Needle rollers
7. Pump gears	15. Separator plates	24. Thrust washer	33. Planet pinion
8. Lockout pin	16. Thrust washer	25. Steel ball	34. Spacer
	17. Snap ring	26. Planet carrier	35. Planet pinion shaft

Fig. 88—Transmission oil pump, reverse brake and planetary unit can be removed as an assembly after detaching transmission from clutch housing.

shown; unseat the snap ring then disassemble piston unit.

Square section sealing rings are used on clutch and brake units. Lubricate thoroughly and use care not to cut rings when installing. Position spring washers as shown in Fig. 92.

Four clutch discs (14—Fig. 87 or 89) are used in reverse brake; three in forward clutch. An equal number of separator plates (15) are used. Separator plates are slightly wavy and should not be flattened. Clutch (brake) pack begins with a separator plate next to actuating piston and ends with a clutch disc next to pressure plate. Clutch discs are 0.112-0.118 in thickness; separator plates 0.090. If clutch discs measure less than 0.110, renew discs as a set. The three spring washers for forward clutch have slightly more cup than those for reverse brake.

TRANSMISSION

Transmissions are constant mesh type using helical cut gears. Two shift levers are used, the left lever selecting high, low and reverse ranges as well as a park position. The right lever controls a four step gear arrangement. Thus with the two shift levers, eight forward speeds (four in high range and four in low range) and four reverse speeds are

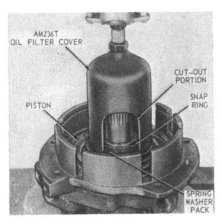

Fig. 91—An oil filter cover properly modified, makes a satisfactory disassembly tool for piston removal.

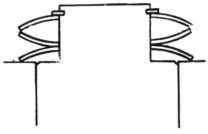

Fig. 92—Belleville spring washers on clutch and brake pistons must be positioned as shown.

available. The four reverse speeds approximate, in mph, the four forward speeds obtained in low range.

TOP (SHIFTER) COVER

All Models

94. **REMOVE AND REINSTALL.** To remove the shifter cover remove the shield and work it up over shifter lever boots. Disconnect connector from starting safety switch and bleed line from fitting on shift cover. If necessary, disconnect rear wiring harness, then unbolt and lift shifter cover from clutch housing.

Any further disassembly required will be obvious after examination of the unit. See Fig. 93.

Reinstall by reversing the removal procedure.

NOTE: Shifter rails and forks are an integral part of the transmission and can be serviced after transmission is split from clutch housing as outlined in paragraph 95.

TRACTOR SPLIT

All Models

95. To split tractor between clutch housing and transmission case, disconnect start-safety switch wiring harness, battery ground straps, drain transmission case (both plugs), and remove hydraulic oil filter cover and element. Remove shifter cover as outlined in paragraph 94, then remove the two

clutch housing to transmission case cap screws located at rear of shifter cover opening under mounting flange. Remove spring from the high range/park shifter shaft (22—Fig. 93) guide. Remove left platform and unhook clutch return spring. Remove right platform, disconnect the two brake pressure lines from brake valve and the main hydraulic pressure line from pressure control valve. Remove hydraulic pump inlet line and reservoir return line at lower right side of transmission case. Disconnect tail light wires and if so equipped, disconnect hydraulic outlet mid-couplers. Support transmission case, place a rolling floor jack under front section and block front axle to prevent front end from tipping. Remove remaining clutch housing to transmission cap screws and separate tractor. See Fig. 94.

NOTE: DO NOT lose the check valve, located in rear end of the main hydraulic pump inlet line, which will probably fall out as line comes out of transmission case.

When rejoining tractors equipped with mid-pto, be sure spring is in place in forward end of pto shaft and ball is in place in rear end of mid-pto shaft. See Fig. 94.

SHIFTER SHAFTS AND FORKS

All Models

96. **REMOVE AND REINSTALL.** To

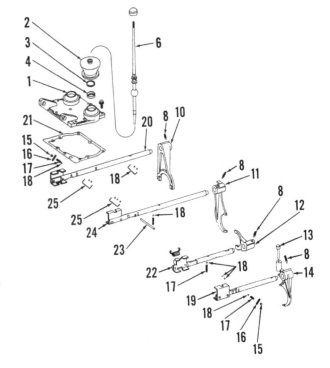

Fig. 93—Exploded view of shifter cover, shifter shafts and shifter forks, Shifter shafts are an integral part of transmission. See Fig. 94.

1. Shifter cover
2. Boot
3. Snap ring
4. Retainer
6. Shift lever
8. Set screw
10. Fork
11. Fork
12. Fork
13. Pin, starter safety switch
14. Fork
15. Plug
16. Spring pin
17. Detent spring
18. Detent ball
19. Shifter shaft, low range & reverse
20. Shifter shaft, 1st, 5th, 2nd, 6th
21. Gasket
22. Shifter shaft, high range
23. Interlock pin
24. Shifter shaft, 3rd, 7th, 4th, 8th
25. Interlock balls

remove shifter rails (shafts) and forks it will first be necessary to split tractor as outlined in paragraph 95. Remove seat and disconnect lift links from rockshaft arms. If so equipped, disconnect all lines from selective control valves to rockshaft housing. Attach a hoist to rockshaft housing, place load selector lever in "L" position, then unbolt and remove rockshaft housing from transmission case.

Remove transmission lubricating oil cup and lines. Remove the starter safety switch pin from low range shifter fork and remove set screws from shifter forks.

NOTE: Before removal and installation of shifter shafts and forks, refer to Fig. 95 to determine location of detent and interlock mechanisms as well as for identification of shifter shafts. Also

note that the five interlock balls are ¼-inch diameter whereas the remaining eight detent balls are 5/16-inch diameter.

DO NOT turn shifter shafts while withdrawing from housing. Detent ball could drop into set screw hole, making shaft impossible to remove.

Place the two left hand shifter shafts in neutral, then pull the outer range shift shaft (L N R) from its bore and use caution not to lose the three detent balls. Shift the inner speed shifter shaft into gear to release interlock pin, then withdraw the inner range shifter shaft (P N H) from its bore and catch detent ball (and spring). Lift shifter forks from transmission case.

Pull inner speed shifter shaft out of its bore until the three interlock balls can be removed from shaft, then complete removal of shaft and remove the three rear detent balls. Remove the remaining front two interlock balls from bore in housing. Pull the right hand speed shifter shaft from its bore and remove the remaining front detent ball. Lift shifter forks from transmission case.

If interlock pin requires renewal, drive out spring pin at right side of transmission case and slide interlock pin out right side.

NOTE: Detent bores are not aligned and interlock pin will not enter left detent bore.

Clean and inspect all parts and renew any that are bent, worn or otherwise damaged. Be sure to inspect all balls and renew any which have flat spots that would prevent them from rolling freely. To aid in the installation of detent balls in case, a simple tool may be made from ½-inch electrical conduit as shown in Fig. 96. Insert tool in shaft bore, place ball in groove in tool and rotate until ball aligns with hole. Push

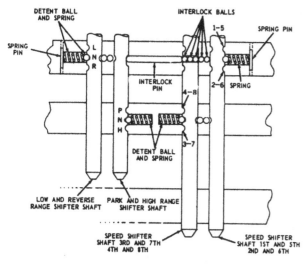

Fig. 94—View showing front of transmission case after tractor split. Tractor shown is 2440 model with reverser and is equipped with 540-1000 rpm continuous running pto with mid-pto.

H. 1000 rpm gear
L. 540 rpm gear
S. Spring
SS. Shifter shafts
TF. Transmission filter

Fig. 95—Schematic view showing shifter shafts and the detent and interlock mechanisms. The rear detents are below shifter shafts instead of sides as indicated. Refer to text for service procedure.

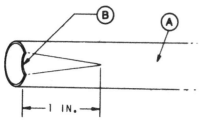

Fig. 96—A ½-inch electrical conduit 7 inches long makes an installation tool for detent balls.

A. Conduit   B. ¼-inch indent

Fig. 97—Front of transmission case with pto shaft and pto gears removed. Countershaft support (S) must be removed before input shaft or pinion shaft can be removed. (Model without independent pto shown.)

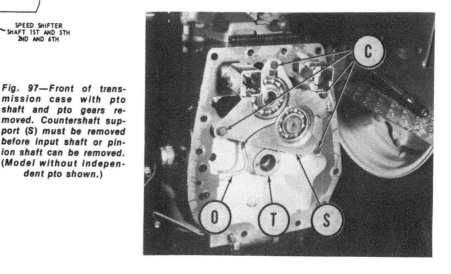

tool out with shifter shaft.

Reinstall shifter shafts and forks by reversing removal procedure. Start with right hand (outer) speed shifter shaft and shift inner speed shaft into a gear position before attempting to install right hand (inner) range shifter shaft so that interlock pin will move to the right.

COUNTERSHAFT

All Models

97. R&R AND OVERHAUL. To remove the countershaft, split tractor as outlined in paragraph 95 and remove shifter shafts and forks as outlined in paragraph 96. Remove the pto gear, or gears, from front of transmission. Remove cap screws from countershaft bearing support (13—Fig. 98). Remove snap ring (33) from its groove at rear end of countershaft, then use a screwdriver and turn locking washer (29) until splines of washer index with splines of countershaft. Pry bearing support off dowels (12), pull assembly

forward, notice that shoulder on shift collar (28) is toward front and lift gears from transmission case as they come off shaft. See Fig. 99.

Inspect all gears, thrust washers and shift collar for broken teeth, excessive wear or other damage and renew as necessary. If support assembly bearings or shafts require service, the shafts and bearings can be pressed out after removing retaining snap rings; however, the transmission drive gear must be removed before countershaft can be removed. Snubber brake plugs (21—Fig. 98) should be renewed if worn. Snubber brake springs (22) should test 63-77 lbs. when compressed to a length of 1½ inches. Needle bearing (35) can be removed from its bore after removing snap ring (34).

On models equipped with reverser, plug (20) is used to plug oil hole in countershaft (24) where reverse pinion (27) was installed on models without reverser. Also, brake snubber spring shims (23) are used to increase spring pressure behind snubber brake plugs (21).

INPUT SHAFT

All Models

98. R&R AND OVERHAUL. To remove input shaft it is first necessary to remove countershaft as outlined in paragraph 97.

With countershaft removed, the transmission input shaft is removed as follows: Remove transmission oil cup and lines. Straighten lock plates and remove input shaft bearing quill (8—Fig. 98) and shims (7) from front of input shaft. Bump input shaft forward until front bearing cup (6) clears its bore, then move input shaft forward, lift rear end of shaft and remove input shaft from transmission case.

With input shaft removed, inspect all gears for chipped teeth or excessive wear. Inspect bearings and renew as necessary. Bump bearing cup (1) forward if removal is required. Inspect needle bearing (4) and renew if necessary.

Install and adjust end play of input shaft as follows: Be sure bearing cup (1) is bottomed in bore and place input shaft in position. Use original shim pack (7), or use a shim pack approximately 0.030 thick, install front bearing quill (8) and tighten cap screws to 35 ft.-lbs. torque. Use a dial indicator to check the input shaft end play which should be 0.004-0.006. Vary shims as required. Shims are available in thicknesses of 0.003, 0.005 and 0.010. Do not forget front oil line clamp when making final installation.

PINION SHAFT

All Models

99. R&R AND OVERHAUL. To remove the transmission pinion shaft, remove the input shaft as in paragraph

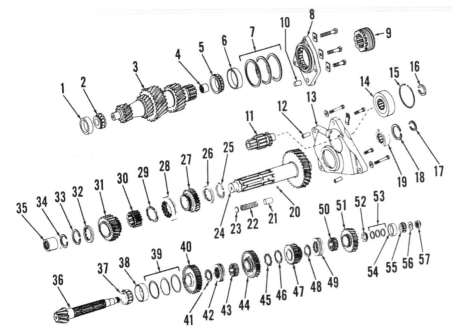

Fig. 98—Exploded view of transmission shafts and gears. Note that transmission drive gear (11) is mounted in countershaft support (13). Items (20 and 23) are used only in models with reverser. Item (27) is not used with reverser.

1. Bearing cup	16. Snap ring	30. Shift collar sleeve
2. Bearing cone	17. Snap ring	31. Low range pinion
3. Input shaft	18. Snap ring	32. Thrust washer
4. Needle bearing	19. Ball bearing	33. Snap ring
5. Bearing cone	20. Plug	34. Snap ring
6. Bearing cup	21. Brake plug	35. Needle bearing
7. Shims	22. Spring	36. Pinion shaft
8. Bearing quill	23. Shim	37. Bearing cone
9. Shifter collar	24. Countershaft	38. Bearing cup
10. Dowel	25. Snap ring	39. Shims
11. Drive gear	26. Thrust washer	40. 1st & 5th gear
12. Dowel	27. Reverse pinion	41. Thrust washer
13. Support	28. Shift collar	42. Shift collar
14. Ball bearing	29. Thrust washer	43. Shift collar sleeve
15. Snap ring	(locking)	44. 2nd & 6th gear

45. Thrust washer	
(outer tangs)	
46. Retaining washer	
47. 4th & 8th gear	
48. Thrust washer	
49. Shift collar	
50. Shift collar sleeve	
51. 3rd & 7th gear	
52. Spacer	
53. Shims	
54. Bearing cup	
55. Bearing cone	
56. Special washer	
57. Nut	

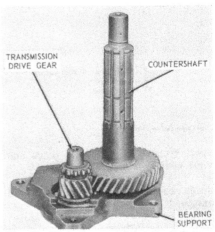

Fig. 99—View of removed countershaft bearing support assembly.

98 and the differential as in paragraph 103.

With differential removed, remove oil line, nut (57—Fig. 98), bearing (55), shims (53) and spacer (52). Use a screwdriver and turn thrust washers (41, 45 and 48) until splines of thrust washers are indexed with splines of pinion shaft. Pull pinion shaft rearward and remove parts from transmission case as they come off shaft. Bearing cup (38) and shims (39) can be removed from housing by bumping cup rearward. Be sure to keep shims (39) together as they control the bevel gear mesh position. Bearing cup (54) can be removed from housing by bumping cup forward.

Check all gears and shafts for chipped teeth, damaged splines, excessive wear or other damage and renew as necessary. If pinion shaft is renewed, it will also be necessary to renew the differential ring gear and right hand differential housing as these parts are not available separately. Bearing (37) is installed with large diameter toward gear end of shaft.

NOTE: Mesh (cone point) position of the pinion shaft and main drive bevel pinion gear is adjusted with shims (39) located between rear bearing cup (38) and housing. If new drive gears or bearings are installed, the mesh position must first be checked and adjusted as outlined in paragraph 106. If same pinion shaft and bearing are installed, reinstall the same shims (39) and check the bearing preload as in paragraph 100.

Install pinion shaft and adjust shaft bearing preload as follows: Use Fig. 98 as a guide and with bearing (37) on pinion shaft, start shaft into rear of housing. With shaft about half-way into housing, place 1st and 5th speed gear (40) on shaft with teeth for shift collar (42) toward front. Place thrust washer

(41) on shaft, then install coupling sleeve (43) and shift collar (42). Move shaft forward slightly and install 2nd and 6th speed gear (44) with teeth for shift collar toward rear. Place thrust washer with outer tangs (45) over shaft, then slide retaining washer (46) over thrust washer (45). Move shaft slightly forward and install 4th and 8th speed gear (47) on shaft with teeth for shift collar toward front. Place thrust washer (48) on shaft and install shift collar sleeve (50) and shift collar (49). Install 3rd and 7th speed gear (51) on shaft with teeth for shift collar toward rear. Push shaft forward until rear bearing cone (37) seats in bearing cup (38) and use screwdriver to turn thrust washer until splines on thrust washers lock with splines of pinion shaft. Install spacer (52), shims (53), bearing (55), washer (56) with tang facing forward, and nuts (57), then adjust pinion shaft bearing preload as outlined in paragraph 100.

100. **PINION SHAFT BEARING ADJUSTMENT.** The pinion shaft bearings must be adjusted to provide a bearing preload of 0.006. Adjustment is made by varying the number of shims (53—Fig. 98).

To adjust the pinion shaft bearing preload, proceed as follows: Mount a dial indicator with contact button on front end of pinion shaft and check for end play of shaft. If shaft has no end play, add shims (53) to introduce not more than 0.002 shaft end play.

NOTE: Do not exceed more than 0.002 shaft end play when beginning adjustment as increased end play increases the possibility of inaccuracies due to parts shifting.

If original shims (53) are not being

used, install a preliminary 0.035 thick shim pack. Shims are available in thicknesses of 0.002, 0.005 and 0.010. Tighten nut (57) to 160 ft.-lbs. torque and measure shaft end play, then remove shims from shim pack (53) equal to the measured shaft end play, PLUS an additional 0.005. This will give the recommended bearing preload of 0.006. Retighten nut to 160 ft.-lbs. torque and stake in position.

TRANSMISSION OIL PUMP

All Models

The transmission oil pump is a gear type pump with a capacity of 9 gpm for 2240 models, 6 gpm for other models. Refer to Fig. 100. Except for the Hi-Lo shift unit or reverser requirements, all oil from transmission pump is routed to inlet side of main hydraulic pump. Oil not used by main pump flows to the hydraulic reservoir at front of tractor, then to oil cooler and from cooler to brake control valve and transmission sump. Oil cooler is shown in Fig. 45.

Transmission oil pump system is protected by a filter relief valve, which is located above filter in transmission housing and operates with a 50 psi pressure differential. If filter becomes clogged, or oil is cold, relief valve will open and route oil back to sump.

101. **TESTING.** Use necessary adapters to attach flow meter inlet hose to oil filter relief valve bore as shown at (1—Fig. 100A). Attach outlet hose from tester to the oil filler opening in transmission housing. Start engine and operate at 2500 rpm, with flow meter open and oil moving through hoses. Meter should indicate correct amount of flow after oil has reached 100-110 degrees F. The transmission pump on 2240 models should flow 9 gpm and other models should flow 6 gpm. On all models, increase pressure gradually until volume of flow decreases and notice the pressure. This pressure is regulated by

Fig. 100—View of transmission oil pump (P) with clutch shaft and pto powershaft removed.

I. Inlet line
O. Outlet line
P. Transmission pump

Fig. 100A—Refer to text for checking the transmission oil pump. Inlet to tester is attached at (1). Inlet screen is under plug (2).

the surge relief valve or the transmission oil regulating valve. Oil pressure at time of flow reduction will be about 165 psi for models with reverser, about 90-140 for other models.

If pump will meet above conditions it can be considered satisfactory. If pump will not meet above conditions, clean the transmission oil filter and the inlet screen, which is located under the squared headed plug (2—Fig. 100A) just behind transmission oil filter, then retest pump. If pump operation is still not satisfactory, check to see that the surge relief valve, located in clutch housing directly back of brake control valve (Fig. 101), is not stuck in the open position. Also check relief valve spring which should test 30-37 lbs. when compressed to a length of 1-1/8 inches. If relief valve seat is damaged, seat can be pried out and a new one driven in.

If performing the above operations does not produce proper pump operations, remove and overhaul pump as outlined in paragraph 102.

102. R&R AND OVERHAUL. To remove the transmission oil pump, first split clutch housing from transmission housing as outlined in paragraph 95.

With clutch housing separated from transmission case, pull clutch shaft, or clutch shaft and pto power shaft, from clutch housing. Remove pump inlet and outlet lines from pump, then remove pump from wall of clutch housing and separate pump body from adapter. See Fig. 102.

Clean and inspect all parts for chipping, scoring or excessive wear. If bearing in pump body requires renewal, press new bearing in bore until it bottoms. Pump gears are available as a matched set only. Pump idler shaft is renewable and diameter of new shaft is 0.624-0.625 inch. Idler gear bushing I.D.

Fig. 102—View showing transmission oil pump separated. Note drive lugs in I.D. of pump drive gear.

is 0.626-0.627 inch. Thickness of new pump gears is 0.508-0.509 inch.

When reassembling pump, coat gears with oil and tighten adapter mounting cap screws to 35 ft.-lbs. torque and housing to adapter cap screws to 23 ft.-lbs. torque. Align slots of clutch shaft, or pto powershaft, with lugs of pump drive gear when installing shafts and be sure seals are on ends of inlet and outlet tubes before rejoining tractor.

DIFFERENTIAL AND FINAL DRIVE

Differential units are equipped with a differential lock. Removal procedure will be the same for all tractors.

Final drives incorporate a planetary gear reduction unit at inner end of the housing and final drive units for all models are similar. See Fig. 105.

DIFFERENTIAL

All Models

103. REMOVE AND REINSTALL. To remove differential, drain transmission case, then remove final drives as outlined in paragraph 109 and the rockshaft housing as outlined in paragraph 143.

The differential lock assembly must be removed as follows: Remove clamp screw from lever (1—Fig. 103), or pedal (12) on some models, and remove the square key (9). Hold the yoke (4) in place, bump shaft (3) rearward and remove Woodruff key (8) when it clears yoke. Continue to bump shaft rearward until plug at rear end of shaft bore comes out, then remove shaft, yoke and collar (7).

Disconnect load control arm spring, right load control shaft cotter pin, clevis pin and retainer, slide load control pivot shaft to the left and lift out load control arm. Remove transmission oil cup and rear oil line. Support the

differential, remove both bearing quills (2 and 14—Fig. 104) and keep shims (3) with the correct quill.

The shims (3) located between bearing quills and transmission case control the differential bearing preload and the backlash of main drive bevel gears. Recommended bearing preload is 0.002-0.005 and recommended backlash is 0.006-0.012. Refer to paragraph 105 for adjustment procedure.

104. OVERHAUL. On all models the bevel ring gear, right hand differential housing and pinion shaft are available as a matched set.

To disassemble unit, remove the eight differential housing bolts and separate housing.

Removal of bearings (5 and 12—Fig. 104), and bearing cups (4 and 13) is obvious. If any of the pinions (9) or pinion shafts (10) are damaged, all mating parts should also be renewed. If axle (side) gears (8) are damaged or excessively worn, closely examine bores in differential housing as they may also be damaged.

Reassemble by reversing the disassembly procedure and tighten differential housing cap screws to 35 ft.-lbs. Cap screws are self-locking.

MAIN DRIVE BEVEL GEARS

All Models

105. ADJUSTMENT. If differential is removed for access to other parts and

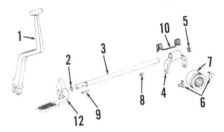

Fig. 103—Exploded view showing parts of differential lock. Some models use pedal (12) instead of lever (1).

1. Lever		7. Collar
2. "O" ring		8. Woodruff key
3. Shaft		9. Square key
4. Yoke		10. Spring
5. Plug		12. Pedal
6. Shoes		

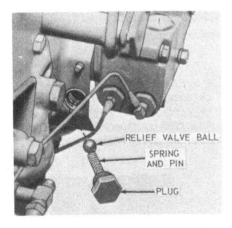

Fig. 101—Surge relief valve assembly for transmission oil pump is located as shown.

RELIEF VALVE BALL

SPRING AND PIN

PLUG

no defects in the adjustments are noted, the shim packs should be kept intact and reinstalled in their original positions. However, if bevel gears, bearings, bearing quills or transmission case are renewed, the main bevel gears should be checked for mesh (cone point) position, differential carrier bearing preload and gear backlash, and in the following order.

106. MESH (CONE POINT) POSITION. The fore and aft position of the transmission pinion shaft is controlled by shims located between pinion shaft rear bearing cup and front wall of differential compartment as shown in Fig. 106. Usually, two 0.010 inch thick shims should be installed; however, the following procedure can be followed to determine the correct thickness of shim stack to be installed.

Use 1.446 inches as the nominal width of an assembled rear pinion shaft bearing and to the 1.446 inches, add the dimension found etched on aft end of pinion shaft gear. Next observe the actual cone point dimension stamped on top rear of transmission case, then subtract the previously determined value from the cone point dimension. This will give the required thickness of shim pack. The bearing assembly width (1.446), plus dimension on end of pinion shaft, plus shim pack, should equal the dimension stamped on top rear of transmission case.

As an example, assume the following values: 6.643 stamped on rear of transmission case and 5.183 etched on aft end of pinion shaft gear. Compute shim pack thickness as follows: add 5.183 and 1.446 which equals 6.629, then subtract 6.629 from 6.643 which leaves

0.014 **which will be the shim pack thickness required for correct mesh position of the main drive bevel gears. Follow this procedure whenever any new parts are installed that affect mesh position of the bevel gears.**

107. DIFFERENTIAL BEARING ADJUSTMENT. The differential carrier bearings should have a preload of 0.002-0.005 and adjustment is made as follows: Install differential and bearing quills with original shim packs, then check differential end play using a dial indicator.

NOTE: When making this adjustment, be positive that clearance exists between the main drive bevel ring gear and pinion shaft at all times.

If no differential end play exists, add shims under right bearing quill to introduce not more than 0.002 end play. If more than 0.002 end play existed on original check, subtract shims.

Measure end play of differential, then subtract shims equal to the measured end play PLUS an additional 0.003 to give the desired 0.002-0.005 bearing preload. Shims are available in thicknesses of 0.003, 0.005 and 0.010.

108. BACKLASH ADJUSTMENT. With differential carrier bearing preload adjusted as outlined in paragraph 107, adjust backlash between main drive bevel gear and pinion shaft to 0.006-0.012 by transferring bearing quill shims from one side to the other as required. Moving shims from left to right will increase backlash. Do not change the total thickness of all shims during backlash adjustment or the previously determined preload adjustment will be changed.

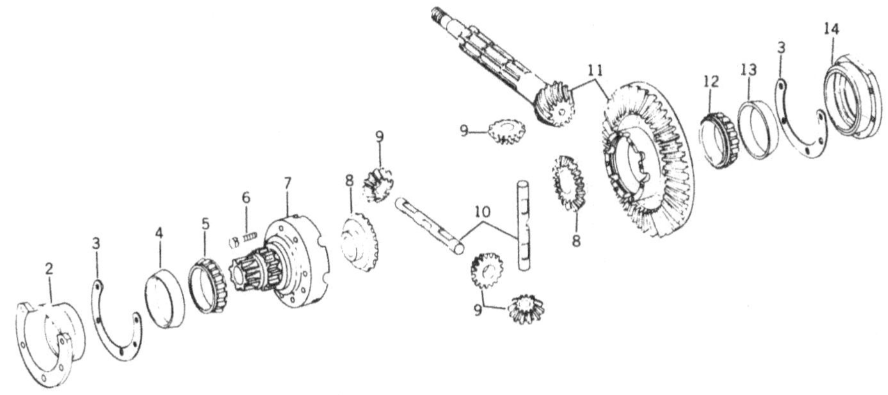

Fig. 104—Exploded view of differential used on all models.

2. Quill, LH	5. Bearing cone	9. Pinions	12. Bearing cone
3. Shims	7. Housing, LH	10. Pinion shafts	13. Bearing cup
4. Bearing cup	8. Bevel (side)gears	11. Bevel gear set	14. Quill, RH

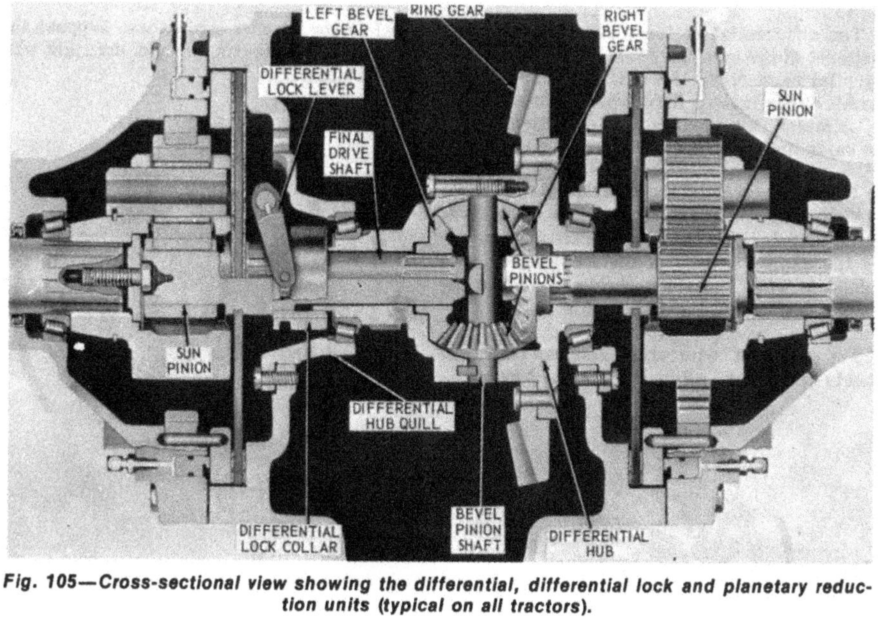

Fig. 105—Cross-sectional view showing the differential, differential lock and planetary reduction units (typical on all tractors).

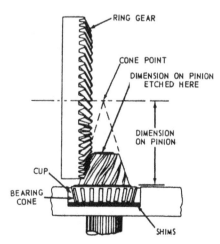

Fig. 106—Simplified view of main drive bevel gears showing location of shims which control gear mesh position. Refer to text for adjustment procedure.

FINAL DRIVE

All Models

109. REMOVE AND REINSTALL. To remove final drive, disconnect battery ground cables and drain transmission. Support rear of tractor and remove wheel and tire. Disconnect fender lights and free wiring harness from clamp on final drive housing, then remove fender. If right hand final drive is being removed and tractor has selective control valve, disconnect pressure line, coupler lines and return hose between valve and rockshaft housing and remote control valve. Disconnect the brake line from final drive housing. Attach hoist to final drive, remove attaching cap screws and pull final drive from transmission case.

Reinstall by reversing removal procedure and tighten attaching cap screws to 85 ft.-lbs. torque.

NOTE: If brake disc came off with final drive, inspect facing on both sides of disc. If facing is thicker on either side, install disc so that thickest facing is toward transmission case.

110. OVERHAUL. To overhaul the removed final drive unit, refer to Fig. 107 and proceed as follows:

Remove lock plate (23), cap screw (24) and retainer washer (25), then pull planet carrier assembly (26) from axle. Support outer end of final drive housing (11) so oil seal (2) will clear and press or drive axle out of housing. The axle bearings, bearing cups and oil seals are now available for inspection or renewal. If outer bearing (4) is renewed, heat it to approximately 300 degrees F and drive it into place while hot.

NOTE: If axle is flanged, be sure oil seal is on axle, metal side out, before installing outer bearing.

Bearing cups are pressed in bores until they bottom. Seal cup (3) will be pushed out when outer bearing cup is removed. Be sure to reinstall seal cup after bearing cup is installed. If ring gear and/or final drive housing is damaged, renew complete unit.

To remove planet pinions (20), expand snap ring (27), lift it from groove of carrier (26) and pull pinion shafts (18). (Snap ring is located on outer side of carrier on 2440 models and on inner side of carrier on other models.) Check carrier, pinions and rollers for pitting, scoring or excessive wear and renew parts as required. If any of the planet pinion rollers are defective, renew complete set.

The final drive on 2440 models uses either a three planetary or four planetary reduction unit. Only three planet units are used on all other models. Each planet gear on 2240 and 2440 models has a single row of 23 individual needle rollers. Other models use three gear planetary units, with double rows of 23 roller bearings each, or 46 rollers in each gear and a spacer between the double rows of rollers.

Reassemble final drive and adjust axle bearings as follows: Coat bores of planet pinions with grease and position rollers in pinions. Place a thrust washer on each side of pinions (with a spacer between double rows, if so equipped), then place pinions in carrier and insert pinion shafts only far enough to retain rollers and thrust washers. Install snap ring (27) in slots of pinion shafts, then complete insertion of pinion shafts and be sure snap ring seats in groove in carrier. Coat inner seal (13) with grease and install axle in housing. Heat inner bearing (15) to approximately 300 degrees F and install bearing on inner end of axle. Place carrier assembly on axle, install retaining washer (25) and cap screw (24) and tighten cap screw until bearing is pulled into place and a small amount of axle end play remains. Now while bearing is still hot, check the amount of torque required to turn the axle with the existing axle end play, then tighten the cap screw to increase the rolling torque 50-80 in.-lbs. for old bearings, or 90-140 in.-lbs. if new bearings are used. Install lock plate (23). Fill axle outer bearing opening with multi-purpose grease and install oil seal with metal side out.

Use new gasket (12) when reinstalling final drive to transmission case. However, before installing final drive, pull final drive shaft (22) and brake disc and inspect. Brake disc is installed with thickest facing next to transmission case.

Refer to paragraph 118 for information on brake pressure plate and pressure ring.

BRAKES

The brakes on all models are hydraulically actuated and utilize a wet type disc controlled by a brake operating valve located on right side of clutch housing. See Fig. 108. Brake discs are splined to the final drive shafts and the brake pressure ring is fitted in inner end of final drive housing. Except for a pedal adjustment, no other brake adjustments are required.

BLEED AND ADJUST

All Models

111. BLEEDING. Brakes must be bled when pedals feel spongy, pedals

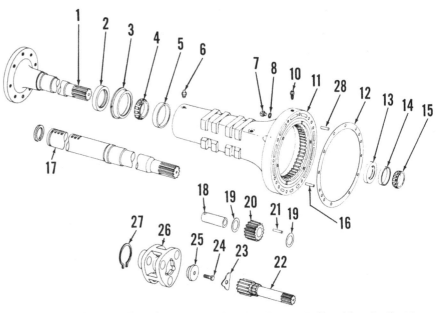

Fig. 107—Exploded view of final drive assembly. All units are similar although all parts are not interchangeable.

1. Axle, flanged	8. "O" ring	16. Dowel pin	23. Lock plate
2. Oil seal	10. Bleed screw	17. Axle	24. Cap screw
3. Oil seal cup	11. Housing	18. Pinion shaft	25. Retaining washer
4. Bearing cone	12. Gasket	19. Thrust washer	26. Carrier
5. Bearing cup	13. Oil seal	20. Pinion	27. Snap ring
6. Grease fitting	14. Bearing cup	21. Rollers (69 used)	28. Dowel (replacement)
7. Plug	15. Bearing cone	22. Final drive shaft	

bottom, or after disconnecting or disassembling any portion of the braking system.

To bleed brakes, start engine and run for at least two minutes at 2000 rpm to insure that brake control valve reservoir is filled.

Attach a bleeder hose (preferably clear plastic) to brake bleed screw located on top side of final drive housing and place opposite end in filler hole of rockshaft housing. Slowly depress and release brake pedal until oil flowing from bleeder hose is completely free of air bubbles, then depress brake pedal and tighten bleed screw. If brake pedal is released rapidly, internal parts of brake valve may be damaged.

Repeat bleeding operation for opposite side brake.

NOTE: Brakes can also be bled without engine running if necessary. Remove right platform and fill brake control valve reservoir by removing filler plug (11—Fig. 109). Follow same bleeding procedure used when engine is running except brake valve reservoir will need to be refilled after each 15 strokes of the brake pedal.

112. **ADJUSTMENT.** Whenever brake control valve has been disassembled, a brake pedal and equalizing valve adjustment must be made to prevent mechanical interference between brake valve pistons and reservoir check valves.

Before making this adjustment, bleed brakes as outlined in paragraph 111.

113. RIGHT PEDAL. Adjust right hand pedal stop screw so brake valve piston is fully extended and arm of brake pedal is snug against end of piston without piston being depressed (zero clearance). Apply a force of about 10 lbs. to LEFT brake pedal and if left brake pedal settles, turn pedal stop screw for right brake pedal counterclockwise about ⅓-turn at which time left brake pedal should stop settling. If left brake pedal does not stop settling, a leak in the braking system is indicated and must be isolated and corrected. Refer to paragraph 116.

114. LEFT PEDAL. Adjust left hand pedal stop screw so brake valve piston is fully extended and arm of brake pedal is snug against end of piston without piston being depressed (zero clearance). Apply a force of about 10 lbs. to RIGHT brake pedal and if right pedal settles, turn pedal stop screw for left brake pedal counter-clockwise about ⅓-turn at which time right brake pedal should stop settling. If right brake pedal does not stop settling, a leak in the braking system is indicated and

must be isolated and corrected. Refer to paragraph 116.

115. PEDAL HEIGHT. If brake pedal height is not aligned after equalization valves are adjusted as outlined in paragraphs 113 and 114, align pedals by turning stop screw on highest pedal about 1/6-turn counter-clockwise.

BRAKE TEST

All Models

116. **PEDAL LEAK-DOWN.** With a 60 lb. pressure applied continuously to each pedal for one minute, the pedal leak-down should not exceed one inch. Excessive brake pedal leak-down can be caused by air in the brake system, faulty brake control valve piston and/or "O" rings, faulty brake pressure ring seals, or faulty brake control valve

Fig. 108—View showing brake control valve. Right platform has been removed.

A. Pedal adjusting screws
B. Brake control valve
C. Mounting cap screws
P. Brake valve pistons

Fig. 109—Exploded view of brake control valve. Only half of component parts are shown. Other half of parts are identical. Two cup plugs (17) are used on some models in place of plug (11).

1. Connector
2. "O" ring
3. Check valve seat
4. "O" ring
5. Spring
6. Ball
7. Retainer
8. Spring
9. Piston
10. Housing
11. Filler plug
12. "O" ring
13. Valve seat
14. "O" ring
15. Check valve
16. Spring
17. Cup plug
18. "O" ring
19. Oil seal

20. Pedal shaft
21. Retainer ring
22. Ball

23. Spring
24. "O" ring
25. Plug

equalizing valves or reservoir check valves.

Brakes should always be bled as outlined in paragraph 111 before any checking or adjusting of braking system is attempted. Faulty brake control valve pistons or "O" rings, as well as a leaking brake line, will be indicated by external leakage around the brake control valve pistons or brake line connections.

Faulty brake pressure ring seals, or brake control valve, can be determined as follows: Isolate brake from brake control valve by plugging brake line. If leak-down stops, the brake pressure ring seals are defective. If leak-down continues, the brake control valve is faulty and can be checked further by depressing brake pedals individually, then simultaneously. If leak-down occurs in both cases, a defective reservoir check valve is indicated. If leak-down

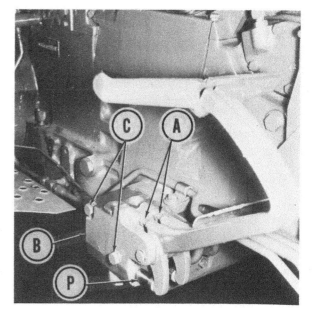

occurs during individual pedal operation but not on simultaneous pedal operation, a faulty equalizer valve is indicated.

Refer to paragraph 117 for brake control valve information and to paragraph 118 for brake pressure ring information.

OVERHAUL

All Models

117. **BRAKE CONTROL VALVE.** To remove brake control valve, remove right platform and thoroughly clean valve and surrounding area. Disconnect brake lines from rear of control valve, remove the mounting cap screws (C—Fig. 108) and remove control valve from clutch housing. Discard gasket located between control valve and clutch housing. Remove "E" ring (21—Fig. 109), pull shaft (20) and remove pedals from control valve. Remove connectors (1), check valve springs (5) and balls (6). Remove seats (3) and ball retainers (7), then push pistons (9) and springs (8) out rear of valve body. Remove filler plug (11) and cup plug (17), then using a screwdriver with proper sized bit, remove reservoir check valve assemblies (items 13, 14, 15 and 16). Remove equalizer valve assemblies (items 22, 23, 24 and 25). "O" rings (18) and oil seals (19) can be removed from piston bores.

Clean and inspect all parts. Piston spring (8) should test 20 lbs. when compressed to a length of 5¾ inches. Renew housing (10) if seats for equalizer balls (22) are damaged. Oil seals (19) are installed with lips toward outside. Pay particular attention to area of reservoir check valve (15) where contact is made with valve piston and renew valve if any doubt exists as to its condition. Brake pedals are fitted with bushings for brake pedal shaft (20) and bushings and/or shaft should be renewed if clearance is excessive.

Lubricate lips of oil seals (19) and all other parts. Use a new cup plug (17) and reassemble by reversing disassembly procedure. Use a new gasket when installing valve on tractor. Bleed brakes as outlined in paragraph 111 and adjust pedals as outlined in paragraphs 113, 114 and 115.

118. **BRAKE PRESSURE PLATE, RING AND DISC.** To remove brake pressure plate, pressure ring and brake disc, remove final drive housing as outlined in paragraph 109. Pull final drive shaft from differential and remove brake disc from final drive shaft. See Fig. 110. Lift brake pressure plate from

dowels in final drive housing by prying it out evenly. If pressure plate is difficult to remove, use moderate air pressure, or attach a small hydraulic pump to brake line connections, be sure bleed valve is closed, then pump oil behind pressure plate to force it from cylinder (groove). Dowels can be removed from final drive housing, if necessary.

Inspect brake disc for worn or damaged facing or damaged splines. If facings require renewal, renew complete disc assembly as facings are not available separately. Inspect pressure plate for scoring, checking, or other damage and renew if necessary. Remove and discard seals from pressure plate and inspect plate for cracks or other

damage.

To reassemble brake assembly, proceed as follows: Place brake disc on final drive shaft so thickest facing is next to transmission case and insert final drive shaft into differential. Place new inner and outer seals on pressure plate and lubricate assembly liberally. Start pressure plate into its cylinder (groove), making sure dowel aligns with dowel in final drive housing and press into cylinder until it bottoms. Be absolutely sure that neither seal is cut or rolled during installation. Install a new gasket, then install final drive housing.

Bleed brakes as outlined in paragraph 111.

Fig. 110—Final drive shaft and brake disc being removed. Thickest disc facing is next to transmission case.

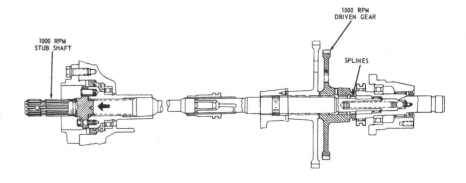

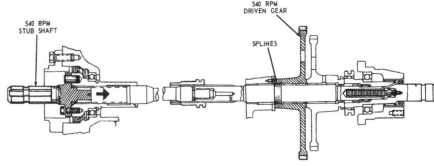

Fig. 111—Schematic view showing arrangement of pto shafts and gears used with dual 540-1000 rpm pto with mid-pto attachments. Note fore and aft movement of pto shafts.

POWER TAKE-OFF (Continuous Or Transmission)

Tractors may be equipped with 540 rpm transmission driven pto, a 540 rpm continuous running pto or a dual 540-1000 rpm continuous running pto. The dual pto is available only on reverser equipped tractors. In addition, a mid-pto is also available with only the dual speed pto and reverser of the continuous running pto units, however, the mid-pto operates only at 1000 rpm.

Changing speeds of the dual speed pto is accomplished by using either a six spline 540 rpm stub (output) shaft or a 21 spline 1000 rpm stub (output) shaft. The 540 rpm stub shaft incorporates a pilot on its forward end which moves the pto shaft assembly forward against spring pressure to engage splines on pto shaft with the splines of the 540 rpm pto driven gear. When the 1000 rpm pto stub shaft, with no pilot, is installed, the spring in forward end of pto shaft assembly moves the pto shaft rearward and the splines of the pto shaft engage with the splines of the 1000 rpm pto driven gear. See Fig. 111.

Shifter couplings are provided to engage and disengage rear pto and mid-pto shafts.

When service is required on the pto system, the following should be taken into consideration. Work involving rear pto shaft can be done by working from rear of tractor. Work involving the pto driven gears, powershaft clutch shaft, mid-pto shaft or mid-pto shifter assembly will require that the clutch housing be separated from the transmission case. Work involving the rear pto shaft shifter assembly will involve removing the rockshaft housing, high and low range shifter shafts and countershaft assembly in addition to separating the clutch housing from transmission case.

REAR PTO SHAFT

119. R&R AND OVERHAUL (540 RPM). To remove rear pto shaft (12—Fig. 112), drain transmission and remove pto shield and shaft guard, if so equipped. Place rear pto shaft control lever in "OFF" position, then remove cap screws from pto shaft bearing quill (17) and pull shaft and quill assembly from transmission case. Be careful not to pull shift collar from front drive shaft. Bearing (14) can be renewed after removing snap rings (4 and 13). Press new oil seal (15) in quill with lips toward front until it bottoms. For ser-

vice on remainder of pto assembly, split tractor as outlined in paragraph 95.

Reassemble by reversing removal procedure and mate splines of rear shaft with splines of shift collar as shaft is installed.

120. R&R AND OVERHAUL (540-1000 RPM). To remove rear pto shaft (39—Fig. 113), drain transmission case and remove pto shield and shaft guard, if so equipped. Place rear pto shaft control in "OFF" position, then remove output (stub) shaft (31) from power-

shaft pilot (33). Remove quill (28), then using caution not to damage bore, pry out oil seal (34). Remove snap ring (35), temporarily attach stub shaft (31) to function as a puller and remove powershaft pilot (33) and washer (36). Powershaft (39) can now be removed but be careful not to pull rear pto shaft shift collar from front drive shaft.

Clean and inspect all parts. Bearing (37) can be removed from pilot after removing snap ring (38). Check condition of "O" ring (32).

To reassemble rear pto shaft, insert

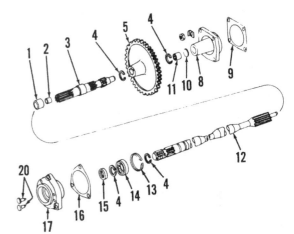

Fig. 112—Exploded view showing gears and shafts of typical transmission driven 540 rpm pto.

1. Bushing
2. Bushing
3. Drive (front) shaft
4. Snap ring
5. Driven gear
8. Cover
9. Gasket
10. Thrust washer
11. Needle bearing
12. Power (rear) shaft
13. Snap ring
14. Ball bearing
15. Oil seal
16. Gasket
17. Bearing quill
20. Drive screw

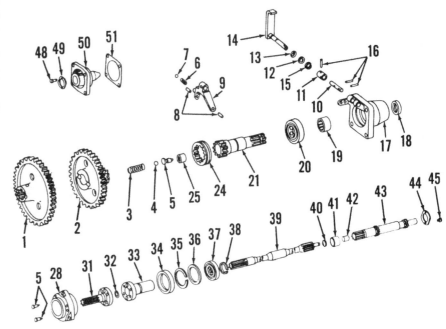

Fig. 113—Exploded view of dual 540-1000 rpm continuous-running pto shafts and gears. Items (6 through 24) are used with mid-pto units whereas items (48 through 51), are used on units with no mid-pto. Driven gears (1 and 2) are fitted with renewable bushings which are not shown.

1. Driven gear (540)	12. Washer	25. Needle bearing
2. Driven gear (1000)	13. "O" ring	28. Quill
3. Spring	14. Shifter lever	31. Rear stub shaft
4. Ball	15. Retaining ring	32. "O" ring
5. Drive screw	16. Spring pins	33. Pilot
6. Detent spring	17. Quill	34. Oil seal
7. Ball	18. Oil seal	35. Snap ring
8. Dowel pin	19. Needle bearing	36. Washer (special)
9. Yoke	20. Ball bearing	37. Ball bearing
10. Shifter shaft	21. Mid-stub shaft	38. Snap ring
11. Coupling	24. Collar	39. Power (rear) shaft

40. Thrust washer	44. Thrust washer
41. Bushing	45. Drive screw
42. Bushing	48. Drive screw
43. Drive (front) shaft	49. Thrust washer
	50. Cover
	51. Gasket

shaft (39) in rear of housing and turn shaft so shaft splines mate with shifter collar splines. Align splines of pilot (33) with splines of pto shaft, bump pilot and bearing assembly into place and install snap ring (35). Install oil seal (34) with lips toward front, then install quill (28), stub shaft (31) and shield and shaft guard, if so equipped.

DRIVEN GEARS, FRONT DRIVE SHAFT AND MID-PTO SHAFT

121. **R&R AND OVERHAUL.** To service the pto driven gears, front powershaft or mid-pto stub shaft, it is first necessary to split tractor as outlined in paragraph 95. Refer also to Fig. 114.

NOTE: The pto clutch power shaft and engine clutch shaft can also be removed at this time by withdrawing them rearward out of clutch housing. Pto drive gear can be pressed from pto clutch power shaft if necessary.

Prior to any disassembly, install rear pto shaft, if removed, and place rear pto shaft shifter lever in "ON" position. This will slide shift collar on rear pto shaft and prevent it from dropping to bottom of transmission case. If shift collar comes off, it will be necessary to remove the rockshaft housing to retrieve it.

Remove spring (3—Fig. 113) from front bore of drive shaft (43), remove driven gears (2 and 1), then pull drive shaft (43). See Fig. 115. Inspect thrust washer (T) at this time. Also inspect driven gear bushings and the drive shaft bushing in front bore of transmission. Renew as necessary.

To disassemble mid-pto stub shaft, remove ball (4—Fig. 113) from bore of stub shaft (21), then remove retaining ring (15), bump out roll pin and pull lever assembly from clutch housing. Bump out roll pin retaining shifter yoke (9), pull yoke shaft (10) from yoke, then remove yoke and shift collar (24) from stub shaft and catch detent ball (7) and spring (6) as yoke is removed. Remove snap ring and bump stub shaft and bearing (20) rearward from quill (17). Straighten locks and remove quill stub nuts them bump quill from clutch housing.

NOTE: On models without mid-pto, only removal of cover (50) is involved and the procedure for doing so will be obvious.

Clean and inspect all parts. Bearing (20) can be pressed from stub shaft after removing snap ring. Needle bearing (25) and drive screw (5) located in bore of stub shaft are renewable. Oil seal (18) is installed in quill with metal

side forward. Renew "O" ring (13) if doubt exists as to its condition.

Reassemble by reversing disassembly procedure and either use a piece of shim stock as a seal protector, or use caution, when inserting stub shaft through seal (18).

REAR PTO SHAFT SHIFTER

122. **R&R AND OVERHAUL.** To remove rear pto shaft shifter assembly, remove the high and low range shifter shafts as outlined in paragraph 96 and the transmission countershaft as out-

lined in paragraph 97. Remove detent assembly located forward of and slightly below pto shift lever. Clip lock wire and remove shifter fork set screw. Remove snap rings from ends of shifter shaft, push shifter shaft out front of transmission case and lift out shifter fork. Pull front pto shaft out and remove shift coupler. Bump roll pin from shifter arm and pull shifter lever from shifter arm and transmission case.

Clean and inspect all parts and renew as necessary. Reinstall shifter mechanism by reversing removal procedure.

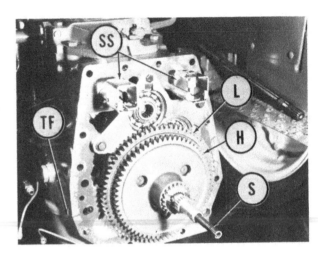

Fig. 114—View showing front end of transmission assembly. Rear gear (L) is 540 rpm driven gear. Front gear (H) is 1000 rpm driven gear.

- H. 1000 rpm gear
- L. 540 rpm gear
- S. Spring
- SS. Shifter shafts
- TF. Transmission filter

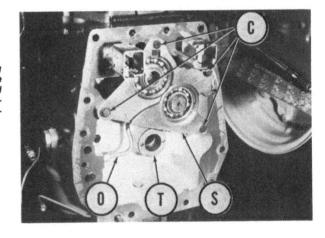

Fig. 115—View showing front of transmission case with pto drive shaft and both pto driven gears removed. Note thrust washer (T).

- C. Cap screws
- O. Oil line
- S. Countershaft support
- T. Thrust washer

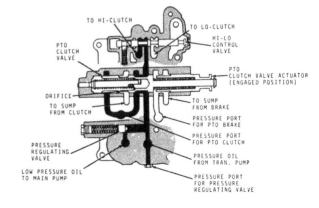

Fig. 116—Cross-sectional view ipto control valve showing fluid flow. Refer to Fig. 121 for exploded view.

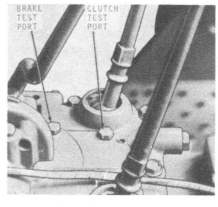

Fig. 117—Cross-sectional view of ipto power train used in 540-1000 rpm models. The 540 rpm unit is similar except that reduction gears (5 and 6) and stub shafts (7 and 8) are not used and shaft (4) extends rearward out of transmission housing. Clutch and front reductions ratio also differ.

1. Pto input shaft	3. Mid pto shaft
2. Ipto clutch	4. Pto output shaft
5. Countershaft gear	7. 1000 rpm stub shaft
6. 540 rpm driven gear	8. 540 rpm stub shaft

INDEPENDENT POWER TAKE-OFF

Some tractors are optionally equipped with an independent power take-off which is driven by a splined hub in the single stage engine clutch cover. Control of the power shaft is entirely independent of transmission and is accomplished by a hydraulically engaged multiple disc clutch contained in clutch housing.

Independent Power Take-Off is available in combination with, or without, Hi-Lo Shift but cannot be used with Reverser Unit.

Fig. 118—View of shift cover showing ipto shift lever installed.

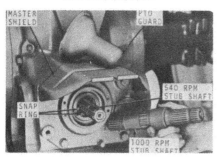

Fig. 119—Stub shafts are interchangeable from rear without draining transmission.

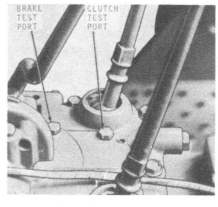

Fig. 120—View of shift cover with test port plugs identified.

Fig. 121—Exploded view of shift cover and valves on models equipped with ipto and Hi-Lo shift unit.

1. Hi-Lo clutch spool
2. Shift cover
3. Regulating valve
4. Test plug
5. Detent assembly
6. Detent pin
7. Cover
8. Ipto shift valve

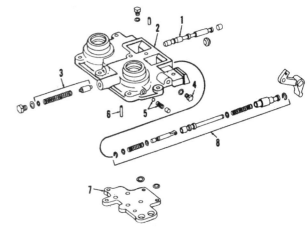

OPERATION

123. Refer to Fig. 116 for cross-sectional view of control valve and to Fig. 117 for cross-section of power train. Hydraulic power to engage the pto clutch and brake is provided by the transmission pump and is combined with the Hi-Lo clutch circuit on models so equipped.

The pto control valve is designed so there is no overlap between clutch and brake circuits. When control lever is moved to the engaged position, pressure fluid flows to the pto clutch and at the same time enters the area behind the valve, preventing rapid opening of the valve and consequent rough clutch engagement. At the same time pressure is ported to the clutch, pto brake pressure is released and brake piston passage opened to the sump.

When control lever is moved to the disenaged position, clutch passage is opened to the sump and pressure applied to a piston located in countershaft bearing support which applies a band-type brake to pto output shaft.

TESTING

124. Test ports for brake and clutch passages are shown in Fig. 120, or circuit pressure can be checked by removing plug (4—Fig. 121). Pressure should be 140-160 psi at any port when that circuit is open, checked at operating temperature and 2100 engine rpm.

If regulated pressure cannot be properly adjusted, refer to paragraph 130 for additional hydraulic system checks and to paragraphs 125 through 128 for overhaul procedure.

OVERHAUL

125. **CONTROL AND PRESSURE REGULATING VALVES.** Control and pressure regulating valves are located in shifter cover as shown in Fig. 121. Regulating valve can be adjusted with-

out removing shifter cover, but overhaul of valve can only be accomplished after cover is removed; proceed as follows:

Remove the four cap screws retaining transmission shield and remove the shield by working it up over shift lever boots. Disconnect Hi-Lo linkage if so equipped. Disconnect rear wiring harness. Shift ipto control lever to engaged position, then remove lever and housing. Remove remaining cap screws and lift off shift cover, valves and shift levers.

Pin (6) is inserted behind spring of valve detent (5). Remove plate (7) and withdraw the pin before attempting to remove Hi-Lo valve spool (1). Valve spool (1), detent (5) and pin (6) will not be present in models not equipped with Hi-Lo shift unit.

Clean and inspect all parts. Front and rear springs in ipto shift valve (8) are interchangeable and three spacer washers are normally used in each spring location. Rear spring should

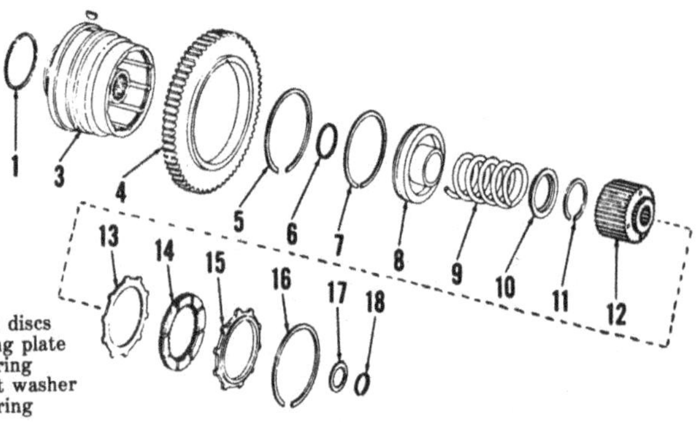

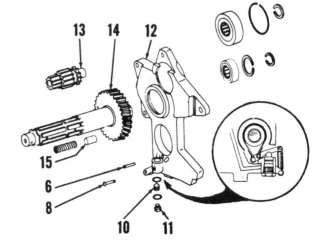

Fig. 123—Exploded view of ipto clutch and associated parts. Refer also to text for additional parts identification.

1. Sealing ring
3. Housing
4. Drive gear
5. Snap ring
6. Sealing ring
7. Piston ring
8. Piston
9. Spring
10. Retainer
11. Snap ring
12. Clutch hub
13. Separator plates
14. Clutch discs
15. Backing plate
16. Snap ring
17. Thrust washer
18. Snap ring

require a pressure of approximately 7 lbs. to move seating washer away from head of guide pin. Add or remove washers to adjust, provided at least one washer is used (three is normal). With valve, front spring and actuator assembled, a pressure of approximately 13 lbs. should be required to move actuator away from front snap ring. Add or remove washers behind front spring to adjust.

Figs. 122 or 122A show an exploded view of ipto valve lever and associated parts. The two detent screws (2) should be tightened until they bottom (springs

solidly compressed); then backed out two full turns to apply the required detent pressure.

126. IPTO CLUTCH AND INPUT GEARS. Refer to Fig. 123 for an exploded view of clutch assembly. To remove the clutch, first split tractor between clutch housing and transmission case as outlined in paragraph 95.

On units equipped with mid-pto, shift rear selector lever to engaged position to prevent the disconnect collar from falling into transmission case and withdraw clutch unit and pto drive shaft as an assembly. On units without mid-pto, remove front snap ring (18) and withdraw clutch unit only, leaving pto drive shaft in transmission housing.

Clutch housings (3) differ on units equipped with 540 rpm single speed pto and 540-1000 rpm models. Gear (4) also differs, the 540 rpm unit having 73 teeth and the dual speed unit 62 teeth. On 540 rpm units the clutch turns at 540 rpm at rated speed and higher torque is transmitted. Ten clutch discs (14) and separator plates (13) are used. On dual speed units the clutch turns at 1000 rpm and speed reduction takes place at output end. Six each clutch discs (14) and separator plates (13) are used on dual speed units. Drive gear (4) is installed on clutch drum with offset forward on models equipped with Hi-Lo shift; or offset rearward on other

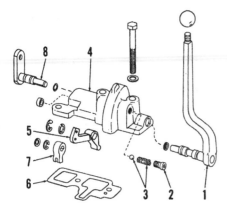

Fig. 122—Exploded view of early ipto valve lever and associated parts.

1. Ipto shift lever
2. Detent plug
3. Detent assembly
4. Cover
5. Lever arm
6. Gasket
7. Hi-Lo Shift valve arm
8. Valve shaft

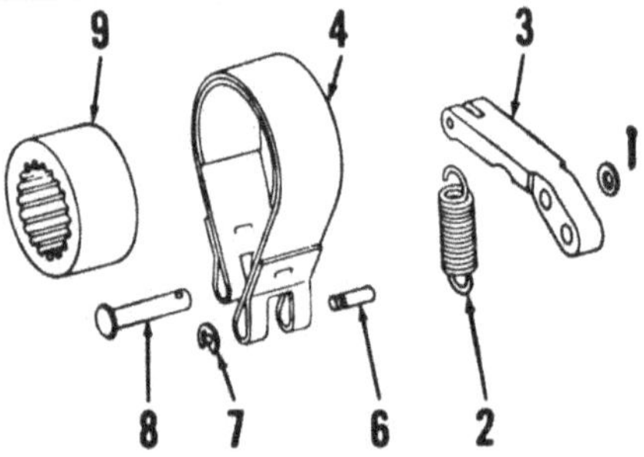

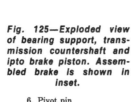

Fig. 124—Exploded view of ipto shaft brake and associated parts. Refer to Fig. 125 for assembled view and for actuating piston and cylinder assembly.

2. Spring
3. Lever
4. Band
6. Pivot pin
7. Retainer
8. Anchor pin
9. Brake drum

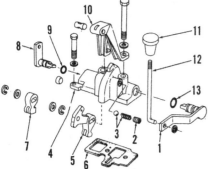

Fig. 122A—Exploded view of late ipto valve lever and associated parts.

1. Ipto shift lever
2. Detent plug
3. Detent assembly
4. Cover
5. Lever arm
6. Gasket
7. Hi-Lo shift valve arm
8. Valve shaft
9. "O" ring
10. Support
11. Knob
12. Rod
13. "O" ring

Fig. 125—Exploded view of bearing support, transmission countershaft and ipto brake piston. Assembled brake is shown in inset.

6. Pivot pin
8. Anchor pin
10. Piston
11. Plug
12. Bearing support
13. Transmission drive gear
14. Countershaft
15. Friction plug

models. Check the gear before removal, for proper installation.

Input gear can be withdrawn from rear of clutch housing on models without Hi-Lo shift, or removed with oil manifold (or transmission oil pump) on models with Hi-Lo shift.

To remove clutch piston (8) compress spring (9) using a press and suitable fixture, then remove snap ring (11). Gear (4) should be heated to approximately 360 degrees before pressing on pto clutch drum.

127. IPTO BRAKE. The band type ipto brake mounts on transmission countershaft bearing support and acts on output shaft to stop the pto shaft when clutch is disengaged and brake applied. Refer to Figs. 124 and 125 for exploded views.

To remove the ipto brake, first remove clutch as outlined in paragraph 126; then remove countershaft and bearing support as in paragraph 97. Brake band is mounted on bearing support as shown, and actuating piston is carried in bearing support cylinder bore. Renew brake band if total thickness (band and lining) is 1/8-inch or less.

128. OUTPUT SHAFT AND GEARS. Refer to Fig. 126 for exploded view of dual ipto shafts and to Fig. 127 for cross-section of reduction gears and associated parts.

To remove the rear pto shaft, reduction gears and associated parts, drain transmission and remove pto shield and shaft guard. Move rear pto lever to "OFF" position. Remove stub shaft, if installed. Remove the cap screws retaining pto shaft bearing quill, remove the quill then withdraw gears, shafts and associated parts. When installing bearing quill on dual ipto models, adjust draft control negative stop screw as outlined in paragraph 134.

On models with 540 rpm single speed ipto, shaft and rear bearing quill will be removed as a unit. On models without mid-pto, it will be necessary to remove rockshaft housing as outlined in paragraph 143 and install coupling sleeve between pto output shaft and pto drive shaft.

HYDRAULIC LIFT SYSTEM

The hydraulic lift system is a closed center, constant pressure type. The stand-by 2200-2300 psi pressure is furnished by an eight piston, constant

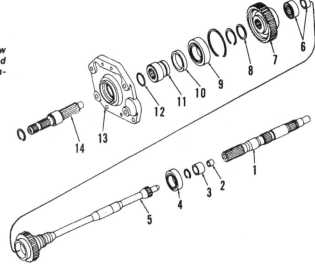

Fig. 126—Exploded view of ipto drive gears and shaft units showing component parts.

1. Ipto drive shaft
2. Bushing
3. Bushing
4. Bearing
5. Ipto output shaft
6. Bearing
7. 540 rpm gear
8. Spring washer
9. Bearing
10. Oil seal
11. Shaft pilot
12. "O" ring
13. Bearing quill
14. Stub shaft

running, variable displacement pump which is mounted in the tractor front support and is driven by a coupling from front end of engine crankshaft. Charging oil for the hydraulic main pump is supplied by the transmission oil pump and oil not used by the main hydraulic pump is routed to the auxiliary hydraulic oil reservoir which provides an auxiliary supply of oil when transmission oil pump is unable to meet the demand of main hydraulic pump. When there is little or no demand by the main hydraulic oil pump, the overflow from auxiliary reservoir is returned to the clutch housing through an oil cooler where part of it fills the brake control valve reservoir and the remainder lubricates the transmission shafts and gears.

TROUBLESHOOTING

All Models

129. The following are symptoms which may occur during the operation of the hydraulic lift system. By using this information in conjunction with the Test and Adjust information, no trouble should be encountered in servicing the hydraulic system.

1. Slow system operation. Could be caused by:
   a. Clogged transmission oil filter.
   b. Transmission oil pump inlet screen plugged.
   c. Faulty transmission oil pump.
   d. Transmission oil pump relief stuck open.
   e. Hydraulic pump stroke control valve not seating properly.
   f. Hydraulic pump crankcase out-filter plugged.
   g. Oil leak on low pressure side of system.
2. Erratic pump operation. Could be caused by:

a. Pump stroke control valve not seating properly.
   b. Leaking pump inlet or outlet valves or valve "O" rings.
   c. Broken or weak pump piston springs.
3. Noisy pump. Could be caused by:
   a. Worn drive parts or loose cap screws in drive coupling.
   b. Air trapped in oil cavity of pump stroke control valve.
4. No hydraulic pressure. Could be caused by:
   a. Pump shut-off valve closed (if so equipped).
   b. No oil in system.
   c. Faulty pump.
5. Rockshaft fails to raise or raises slowly. Could be caused by:
   a. Excessive load.
   b. Low pump pressure or flow.

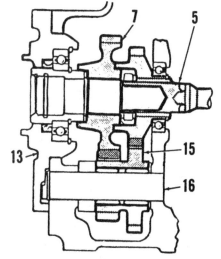

Fig. 127—Cross-sectional view of ipto reduction gears and associated parts. Refer to Fig. 126 for parts identification except for the following.

15. Countershaft gear   16. Countershaft

c. Rockshaft piston "O" ring failed.

d. Flow control valve maladjusted.

e. Thermal relief valve defective.

f. Cam follower adjusting screw maladjusted.

g. Transmission oil filter plugged.

h. Defective seals between cylinder and rockshaft housing, or between rockshaft housing and transmission case.

6. Rockshaft settles under load. Could be caused by:

  a. Leaking discharge valve.

  b. Leaking rockshaft cylinder check valve.

  c. Leaking cylinder pipe plug.

  d. Faulty rockshaft cylinder valve housing.

7. Rockshaft valves hunt. Could be caused by:

  a. Control valves maladjusted.

  b. Rockshaft piston "O" ring faulty.

  c. Discharge valve leaking.

  d. Thermal relief valve leaking.

8. Rockshaft lowers too fast to too slow. Could be caused by:

  a. Throttle valve maladjusted.

  b. Broken or disconnected valve linkage.

9. Rockshaft raises too fast. Could be caused by:

  a. Flow control valve incorrectly set.

10. Insufficient load response. Could be caused by:

  a. Control valve clearance excessive.

  b. Control valves sticking.

  c. Control lever not positioned correctly on quadrant.

  d. Worn load control shaft or bushings.

  e. Negative stop screw turned too far in.

11. Hydraulic oil too hot. Could be caused by:

  a. Control valves adjusted too tight and held open.

  b. Control valves leaking.

  c. Control valve "O" rings faulty.

  d. Thermal relief valve faulty (leaking).

HYDRAULIC SYSTEM TESTS

All Models

Before making any tests on the main hydraulic pump or lift system, be sure the transmission oil pump is satisfactory as the performance of main hydraulic pump is dependent upon being charged by the transmission oil pump. For information on testing of transmission oil pump, refer to paragraph 101.

130. **MAIN HYDRAULIC PUMP TEST.** The main hydraulic pump can be tested for standby pressure and

Fig. 128—Install hydraulic test unit as shown. Refer to text.

A. Return hose
B. Fitting Y3018
C. Fitting Y3021
D. Pressure hose
E. Fitting Y3005
F. Fitting JDH-38
G. Fitting R34063
H. Plug open line

flow rate by using a hydraulic test unit. Pump standby pressure can also be checked by using a 3000 psi gage. Refer to paragraph 131 for procedure using hydraulic test unit and to paragraph 132 for procedure using pressure gage only.

NOTE: On models equipped with selective remote control valves, main hydraulic pump pressure can also be checked as outlined in paragraph 146.

131. When testing pump using the hydraulic test unit, refer to Fig. 128 for connecting the test unit.

NOTE: If flow rate of main hydraulic pump is to be checked, it is necessary to connect hydraulic test unit so that discharged oil will be directed back to the pump inlet to insure an adequate oil supply for the main hydraulic pump.

Start the engine and run at 2500 rpm. Close the hydraulic test unit control valve and note the pressure gage which should read 2220-2280 psi. This is the pump standby pressure. If pump standby pressure is not as stated, loosen jam nut and turn pump stroke control valve adjusting screw (Fig. 129) in to increase pressure or out to decrease pressure.

With the correct pump standby pressure established, close tester control valve so that pressure gage indicates 2050 psi for 2240 models; 2000 psi for other models. At this (system working) pressure, the tester flow meter should show a minimum flow of 12.5 gpm for 2240 models, 9.5 gpm for other models with the 1.38 cu.-in. pump, or 18.5 gpm minimum flow for the 2.40 cu.-in. pump. If main hydraulic pump will not meet both of the above conditions it must be removed and overhauled as outlined in paragraphs 139 or 141.

132. When no hydraulic test unit is available, test main hydraulic pump standby pressure by installing a 0-3000 psi pressure gage as shown in Fig. 128A.

Start engine and run at 2500 rpm. With transmission in park, move the selective control valve lever forward and note the gage reading which should be 2220-2280 psi. This is the main hydraulic pump standby pressure. If pump standby pressure is not as stated, loosen jam nut and turn pump stroke control valve adjusting screw (Fig. 129) in to increase pressure or out to decrease pressure. If pump will not produce the 2220-2280 psi pressure, it must be removed and overhauled as outlined in paragraph 139 or 141.

133. **FLOW CONTROL VALVE TEST.** Tractors are equipped with a

Fig. 128A—If no flo-rater is available to test pump stand-by pressure, install gage as shown.

Fig. 129—Hydraulic pump stroke control valve is located as shown at arrow.

flow control valve that is incorporated into the rockshaft valve circuit to reduce the oil flow. See Fig. 130. Check the flow control valve setting with a hydraulic test unit as follows: Connect the hydraulic test unit outlet line as outlined in paragraph 131. Place the rockshaft load control selector lever in "L" position, start engine and run at 2500 rpm, then move the rockshaft control lever to maximum lift position. Close the tester control valve until tester pressure gage shows 1750 psi, then check flow meter gage which should show a 4¾-5¾ gpm flow. If oil flow is not as stated, refer to Fig. 130 and remove plug, valve and spring, then vary washers (shims) as required. Also check spring which should test 7½-9 lbs. when compressed to a length of 63/64-inch.

If tractor is equipped with selective (remote) control valves, refer to paragraph 146 for testing information.

HYDRAULIC SYSTEM ADJUSTMENT

All Models

The following paragraphs outlined the adjustments that can be made when necessary to correct faulty hydraulic operation, or that must be made when reassembling a hydraulic lift system that has been disassembled for service.

However, because of the interaction of component parts, the adjustments must be made in the following order.
A. Negative stop screw
B. Rockshaft control lever neutral range
C. Control lever position
D. Load control
E. Rate-of-drop

134. NEGATIVE STOP SCREW ADJUSTMENT. The negative stop screw is located on the right rear of the transmission case as shown in Fig. 131, on models without dual independent power take-off; or as shown at arrow—Fig. 132 on models with dual ipto.

To adjust the negative stop screw, loosen jam nut and turn stop screw in until it just contacts the load control arm, then turn (back) screw out 1/6-turn and tighten jam nut.

NOTE: Contact of stop screw with load control arm can be more easily felt if rockshaft housing filter plug is removed and a screwdriver is held against upper end of the load control arm.

135. CONTROL LEVER NEUTRAL RANGE ADJUSTMENT. To adjust the control lever neutral range, first remove the pipe plug, located directly in front of control lever tube which will expose the control valve adjusting screw. See Fig. 133. Place selector lever in upper (D) notch, start engine and run at 2500 rpm. Start with rockshaft control lever at rear of quadrant, move lever slowly forward until rockshaft just starts to lower and mark this point on upper edge of quadrant. Now move lever slowly rearward until rockshaft just begins to raise and mark this point on upper edge of quadrant. Distance between these two marks should be 1/8 to 3/16-inch.

If neutral range of rockshaft control lever is not as stated, turn the control valve adjusting screw clockwise to increase, or counter-clockwise to decrease, the control lever neutral range.

136. CONTROL LEVER POSITION ADJUSTMENT. Place load selector in upper (D—Fig. 134) position, start engine and run at 2500 rpm, then move rockshaft control lever fully forward to completely lower rockshaft.

Loosen rockshaft control lever adjusting nut and move control lever rearward until there is 5/16-inch clearance between control lever friction pin and bottom end of quadrant slots.

Then rotate control shaft arm clockwise as far as possible, then rotate control shaft arm counter-clockwise to the point where rockshaft just starts to raise. Tighten control lever adjusting nut.

137. LOAD CONTROL ARM ADJUSTMENT. Remove the rockshaft filler hole plug to expose the load control arm cam follower adjusting screw. See Fig. 135. Place selector lever in lower position, start engine and run at 2500 rpm. Move rockshaft control lever

Fig. 133—Control valve adjusting screw is located under plug (B) which is in front of control shaft tube.

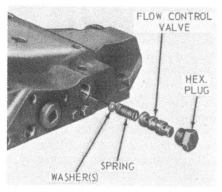

Fig. 130—Flow control valve is installed in right side of rockshaft housing.

Fig. 131—Negative stop adjusting screw is located on right rear of transmission case without dual ipto.

Fig. 132—Arrow shows negative stop screw on models with dual ipto.

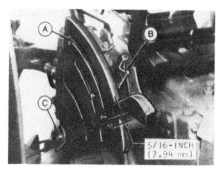

Fig. 134—Set rockshaft control lever as shown when making control lever adjustment.

A. Quadrant slot   upper "D" notch
B. Selector lever in   C. Adjusting nut

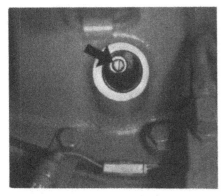

Fig. 135—Cam follower adjusting screw is located behind rockshaft filler plug hole.

Fig. 136—Arrow shows location of the rate-of-drop adjusting screw.

fully forward to completely lower rock-shaft, then move control lever rear-ward until distance between control lever friction pin and top end of quadrant slot is 3-5/8 inches from upper end of slot. Hold rockshaft control lever in this position, loosen jam nut and turn cam follower adjusting screw counter-clockwise until rockshaft is fully lowered, then turn adjusting screw clockwise until rockshaft starts to raise. Tighten jam nut.

NOTE: If rockshaft begins to raise before rockshaft control lever reaches the 3-5/8 inch setting, it will be necessary to turn cam follower adjusting screw counter-clockwise to allow the lever to be positioned while rockshaft remains in lowered position.

138. **RATE-OF-DROP ADJUSTMENT.** The rate-of-drop (throttle) adjusting screw is located on top side of rockshaft housing as shown in Fig. 136. Turn adjusting screw clockwise to decrease rate-of-drop, or counter-clockwise to increase rate-of-drop. Tighten jam nut after adjustment is made. Rate-of-drop will vary with the weight of the attached implement.

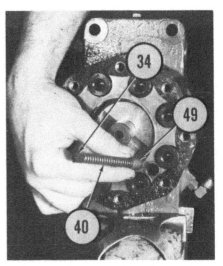

Fig. 137A—Crankcase outlet valve is removed from bore as shown. Valve pin is still in bore.

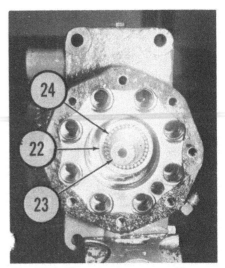

Fig. 137B—View showing pump with cover, inlet valves and pistons removed. Pump shaft (23), rollers (24) and cam race (22) can be removed as a unit, if desired.

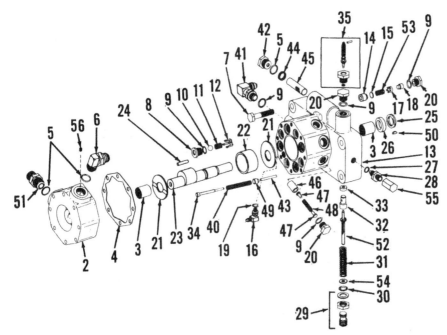

Fig. 137—Exploded view of the 1.38 cu. in. main hydraulic pump.

2. Cover	16. Adjustable elbow	28. Connector	44. Seal (2 used)
3. Needle bearing	17. Guide	29. Adjusting screw	45. Filter screen
4. Gasket	18. Stop	assembly	46. Pump piston
5. "O" ring	19. "O" ring	30. "O" ring	47. Guides
6. Adjustable elbow	20. Plug	31. Spring	48. Piston spring
8. Inlet valve seat	21. Thrust washer	32. Stroke control valve	49. Crankcase outlet
9. "O" ring	22. Cam race	33. Valve seat	valve
10. Inlet valve ball	23. Pump shaft (cam)	34. Spring guide	50. Woodruff key
11. Spring	24. Roller bearing (33	35. Pump shut-off assy.	51. Connector
12. Guide	used)	40. Spring	52. Valve guide
13. Housing	25. Quad ring	41. Adjustable elbow	53. Spring
14. Outlet valve seat	26. Oil seal	42. Plug	55. Cap
15. Outlet valve	27. "O" ring	43. Pin	56. Orifice

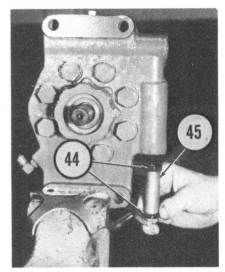

Fig. 137C—Remove pump filter (45) as shown. Two seals (44) are used.

MAIN HYDRAULIC PUMP

All Models

The main hydraulic pump may be either 1.38 cu. in. or a 2.40 cu. in. piston type. While pump assemblies and parts will differ, operation of the two is basically the same.

Pumps maintain a 2000-2050 psi working pressure with a standby pressure of 2220-2280 psi. Standby operation of the pump occurs when pressure in the pump crankcase builds to 2220-2280 psi thereby holding the pump pistons away from the pump cam. The pump crankcase (standby) pressure is controlled by stroke control valve located in a bore in the pump housing.

A pump off (destroking) screw (35—Fig. or 32—Fig. 138A) is optionally available which will make pump inoperative and will act as an aid during cold weather starts.

139. R&R AND OVERHAUL 1.38 CU. IN. PUMP. The hydraulic pump can be removed without radiator being removed, however, removal of radiator provides additional working room. The following procedure is based on radiator not being removed.

To remove hydraulic pump, first remove hood and side panels. Drain radiator and remove lower hose. Disconnect reservoir to pump line and drain reservoir. Disconnect inlet and pressure lines from pump. Remove air cleaner assembly. Remove engine timing hole plug and turn engine until the clamping cap screw at forward end of pump drive coupling is at about the 10 o'clock position. Use a long extension and remove cap screw. Remove pump mounting bolts, pull pump forward to remove pump shaft from drive coupling, then disconnect bleed line and

remove pump from left side of tractor.

With pump removed, clean unit and prior to any disassembly, use a dial indicator and check pump shaft end play. Shaft end play should be 0.004-0.038 and if end play is excessive, renew thrust washers (21—Fig. 137) when reassembling. Clamp pump in a vise and remove cover (2), then remove crankcase outlet valve (items 34, 40, 43

and 49) from pump housing. Also see Fig. 137A. Remove outer thrust washer (21—Fig. 137).

NOTE: At this time, shaft (23) can be removed if desired by removing Woodruff key (50) and pushing shaft out of body and cam ring (22). Be sure not to lose any of the 33 rollers (24) which will be loose. However, if pump requires

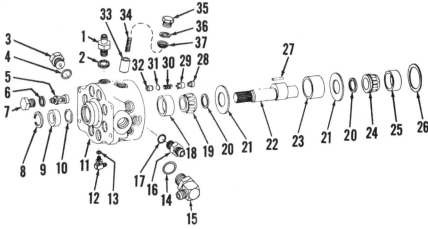

Fig. 138—Exploded view of the 2.40 cu. in. main hydraulic pump.

1. Connector	11. Housing	21. Thrust washer (2 used)
2. Special washer	12. Elbow	22. Shaft
3. Plug	13. "O" ring	23. Race
4. "O" ring	14. "O" ring	24. Bearing cone
5. Inlet valve (8 used)	15. Elbow	25. Bearing cup
6. "O" ring (8 used)	16. Connector	26. Shim
7. Plug (8 used)	17. "O" ring	27. Bearing (36 used)
8. Snap ring	18. Bearing cup	28. Stop (8 used)
9. Oil seal	19. Bearing cone	29. Guide (8 used)
10. Quad-ring	20. Spacer (2 used)	
		30. Spring (8 used)
		31. Discharge valve (8 used)
		32. Seat (8 used)
		33. Piston (8 used)
		34. Spring (8 used)
		35. Plug (8 used)
		36. "O" ring (8 used)
		37. Sheath

Fig. 137D—Pump stroke control valve being removed. Valve guide (52—Fig. 137) is in inside spring (31).

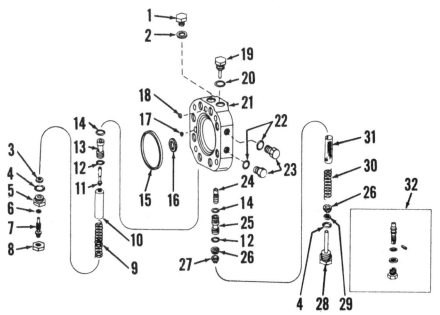

Fig. 138A—Exploded view of the stroke control valve for the 2.40 cu. in. pump.

1. Plug	9. Spring	17. "O" ring	25. Sleeve
2. "O" ring	10. Guide	18. "O" ring	26. Packing (2 used)
3. Washer	11. Stroke control valve	19. Plug	27. Guide
4. "O" ring	12. "O" ring	20. "O" ring	28. Plug with pin
5. Bushing	13. Sleeve	21. Housing	29. Washer
6. "O" ring	14. "O" ring	22. "O" rings	30. Spring
7. Adjusting screw	15. "O" ring	23. Plugs	31. Filter
8. Nut	16. Packing (2 used)	24. Outlet valve	32. Pump shut-off

disassembly, it is good policy to completely disassemble pump and inspect all parts and to remove pump pistons before removing pump shaft. See Fig. 137B.

Remove pistons assemblies (20, 46, 47 and 48—Fig. 137), then if not already removed, remove shaft (23), rollers (24) and cam ring (22). Remove inlet valves (8, 10, 11 and 12) and outlet valves (14, 15, 17, 18 and 20). Identify all pistons, valves, springs and seats so they can be reinstalled in their original positions. Remove plug (42) and remove filter screen (45). Also see Fig. 137C. Loosen jam nut and remove stroke control valve adjusting screw (29—Fig. 137), spring (31), spring guide and valve (32). Also see Fig. 137D.

Clean and inspect all parts. Use data in paragraph 140 as a guide for renewal of parts. Discard all "O" rings and use new during assembly. On some early pumps, split bushings are used instead of shaft needle bearings (3—Fig. 137). Bushings are renewable and are installed flush with inner ends of bushing bores. Pay particular attention to shaft bore around the quad ring seal groove as leakage at this point can cause the pump to be slow in going out of stroke. If outlet valve seats (14) are damaged, drive them out and install new seats with large chamfered end toward bottom of bore. Press seats in to 1.171 below spot face surface of bore. O.D. of new outlet valve (15) is 0.609-0.611 and any valve that is distorted, scored or worn should be renewed. If stroke control valve seat (33) is damaged, remove plug (20), or shut-off assembly (35) if so equipped, and drive out seat (33). Install new valve chamfered end first and drive it in bore until it bottoms. Spring (31) should test 160-190 lbs. when compressed to a length of 2½ inches. Renew any pistons (46) which are scored or pitted. Piston springs (48) should exert 18-22 psi when compressed to 1.26 inches and springs MUST test within 1.5 lbs. of each other when compressed to a length of 1.26 inches.

When reassembling, dip all parts in oil. Use grease to hold roller in I.D. of cam ring during assembly. Seal (26) is installed with printed side toward outside. Thrust washers are installed with grooved sides away from pump shaft cam.

140. The following data applies to 1.38 cu. in. pumps and can be used as a guide for parts renewal.

Thrust washer thickness....0.087-0.091
Shaft bushings bore I.D.
(early models) 1.1245-1.1255
Piston bore I.D. 0.680-0.681
Cam race O.D.............2.235-2.245

Cam race I.D. 1.800-1.801
Crankcase bore depth 2.660-2.666
Cover to body screws—
  First torque ....14 ft.-lbs.
  Final torque ....35 ft.-lbs.
Torque, piston plugs........90 ft.-lbs.
Torque, pump mounting
  screws ....85 ft.-lbs.

141. R&R AND OVERHAUL 2.40 CU. IN. PUMP. The hydraulic pump can be removed without radiator being removed; however, removal of radiator provides additional working room. The following procedure is based on radiator not being removed.

To remove hydraulic pump, first remove hood and side panels. Drain radiator and remove lower hose. Disconnect reservoir to pump line and drain reservoir. Disconnect inlet and pressure lines from pump. Remove air cleaner assembly. Remove engine timing hole plug and turn engine until the clamping cap screw at forward end of pump drive coupling is at about the 10 o'clock position. Use a long extension and remove cap screw. Remove pump mounting bolts, pull pump forward to remove pump shaft from drive coupling, then disconnect bleed line and remove pump from left side of tractor. With pump removed, clean unit and prior to disassembling the pump, check pump shaft end play using a dial indicator, and record the measurement for convenience in reassembling. End play should be 0.001-0.003 for the large pump. End play is adjusted by adding

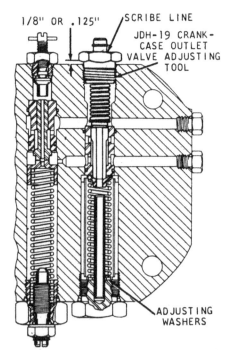

Fig. 139—Special adjusting tool (JDH-19) can be used to determine stroke control valve setting. Refer to text.

or removing shims (26—Fig. 138) which are available in thicknesses of 0.006 and 0.010. Bearing wear, or wrong number of adjusting shims can cause excessive end play.

To disassemble the pump, remove the four cap screws retaining the stroke control valve housing (21—Fig. 138A) to front pump, and remove the housing. Withdraw discharge valve plugs, guides, springs and valve (28 through 31—Fig. 138). Remove all piston plugs (35), springs (34) and pistons (33); then carefully withdraw pump shaft (22) together with bearing cones, thrust washers (21), cam race (23) and the 36 loose needle bearings (27).

Remove the plug retaining inlet valve assembly (5) and check inlet valve lift using a dial indicator. Lift should be 0.060-0.080 inch. If lift exceeds 0.080, spring retainer is probably worn and valve should be renewed. Also check for apparent excessive looseness of valve stem in guide. Do not remove inlet valve assembly unless removal is indicated or discharge valve seat (32) must be renewed. To remove the inlet valve, use a small pin punch and drive valve out, working through discharge valve seat (32). If inlet valve is to be reused, place a flat disc on inlet valve head from the inside, so that all the driving force will not strike valve head in the center. Be sure the disc will drive through the hole without touching. Discharge valve seat can be driven out after inlet valve is removed. Be sure to reinstall pistons, springs, valves and seats to their own respective bores. The piston bores are lined with a Teflon sheath, so all bores should be carefully inspected. Scored pistons or bores could cause pistons to stick.

The manufacturer recommends that the eight piston springs (34) test within 1½ lbs. of each other at 34 to 40 lbs. when compressed to a height of 1-5/8 inches. Install seal (9) only deep enough to allow snap ring to enter groove, to avoid blocking the relief hole in body.

Valves located in stroke control valve housing control pump output as follows: The closed hydraulic system has no discharge except through the operating valves or components. Peak pressure is thus maintained for instant use. Pumping action is halted when line pressure reaches a given point by pressurizing the camshaft reservoir of pump housing, thereby holding pistons outward in their bores.

The cutoff point of pump is controlled by pressure of spring (9—Fig. 138A) and can be adjusted by turning adjusting screw (7). When pressure reaches the standby setting, valve (11) opens and meters the required amount of fluid at reduced pressure in crankcase section of pump. Crankcase outlet valve (24) is held closed by hydraulic pressure and blocks the outlet passage. When pressure drops as a result of system demands, crankcase outlet valve is opened by the pressure of spring (30) and a temporary hydraulic balance on both ends of valve, dumping the pressurized crankcase fluid and pumping action resumes. Stroke control valve spring (9) should test 125-155 lbs. pressure when compressed to 3.3 inches, and crankcase outlet springs (30) should test 45-55 lbs. at 2.2 inches.

Cutoff pressure is regulated by the setting of adjusting screw (7) and adjustment procedure is given in paragraph 130. Cut-in pressure is determined by the thickness of shim pack (29) and pressure of spring (30). A special tool (JDH-19) is available to determine thickness of shim pack; refer to Fig. 139 and proceed as follows:

Assemble outlet valve (24—Fig. 138A) and all components using existing shim pack (29). Install special tool (JDH-19) in place of plug (19) using one 1/8-inch thick washer as shown in Fig. 139. If shim pack thickness is correct, scribe line on tool plunger should align with edge of tool plug bore as shown; if it does not, remove top plug (28—Fig. 138A) and add or remove shim washers (29) as required. Shims (29) are available in 0.030 thickness. If special tool is not available use shim washers of same thickness as those removed, then add shims to raise cut-in pressure, or remove shims to lower pressure.

When installing stroke control housing, add or remove shims (26—Fig. 138) as necessary to obtain specified pump shaft end play of 0.001-0.003 inch. Always use new "O" rings, packings and seals. Oil all parts liberally with clean hydraulic system oil. Tighten stroke control valve housing retaining cap screws to 85 ft.-lbs. and tighten piston cap plugs to 100 ft.-lbs. torque. Adjust standby pressure as outlined in paragraph 130 after tractor is reassembled.

142. The following data applies to 2.40 cu. in. pumps and can be used as a guide for parts renewal.

Thrust washer thickness . . 0.1235-0.1265
Piston O.D. 0.8740-0.8744
Piston bore I.D. 0.8747-0.8753
Torque, main pump to stroke
   control valve housing . . . . . . 85 ft.-lbs.
Torque, pump to pump
   support . . . . 85 ft.-lbs.
Torque piston plugs . . . . 90 ft.-lbs.

ROCKSHAFT HOUSING, CYLINDER AND VALVE

All Models

143. **REMOVE AND REINSTALL.** To remove rockshaft housing, disconnect battery ground straps, remove transmission shield and disconnect wires from starter safety switch. Disconnect

lift links from rockshaft arms. Remove seat assembly. If tractor is equipped with remote hydraulic system, disconnect lines from rear of selective (remote) control valves and remove return hose between valve and rockshaft housing. Attach hoist to rockshaft housing, place selector lever in lower (L) position, then unbolt and lift rockshaft housing from transmission case. See Fig. 140.

Installation is the reverse of removal but be sure dowels are in place and that roller link mates properly with the load control arm cam follower.

144. **OVERHAUL.** With rockshaft housing assembly removed, disassemble as follows: Remove rockshaft arms from rockshaft. Remove selector lever from load control arm. Turn unit so bottom side is accessible, then drive out spring pin (SP—Fig. 140) which retains control shaft to pivot block. Unbolt plate from rockshaft housing and remove quadrant and control shaft. Remove remote cylinder outlet adapter from left side of housing and unhook valve spring from linkage. Remove front outside cylinder mounting cap screw, then remove remaining cylinder mounting cap screws, lift cylinder assembly from housing and disengage selector arm from roller link as cylinder is removed. Do not lose throttle

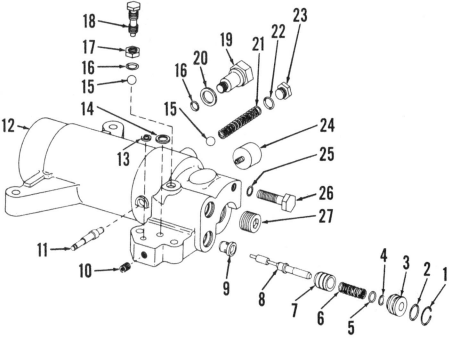

Fig. 140—View showing bottom side of rockshaft housing. Spring pin (SP) retains control lever shaft to pivot block.

Fig. 141—Exploded view showing typical control valve components.

1. Snap ring (2 used)	8. Control valve (2 used)	15. Ball
2. "O" ring (2 used)	9. Seat (2 used)	16. "O" ring
3. Plug (2 used)	10. Plug	17. Nut
4. Backup ring (2 used)	11. Pin	18. Throttle valve
5. "O" ring (2 used)	12. Housing	19. Special plug
6. Spring (2 used)	13. Gasket	20. Gasket
7. Sleeve (2 used)	14. Packing	21. Spring
		22. "O" ring
		23. Plug
		24. Thermal relief
		25. "O" ring
		26. Cap screw
		27. Plug

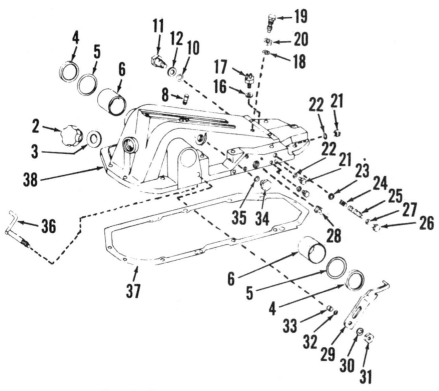

Fig. 142—Exploded view of typical rockshaft housing.

2. Oil filler cap	12. Gasket	22. "O" ring
3. Gasket	16. Aluminum washer	23. Washer (shim .048)
4. Retainer	17. Starter safety	24. Spring
5. "O" ring	switch	25. Flow control valve
6. Bushing	18. "O" ring	26. Plug
8. Pin	19. Throttle valve	27. "O" ring
10. "O" ring	adjusting screw	28. Plug
11. Remote cylinder	20. Jam nut	29. Load selector lever
outlet adapter	21. Plug	30. Washer

31. Nut	
32. "O" ring	
33. Bushing	
34. Plug	
35. "O" ring	
36. Load control arm	
37. Gasket	
38. Rockshaft housing	

wear, bends or other damage. Use Figs. 141, 142, 143 and 145 to identify parts. Check valve spring (21—Fig. 141) should test 22½-27½ lbs. when compressed to a length of 2-1/8 inches. Control valve springs (6) should test 8-5/16 lbs. when compressed to a length of 7/8-inch. New piston diameter is 3.244-3.246. If so equipped, flow control valve spring (24—Fig. 142) should check 7½-9 lbs. when compressed to a length of 53/64-inch. Pay particular attention to seating area of the two control valves (8—Fig. 141). Leakage in this area could cause rockshaft settling if it occurs in the discharge (upper) valve, or upward rockshaft creep if it occurs in the pressure (lower) valve. Also pay close attention to the valve seats (9) and sleeves (7). Inspect rockshaft assembly for fractures, damaged splines or other damage. Check control linkage for worn or bent condition. A worn adjusting cam (10—Fig. 144) will cause difficulty in adjusting lever neutral range.

When reassembling the rockshaft and cylinder unit, use sealant on threads of any pipe plugs that were removed and coat all "O" rings with oil.

Start rockshaft into rockshaft housing, align master splines of rockshaft and control arm, then push rockshaft into position. Install rockshaft bushings, which will be free in rockshaft bores, then install "O" rings and retainers with cupped side outward. Install cam on rockshaft and connecting rod to control arm. Insert valve seats (9—Fig. 141) in their original bores small end first, then place valves (8) in seats with smaller diameter of valves in seats. Install sleeves (7), chamfered end first, and valve springs (6). Install "O" rings (5) and back-up rings (4) in I.D. of plugs (3) with back-up rings toward outer end of valves. Install "O" rings (2) on O.D. of plugs, then carefully install plugs over ends of valves and install snap rings (1). Install "O" ring (9—Fig. 143) and back-up ring (10) on

valve ball as it will fall free.

Remove cam from rockshaft, then pull rockshaft from control arm and housing. Bushing, "O" ring and retainer will be removed by the rockshaft on the side rockshaft is removed from. If necessary, separate piston rod from control arm by driving out spring pin. Load selector shaft can also be removed if necessary. Remove flow control valve assembly from right front of hydraulic housing.

With components removed, disassemble cylinder and valve unit as follows: Remove plug (23—Fig. 141), spring (21) and check ball (15). Remove thermal relief valve (24). Remove retainer ring and pull linkage from linkage pivot pin. Remove snap rings (1), valve plugs (3), spring (6), sleeves (7), valves (8) and seats (9) from bores in cylinder housing. Keep valve assemblies identified so they can be reinstalled in original bores. Remove piston from cylinder by bumping open end of cylinder against a wood block or remove plug (27) and push piston out. Remove and discard piston "O" ring and back-up ring.

Clean and inspect all parts for undue

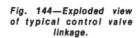

Fig. 143—View of typical rockshaft, piston and lift arms.

1. Cap screws	7. Piston rod
2. Washer	8. Piston
3. Lift arm	9. "O" ring
4. Rockshaft	10. Backup ring
5. Control arm	11. Cam
6. Spring pin	12. Cap screw

Fig. 144—Exploded view of typical control valve linkage.

1. Retaining rings
2. Roller (load selector) link
3. Pivot block
4. Link
5. Washer
6. Spring
7. Special nut
8. Link
9. Spring

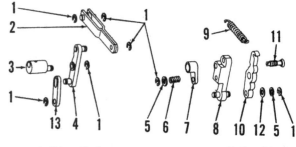

10. Valve adjusting cam	12. Bowed washer
11. Adjusting screw	13. Link

piston (8) with "O" ring toward closed end of piston. Lubricate piston assembly and install it in cylinder. Install check valve and thermal relief valve. Assemble linkage and install it on linkage pivot pin. If unit is so equipped, install flow control valve. If not already done, remove the throttling valve screw (18—Fig. 141) from rockshaft housing. Install selector control lever shaft in rockshaft housing. Place "O" ring (25) in front cap screw hole of cylinder and valve unit and the gasket (13) and seal (14) on flange of valve housing, then install cylinder and valve assembly in rockshaft housing and start slot of roller link over selector control lever shaft and enter connecting rod in piston as cylinder valve unit is positioned. Install and tighten the front cap screw (26) before tightening the other cylinder mounting cap screws. Connect the linkage spring and install the throttling valve ball (15) and adjusting screw (18). Install control quadrant assembly and selector lever.

Install rockshaft housing assembly by reversing the removal procedure and adjust assembly as outlined in paragraph 134 through 138.

LOAD CONTROL (SENSING) SYSTEM

All Models

The load sensing mechanism is located in the rear of the transmission case. See Fig. 145 for an exploded view showing component parts.

The load control shaft (9) is mounted in tapered bushings and as load is applied to the shaft ends from the hitch draft links, the shaft flexes forward and actuates the load control arm (20) which pivots on shaft (19). Movement of the load control arm is transmitted to the rockshaft control valves via the roller link (2—Fig. 144) and control linkage and control valves are opened

or closed permitting oil to flow to or from the rockshaft cylinder and piston.

145. **R&R AND OVERHAUL.** To remove the load sensing mechanism, remove the rockshaft housing assembly as outlined in paragraph 143, the left final drive assembly as outlined in paragraph 109, and the three-point hitch.

Remove cam follower spring (28—Fig. 145), then slide pivot shaft (19) to the left and lift out control arm (20) assembly. Removal of pivot shaft (19) can be completed if necessary by removing snap ring. Remove pin (1), retainer (2) and bushing (4) from right end of load control shaft and bump shaft from transmission case. The negative stop screw (10 or 16), can also be removed if necessary.

Inspect bushings (8) in transmission case and renew if necessary. Drive old bushings out by inserting driver through opposite bushing. New bushings are installed with chamfer toward inside. Check the special pin (15) for damage in area where it is contacted by load control shaft and renew if necessary. Also check contact areas of negative stop screw (10 or 16) and load control arm (20). Check load control shaft (9) to be sure it is not bent or otherwise damaged. Wear or damage to any other parts will be obvious.

Reassemble load sensing assembly by reversing the disassembly procedure and when installing hitch draft links, tighten retaining nuts until end play is removed between link collar and retaining ring, then tighten nut until next slot aligns with cotter pin hole and install cotter pin. After assembly is completed, adjust the negative stop screw as outlined in paragraph 134.

REMOTE CONTROL SYSTEM

Tractors may be equipped with single or dual selective (remote) control valves, mounted on a bracket attached to right final drive housing. Each control valve will operate a single or double acting remote cylinder and contained within the control valves are a combination flow control valve and check valve, a metering valve and two sets of operating valves. See Fig. 146 and Fig. 146A for exploded views of the selective control valves used.

SELECTIVE CONTROL VALVE

All Models So Equipped

146. **TEST.** The selective control valve can be used to check the main hydraulic pump pressure and in addition, the operation of the selective control valve can be tested either by checking flow or by making a time cycle of the remote cylinder.

To check the main hydraulic pump pressure via the selective control valve, proceed as follows: Use a gage capable of registering at least 3000 psi connected to a line fitted with a male disconnect fitting. Connect line fitting to disconnect coupling, start engine and run at 2500 rpm (clutch engaged). Move control valve lever to pressurize line, hold lever in this position and check the gage reading which should be 2220-2280 psi. If main hydraulic pump pressure is not as stated, adjust pump stroke control valve as outlined in paragraph 131 or overhaul pump as outlined in paragraphs 139 or 141.

147. **R&R AND OVERHAUL.** All models equipped with remote control, use a selective control valve of the type shown in Fig. 146 or Fig. 146A. To remove the valve or valves, disconnect breakaway coupling lines, inlet lines and return lines from valve, then unbolt valve bracket from final drive housing. Lift the valve and bracket as a unit from tractor.

To overhaul the removed valve, unbolt valve lever (61) and side cover. Remove bracket and end cap (37) which will contain spring pin (45) and metering valve (50) as shown. Note while loosening that the four valve guides (33) are spring loaded and retained by the cap. The four guides should move out with the cap as cap screws are loosened; if they do not, protect them from flying as cap is removed, thus becoming lost or damaged. Withdraw guides, springs (33 and 32), valves (31)

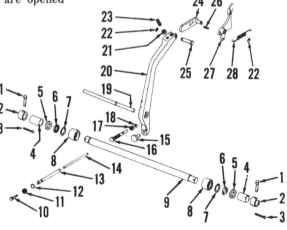

Fig. 145—Load control mechanism is located in rear of transmission case. Flexing of control shaft (9) actuates load control arm (20). Items 16 through 18 used without ipto and 10 through 14 used with ipto.

1. Pin
2. Retainer
3. Cotter pin
4. Bushing
5. Thrust washer
6. Sealing ring
7. "O" ring
8. Bushing
9. Load control shaft
10. Cap screw
11. Jam nut
12. "O" ring
13. Rear rod
14. Front rod
15. Special pin
16. Special screw
17. Jam nut
18. "O" ring
19. Control arm shaft
20. Load control arm
21. Jam nut
22. Retaining ring
23. Adjusting screw
24. Control arm extension
25. Pin
26. Pin
27. Cam follower
28. Spring

and associated parts, keeping them together and in proper order. Remove flow control valve (46) and spring (47). Remove snap ring (25) and detent piston outer guide (23), then withdraw detent spring (21), piston (20) and pin (19).

Invert valve body in vise, rocker assembly end up as shown in Fig. 147. Drive out the spring pin (8—Fig. 146 or Fig. 146A) securing rocker (10) to lever shaft (60), withdraw the shaft and lift out rocker.

Clean all parts and inspect housing side cover and end caps for cracks, nicks or burrs. Small imperfections can be removed with a fine file, renew the

parts if their condition is questionable. Inspect poppet valves (31) and their seats in housing (14) for grooves, scoring or excessive wear and renew or recondition parts as necessary.

Selective control valve is adjustable and if disassembled, valve must be adjusted during assembly. Adjustment requires the use of a special adjusting cover (JDH-15C) or equivalent, and a dial indicator as shown in Fig. 148. Install adjusting cover (A—Fig. 149) instead of valve end cap (37—Fig. 146 or 73—Fig. 146A). Snug up the retaining cap screws and tighten the four screws (S—Fig. 149) gently but firmly to hold the operating poppet valves seated. Install dial indicator with contact point touching lever two inches from rocker shaft as shown in Fig. 148.

NOTE: Lever may be reversed, if necessary for convenience in mounting dial indicator.

Remove the two plugs (62—Fig. 146 or Fig. 146A) and, reaching through plug holes, loosen the cam holding screws (H—Fig. 149) on each side of valve body.

Gently rock valve lever to be sure it

Fig. 147—View of valve rocker with top cover removed, showing adjusting screws and rubber keepers.

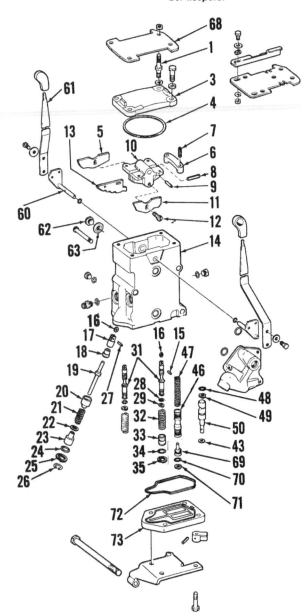

Fig. 146—Exploded view of typical early selective (remote) control valve. Two valves may be used which differ only in control lever location. Refer to Fig. 146A for later type.

Fig. 146A—Exploded view of selective (remote) control valve typical of late type. Refer to Fig. 146 for legend except the following:

69. Flow control valve stop
70. "O" ring
71. Backup ring
72. Packing
73. End cap

3. Cover	23. Outer guide
4. Packing	25. Backup ring
5. Operating cam	26. Snap ring
6. Rubber keeper	27. Pin
7. Adjusting screw	28. "O" ring
8. Spring pin	29. Backup ring
9. Drive pin	31. Poppet valves
10. Rocker	32. Springs
11. Operating cam	33. Valve guides
12. Special screw	37. End cap
13. Detent cam	42. Metering lever
14. Housing	43. Thrust washer
15. Spring pin	44. "O" ring
16. Roller	45. Stop pin
17. Detent follower	46. Flow control valve
18. Inner guide	47. Spring
19. Detent pin	48. "O" ring
20. Detent piston	49. Backup ring
21. Spring	50. Metering valve
22. "O" ring	68. Stop plate

is centered in neutral detent, then zero the dial indicator against lever arm.

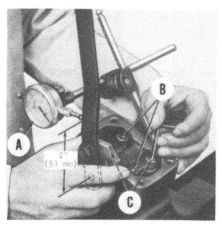

Fig. 148—Dial indicator contact point should touch control lever 2 inches from rocker axis as shown.

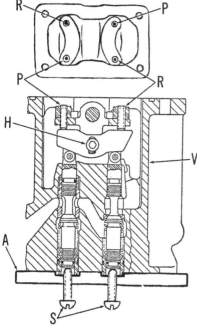

Fig. 149—Cross-sectional view of control valve with adjusting cover installed, showing adjusting points.

A. Adjusting cover
H. Holding screw
P. Pressure adjusting screws
R. Return adjusting screws
S. Seating screws
V. Valve body

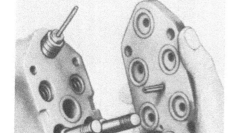

Fig. 150—Removing valve end cap.

Carefully tighten adjusting screws (R and P) equally a little at a time until cams are solid against cam followers, adjusting screws are all touching cam, and dial indicator still reads zero.

Tighten cam holding screws (H) at this time. Back the two return valve adjusting screws (R) out 1/8-turn and the two pressure valve adjusting screws (P) out ¼-turn.

With adjustment completed as outlined, rock the valve lever both ways

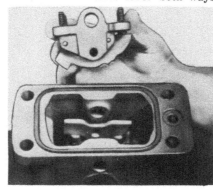

Fig. 152—Installing control rocker.

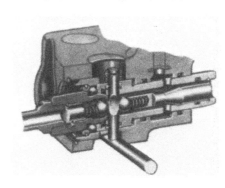

Fig. 153—Cross-sectional view of breakaway coupling showing component parts. Refer to Fig. 154 for exploded view.

while watching indicator dial. Indicator reading should be 0.016-0.032 inch in either direction, adjust if necessary by turning the appropriate return adjusting screw (R) a slight amount. With return adjusting screws correctly adjusted, back out the two adjusting cover screws (S) which contact return valves, then again check valve lever movement which should now be 0.056-0.071 in either direction. Adjust if necessary, by turning the appropriate pressure adjusting screw (P).

Remove adjusting cover (A) and reassemble using new seals and gaskets. Tighten the cap screws retaining control valve cap (37—Fig. 146 or 73—Fig. 146A) to a torque of 35 ft.-lbs. and cap screws containing valve cover (3) to a torque of 23 ft.-lbs.

BREAKAWAY COUPLER

All Models

148. Fig. 153 shows a cross-sectional view of coupler assembly and Fig. 154 shows an exploded view. When handle of lever (30) is crosswise to hose centerline as shown in Fig. 153, ball checks are seated, hose ends blocked and hoses can be disconnected. Also, check balls will seat if hoses are pulled from receptacle.

To disassemble the coupler, first remove unit from tractor. Punch a hole in expansion plugs (21—Fig. 154) and pry out the plugs. Remove "E" clips (22) and springs (23), then withdraw operating levers (30). Use a soft drift of the appropriate size and drive receptacle (13) rearward out of housing (24). Remove snap ring (9) and the six balls (10). Invert the receptacle (13) and insert large end of receptacle into large

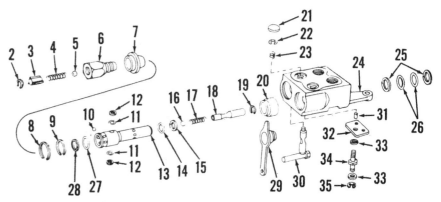

Fig. 154—Exploded view of breakaway coupler of the general type used on most models. Refer to Fig. 153 for cross-section.

2. Snap ring
3. Guide
4. Spring
5. Check ball
6. Plug
7. Dust cover
8. Snap ring
9. Snap ring
10. Steel ball
11. "O" ring
12. Backup ring
13. Receptacle
14. "O" ring
15. Backup ring
16. Check ball
17. Spring
18. Plug
19. Snap ring
20. Sleeve
21. Expansion plug
22. "E" ring
23. Spring
24. Body
25. Backup ring
26. "O" ring
27. "O" ring
28. Backup ring
29. Dust cover
30. Lever
31. Pin
32. Cam
33. Washer
34. Special cap screw
35. Retainer

bore of housing, and using housing as a holder, push down on exposed end of plug (18) and unseat and remove snap ring (19). Withdraw plug (18), spring (17) and ball (16).

"O" rings and back-up rings can be renewed at this time. It is recommended that all be renewed if normal wear is the cause of failure. Assemble coupling by reversing the disassembly procedure, using Figs. 153 and 154 as a guide.

REMOTE CYLINDER

All Models

149. To disassemble the remote cylinder, remove oil lines and end cap (18—Fig. 155). Remove stop valve (14) and bleed valve (13) by pushing stop rod (9) completely into cylinder. Withdraw stop valve from bleed valve being careful not to lose the small ball (12). Remove nut from piston rod and remove piston and rod. Push stop rod (9) all the way into cylinder and drift out pin (27). Remove piston rod guide (26).

Renew all seals and examine other parts for wear or damage. Wiper seal (35) should be installed with lip toward outer end of bore. Install stop rod seal

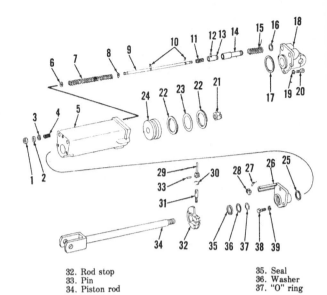

Fig. 155—Exploded view of remote cylinder.

1. Adapter
2. Packing
3. Adapter
4. Spring
5. Cylinder
6. Washer
7. Spring
8. Washer
9. Stop rod
10. Snap ring
11. Spring
12. Ball
13. Bleed valve
14. Stop valve
15. Spring
16. Gasket
17. Gasket
18. Cap
21. Nut
22. Backup ring
23. "O" ring
24. Piston
25. Gasket
26. Guide
27. Pin
28. Stop rod arm
29. Stop lever
30. Washer
31. Stop screw

32. Rod stop
33. Pin
34. Piston rod

35. Seal
36. Washer
37. "O" ring

assembly (1, 2 and 3) with sealing edge toward cylinder. Complete the assembly by reversing the disassembly procedure and tighten the end cap screws to 85 ft.-lbs. torque and the piston rod guide screws to 35 ft.-lbs. torque.

To adjust the working stroke, lift the piston stop lever (29), slide the adjustable stop (32) along piston rod to the desired position and press the stop lever down. If clamp does not hold securely, lift and rotate stop lever ½-turn clockwise and reset. Make certain that adjustable stop is located so that the stop rod contacts one of the flanges on adjustable stop.

JOHN DEERE

Models ■ 4040 ■ 4240 ■ 4440 ■ 4640 ■ 4840

Previously contained in I&T Shop Manual No. JD-51

SHOP MANUAL

JOHN DEERE

SERIES 4040, 4240, 4440, 4640 & 4840
Tractor serial number located on rear of transmission case. Engine serial number located on front right side of engine block.

INDEX (By Starting Paragraph)

CONDENSED SERVICE DATA

GENERAL	4040 Diesel	4240 Diesel	4440 Diesel	4640 Diesel	4840 Diesel
Engine Make			OWN		
Engine Model	6404D	6466D	6466T	6466A	6466A
Number of Cylinders			6		
Bore-Inches	4.25	4.56	4.56	4.56	4.56
Stroke-Inches	4.75	4.75	4.75	4.75	4.75
Displacement-Cu. In.	404	466	466	466	466
Compression Ratio	16.7:1	17.0:1	14.9:1	14.9:1	14.9:1
Induction*	N-A	N-A	T	T-I	T-I
Cylinder Sleeves			WET		

TUNE-UP					
Firing Order			1-5-3-6-2-4		
Valve Tappet Gap–					
Exhaust-Inch	0.028	0.028	0.028	0.028	0.028
Inlet-Inch	0.018	0.018	0.018	0.018	0.018
Injection Timing–Static	TDC	TDC	TDC	TDC	TDC
Governed Speeds-Engine RPM –					
Low Idle	800	800	800	800	800
High Idle	2400	2400	2375	2375	2375
Loaded	2200	2200	2200	2200	2200
Horsepower at Pto Shaft**	90.80	110.94	130.58	156.30	180.63
Battery:					
Volts			12		
Ground Polarity			Negative		

SIZES-CAPACITIES-CLEARANCES					
Cooling System (Quarts)+	24	30	36	38	40
Crankcase Oil (Quarts)–					
Including Filters	17	17	16	20	20
High Clearance Final Drive Housing (Quarts)	2¼	2¼	2¼	.....	.....
Transmission & Hydraulic System (Gallons)+ +					
Syncro-Range	15½	15½	.....	.....	.....
Power Shift	13½	13½	13½	20	20
Quad Range	15½	15½	15½	25	.....
Fuel Tank (Gallons)	37	56	65	83	103
Crankshaft Sizes and Clearances			See Paragraph 65		
Piston, Rings & Sleeves			See Paragraph 60		

TIGHTENING-TORQUES-Ft.-Lbs.					
Cylinder Head-Final	135-165	135-165	135-165	135-165	135-165
Main Bearing Caps			See Paragraph 65		
Connecting Rod Caps			See Paragraph 64		

*N-A = Naturally Aspirated; T = Turbocharged; T-I = Turbocharged and Intercooled

**Horsepower is according to Nebraska test.

+ Add approximately 2 quarts if equipped with heater.

+ +Approximate capacity only. When draining, 3 to 6 gallons may remain in case. Add 4½-5 gallons if equipped with Power Front Wheel Drive.

FRONT SYSTEM

Three different tricycle front end units have been used on 4040 models: Single front wheel, Dual front wheels and "Roll-O-Matic" dual wheel tricycle units. Tricycle front end units require a special front support and steering motor assembly. Some units are convertible to adjustable axle models.

Adjustable front axle is available on all models. On some models, the adjustable front axle may be High Clearance or standard height in narrow, regular or wide widths. The adjustable front axle on tractors with Power Front Wheel Drive is regular width only.

TRICYCLE FRONT END UNITS

1. **REMOVE & REINSTALL.** The spindle extension (pedestal) attaches directly to steering motor spindle by four cap screws. To remove the unit, support front of tractor with a hoist using the special John Deere lifting plate or other suitable support attached to tractor front support. Tighten the retaining cap screws to a torque of 300 ft.-lbs. when unit is reinstalled.

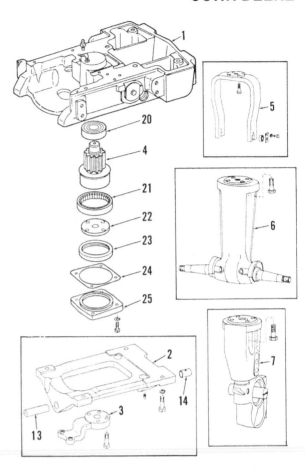

Fig. 2 — Partially exploded view of convertible front support assembly. The unit can be used with tricycle or adjustable tread axles. Axle pivot bracket (2) must be attached to support (1) to mount axle and steering arm (3) is bolted to the spindle (4). Tricycle fork (5) or pedestal (6 & 7) is bolted directly to spindle (4).

1. Front support
2. Axle pivot bracket
3. Steering arm
4. Spindle
5. Single wheel tricycle fork
6. Dual wheel tricycle pedestal
7. Roll-O-Matic pedestal
13. Pivot pin
14. Bushing
20. Ball bearing
21. Roller bearing
22. Spacer (some models)
23. Oil seal
24. Gasket
25. Steering motor spindle quill

2. **SINGLE WHEEL TRICYCLE.** The fork mounted single wheel is supported on taper roller bearings as shown in Fig. 3. Bearings should be adjusted to provide a slight rotational drag by means of adjusting nut (4). The one-piece wheel & rim assembly (5) accommodates a 7.50"-16" tire.

3. **DUAL WHEEL TRICYCLE.** An exploded view of the dual wheel tricycle pedestal and hub is shown in Fig. 4. Horizontal axles are not renewable. Service consists of renewing the complete pedestal assembly.

4. **ROLL-O-MATIC UNIT.** The "Roll-O-Matic" front wheel pedestal and associated parts are shown exploded in Fig. 5. The unit can be overhauled without removing the assembly from tractor.

Support front of tractor and remove wheel and hub units. Remove knuckle caps (6) and thrust washers (5), then pull knuckle and gear units (4) from housing.

The "Roll-O-Matic" unit is equipped with a lock (2) and lock support (3) which may be installed for rigidity when desired. Check the removed parts against the values which follow:

Knuckle Bushing ID ....2.127-2.129 in.
Knuckle Shaft OD ......2.124-2.126 in.

Renew thrust washers (5) if worn.

Bushings are presized and contain a spiral oil groove which extends to one

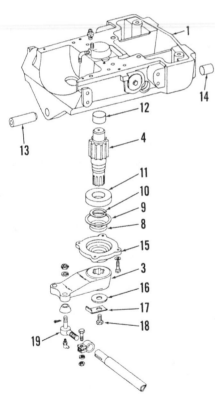

Fig. 1 — Partially exploded view of front support that can only be used with adjustable and fixed tread front axles. Notice that axle pivot bosses are cast as part of support.

1. Front support
3. Center steering arm
4. Spindle
8. Oil seal
9. "O" seal
10. Snap ring
11. Ball bearing
12. Bushing
13. Pivot pin
14. Bushing
15. Bearing quill
16. Washer
17. Lock plate
18. Cap screw
19. Rod end

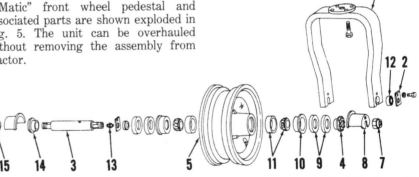

Fig. 3 — Exploded view of front wheel fork and axle assembly used on single wheel tricycle models. The one-piece wheel (5) is for 7.50-16 inch tire.

1. Fork
2. Lock plate
3. Axle
4. Bearing adjusting nut
5. One-piece wheel
7. Nut
8. Dust shield
9. Felt washers
10. Felt retainer
11. Bearing cup and cone
12. Washer
13. Grease fitting
14. Spacer
15. Nut

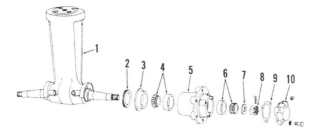

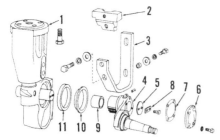

Fig. 4—Dual wheel tricycle pedestal showing one wheel hub and associated parts.

1. Pedestal
2. Oil seal
3. Oil seal cup
4. Bearing cup and cone
5. Hub
6. Bearing cone and cup
7. Washer
8. Nut
9. Gasket
10. Cap

ADJUSTABLE AND FIXED TREAD AXLES

5. HOUSING & PIVOT BRACKET. The front axle attaches directly to the front support (1–Fig. 1) or to the pivot bracket (2–Fig. 2). Install pivot pin bushings flush with the bottom of chamfer, and be sure that lubrication channel is 180 degrees from grease fittings. Install front pivot pin flush with rear edge of pin hole in axle. Install shims, if necessary, between large washer and pivot pin to provide some clearance between washer and front support. Heads of retaining screws should be toward front. Tighten nuts on pivot retaining screws to 220 ft.-lbs., then tighten slightly if necessary to align castellation for cotter pin.

6. SPINDLES & BUSHINGS. Steering arm (6–Fig. 8) is splined to spindle (5) and retained by a bolt. Spindle bushings are presized. Shim washers (8) should be used to adjust end play of spindle to 0.010-0.040 inch.

Refer to paragraph 35 for service to

edge of bushing. When installing new bushings, use a piloted arbor and press bushings into knuckle arm so that OPEN end of spiral grooves are together as shown in Fig. 6. Bushing at

spindle end of "Roll-O-Matic" unit should be pressed into arm so that outer edge (B) is 1/32-inch below edge of bore. There should be a gap (C) of 1/32 to 1/16-inch between bushings and distance (A) from edge of inner bushing to edge of bore should measure 3/16-inch. Soak felt washers in engine oil prior to installation. Install one of the knuckles so that wheel spindle extends behind vertical steering spindle. Pack the "Roll-O-Matic" unit with multipurpose type grease and install the other knuckle so that timing marks on gears are in register as shown at (M–Fig. 7). Tighten the thrust washer attaching screws to a torque of 55 ft.-lbs., and bend up corners of lock plates.

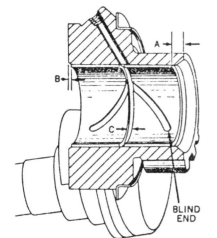

Fig. 5—"Roll-O-Matic" Spindle extension showing component parts. Lock (2) and support (3) may be installed to increase rigidity.

1. Pedestal extension
2. Lock
3. Support
4. Knuckle
5. Thrust washer
6. Cap
7. Gasket
8. Lock plate
9. Bushing
10. Felt retainer
11. Felt washer

Fig. 8—Exploded view of typical, standard, adjustable tread front axle.

1. Steering motor arm
2. Pivot pin
3. Bushing
4. Axle housing
5. Spindle
6. Steering arm
7. Knee
8. Shim washer
9. Bushing
10. Thrust washers
11. Dowel pins
12. Clamp

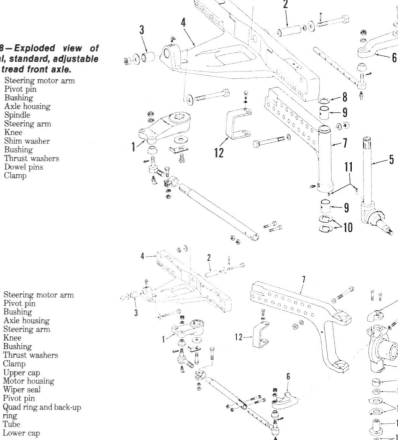

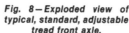

Fig. 6—Cross-sectional view of knuckle showing details of bushing installation. Refer also to paragraph 4.

Fig. 7—Make sure timing marks (M) are aligned when installing knuckles in "Roll-O-Matic" unit.

1. Steering motor arm
2. Pivot pin
3. Bushing
4. Axle housing
6. Steering arm
7. Knee
9. Bushing
10. Thrust washers
12. Clamp
14. Upper cap
15. Motor housing
16. Wiper seal
17. Pivot pin
18. Quad ring and back-up ring
19. Tube
20. Lower cap

Fig. 11—Exploded view of adjustable tread front axle used on models with power front wheel drive. Refer to paragraph 30 and following for service to the drive motors and related parts.

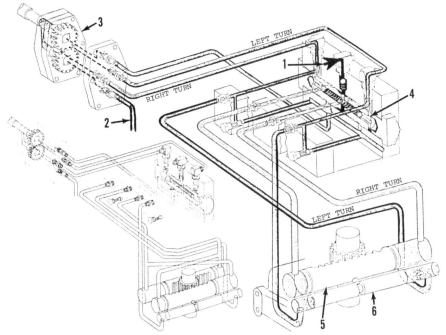

Fig. 30 — Schematic view of steering system. The view at lower left is of type used on 4640 and 4840 models.

All Models

21. Be sure that fluid is at correct level and that all hydraulic system filters are clean before suspecting the steering system. Operate another hydraulic system function such as the rockshaft to determine whether system has pressure. Refer to paragraph 276 if hydraulic system fails to operate satisfactorily.

SLOW RESPONSE OR HARD STEERING
　Tractor heavily weighted and/or not in motion.
　Low hydraulic system pressure. Refer to paragraph 276.
　Filters screens (17 & 23 – Fig. 36) in control valve plugged or damaged.
　Steering feed back orifices plugged, eroded or otherwise damaged. Remove filters (17 – Fig. 36) and check orifices in valve body.
　Cold oil in control circuit because of lack of circulation through metering pump.
　Steering valve spool and/or body scored or damaged. Examine spool and bore of body for signs of sticking.
　Metering pump leaking internally. Check pump for wear and/or damage.
　Steering motor piston seal damaged. Check piston, seal and cylinder.
　Feed back piston pin (40 – Fig. 48) or spring damaged or stuck. Remove pin and check.

STEERING WHEEL CREEPS
　The steering wheel may require constant movement to maintain straight travel, may creep at locks and/or creep excessively when steering manually.
　Steering motor feed back piston rings damaged.
　Metering pump shims, wrinkled, folded or torn.

FRONT WHEELS LOCK TO ONE SIDE
　The steering valve manual steering check valve seat (3 – Fig. 36), ball (5) or spring (6) damaged. Inspect and install new parts as necessary.
　Steering valve spool and/or body scored or damaged.

STEERING WHEEL CONTINUES TO TURN WITH WHEELS IN LOCK POSITION
　Check make up valve (12 and 13 – Fig. 36) for damage. Install new parts as necessary.

STEERING WANDERS
　Steering motor feedback piston pin (40 – Fig. 48) stuck or spring (39) broken. Remove pin and inspect.

front axle used with power front wheel drive.

7. **TIE RODS AND TOE-IN.** The outer tie rod ends on axles with adjustable tread are adjustable with the several holes provided for changing axle width. Tie rod end at the inner ends are threaded to provide finer adjustment. Make sure that tie rods are equal length so the tractor will turn the same in either direction. Adjust toe-in to ⅛ to ⅜-inch, and tighten clamp bolts to 35 ft.-lbs. torque. If tie rod ends are renewed, tighten nuts to 100 ft.-lbs. torque.

STEERING SYSTEM

Some differences will be noted between certain of these units, but all are equipped with full power steering. No mechanical linkage exists between the steering wheel and front unit; however, steering can be manually accomplished by hydraulic pressure without tractor normal hydraulic pressure.

OPERATION

All Models

20. The power steering system consists of a metering pump (3 – Fig. 30), a control valve (4) and a steering motor assembly (5 & 6). The power steering system, in normal operation, uses the tractor hydraulic system fluid which has been pressurized by the main hydraulic pump to turn the front wheels.

The metering pump (3 – Fig. 30) is located at the lower end of the steering wheel shaft. The metering circuit is filled (2) with oil from the transmission and charging oil circuit. Turning the steering wheel and metering pump (3) causes the oil trapped in the metering circuit to move and apply pressure to one end of the control valve spool (4).

The control valve spool will be moved by the metering oil to one end of the valve body. When the control valve spool is not centered, a passage is opened allowing pressurized tractor hydraulic fluid to flow from inlet (1) through control valve and into one end of steering motor cylinder (5). During normal operation, oil from the feed back cylinders (6) is used to center the control valve spool after steering wheel is stopped or after turn is completed.

The steering system is equipped with check valves which trap oil in the system. In case of hydraulic system failure, the trapped oil is circulated within the steering system by the metering pump and pressure is exerted against both the steering motor piston and the feed back piston to turn the steering wheels.

Some differences will be noticed between control valves and steering motors used. For instance, the steering motor used on 4640 and 4840 models contains two operating pistons. The additional piston is identical to the one in the usual location, but is on the opposite (rear) side of the spindle.

FRONT WHEELS TURN SHARPER IN ONE DIRECTION THAN THE OTHER

Steering motor spindle incorrectly indexed with front support. Refer to paragraph 28 and Fig. 49.

Tie rods not adjusted equally. Refer to paragraph 7.

ERRATIC STEERING EFFORT

Metering pump return oil check valve in lower fitting (34L–Fig. 33) does not seat. Check and renew if damaged.

Steering control valve manual steering valves (3 and 5–Fig. 36) leaking. Install new parts if damaged.

FRONT WHEELS TWITCH WHEN ENGINE IS STARTED OR JERKS WHEN TURNING

Air in steering system. Check for loose or damaged connections and bores.

NO STEERING FEEL

Metering pump friction spring (32–Fig. 33) damaged. Install new spring.

METERING PUMP HOSE FAILURE

Return oil check valve in lower fitting (34L–Fig. 33) missing or damaged. Install new valve. Damage to check valve could be caused by pump friction spring (32) damage.

OIL LEAKAGE IN CONTROL SUPPORT OR OUT OF STEERING COLUMN

Metering pump oil seal (27–Fig. 33) damaged. Install new seal.

Metering pump shims torn, wrinkled or folded. Install new shims.

LEAKAGE THROUGH STEERING MOTOR BLEED LINE

Leakage past piston seal. Install new seals.

STEERING WHEEL MOVES UP AND DOWN WITH ENGINE RUNNING

Return oil check valve in lower fitting (34–Fig. 33) leaking. Install new parts as necessary.

LOSS OF MANUAL STEERING

Return oil check valve in lower fitting (34L–Fig. 33) of metering pump leaking or missing.

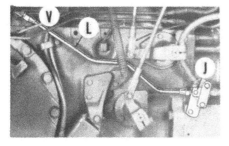

Fig. 31 – View of tractor showing location of junction block (J), line (L) and pressure control valve (V).

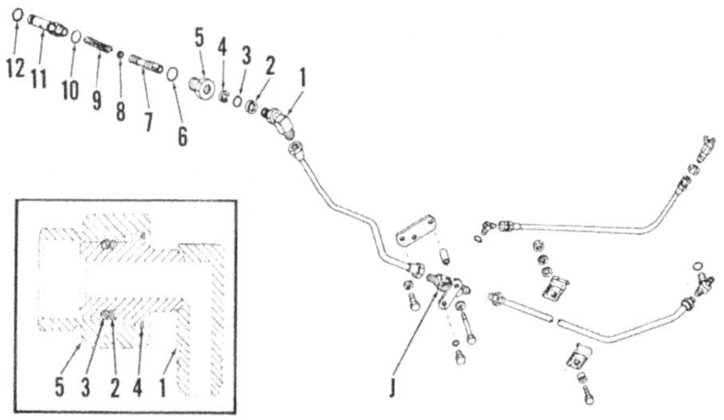

Fig. 32 – Exploded drawing of pressure control valve. Inset shows cross-section of retaining plug and elbow assembled.

J. Junction block
1. Elbow
2. Back-up ring
3. "O" ring
4. Snap ring
5. Plug
6. "O" ring
7. Pressure control valve
8. Shims
9. Spring
10. "O" ring (same as 12)
11. Valve body
12. "O" ring (same as 10)

Metering pump leaking internally. Inspect and repair or renew as necessary.

Steering valve inlet check valve (19, 20, 21 and 22–Fig. 36) damaged.

Manual steering valves (3, 5 and 6–Fig. 36) leaking.

Make up valves (12 and 13–Fig. 36) leaking.

BLEEDING

All Models

22. The power steering will normally not require bleeding. If air is present in the power steering hydraulic system after service, cycling the steering wheel several times should purge air from lines and components. Air in operating system can be detected by twitch or jerk while turning, especially after just starting engine. Cycling the steering wheel several times may remove air, but check all lines and fittings for air and hydraulic leaks. Repair may be necessary to prevent reoccurrence.

PRESSURE CONTROL (PRIORITY) VALVE

All Models

23. The pressure control valve stops hydraulic fluid flow to the tractor hydraulic lift system when system pressure drops below 1600-1700 psi, thus giving priority to the steering and brake units. The pressure control valve is located in a bore of rockshaft housing.

To check operation, clean the area around the junction block (J–Fig. 31) and around the pressure control valve (V). Attach a 0-5000 psi pressure gage to the junction block and attach a hose to the selective control valve breakaway couplings so that oil can be circulated through the remote system. Move rate of operation lever, located just above breakaway couplings, to maximum (clockwise) position. Start engine and operate at 800 rpm then move the selective control valve lever to circulate hydraulic fluid through that system. The pressure indicated by gage attached to the junction block should be within range of 1600-1700 psi.

If pressure is incorrect, disconnect line (L–Fig. 31) from junction block (J) and control valve (V). Remove adapter (5–Fig. 32), then withdraw valve (7), shims (8) and spring (9). Add or deduct shims (8) as necessary to change pressure to within limits. Each shim represents about 50 psi. Reassemble and recheck pressure. Inability to adjust the pressure could be caused by damaged valve (7) and/or body (11), restriction in pressure or return passages in rockshaft housing or restriction in selective control valve circuit.

Spring (9–Fig. 32) should exert 45-55 lbs. when compressed to height of 3½ inches.

METERING PUMP AND STEERING COLUMN

All Models

24. Some repairs can be accomplished without removing the steering column and metering pump from tractor. Use extreme care to be sure that dirt is not permitted to enter the steering system lines or ports.

Disconnect battery ground cables, relieve hydraulic system pressure and remove steering wheel. Remove the left and right cowl panels and the rear control support cover. Remove snap ring from upper end of steering shaft and withdraw the sleeve from around top of steering shaft. Disconnect the four hydraulic lines from metering pump and cover all openings. Detach spring from lug on steering column, remove the two pivot screws, then tilt the metering pump and steering column up and out toward front of tractor. The metering pump should be unbolted and separated from lower end of steering column before servicing.

To disassemble the removed metering pump, loosen the two remaining screws, break the seal between pump body and pump cover then remove screws gradually. Examine parts of pump carefully, especially the two 0.0005 inch

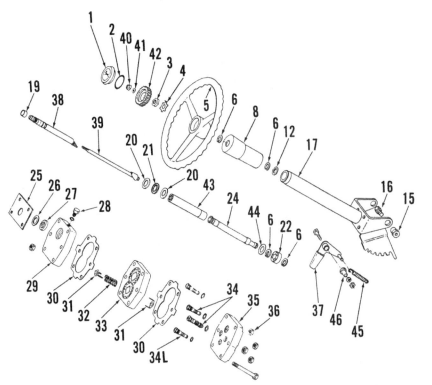

Fig. 33 — Exploded view of metering pump and steering column used.

1. Emblem	19. Bushing	30. Shim	38. Upper steering shaft
2. "O" ring	20. Thrust races	31. Friction plates	39. Column release rod
3. Nut	21. Thrust bearing	32. Spring	40. Special jam nut
4. Lockwasher	22. Collar	33. Pump body and gears	41. Special washer
5. Steering wheel	24. Steering shaft	34. Special hose	42. Release knob
6. Snap ring	25. Plate	connectors (not all	43. Steering shaft
8. Sleeve	26. Washer	the same)	coupling
12. Oil seal	27. Oil seal	35. Pump cover	44. Washer
15. Bushing	28. Plug	36. Special nut	45. Spring
16. Cap screw	29. Metering pump base	37. Tilt release	46. Spacer
17. Steering column			

in body bores. Make sure that marks on gears are up and that splined top gear is in correct (upper) bore. Install friction spring (32) and plates (31) in idler gear. Carefully position the second shim (30) onto pump body and lower pump base (29) over dowels onto remainder of pump. Use care to not damage friction plate or shims when positioning the pump base. Install nuts on the two lower retaining screws and tighten nuts to 35 ft.-lbs. torque, then remove alignment dowels.

Assemble remaining parts and attach to column. Tighten the four remaining screws to 35 ft.-lbs torque. Check unit for leaks and proper operation after reinstalling and bleeding (cycling) the system.

STEERING VALVE

All Models

25. The control circuit is operated by the metering pump which moves the control valve which in turn directs pressurized hydraulic fluid to the steering motor cylinders to move the front wheels. The actual location of the control valve will depend upon specific model of tractor. It will not be necessary to remove the valve for some service such as cleaning filters. The valve spool (7 – Fig. 36) and body (24) are available only as a matched set. Specific service will depend upon difficulties encountered. Be sure that all parts are clean when assembling.

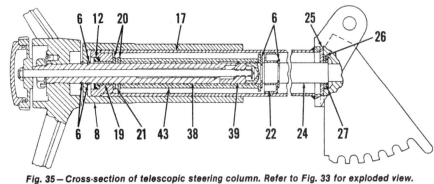

Fig. 35 — Cross-section of telescopic steering column. Refer to Fig. 33 for exploded view.

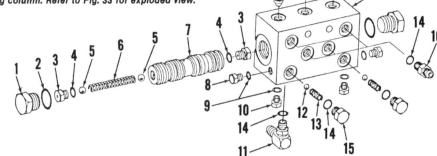

Fig. 36 — Exploded view of typical power steering control valve. The additional parts shown in Fig. 37 are used on some models with power front wheel drive.

thick shims (30 – Fig. 33). Internal or external leakage can be caused by wrinkled, torn or otherwise damaged shims. The special fittings (34) can be removed after removing special nuts (36). Be sure to reinstall fittings in proper ports.

Free length of friction spring (32) should be 0.735 in. and the spring should exert 81-99 lbs. force when compressed to height of 0.52 inch. Locate JDH-42-2 or equivalent dowels in the two center holes of cover (35) and four cap screws in corner holes to align pump while assembling. Carefully install one shim (30) over cover (35), then install body (33) onto cover and shim. Position gears

1. Plugs	7. Control valve spool	13. Springs	20. Spring
2. "O" rings	8. Plug	14. "O" rings	21. Inlet check valve
3. Seats	9. "O" rings	15. Plug	ball (⅜-inch steel)
4. "O" rings	10. Plugs	16. Connector fittings	22. Pressure inlet fitting
5. Steel balls (5/16-in.)	11. Elbow	17. Control circuit filters	23. Inlet filter
for manual steering	12. Steel balls (⅜-inch)	18. Elbows	24. Valve body
6. Spring	for make up valves	19. Guide	

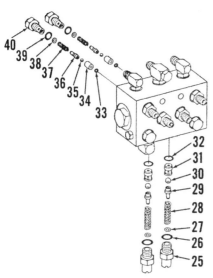

Fig. 37—View of additional parts installed in power steering control valve of 4640 tractors with power front wheel drive.

25. Special plugs	
26. "O" rings	33. "O" rings
27. Shims	34. Valve seats
28. Springs	35. Steel balls (3/16-inch)
29. Guides	36. Guides
30. Steel balls same as	37. Springs
(5–Fig. 36)	38. Shims
31. Valve seats	39. "O" ring
32. "O" rings	40. Special plugs

STEERING MOTOR

All Models

The steering motor is contained in the front support assembly and most repair can be accomplished without removing the front support from the tractor. Some differences may be noticed between standard two wheel drive tractors and models with power front wheel drive. Model 4640 and 4840 tractors use two racks in steering motor to turn the spindle. Where applicable, differences will be noted.

Fig. 40—View of tractor with front support (1) moved forward in frame rails (F) and held in position by screws (S). The piston plugs (37) and other components of the steering motor are accessible for service.

Fig. 41—View of bar (B) installed and spacers (S) being used to push piston plugs (35) in so that snap ring (36) can be removed. Procedure is similar for 4640 and 4840 models.

26. STEERING MOTOR ACCESS. Most service to the steering motor can be accomplished after sliding the front support forward in the frame rails.

Support tractor at sides of frame and support the fuel tank and steering motor (front support housing). Remove muffler, air intake tube, hood side shields, side grille screens and hood. Relieve hydraulic pressure and disconnect battery ground cables. Remove fuel tank leak-off line, shut-off fuel valve at bottom of tank and disconnect steering motor bleed line from connector above the steering motor. Remove fuel supply line from between fuel tank and fuel filters. Detach the four steering lines from steering control valve and cover all of the openings and ends of lines. Remove oil cooler return hose and

clamps securing hose to fuel tank. Disconnect radiator support rod and the hydraulic line clamp from radiator. Disconnect wire from fuel gage sending unit in tank, also disconnect horn and air conditioner control wires if so equipped. Remove lower frame plate or hydraulic line guard. Remove radiator mounting screws and the upper frame plates over the main hydraulic pump. Loosen the baffle plate. Attach hoist to front support using special lift bracket or equivalent. Use frame side stands or equivalent to securely hold rear of tractor while removing front support (steering motor). Remove screws which attach the front support to the frame side rails, then carefully slide the front support forward until two bolts can be installed through front holes in frame rails and into the rear threaded holes in support as shown at (S–Fig. 40).

Most repair can be accomplished as outlined in paragraph 28 with the front support attached to the tractor.

Tighten screws attaching frame rails to steering motor to 275 ft.-lbs. Bleed system after installation is complete by turning steering wheel from lock to lock several times. Check for leaks and proper operation.

27. REMOVE AND REINSTALL. To remove the complete steering motor and front support assembly, proceed as

Fig. 42—Partially exploded drawing of convertible steering motor used on some 4040 models showing steering spindle (4) and related parts.

1. Front support
2. Axle pivot bracket
3. Steering arm
4. Spindle
5. Single wheel tricycle fork
6. Dual wheel tricycle pedestal
7. Roll-O-Matic pedestal
13. Pivot pin
14. Bushing
20. Ball bearing
21. Roller bearing
22. Spacer (some models)
23. Oil seal
24. Gasket
25. Steering motor spindle quill

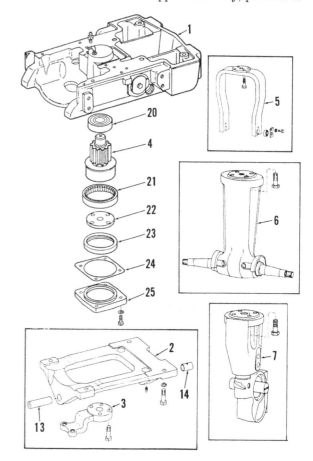

follows: Remove hood and support front of tractor from a hoist by installing engine sling or other suitable support. Remove fuel tank and air cleaner from front support. On tricycle models, remove wheels and pedestal assembly. On axle models, remove front axle and support as a unit if so desired.

Place a rolling floor jack or a hoist under steering motor to support the motor. Relieve hydraulic system pressure. Disconnect and plug fluid lines and remove any brackets and covers. Remove the cap screws securing steering motor to side frames and roll or hoist steering motor away from tractor.

To install the steering motor, reverse the removal procedure then bleed system by turning the steering wheel to each extreme several times. Tightening torques are as follows:

Steering motor to
side frames . . . . 275 ft.-lbs.
Steering spindle arm bolt . . . 300 ft.-lbs.

28. **OVERHAUL.** The following overhaul procedure can be accomplished with the steering motor and front support moved forward in the frame rails as outlined in paragraph 26 or with unit removed from tractor as outlined in paragraph 27.

Attach a bar (B–Fig. 41) between frame mounting holes and remove the piston caps (cover washers) if used. Use a spacer (S) to push piston plugs (35) in, then remove the retaining snap rings (36). Remove piston plugs, backup rings, "O" rings and spring washers.

On tricycle models, unbolt and remove the tricycle pedestal (6 or 7–Fig. 42) or yoke (5) from spindle (4). On adjustable axle models with convertible steering gear, detach tie rods from steering arm (3) then unbolt and remove steering arm from spindle (4). On standard adjustable axle steering motor, detach tie rods from steering arm (3–Fig. 44 or Fig. 45), remove screw (18) then pull steering arm (3) from spindle (4).

Turn the spindle to one extreme (either left or right), then unbolt quill (25–Fig. 43, 15–Fig. 44 or 15–Fig. 45). The quill, seal, steering spindle and lower bearing can now be removed. Align the feed back piston pin (40–Fig. 43, Fig. 44 or Fig. 45) with notch in housing as shown in Fig. 46, then turn piston so that pin and spring can be removed. A special tool (T–Fig. 46) is available, which engages the two depressions in top of piston to facilitate turning the piston. Withdraw steering motor pistons (31–Fig. 43, Fig. 44 or Fig. 45). Remove snap rings (47) then withdraw plugs (46) and feed back piston (43).

Check pistons (31) and sleeves (28) for wear or scoring. Sleeves can be pulled from housing using appropriate sized piloted puller. Examine parts against the following values.

Piston (31) O.D. . . . . 2.621-2.623 in.
Sleeve (28) I.D. . . . . 2.625-2.626 in.
Pin (40) O.D. . . . . 0.7357-0.7363 in.
Bore for pin (40) in
   piston (31) . . . . 0.7365-0.7375 in.

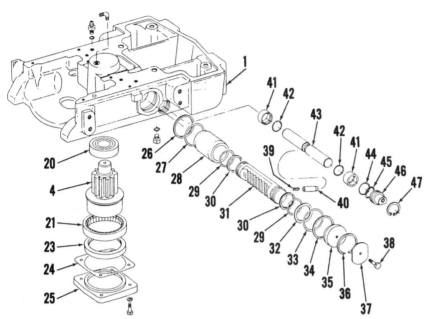

Fig. 43—Exploded view of convertible steering motor also shown in Fig. 42. Refer to Fig. 42 for legend except the following.

26. Back-up rings		37. Washers	42. "O" rings
27. "O" rings	32. Spring washers	38. Retaining screws	43. Feedback piston
28. Cylinder sleeves	33. "O" rings		44. "O" rings
~~29. "O" rings~~	~~34. Back-up rings~~	~~39. Spring~~	~~45. Back-up rings~~
30. Back-up rings	35. Piston plugs	40. Special pin	46. Plugs
31. Piston	36. Snap rings	41. Sleeves	47. Snap rings

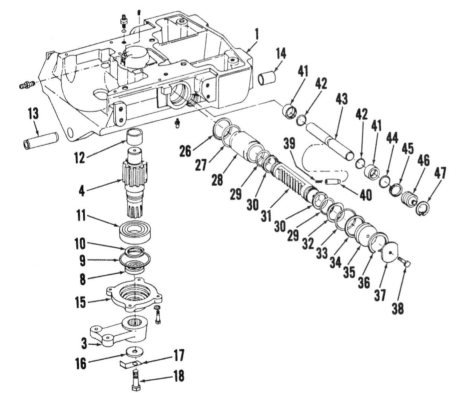

Fig. 44—Exploded view of standard front support and steering motor used on 4040, 4240 and 4440 models without convertible steering motor. The unit shown above cannot be used with tricycle front wheel or wheels. Refer to Fig. 43 for legend except the following.

1. Front support			15. Bearing quill
3. Center steering arm	9. "O" ring	12. Bushing	16. Washer
4. Spindle	10. Snap ring	13. Pivot pin	17. Lock plate
8. Oil seal	11. Ball bearing	14. Bushing	18. Cap screw

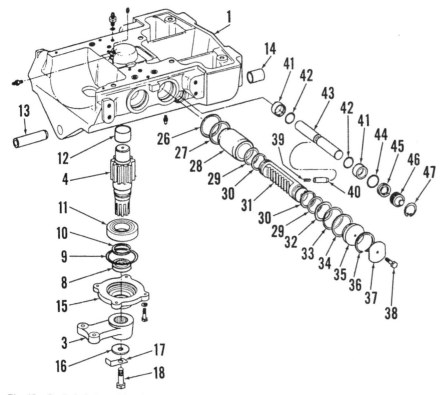

Fig. 45 — Exploded view of the front support and steering motor used on 4640 and 4840 models. Two pistons (31) are used and are located in bores located in front of and behind the spindle (4). Refer to Fig. 43 and Fig. 44 for legend and to Fig. 48 for cross-section of this unit.

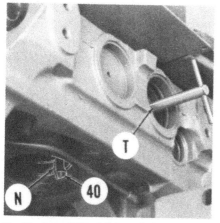

Fig. 46 — A special tool (T) is available for turning the front piston permitting pin (40) to be removed through notch (N).

Fig. 47 — View showing installation of plug (46) and snap ring (47) in one end of feedback piston bore.

Pin (40) length . . . . .2.58 in.
Spring (39) pressure at
    0.74 in. . . . . .45-55 lbs.
Feed back piston (43)
    O.D. . . . . .1.499-1.500 in.
Bushing (41) I.D. . . . . . . .1.501-1.506 in.
Bushing (41) O.D. . . . .1.8515-1.8525 in.

Feed back bushings (41) should only be removed if new bushings are to be installed. Use a blind hole puller to remove bushings (41) and be careful not to catch lip of casting while attempting to remove. See Fig. 48. Press new bushings (41) in until seated against inner shoulder in bore.

Check condition of bushing (12 – Fig. 44 and Fig. 45). Excessive wear could be caused by leakage or inadequate lubrication. Drive new bushing (12) into bore

until flush to 0.015 inch below flush with casting bore. Identification numbers on bearing (11) should be toward outside (down).

Dip all parts in John Deere Hy-Gard Transmission oil or equivalent while assembling. Soak fiber back-up rings in oil for 30 minutes before assembling.

If sleeves (28 – Fig. 43, Fig. 44 or Fig. 45) were removed, install fiber back-up ring (26) in groove, then install "O" ring (27) in same groove, but toward outside of back-up ring. Use a piloted driver and carefully drive sleeves into position until bottomed in casting bore. Repeat procedure for remaining sleeves. A total of four sleeves are used on each 4640 and 4840 model tractor. Two sleeves are used on other models.

Install "O" rings (42), then slide feed back piston (43) into bore. Center the feed back piston so that pin groove is aligned with notch (N – Fig. 46) in spindle bore.

Install back-up rings (30 – Fig. 43, Fig. 44 or Fig. 45) and "O" rings (29) in grooves of piston (31). The back-up rings should be toward center of piston and the concave (hollowed) side of back-up ring should be against "O" ring. Slide the steering motor piston (31) into cylinder, with the two counter sunk holes toward right side. Turn the piston to align hole in piston with notch (N – Fig. 46), grease the spring (39 – Fig. 43, Fig. 44 or Fig. 45) and pin (40), then slide the spring and pin into hole in piston. Turn the steering motor piston (31) to position the pin (40) in groove of feed back piston (43). On 4640 and 4840 models, install the rear piston using similar procedure.

Install back-up ring (34) and "O" ring (33) in groove, position spring washer (32) then slide plug (35) in bore. Use a bridge and spacer as shown in Fig. 41 to compress spring washer enough to install retaining snap ring (36). Install back-up rings, "O" rings, spring washer and plugs (33, 34, 35 & 36 – Fig. 43, Fig. 44 or Fig. 45) in the remaining cylinder bores.

Install oil seal (8 – Fig. 44 and 45) with

Fig. 48 — Cross-section of steering motor typical of 4640 and 4840 models. Refer to Fig. 43 and Fig. 44 for legend. Other models are similar except that rear piston and associated parts are not used.

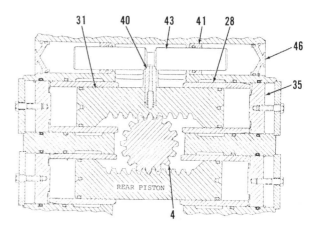

POWER FRONT WHEEL DRIVE

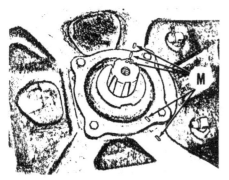

Fig. 49 — Align the two marks (2M) when installing spindle. Align marks on bearing quill as shown at (M). Refer to text.

spring loaded lip up toward spindle gear (4). Push the (front on 4640 and 4840 models) steering motor piston (31) to the right extreme of travel. On 4640 and 4840 models, push the rear steering motor piston to the extreme left of travel. On all models, slide the spindle into position with "V" mark on spindle aligned with "I" mark at right front as shown at (2M—Fig. 49). Install "O" ring (9—Fig. 44 or Fig. 45) in groove, then install quill (15) with mark on quill aligned with other two marks as shown at (M—Fig. 49). Assemble models with convertible steering motor (Fig. 43) using a similar procedure.

All models except 4840 may be equipped with a Power Front Wheel Drive unit. The front wheels are driven by hydraulic fluid supplied by the tractor main hydraulic system.

OPERATION

All Models So Equipped

30. The hydraulic power Front Wheel Drive unit consists of an axial piston hydraulic motor in each front wheel hub which turns the wheel through a planetary gear reduction unit.

Power is supplied by the constant pressure, variable volume Main Hydraulic System Pump, thus providing a specified amount of torque to front wheels without regard to tractor ground speed. This design permits a "Power Assist" traction boost which need not be precisely synchronized with tractor drive gears, and which eliminates possibility of damage or overload due to wheel slippage, altered tire size or tread wear.

The control valve is electrically operated by switches contained in Shift Lever Quadrant, and automatically coupled to the directional control solenoid. On Syncro-Range and Power Shift tractors, power front drive unit is automatically disengaged above 6th gear in "Low Torque" range, above 4th gear in "High Torque" range; or whenever Inching Pedal (Power Shift Models) or Clutch Pedal (Syncro-Range and Quad-Range Models) is depressed.

On Quad-Range models "High Torque" is limited to the 4 speeds in "A" range and 1st speed in "B" range. When speed selector lever is placed in 2nd or 3rd in

"B" range, and 1st, 2nd or 3rd in "C" range, the high torque is automatically switched to low torque drive if operating switch is in the engaged position. Fourth gear disengages front wheel drive unit in all but "A" range in both high and low torque positions, and no front wheel drive is possible in "D" range.

The panel mounted operating (selector) switch has three positions: Off, High Torque and Low Torque.

With operating switch in Off position, the wheels do not drive, but are allowed to rotate freely.

When the operating switch is in High Torque position, oil pressure from the main hydraulic pump is directed to the drive motors in both front wheel motors and to the front wheel drive brake.

With the operating switch in Low Torque position, the control valve directs pressurized oil to only one of the front wheel drive motors. Oil returning from this front wheel drive motor flows back to the control valve and is directed to the other front wheel drive motor. The drive motors in Low torque drive are connected hydraulically in series and since both drive wheels displace the same volume, both wheels will rotate a similar amount with a minimum amount of slippage.

31. **CONTROL VALVE.** Refer to Fig. 52 for a cross-sectional view of front wheel control valve. The unit contains three solenoid-operated pilot valves; F (Forward), R (Reverse) and T (Torque), and three hydraulically actuated operating valves; (Pressure Control Valve) which gives priority to other hydraulic functions if necessary, (Direction Valve) and (Torque Valve). A pressure relief valve is mounted near pressure inlet.

The three pilot valves are spring loaded in the closed position and direction valve is spring centered, blocking fluid passage to wheel motors whenever electric power is cut off to control valve. When forward or reverse solenoid is electrically energized, the pilot valve opens and admits pressure to one end of direction valve, shifting the valve and routing pressure fluid to front wheel motors in forward or reverse direction depending on which way valve is moved.

Torque valve is spring loaded in High Torque position. When Torque Pilot Valve (T) is not energized, flow through the direction valve is delivered at equal pressure to each front wheel motor. In Forward position, when torque pilot valve is energized, pressure fluid enters piston end of torque valve and moves it

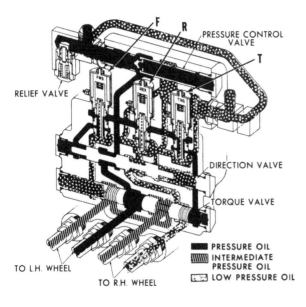

Fig. 52 — Cross-sectional view of Power Front Wheel Drive control valve showing component parts. Shaded areas indicate pressurized operating fluid with valve in forward, low torque position.

12

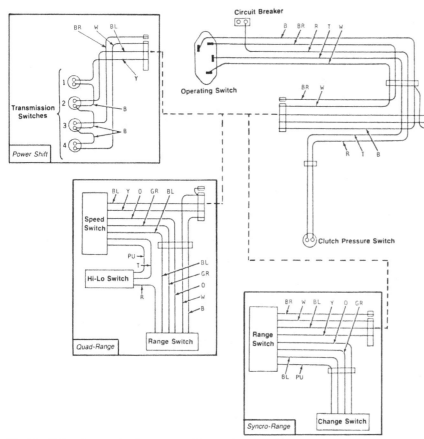

Fig. 53 — Wiring diagram for Power Front Wheel drive. The wiring and switches which are different with various transmissions are shown in the insets.

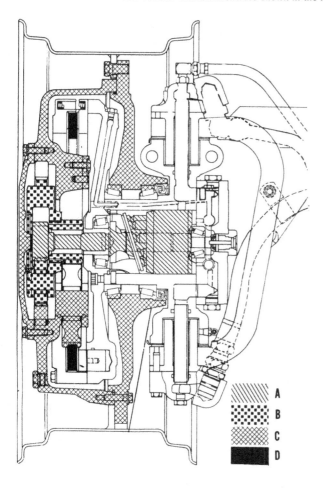

Fig. 54 — Schematic view of wheel drive unit showing the double reduction principle. "A" shows the hydraulic motor, shaft, and primary sun gear which turns at the fastest speed. "B" shows the primary planet carrier and secondary sun gear, which turns at intermediate speed. "C" shows fluid housing and wheel hub which turns at the slowest speed. "D" shows secondary planetary ring gear which is stationary when unit is engaged.

against spring pressure to route full volume to flow to left wheel motor. Discharge port of left wheel motor is connected through control valve to inlet port of right wheel motor and discharge port of right wheel motor to discharge port of valve. The two wheel motors, being series connected, now deliver half the torque on half the volume of fluid.

In reverse position, when torque pilot valve is energized, pressure fluid enters piston end of torque valve and moves it against spring pressure to route full volume flow to right wheel motor. Discharge port of right wheel motor is connected through control valve to inlet port of left wheel motor and discharge port of left wheel motor to discharge port of valve.

Electrical switches control current to the solenoids of control valves. Refer to Fig. 53 for schematic.

32. **WHEEL DRIVE UNIT.** Refer to Fig. 54. The front wheel drive unit consists of a fixed displacement axial piston motor and primary sun gear (A), which drives a dual reduction planetary gear unit (B). The primary planet carrier is splined to a secondary sun gear which drives hub (C).

The planetary secondary ring gear (D) floats in its housing and is coupled to spindle carrier by hydraulic pressure whenever power is applied. The wheel manifold contains check valves to admit pressure to brake pistons and wheel motor pistons simultaneously. One planet brake piston contains a relief valve to prevent damage due to pressure buildup during neutral operation. In forward direction, pressure oil is routed through the upper hoses to front wheel assembly, and return oil is directed through lower hoses to control valve and then to the reservoir. For reverse direction, pressure oil is directed to the lower hoses and return is through upper hoses. Refer to Fig. 55 for oil pressure flow through manifold. The upper check valve body has a 0.031 inch orifice which allows brake piston apply pressure to bleed to low pressure passage after controls are moved to the neutral, or off position. A relief valve is provided in the center of manifold to prevent damage due to excessive pressure buildup.

TROUBLESHOOTING

All Models So Equipped

CAUTION: If engine is to be run while checking Power Front Wheel Drive, shut off the hydraulic pump and disconnect the **wiring harness near the solenoids at the electrical connectors. This will prevent tractor movement if a short circuit or accidental switch movement should attempt to engage the front wheel drive.**

33. On all models, electrical circuits can be checked whenever key switch is turned on, but on models with Perma-Clutch or Power Shift the clutch pressure switch must be by-passed using a jumper wire. Solenoids close with an audible click. With key switch on and clutch switch jumper wire installed, test as follows:

(1) Move selector switch from "OFF" to "LOW TORQUE" (lower) position; torque solenoid should engage with an audible click.

(2) Move selector switch back to "OFF" position and Gear or Speed selector lever to 1st or 2nd forward gear; then, move selector switch to "HIGH TORQUE" (upper) position. Forward solenoid should engage with an audible click.

(3) Repeat test 2 except with Gear or Speed selector lever in 1st or 2nd reverse gear. Reverse solenoid should engage with an audible click.

NOTE: **HIGH TORQUE (upper) position of selector switch is used to test forward and reverse solenoids to keep from energizing torque solenoid.**

If any of the solenoids fail to engage at the proper time, continuity checks of the circuits should be conducted to determine the cause. Refer to Figs. 53, 56, 57 or 58 for wiring diagrams of power front drive units. Check red wire from clutch pressure switch to circuit breaker and make sure circuit breaker is not open.

Internal hydraulic leakage or failure, because of the closed center system, will usually be signalled by heat or noise.

ELECTRICAL SYSTEM

All Models So Equipped

34. ADJUSTMENT. Except for circuit breaker terminal, terminals for solenoid connections and rotary switch connections, quick disconnect terminals are used.

On Syncro-Range models, the two rotary switches must be synchronized as follows:

Remove console cover and be sure key is off, or remove from switch. Place shift lever in Fifth Gear.

Disconnect rod from Speed Range Switch and turn switch counter-clockwise against internal stop. Turn switch clockwise THREE detents, then adjust control rod yoke if necessary until rod can be reconnected to switch arm.

NOTE: **Be sure all shift lever supports and switch brackets are tight before making the above adjustments.**

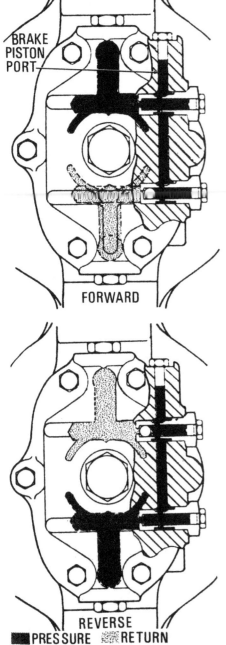

Fig. 55—Schematic view of pressure and return oil through wheel oil manifold. Brake piston port is pressurized in both directions.

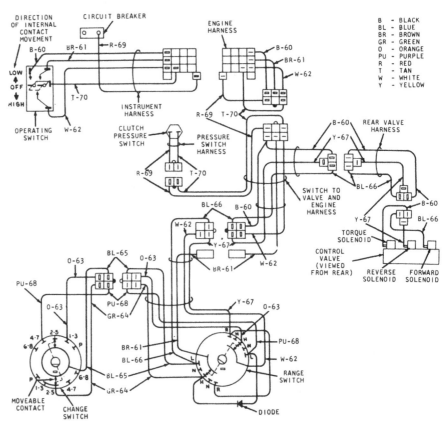

Fig. 56—Wiring diagram for Power Front Drive electrical system used on Syncro Range models.

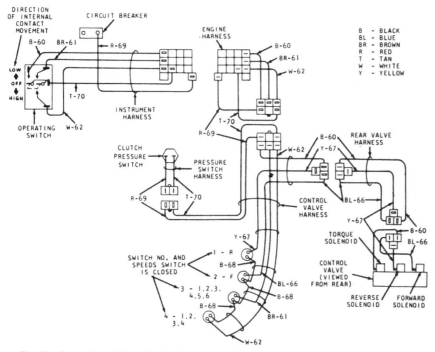

Fig. 57 – Power Front Drive electrical system wiring diagram used on Power Shift Models.

nect yoke from Range Switch arm (right lever) and turn switch counter-clockwise to internal stop. Turn clockwise THREE detents and adjust yoke if necessary, so that clevis pin will line up with hole in switch arm, and reconnect yoke. Move both levers to all positions and check for proper adjustment.

On Power Shift models, disconnect the wiring harness from the 4 transmission switches and check for continuity as follows:

TOP switch (No. 1) – closed in all reverse speeds.

2nd switch – closed in all forward speeds.

3rd switch – closed in 1st through 6th speeds.

BOTTOM switch (No. 4) – closed in 1st through 4th speeds. Refer to Fig. 57 for wiring diagram. If adjustment is necessary, remove switches and add or remove shim washers to make switches close ONLY as outlined above.

The clutch pressure switch should be checked for continuity, and should make contact as the clutch engages.

OVERHAUL

All Models So Equipped

35. **PILOT VALVES.** Solenoid coil (2 – Fig. 59) can be removed from pilot valve after removing nut (1), without disturbing hydraulic circuits. Pilot valves can be removed for inspection,

On Quad-Range models, remove console cover and be sure key is off, or removed from switch. Place shift levers in C-1 position and disconnect yoke from Speed Switch arm (left lever). Turn speed switch counter-clockwise to internal stop. Turn clockwise THREE detents and adjust yoke, if necessary, so that clevis pin will line up with hole in switch arm, and reconnect yoke. Discon-

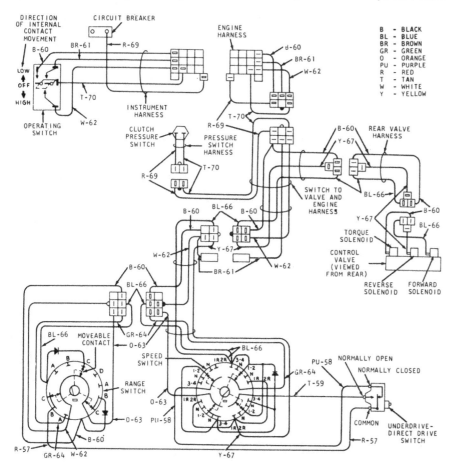

Fig. 58 – Power Front Drive electrical system Wiring Diagram used on Quad-Range Tractors.

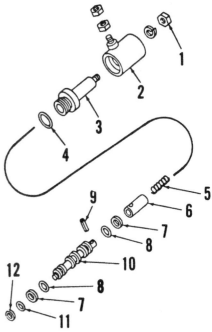

Fig. 59 – Exploded view of solenoid and pilot valve assembly, interchangeable in all three locations.

1. Nut
2. Solenoid
3. Armature
4. "O" ring
5. Spring
6. Plunger
7. Back-up ring
8. "O" ring
9. Spring pin
10. Bushing and valve
11. "O" ring
12. Back-up ring

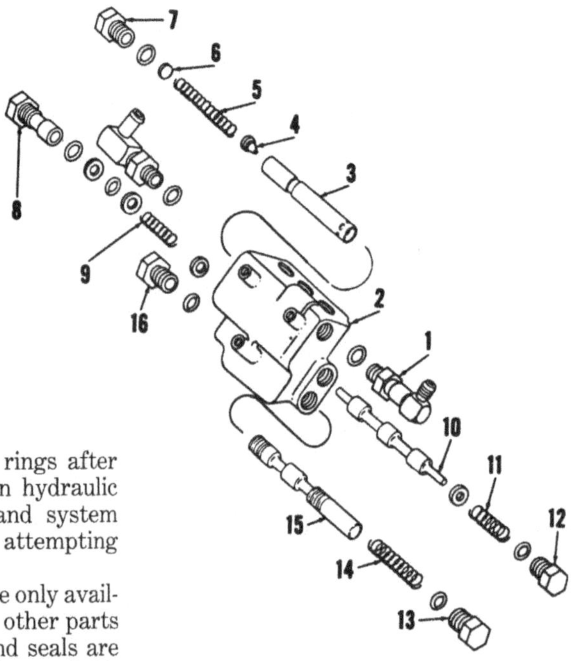

Fig. 60 — Exploded view of Power Front Wheel Drive control valve showing component parts.

1. Inlet elbow
2. Valve body
3. Priority valve
4. Orifice
5. Spring
6. Shim
7. Plug
9. Centering spring
10. Directional valve
11. Centering spring
12. Plug
13. Plug
14. High-torque spring
15. Torque valve
16. Plug

renewal, or renewal of "O" rings after removing solenoid coil. Main hydraulic pump should be shut off and system pressure relieved before attempting removal of valve unit.

Valve spool and bushing are only available as an assembly (10). All other parts are available individually, and seals are provided in the control valve seal kit assembly.

36. **CONTROL VALVE.** Refer to Fig. 60 for an exploded view of control valve assembly. Valve spools (3, 10 and 15) or body (2) are not available separately but spools may be removed for inspection or renewal of other parts. All seals including those on pilot valves (Fig. 59) are available in a valve sealing kit. Shims (6 – Fig. 60) control setting of priority valve (3). Operating pressure should be 1930-1970 psi with fluid at operating temperatures and engine at 2150 rpm.

Refer to Figs. 52 and 60 and assemble by reversing disassembly procedure, making sure valves are installed correct end forward. New sealing rings should always be used when assembling control valve unit.

36A. **WHEEL MANIFOLD.** Refer to Fig. 61 for an exploded view of wheel manifold and associated parts, and Fig. 55 for a schematic view of oil flow. Valves (7-12 and 13-17 – Fig. 61) can be

removed for inspection or service without removing manifold plate. Plate can be removed without taking weight off wheel; proceed as follows:

Drain wheel housing by turning plug (30 – Fig. 68) downward and removing plug. Housing capacity is approximately 9 quarts. Turn wheel for access to manifold cap screws and loosen screws evenly. Wheel motor spring should push manifold approximately 1/8-inch away from motor housing (spindle) as screws are loosened. Valve plate (1 – Fig. 61) and/or bearing plate (13 – Fig. 62) may remain with motor or be removed with manifold. Do not allow plates to drop as manifold is removed. Upper check valve body (17 – Fig. 61) has an orifice, and is longer than the lower, which has no orifice.

Install by reversing the removal procedure. Manifold plate should easily slide on attaching bolts to within 1/8-inch of motor housing. Considerable pressure is required to compress motor spring, but compression should be even without

binding. Tighten manifold cap screws to a torque of 35 ft.-lbs.

37. **WHEEL MOTOR.** The hydrostatic wheel motor is axial piston type shown exploded in Fig. 62. Wheel motor can be removed as outlined in paragraph 38 and overhauled as in paragraph 39.

38. **REMOVE AND REINSTALL.** To remove the hydrostatic wheel motor, first remove manifold as outlined in paragraph 36A and primary planetary unit as in paragraph 40. If brass bearing plate (13 – Fig. 62) remained with motor, insert a suitable wood dowel in one of the small round drain holes to use as a pry. NEVER use a screwdriver or steel wedge of any kind.

Grasp inner bearing cone (3 – Fig. 61) and withdraw wheel motor (2 – Fig. 68), swash plate (3), motor shaft (7) and outer bearing cone (8) as a unit from motor housing. If swash plate binds in housing or drags on locating pin, pressure can be applied at outer end by inserting a wood dowel or brass drift in spline end of shaft (7).

NOTE: Do not rotate assembly when out of housing, as damage may occur to motor parts while not supported by housing.

Install by reversing the removal procedure, making sure thick side of swash plate ramp is toward rear of tractor and that locating dowel pin enters appropriate notch in swash plate.

39. **OVERHAUL.** To overhaul the removed wheel motor unit, use specially cut pieces of clean cardboard, sheet gasket material or plastic to protect the polished inner face of cylinder block (5 – Fig. 62) then use suitable pullers to remove inner bearing cone (3 – Fig. 61).

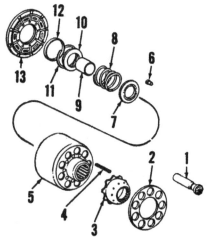

Fig. 62 — Exploded view of front wheel drive motor.

1. Piston
2. Slipper retainer
3. Ball guide
4. Springs
5. Cylinder block
6. Aligning dowel
7. Spring seat
8. Spring
9. Spring guide
10. Spring retainer
11. Retaining ring
12. Centering ring
13. Brass bearing plate

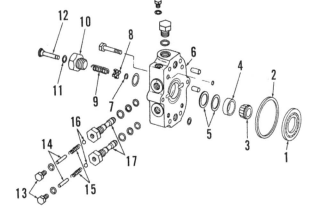

Fig. 61 — Exploded view of Power Front Drive Wheel manifold showing component parts.

1. Valve plate
2. Packing
3. Bearing cone
4. Bearing cup
5. Shims (0.003 & 0.010)
6. Manifold
7. Snap ring
8. Retainer
9. Spring
10. Valve plug
11. "O" ring
12. Valve
13. Plug
14. Spring pin
15. Spring
16. Check valve ball
17. Valve body

Motor shaft, outer bearing cone and swash plate can now be removed. (see preceding NOTE.)

Before disassembling motor, note the splines on ball guide and cylinder block. If splines do not have a master spline, mark both surfaces for reassembly.

To disassemble the motor, grasp outer diameter of slipper retainer (2 – Fig. 62) and remove retainer and nine pistons (1). Remove ball guide (3) and withdraw nine slipper retainer springs (4). See Fig. 63 for method of removal and stacking of pistons for maximum cleanliness.

Wheel motor spring (8 – Fig. 62) must be slightly compressed to remove or install retaining ring (11). Either of two methods is acceptable:

(1) Select a ⅜-inch bolt long enough to extend through cylinder block, nut and large flat washers. Install bolt through cylinder block as shown in Fig. 64 (washer on each end) and compress the spring by tightening nut, then unseat and remove retaining ring (11 – Fig. 62).

NOTE: Retaining ring should be pulled (not pried) out of its groove. Prying will probably cause a burr to be turned up on the lapped surface, resulting in oil leakage.

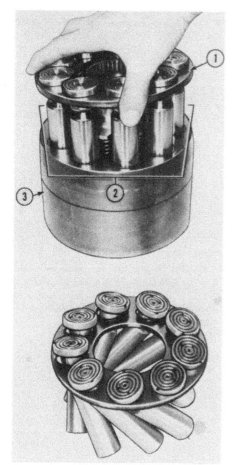

(2). As an alternate method, place cylinder block in a press on wood blocks, retaining ring up. Use a step plate on spring retainer and slightly compress the spring to remove retaining ring.

When using either method, release pressure slowly after removing retaining ring using extreme care to keep from scratching or scoring polished, machined face of block.

Clean all parts and inspect for excessive wear or other damage. Check pistons and bores in cylinder block for linear scratches and excessive wear. Check cylinder block face for shiny streaks, indicating high pressure leakage between cylinder block and brass bearing plate. Inspect piston slippers for scratches, imbedded material or other damage. Light scratches in slippers can be removed by lapping. All nine slippers must be within 0.002 inch of each other in thickness.

Lubricate all parts with hydraulic fluid and reassemble by reversing the disassembly procedure. Spring seat (7 – Fig. 62) must be installed with beveled side away from cylinder block spring (8) and spring retainer (10) installed with retaining groove away from spring. When installing ball guide (3), align master spline of ball guide and cylinder block. On some motors, a punch mark is used on one outer tang of ball guide which aligns with a corresponding mark on cylinder block face.

Inspect the steel valve plate (1 – Fig. 61) and brass bearing plate (13 – Fig. 62) for excessive wear or scoring in areas shown in Figs. 65 and 66. Although both plates are available individually, it is good shop practice to renew both plates whenever one is damaged, to assure proper sealing necessary for efficient operation.

Before installing motor assembly and shaft in motor housing (4 – Fig. 68), check wheel hub bearings (10 and 11) for proper adjustment. Shims (13) behind outer bearing cup are available in 0.003 inch and 0.010 inch thickness to allow for an adjustment range of from 0.004 inch preload to 0.002 inch end play.

Preload may be considered correct when a pull of 3 pounds at wheel hub is required to turn the hub.

Install outer bearing cone (8) on motor shaft (7). Install motor assembly and swash plate (3) on motor shaft and press inner bearing cone (3 – Fig. 61) onto motor shaft. Thoroughly lubricate motor and shaft assembly and carefully install into motor housing (4 – Fig. 68), with swash plate (3) installed thick side toward rear of tractor and brass bearing plate (13 – Fig. 62) protecting lapped surface of cylinder block. Refer to Fig. 61. Lubricate valve plate (1) and install before oil manifold is bolted to housing. Spring (8 – Fig. 62) will resist as the manifold is installed, so tighten evenly to avoid binding. Tighten manifold cap screws to a torque of 35 ft.-lbs. Motor shaft end play should be 0.006-0.012 inch and can be measured with a dial indicator by removing relief valve plug (10 – Fig. 61) and associated parts. Shims (5) are available in thicknesses of 0.003-0.010 inch and are used for adjustment if necessary. Remove bearing cup (4) and adjust by adding or removing shims as required.

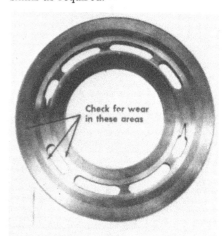

Fig. 65 – Inspect steel valve plates for wear or scoring in areas shown.

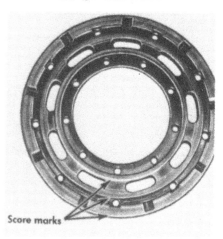

Fig. 66 – Inspect brass bearing plates as indicated.

Fig. 63 – Method of removal and convenient stacking of pistons.

1. Retainer   2. Pistons   3. Cylinder block

Fig. 64 – Use a ⅜-inch, full threaded bolt, flat washers and nut to disassemble motor spring as shown.

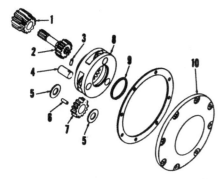

Fig. 67—Exploded view of front wheel drive cover, primary planet carrier and associated parts.

1. Secondary sun gear	6. Bearing roller
2. Primary sun gear	7. Primary pinion
3. Dowel	8. Planet carrier
4. Pinion shaft	9. Snap ring
5. Thrust washer	10. Cover

40. PRIMARY PLANETARY UNIT. The primary planetary unit and secondary sun gear are shown exploded in Fig. 67. Primary ring gear is carried in planetary housing (29 – Fig. 68).

Each planet pinion (7 – Fig. 67) rides on seventeen (17) loose needle rollers (6). Secondary sun gear (1) splines into hub of planet carrier (8), and serves as pilot for primary sun gear (2). Primary planetary unit may be withdrawn after removing retaining cover (10). Tighten cover retaining cap screws to a torque of 35 ft.-lbs. when installing.

41. SECONDARY PLANETARY AND DRIVE BRAKE UNITS. Refer to Fig. 68 for an exploded view of wheel drive unit. Secondary planet gears (28) are carried in planetary housing (29) which also contains primary ring gear. Secondary ring gear (22) floats in housing until hydraulic pressure is applied by the nine brake pistons (18).

Each planet pinion (28) rides on twenty-two loose needle rollers on shaft (27). Shaft (27) is keyed to housing (29) and carrier plate (24) by steel balls (26). Tighten cap screws retaining carrier plate (24) to a torque of 35 ft.-lbs. when unit is reassembled.

Brake backing plate (23) attaches to planetary piston housing (15) with 18 cap screws and encloses planetary ring gear (22), brake facings (21), brake pressure plate (19) and six separator springs (20). Backing plate must be removed for renewal of brake parts, including the nine brake pistons (18).

NOTE: One of the nine brake pistons should contain a bleed valve, and should be installed slotted end out in the second hole counter-clockwise from the word "UP", which is cast into the brake piston housing.

Tighten brake backing plate attaching screws to a torque of 35 ft.-lbs. when unit is assembled. Mount a dial indicator

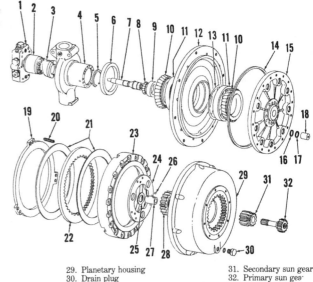

Fig. 68—Exploded view of wheel drive unit showing component parts.

1. Wheel manifold	
2. Motor	
3. Swashplate	
4. Motor housing	
5. Snap ring	
6. Oil seal	
7. Motor shaft	
8. Bearing cone	
9. Bearing cup	
10. Bearing cone	
11. Bearing cup	
12. Wheel hub	
13. Shim	
14. Packing	
15. Piston housing	
16. Quad ring	
17. Back-up ring	
18. Brake piston	
19. Pressure plate	
20. Separator spring	
21. Brake facings	
22. Planetary ring gear	
23. Backing plate	
24. Retainer	
25. Thrust washer	
26. Locking ball	
27. Pinion shaft	
28. Planet pinion	

29. Planetary housing	31. Secondary sun gear
30. Drain plug	32. Primary sun gear

on outboard side of brake backing plate (23) and check free play of planetary ring gear (22) at three places around internal teeth (inside) circumference. Free play should be at least 0.014 inch.

Tighten planetary brake piston housing retaining cap screws to a torque of 45 ft.-lbs. when unit is reassembled.

ENGINE

All models are equipped with a six cylinder engine; however, many differences will be noted.

4040 – 404 CID Diesel
4240 – 466 CID Diesel
4440 – 466 CID Turbocharged Diesel
4640 – 466 CID Turbocharged/intercooled Diesel
4840 – 466 CID Turbocharged/intercooled Diesel

REMOVE AND REINSTALL

All Models

45. To remove the engine and clutch assembly as a unit, first drain cooling system and, if engine is to be disassembled, drain oil pan. Remove vertical air stack, muffler, hood, cowl, grille screens and engine side panels. Discharge brake accumulators by opening the right brake bleed screw and depressing right brake pedal for about 60 seconds.

Make a clutch split by disconnecting batteries and cables, tachometer cable, heater hoses, throttle linkage, electric wiring connectors on right cowl and inside control support housing and hydraulic lines on right side of engine. Disconnect fitting to cold starting pipes and disconnect wiring and hydraulic lines on left side. On Perma-Clutch models, place a drain pan under housing to catch the oil as engine is separated from housing. Remove the 3-point hitch center link attaching bracket at rear of

transmission housing, remove the large plug behind bracket, and pull the hex shaft that extends all the way to the crankshaft and drives the transmission pump. Shaft can become bent if not removed. If equipped with Power Front Wheel Drive, disconnect drain pipe on left side. If equipped with air conditioning, break connections in hoses at couplers on left side bracket. Hold coupler body with one wrench and remove coupler with another wrench. If escaping gas can be heard, tighten coupler and loosen again. Suitably support both halves of tractor separately, remove cap screws securing engine to clutch housing and roll rear half of tractor away from engine and front unit.

NOTE: If tractor is equipped with Sound-Gard body, it will be necessary to remove floor mat, floor panel and filler panels for access to clutch housing cap screws.

Remove wiring harness on engine if necessary, and disconnect hydraulic pump coupling and support. Disconnect fuel line to fuel pump after closing shut-off valve and remove fuel return pipe. Remove air intake pipe and disconnect both radiator hoses. Remove air conditioning compressor and hoses as an assembly if so equipped. Install JDG-1 engine sling and lifting brackets to a hoist, remove side frame cap screws to engine and slide engine out of side frames.

NOTE: When engine is removed, front unit may be heavy in front therefore unstable. Remove front end weights, if used, and securely support front frame to prevent forward tipping.

Reassemble tractor by reversing the disassembly procedure. Tightening torques are as follows:

Hydraulic pump drive . . . . . . .30 ft.-lbs.
Hydraulic pump support . . . . .85 ft.-lbs.
Side frame to cyl. block . . . . .275 ft.-lbs.
Cylinder block to clutch
    housing . . . . .300 ft.-lbs.

CYLINDER HEAD

All Models

48. To remove the cylinder head, drain cooling system and remove side panels, grille screens, air stack, muffler and hood. Relieve hydraulic pressure by pumping brake pedals several times. Remove air intake pipes up as far as turbocharger, if so equipped, or to intake manifold. Remove water manifold, by-pass pipe, thermostat housing and upper water hose. Disconnect turbocharger oil pipes and remove intake manifold; then unbolt and remove exhaust manifold (or turbocharger and manifold as a unit if so equipped.)

Remove alternator and fan blades then unbolt water pump and tip it forward out of the way. Remove injector lines and injectors. Remove ventilator tube, rocker arm cover, rocker arm assembly and push rods, then unbolt and remove cylinder head.

NOTE: Make sure cylinder liners are held down with bolts and washers if engine is to be turned.

The cylinder head gasket is available in two different thicknesses for 4040 and 4240 models. The black gasket is 0.044 inch thick and the black and grey striped gasket is 0.050 inch thick. To determine which gasket to use, measure the amount of piston stand out (above the gasket surface of block). A dial indicator with magnetic mount can be used to measure piston stand out. If top of the highest piston is less than 0.009 inch above gasket surface of block, use black gasket. If top of highest piston is between 0.009-0.014 inch above surface of block, use black and grey striped

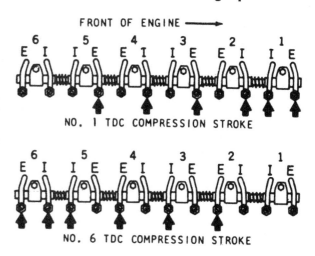

Fig. 71 — Adjust valves indicated by arrows in upper half of Fig. when No. 1 piston is at TDC on compression stroke. Tappet gap should be as listed in text. Refer to lower half of Fig. for remainder of valves.

gasket. If any piston stands out more than 0.014 inch, determine the cause and correct the problem before continuing assembly.

Install the cylinder head gasket dry, but oil threads of retaining screws. Be sure to install a hardened flat washer under the head of all cylinder head retaining screws. Tighten the head retaining cap screws evenly and in sequence shown in Fig. 69 to initial torque of 105 ft.-lbs., then retighten in same sequence to 117-143 ft.-lbs. torque. Tighten the rocker arm clamp screws to 55 ft.-lbs. torque.

NOTE: NEVER run engine with turbocharger oil lines disconnected.

Complete assembly to the point of permitting engine to run, then run engine at 2100 rpm for ½-hour. Retighten cylinder head to 135-165 ft.-lbs. torque, then readjust valve clearance as outlined in paragraph 51.

VALVES AND SEATS

All Models

50. The valve face and seat angles are both 45 degrees on 4040 and 4240 models. Valve face and seat angles are 30 degrees for turbocharged models.

Some models are originally equipped with hardened steel inserts which are pressed into machined bores in cylinder head. On 4040 and 4240 models, the intake valve should be recessed

0.036-0.050 inch and the exhaust should be recessed 0.044-0.058 inch. Turbocharged models should have both intake and exhaust valves protrude beyond gasket surface of cylinder head 0.024-0.038 inch. The manufacturer recommends installation of new valve and seat insert if valve is recessed more than 0.006 inch on turbocharged models.

Intake and exhaust valve stem diameter is 0.3715-0.3725 inch and recommended stem to guide clearance is 0.002-0.004 inch. Guides can be knurled to provide correct clearance if clearance is less than 0.010 inch. If clearance exceeds 0.010 inch, bores in guides should be reamed to correct oversize and new valve with larger (oversize) stem fitted. Be sure to reseat valve after guide has been knurled or reamed oversize. Valve seat width should be 0.083-0.093 inch.

Valve clearance (tappet gap) should be adjusted using procedure outlined in paragraph 51.

51. TAPPET GAP ADJUSTMENT. The two-position method of valve tappet gap adjustment is recommended. Refer to Fig. 71 and proceed as follows:

Turn engine crankshaft by hand using JDE-81 or similar engine rotation tool until number 1 and 6 cylinders are at TDC. This TDC location is determined by using timing pin in flywheel housing to engage hole in flywheel. Check the valves to determine whether front or rear cylinder is at top of compression stroke. (Exhaust valve on adjacent cylinder will be partly open). Use the appropriate diagram (Fig. 71) and adjust the indicated valves; then turn crankshaft one complete turn until "TDC" timing mark is again aligned. Adjust remainder of valves using the other diagram. Recommended valve tappet gap is as follows:

Intake . . . . .0.018 in.
Exhaust . . . . .0.028 in.

Fig. 69 — Use the sequence shown to tighten cylinder head retaining screws.

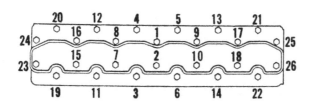

FRONT ▶

VALVE ROTATORS

All Models

52. Positive type valve rotators are originally installed on all (intake and exhaust) valves.

Normal service consists of renewing the complete unit. Rotators can be considered satisfactory if the valve turns a slight amount each time it opens.

VALVE GUIDES AND SPRINGS

All Models

53. Valve guide bores are an integral part of cylinder head. Standard valve guide bore diameter is 0.3745-0.3755 inch and normal stem clearance is 0.002-0.004 inch. If stem to guide clearance is 0.006-0.010 inch, the guide can be knurled; however, if clearance exceeds 0.010 inch, resize guide and install valve with correct oversize stem.

Intake and exhaust valve springs are interchangeable and may be installed either end up. Renew any spring which is distorted, rusted or discolored, or does not meet the test specifications which follow:

Free Length (approx.)........2.12 in.
Lbs. Test at Length (Inches)
   Closed ....54-62 at 1.81 in.
   Open ....133-153 at 1.36 in.

ROCKER ARMS

All Models

54. The rocker arm shaft attaches to bosses which are cast into cylinder head and is held in place by clamps. Shaft rotation is prevented by a spring pin in cylinder head which enters a hole in shaft for positive positioning of lubrication passages.

Rocker arms are right hand and left hand assemblies. Recommended clearance of rocker arms to shaft is 0.0005-0.0035 inch. Bushings are not available; if clearance is excessive,

Fig. 72 — Score oil seal wear sleeve as shown using a blunt chisel, then pry from shaft when renewal is indicated. Timing marks align as shown.

renew rocker arms and/or shaft.

When reassembling, make sure spring pin aligns with locating hole in shaft, tighten clamp screws to 55 ft.-lbs. torque, then adjust tappet gap as outlined in paragraph 51.

CAM FOLLOWERS (TAPPETS)

All Models

55. The mushroom type cam followers can be removed from below after removing camshaft as outlined in paragraph 58. The cam followers operate in unbushed bores in engine block and are available in standard size only.

TIMING GEAR COVER AND CRANKSHAFT FRONT OIL SEAL

All Models

56. To remove the timing gear cover, first drain cooling system and remove hood, grille screens and engine side panels. Remove radiator and fan shroud from left side after disconnecting oil cooler from radiator.

Remove pressure and return lines from hydraulic pump, disconnect pump drive shaft and coupler and remove pump and support. Loosen fan belts, remove crankshaft damper pulley using a suitable puller, remove cover retaining cap screws and lift off the cover. The lip type front oil seal is supplied in a kit which also includes a steel wear sleeve which is pressed on crankshaft in front of gear as shown in Fig. 72. Score old sleeve lightly with a blunt chisel and pry sleeve from shaft. Coat inner surface of new sleeve with a non-hardening sealant and install with a suitable screw-type installer such as JDE-3. Install oil seal in cover from inside. Sealing lip should be toward the rear and seal should bottom in its bore.

When installing timing gear cover, tighten retaining cap screws to a torque of 30 ft.-lbs. and damper pulley retainer cap screw to 150 ft.-lbs. Complete the assembly by reversing the disassembly procedure. Other tightening torques are given in paragraph 45.

TIMING GEARS

All Models

57. The timing gear train consists of crankshaft gear, camshaft gear, injection pump drive gear (which fits on camshaft directly behind camshaft gear) and injection pump shaft gear as shown in Fig. 72. Gears are available in standard size only. If backlash is excessive, renew the parts concerned. The "V" marks on camshaft gear must align with crankshaft gear (Fig. 72). Number 1 cylinder must be at TDC when aligning timing

marks. When installing crankshaft gear, heat gear to approximately 350 degrees F. using a hot plate or oven and install with a press or JDH-7 driver with timing mark to front.

CAMSHAFT AND BEARINGS

All Models

58. To remove the camshaft, first split tractor at front of engine as follows:

Drain cooling system, remove air stack and muffler, grille screens, hood, left tractor step, batteries and boxes. Remove air conditioning compressor if so equipped, and disconnect front couplers by using two wrenches on fittings. If gas escapes, tighten couplers and loosen them again until no gas escapes. Lay compressor and hoses aside as a unit. Disconnect hydraulic pump inlet and outlet pipes, pump coupler and support. Remove air intake pipes to intake manifold or turbocharger. Close fuel shut-off and disconnect supply pipe to fuel pump. Disconnect fuel return line, steering pipes, oil cooler return pipe, radiator and heater hoses. If equipped, disconnect Power Front Wheel drain pipe. Separate front wiring harness at connectors. Support front assembly to prevent tipping, install JDG-1 engine sling, and remove side frame to engine cap screws. Roll rear section away from front end and place a support under clutch housing.

Remove timing gear cover. Remove oil pump as in paragraph 70.

Remove rocker arm cover, rocker arms assembly and push rods. Raise and secure cam followers in their uppermost position using magnetic holders or other suitable means. Remove speed-hour meter drive from right side of engine fuel supply pump. Working through openings provided in camshaft gear, remove the four cap screws securing camshaft thrust plate to front face of engine block, then withdraw camshaft and gear assembly forward out of engine.

Camshaft journals should be 2.3745-2.3755 inches diameter and should have a clearance of 0.002-0.005 inch in bushings. The presized copper lead camshaft bushings are interchangeable. To install bushings after camshaft is out, detach cylinder block from clutch housing, remove clutch, flywheel and camshaft bore plug, then pull bushings into block bores using a piloted puller. Make sure oil supply holes in block are aligned with holes in bushings. The elongated hole in bushing goes to the top.

Camshaft end play of 0.0025-0.0085 inch is controlled by the thickness of camshaft thrust plate. End play can be measured with a dial indicator when camshaft is installed, or with a feeler gage when camshaft unit is out. Thrust plate thickness is 0.187-0.189 inch.

Camshaft gear and injector pump drive gear if so equipped can be removed with a press when camshaft is out, after removing the retaining cap screw and washer. Align key slot in gear with Woodruff key in shaft, make sure thrust plate and spacer are installed and press gear on shaft until it bottoms. Tighten gear retaining cap screw to a torque of 85 ft.-lbs. and thrust plate retaining cap screws to a torque of 20 ft.-lbs. Align timing marks as outlined in paragraph 57.

ROD AND PISTON UNITS

All Models

59. Connecting rod and piston units are removed from above after removing cylinder head, oil pan and rod bearing caps. When reinstalling, correlation numbers, small and large slots and tangs on rod and cap must be in register. Rods and head of piston are stamped "FRONT" for proper installation. Tighten connecting rod cap screws to 55 ft.-lbs. torque plus ¼-turn.

NOTE: Do not rotate crankshaft with head removed nor attempt to remove rod and piston units without first bolting liners down using washers and short cap screws as shown in Fig. 74.

PISTONS, RINGS, PINS AND SLEEVES

All Models

60. **PISTONS AND RINGS.** All models are equipped with aluminum alloy, cam ground pistons which use two compression rings and one oil control ring. All rings are located above the piston pin.

All 4040 and 4240 models have keystone type compression rings in the two top grooves. A conventional oil control

Fig. 74—Lock sleeves in position using a cap screw and washer as shown, when cylinder head is removed for engine service.

ring is used in lower groove. A special ring groove wear gage is necessary for checking the keystone grooves for wear. Special tool JDE-62 is used to check keystone grooves of 4040 models; tool JDE-55 is required for checking grooves of 4240 models. A new oil control ring for 4040 and 4240 models should have 0.002-0.004 inch side clearance in groove and piston should be renewed if side clearance exceeds 0.006 inch.

All 4440, 4640 and 4840 models have keystone compression rings in the top two grooves of pistons. A special ring groove wear gage (JDE-55) should be used to check for wear. A new oil control ring should have 0.0024-0.0040 inch side clearance in groove and piston should be renewed if side clearance exceeds 0.0065 inch.

Pistons and sleeves are selectively fitted on some models. Pistons marked "L" should only be used in sleeves which are stamped "LL" or "LV" on the sleeve flange. Pistons marked "H" should be installed in sleeves with "HH" or "HV" stamped on flange.

The manufacturer recommends cleaning pistons using Immersion Solvent "D-Part," and Hydra-Jet Rinse Gun or Glass Bead Blasting Machine.

61. **SLEEVES, PACKING AND "O" RINGS.** The renewable wet type cylinder sleeves are available in standard size only; however, some sleeves and pistons are selectively fitted. Pistons marked "L" should only be used in sleeves with "LL" or "LV" stamped on flange. Pistons marked "H" should only be used in sleeves stamped "HH" or "HV".

Sleeve flange at upper edge is sealed by the cylinder head gasket. Sleeves are sealed at lower edge by packing shown in Fig. 75. Sleeves normally require loosening using a sleeve puller, after which they can be withdrawn by hand. Out-of-round or taper should not exceed 0.005 inch. If sleeve is to be reused, it should be deglazed using a normal cross-hatch pattern.

When reinstalling sleeves, first make sure sleeve and block bore are absolutely clean and dry. Carefully remove all rust and scale from seating surfaces and packing grooves in block and from areas of water jacket where loose scale might interfere with sleeve or packing installation. If sleeves are being reused, buff rust and scale from outside of sleeve.

Install sleeve without the seals and measure standout. Check sleeve standout at several locations around sleeve. Also check to be sure that sleeve will slip fully into bore without force. If sleeve cannot be pushed down by hand, recheck for scale or burrs. Sleeve stand-out should be 0.000-0.004 inch for used cylinder block, but can be 0.002-0.005

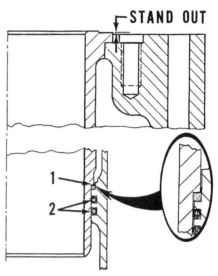

Fig. 75—Cross-section of cylinder sleeve showing square section packing (1) and round "O" rings (2). Refer to text for correct installation of cylinder sleeves.

inch for new cylinder block. Check fit of other sleeves, if standout is less than specified. If stand-out is excessive, check for scale or burrs, then, if necessary, select another sleeve. After matching sleeves to all the bores, mark the sleeves then refer to the appropriate following paragraphs for packing and sleeve installation.

Apply liquid (lubricating) soap such as part number AR54749 to the square section ring (1–Fig. 75) and install over lower end of cylinder liner (sleeve). Slide the square section ring up against shoulder on sleeve, make sure that ring is not twisted and that longer slides are parallel with side of sleeve as shown. Apply the liquid soap to round section "O" rings (2) and install in grooves in cylinder block. If some of the "O" rings are red and some are black, the red "O" rings should be installed in top groove in block and black "O" rings should be in lower groove. Be sure that "O" rings are completely seated in grooves so that installing the sleeve will not damage the "O" rings. Observe the previously affixed mark indicating correct cylinder location, then install sleeves carefully into correct cylinder block bore. Also check the sleeve flange for marks "LL" or "HH". If flange is stamped, install sleeve with stamped "LL" or "HH" mark toward front of engine. Work sleeve gently into position by hand until it is finally necessary to tap sleeve into position using a hardwood block and hammer.

NOTE: Be careful not to damage the packing rings. Check the sleeve standout (Fig. 75) with packing installed. The difference between this measured standout and similar measurement taken earlier for same sleeve in same bore without packing

will be the compression of the packing. If compression is less than 0.0126 inch, the square section packing ring will not seal properly. Remove sleeve from cylinder block, check packing to be sure that installation has not cut the packing ring. If shoulders on sleeve and in cylinder block do not provide proper compression of the square section packing ring, install different sleeve and recheck. If a different sleeve will not provide enough compression of packing sleeve and correct packing is installed, suggested repair is to install new cylinder block.

62. **SPECIFICATIONS.** Specifications of pistons and sleeves are as follows:

4040
Piston Skirt Diameter*—
   Marked "L" ....4.2459-4.2466 in.
   Marked "H" ....4.2466-4.2473 in.
Cylinder Bore Diameter—
   Marked "LL" ......4.2493-4.2500 in.
   Marked "H" ....4.2500-4.2507 in.
Piston Skirt* to Cylinder
   Clearance ....0.0027-0.0041 in.
Piston Pin Diameter ..1.4997-1.5003 in.
Piston Pin Bore in
   Piston ....1.5003-1.5009 in.

4240
Piston Skirt Diameter*—
   Marked "L", green ..4.5577-4.5584 in.
   Marked "H", black ..4.5584-4.5591 in.
Cylinder Bore Diameter—
   Marked "LV" ......4.5615-4.5625 in.
   Marked "HV" ......4.5625-4.5635 in.
Piston Skirt* to Cylinder
   Clearance ....0.0036-0.0053 in.
Piston Pin Diameter ..1.6247-1.6253 in.
Piston Pin bore in
   Piston ....1.6253-1.6259 in.

4440, 4640 & 4840
Piston Skirt Diameter*—
   Marked "L" ....4.5575-4.5582 in.
   Marked "H" ....4.5582-4.5589 in.
Cylinder Bore Diameter—
   Marked "LL" or
   "LV" ....4.5615-4.5625 in.
   Marked "HH" or
   "HV" ....4.5625-4.5635 in.
Piston Skirt* to Cylinder
   Clearance ....0.0036-0.0053 in.
Piston Pin
   Diameter.........1.8739-1.8745 in.
Piston Pin Bore in
   Piston ....1.8748-1.8752 in.
*Measure piston skirt at right angles to piston pin 0.090 inch from bottom of skirt.

PISTON PINS
All Models

63. The full floating type piston pins are retained in piston bosses by snap

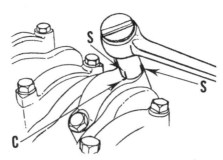

Fig. 75A — Connecting rod bolts should be tightened exactly ¼-turn after tightening to 55 ft.-lbs. Do not depend upon location of tightening handle. Instead, mark the socket at four equally spaced locations (S), mark rod cap (C) as shown in line with one mark on socket, then turn socket until next mark is aligned.

rings. Pins are often available in oversizes as well as standard. Check parts source for availability.

The recommended fit of piston pins is a hand push fit in piston bores and a slip fit in connecting rod bushings. Standard diameter and clearances are as follows:

4040
Piston Pin Diameter—
   Standard ....1.4997-1.5003 in.
Bore in Connecting Rod for Piston
Pin—
   Standard ....1.5010-1.5020 in.
Bore in Piston for Piston Pin—
   Standard ....1.5003-1.5009 in.
Piston Pin to Rod Bushing—
   Desired Clearance ..0.0007-0.0023 in.
   Wear Limit ....0.003 in.

4240
Piston Pin Diameter—
   Standard ....1.6247-1.6253 in.
Bore in Connecting Rod for Piston
Pin—
   Standard ....1.6260-1.6270 in.
Bore in Piston for Piston Pin—
   Standard ....1.6253-1.6259 in.
Piston Pin to Rod Bushing—
   Desired Clearance ..0.0007-0.0023 in.
   Wear Limit ....0.003 in.

4440, 4640 & 4840
Piston Pin Diameter—
   Standard ....1.8739-1.8745 in.
Bore in Connecting Rod for Piston
Pin—
   Standard ....1.8752-1.8762 in.
Bore in Piston for Piston Pin—
   Standard ....1.8748-1.8752 in.
Piston Pin to Rod Bushing—
   Desired Clearance ..0.0007-0.0023 in.

CONNECTING RODS AND BEARINGS
All Models

64. Connecting rod bearings are steel-backed inserts. Bearings are available in standard size as well as undersizes of

0.002, 0.010, 0.020 and 0.030 inch. Refer to paragraph 63 for specifications concerning the piston pin bushing.

Mating surfaces of rod and cap have milled tongues and grooves which positively locate cap and prevent it from being reversed during installation. Connecting rods are marked "FRONT" for proper installation. Check the connecting rods, bearings and crankpin journals for excessive taper and against the values which follow:

4040 & 4240
Crankpin Std.
   Diameter ....2.9980-2.9990 in.
Crankpin to Rod Bearing Diametral
Clearance—
   Desired ....0.0010-0.0040 in.
   Wear Limit ....0.006 in.
Rod Bolt Torque ....55 ft.-lbs.
                   plus ¼-turn.

4440, 4640 & 4840
Crankpin Std.
   Diameter ....2.9980-2.9990 in.
Crankpin to Rod Bearing Diametral
Clearance—
   Desired ....0.0010-0.0040 in.
Rod Bolt Torque ....55 ft.-lbs.
                   plus ¼-turn.

Check crankshaft journals for taper and out-of-round conditions. Limit for journal taper is usually considered to be 0.0001 inch for each one inch of journal length. Out-of-round limit is 0.0040 inch.

Connecting rod bolts should be very carefully tightened as follows: Torque the rod screws evenly to exactly 55 ft.-lbs.; then mark the socket wrench and rod cap as shown in Fig. 75A and tighten each rod screw exactly ¼-turn. It should be noted that final torque is not 55 ft.-lbs.; but that a precision micrometer type adjustment is accomplished using the bolt threads for accurate adjustment. It is therefore important to be sure that the socket is accurately marked at 90 degrees, that the beginning torque is exactly 55 ft.-lbs. and that the rod bolt is turned exactly 90 degrees.

CRANKSHAFT AND BEARINGS
All Models

65. The crankshaft is supported in seven main bearings. Crankshaft end play is controlled by the flanged fifth main bearing. Upper and lower bearing shells may be interchangeable, with both halves containing oil holes; however, if only one half of bearing has a hole, make sure that half is installed in the block, to insure that oil reaches all bearings.

All main bearing caps can be removed from below after removing oil pan and oil pump, and caps are numbered for proper reassembly. When renewing bearings, make sure that locating lug on bearing shell is aligned with milled slot in cap and block bore. After caps are loosely installed, bump crankshaft forward and rearward to align thrust flanges; then tighten main bearing cap screws to a torque of 150 ft.-lbs. Main bearings are available in undersizes of 0.002, 0.010, 0.020 and 0.030 inch. The thrust bearing is available in all undersizes with standard flange width; and in 0.010 inch undersize and 0.007 inch oversize flange width. Thrust flange thickness should be sufficient to permit crankshaft end play within desired limits.

To remove the crankshaft, first remove engine as outlined in paragraph 45 and proceed as follows: Remove flywheel and crankshaft rear oil seal retainer. Remove crankshaft pulley, timing gear cover, oil pan and oil pump. Remove rod and main bearing caps and lift out crankshaft. The hardened crankshaft front and rear wear sleeves are a press fit on flywheel flange and nose of crankshaft, and are both renewable. When installing, refer to paragraph 56 for front wear sleeve, and paragraph 66 for rear sleeve. Check crankshaft and bearings against the values which are given for crankpin in paragraph 64, and for main bearings as follows:

4040, 4240

Crankshaft End Play –
Desired . . . . 0.0015-0.0150 in.
Main Bearing Journal –
Standard Diameter . . 3.3720-3.3730 in.
Desired Clearance in
Bearing . . . . 0.0010-0.0040 in.
Wear Limit . . . . 0.0077 in.
Regrind if Taper
Exceeds . . . . . 0.0001 in./1 in. length
Regrind if Out-of-Round
Exceeds . . . . 0.0004 in.
Connecting Rod Journal –
Standard Diameter . 2.9980-2.9990 in.
Desired Clearance in
Bearing . . . . 0.0010-0.0040 in.
Wear Limit . . . . 0.0060 in.

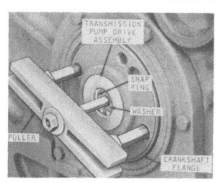

Fig. 76 – Method of removing clutch shaft pilot bushing adapter.

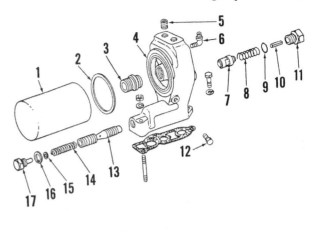

Fig. 81 – Exploded view of filter housing and associated parts showing relief and regulating valves. Fitting (6) is for models with turbocharger.

1. Filter
2. Gasket
3. Threaded adapter
4. Housing
5. Plug
6. Fitting
7. Relief valve
8. Spring
9. "O" ring
10. Roll pin
11. Plug
12. Plug
13. Pressure regulating valve
14. Spring
15. Shims
16. Aluminum gasket
17. Plug

Piston Cooling Orifice
Torque . . . . 85-115 ft.-lbs.
Main Cap Torque . . . . 150 ft.-lbs.
Rod Bolt Torque . . . . 55 ft.-lbs.
plus ¼-turn
Rear Oil Seal Housing Torque . . 20 ft.-lbs.
Flywheel to Crankshaft
Torque . . . . 85 ft.-lbs.

4440, 4640 & 4840

Crankshaft End Play –
Desired . . . . 0.0015-0.0150 in.
Wear Limit . . . . 0.0150 in.
Main Bearing Journal –
Standard Diameter . . 3.3720-3.3730 in.
Desired Clearance in
Bearing . . . . 0.0012-0.0042 in.
Regrind if Taper
Exceeds . . . . . 0.0001 in./1 in. length
Regrind if Out-of-Round
Exceeds . . . . 0.0004 in.
Connecting Rod Journal –
Standard Diameter . . 2.9980-2.9990 in.
Desired Clearance in
Bearing . . . . 0.0010-0.0040 in.
Piston Cooling Orifice
Torque . . . . 85-115 ft.-lbs.
Main Cap Torque . . . . 150 ft.-lbs.
Rod Bolt Torque . . 55 ft.-lbs. plus ¼-turn
Crankshaft Damper to Crankshaft
Torque . . . . 150 ft.-lbs.
Rear Oil Seal Housing
Torque . . . . 20 ft.-lbs.
Flywheel to Crankshaft
Torque . . . . 85 ft.-lbs.

CRANKSHAFT REAR OIL SEAL

All Models

66. The crankshaft rear oil seal is contained in a retainer plate which is attached to rear face of cylinder block by cap screws. Seal is available only in a kit which also includes the steel wear sleeve. To renew the seal, first detach (split) engine from clutch housing as outlined in paragraph 166 and remove clutch and flywheel.

Unbolt and remove oil seal retainer plate and score wear sleeve lightly with a dull chisel so as not to damage flywheel flange, and pry wear sleeve from crankshaft using a screwdriver or pry bar. Drive or press the seal from retainer being careful not to damage retainer plate. Install seal with closed side to rear of engine. Install wear sleeve with a suitable driver so that rear edge is flush with rear face of crankshaft flange. Check sealing face runout with a dial indicator. Runout should not exceed 0.006 inch. Tighten screws attaching seal retainer to 20 ft.-lbs. torque when correctly centered.

FLYWHEEL

All Models

67. Flywheel is doweled to crankshaft flange and retained by four cap screws. Flywheel can be removed by using forcing screws in tapped holes provided.

To install flywheel ring gear, heat gear evenly to approximately 300 degrees F. and position gear so that chamfered end of gear teeth face toward front of engine. When installing flywheel, align dowel hole in flywheel over dowel in flange, coat threads of retaining cap screws with sealant and tighten evenly to a torque of 85 ft.-lbs.

CLUTCH SHAFT PILOT BUSHING ADAPTER

All Models

68. A bearing puller can be used to pull the sealed pilot bearing from the crankshaft bore after torsional damper is removed.

The clutch shaft pilot bushing is carried in an adapter which is pressed into end of crankshaft. The adapter is also the hex drive for the transmission pump drive shaft. The pilot bushing can be removed using a blind hole puller to withdraw bushing from bore in adapter. If necessary, adapter can be pulled from crankshaft as shown in Fig. 76. Install a bolt and washer which will fit into pilot bushing bore, install a H-214-R snap ring in adapter bore groove to prevent

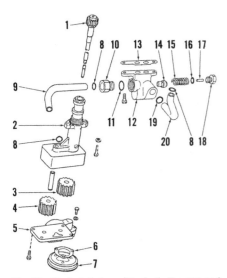

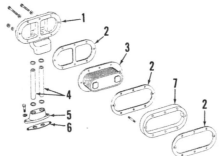

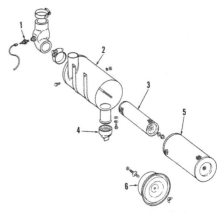

Fig. 84—Exploded view of typical oil pump and oil cooler by-pass valve.

1. Drive gear and shaft	11. "O" ring
2. Pump housing	12. By-pass housing
3. Drive gear	13. Gasket
4. Idler gear	14. By-pass valve
5. Cover	15. Spring
6. Gasket	16. "O" ring
7. Intake screen	17. Roll pin
8. "O" ring	18. Plug
9. Tube	19. "O" ring
10. Gland nut	20. Tube

Fig. 86—Exploded view of engine oil cooler typical of type used on all models. Item 3 is larger on 4440, 4640 and 4840 models, and has a spacer (7) to allow for extra depth.

1. Cover	4. Tubes
2. Gasket	5. Adapter
3. Oil cooler	6. Gaskets

Fig. 87—Exploded view of engine air cleaner.

1. Warning switch	4. Unloader valve
2. Housing	5. Primary element
3. Safety element	6. Cover

washer from coming out of bore and use type puller shown. Protect the face of new adapter and drive it tight against crankshaft flange. New bushing should be driven in until it is 0.020-0.040 inch below adapter face.

OIL PAN

All Models

69. Engine oil pan can be removed without interference from other components. Drain engine oil, remove oil filter access plate and filter from side of oil pan. Use a floor jack or other lifting means to remove and install the cast pan. When installing, tighten the ⅜-inch cap screws to a torque of 35 ft.-lbs. and ½-inch cap screws to a torque of 85 ft.-lbs.

OIL PUMP

All Models

70. **REMOVE AND REINSTALL.** To remove the engine oil pump, first remove oil pan as outlined in paragraph 69, then unbolt and remove oil pump.

When reinstalling, make sure that oil pump and camshaft gears mesh properly, and reinstall in reverse order of disassembly. Tighten pump to block attaching screws to 20-25 ft.-lbs. torque.

72. **OVERHAUL.** To overhaul the removed oil pump, first remove pump cover, remove idler gear and examine pump gears and cover for wear or scoring. Gears are available as a matched set

only. Check for wear or excessive looseness between shaft and housing at drive gear end.

Driven gear (3 – Fig. 84) is retained to drive shaft by a press fit on shaft. Use a press and suitable mandrel to press shaft out of gear, then press drive gear on shaft until gear is flush with mating surface of housing. When reassembling, drive gears and idler gears must spin freely and cover must not bind on gears. Reinstall pump as outlined in paragraph 70.

PRESSURE REGULATOR VALVE

74. The oil pressure regulator valve (13 – Fig. 81) is located in filter housing as is the filter by-pass valve (7).

Oil from the pump is circulated through the oil cooler (or is by-passed if cooler is restricted), then circulated through filter (or is by-passed if filter is restricted), and finally the oil flows past the oil pressure relief valve which limits the maximum oil pressure. Correct regulated pressure is 40-50 psi when engine is at normal operating temperature and at 1900 rpm. Pressure can be adjusted by adding or removing shims (15). Adding or removing one shim will change regulated pressure by about 5 psi. Oil filter by-pass valve (relief valve) operates on a pressure differential of approximately 15 psi. Filter by-pass valve opens for cold starting, or if filter becomes plugged, to assure continuing lubrication. Oil cooler by-pass valve (14 – Fig. 84) operates on a pressure differential of 15 psi and opens for cold starting, or in case of cooler restriction.

OIL COOLER

All Models

75. All models are equipped with an engine oil cooler of the type shown in Fig. 86. Some variation from oil cooler shown may be noted on some models. The oil is cooled by the engine coolant

liquid which circulates through tubes in cooler body. Disassembly for cleaning or other service is normally not required except in cases of contamination of cooling or lubrication systems.

AIR INTAKE SYSTEM

All Models

76. Air used to operate the engine enters through an air stack to the air cleaner. The air then passes through a dual element dry type filter into the manifold or turbocharger; then on into the engine. A vacuum switch (1 – Fig. 87) connects to a warning indicator lamp which lights to warn the operator if the filters are restricted.

Because of the balanced system and the large volume of air demanded (especially by a turbocharged diesel engine), it is of utmost importance that only the approved parts which are in good condition be used.

To check for air cleaner restriction, install a "T" fitting, an elbow, and the vacuum switch where switch was originally installed, and connect a JDST-11 water vacuum gage in "T" fitting.

Normal vacuum at port for switch (1) with clean filters, air stack extension installed and engine operating at full load is as follows:

Model	Inches of water
4040	8 in.
4240	9.5 in.
4440	11 in.
4640	11.5 in.
4840	14.5 in.

Rpm used for test
  All Models 2200

Warning switch closes at a vacuum of 24-26 inches of water (approximately 1.8 inches Hg) and vacuum should never be permitted to be higher than 25 inches of water on any model.

On turbocharged models, intake manifold pressure, checked at the ³/₈-inch pipe plug near the inlet tube on intake manifold at 2200 rpm, full load, should be 29-32 inches of mercury (14-16 psi) for 4440; 32-38 inches of mercury (16-19 psi) for 4640; 36-42 inches of mercury (18-21 psi) for 4840 models. A low manifold air pressure may indicate intake manifold air leaks, restricted air intake or air cleaner, exhaust leaks or a malfunctioning turbocharger. Refer to paragraph 84 for service to intercooler located in intake manifold of 4640 and 4840 models.

TURBOCHARGER

OPERATION

4440-4640-4840 Models

77. The exhaust driven turbocharger supplies air to the intake manifold at above normal atmospheric pressure. The additional air entering the combustion chamber permits an increase in the amount of fuel burned, and increased power output over an engine of comparable size not so equipped.

The use of the engine exhaust to power the compressor increases engine flexibility, enabling it to perform with the economy of a smaller engine on light loads yet permitting a substantial horsepower increase at full load. Horsepower loss because of altitude or atmospheric pressure changes is also largely reduced.

Because a turbocharger compresses the incoming air, the heat of compression causes the air to expand and become less dense than it would be at a lower temperature. Models 4640 and 4840 are equipped with an intercooler which lowers the temperature of the intake air which, in turn, increases the density of the intake air. This permits the use of more fuel and results in more power. Since a greater power output causes more heat, the turbocharger is driven faster, which produces more manifold pressure, and even greater power.

The turbocharger contains a rotating shaft which carries an exhaust turbine wheel on one end and a centrifugal air compressor on the other. The rotating member is precisely balanced and capable of rotative speeds up to 100,000 rpm. Bearings are of the floating sleeve type and the unit is lubricated and cooled by a flow of engine oil under pressure. Exchange turbocharger units are available, or a qualified technician can overhaul the unit if parts are available.

SERVICE

4440-4640-4840 Models

78. In a naturally aspirated diesel engine (without turbocharger) an approximately equal amount of air enters the cylinders at all loads, and only the amount of fuel is varied to compensate for power requirements. Turbocharging may supply up to 3 times the normal amount of air under full load.

All diesel engines operate with an excess of air under light loads. In a naturally aspirated engine, most of the air is used at full load, and increasing the amount of fuel results in a higher smoke level with little increase in power output. Turbocharging provides a variation of air delivery, and a turbocharged engine operates with an excess of air up to and beyond the design capacity of the engine. When more fuel is provided, the turbocharger speed and air delivery pressure increase, resulting in additional horsepower and heat, with little change in smoke level. Smoke cannot, therefore, be used as a guide to safe maximum fuel setting in a turbocharged engine. **DO NOT** increase horsepower output above that given in **CONDENSED SERVICE DATA** at the front of this manual.

Schwitzer turbocharger Models 3LD, 3LDA and 3LM or AiResearch TO-4 units have been used. Be sure to install correct replacement unit if turbocharger is exchanged. The turbocharger consists of the following three main sections. The turbine, bearing housing and compressor.

Engine oil taken directly from the clean oil side of the engine oil filters, is circulated through the bearing housing. This oil lubricates the sleeve type bearings and also acts as a heat barrier between the hot turbine and the compressor. The oil seals used at each end of the shaft are of the piston ring type. When servicing the turbocharger, extreme care must be taken to avoid damaging any of the parts.

CAUTION: DO NOT operate the turbocharger without adequate lubrication. When turbocharger is first installed, or engine has not been run for a month or more, or new oil filter has been installed, turn engine over with starter with fuel cut-off wire disconnected until oil pressure indicator light goes out; reconnect the wire and start engine. Run engine at slow idle speed for at least two minutes before opening throttle or putting engine under load.

Some other precautions to be observed in operating and servicing a turbocharged engine are as follows:

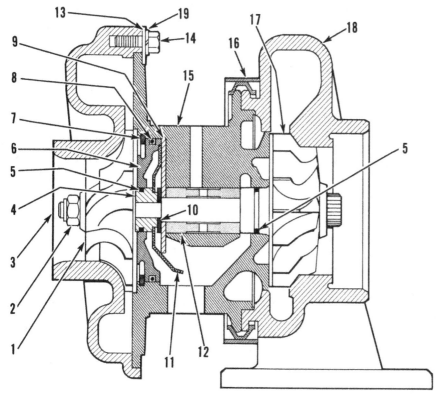

Fig. 90 – Cross-sectional view of Schwitzer 3LD turbocharger showing component parts. Refer to Fig. 91 for legend. Other 3LDA and 3LM models are similar.

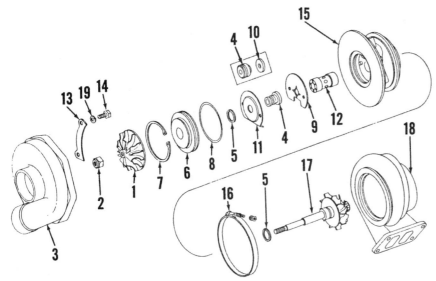

Fig. 91—Exploded view of Schwitzer Model 3LD & 3LDA turbocharger assemblies used on some models. Parts (4 & 10) shown in inset are for early 3LD models.

1. Compressor wheel	5. Seal rings (not	10. Thrust ring
2. Locknut	identical)	11. Oil deflector
3. Compressor cover	6. Flinger plate insert	12. Bearing
4. Flinger (Spacer	7. Snap ring	13. Clamp plate
Sleeve)	8. "O" ring	14. Cap screw
	9. Thrust plate	15. Bearing housing

16. Clamp ring
17. Turbine wheel and shaft
18. Turbine housing
19. Lockwasher

Do not operate at wide-open throttle immediately after starting. Allow engine to idle until turbocharger slows down before stopping engine. This will insure adequate lubrication to the shaft bearings at all times.

Because of increased air flow, care of air cleaner and connections is of added importance. Check the system and condition of restriction indicator whenever tractor is serviced. Make sure exhaust pipe opening is closed and air cleaner connected whenever tractor is transported, to keep turbocharger from turning due to air pressure. If exhaust outlet is equipped with weathercap, tape the cap closed. If weathercap is missing, use tape to close exhaust outlet pipe.

79. REMOVE AND REINSTALL. To remove the turbocharger, first remove muffler, hood, exhaust elbow and adapter. Remove oil lines and intake hose connections, then unbolt and remove turbocharger from exhaust manifold.

To inspect the removed turbocharger unit, examine turbine wheel and compressor impeller for blade damage, looking through housing and end openings. Using a dial indicator with plunger extension, check radial bearing play through outlet oil port while moving both ends of turbine shaft equally. Check shaft end play with dial indicator working from either end. If end or side play exceeds 0.009 inch; or if any of the blades are broken or damaged, renew or overhaul the unit.

When installing, attach turbocharger to manifold using a new gasket and tighten stud nuts to a torque of 35

ft.-lbs. Install inlet oil line, outlet line, adapter and exhaust elbow after first making sure parts are perfectly aligned. Adapter must have a minimum of 1/16-inch end play and be free to rotate. Undue stress on turbocharger at installation may cause bearing failure. Prime turbocharger as outlined in paragraph 78.

Schwitzer Model 3LD and 3LDA

80. OVERHAUL. Remove turbocharger as outlined in paragraph 79. Before disassembling, place a row of light punch marks across compressor cover, bearing housing and turbine housing to aid in reassembly. Clamp turbocharger mounting flange (exhaust inlet) in a vise and remove cap screws (14—Fig. 91), lockwashers and clamp plates (13). Remove compressor cover (3). Remove nut from clamp ring (16), expand clamp ring and remove bearing housing assembly (15) from turbine housing (18).

CAUTION: Never allow the weight of the bearing housing assembly to rest on either the turbine or compressor wheel vanes. Lay the bearing housing assembly on a bench so that turbine shaft is horizontal.

Remove locknut (2) and slip compressor wheel (1) from end of shaft. Withdraw turbine wheel and shaft (17) from bearing housing. Place bearing housing on bench with compressor side up. Remove snap ring (7), then using two screwdrivers, lift flinger plate insert (6) from bearing housing. Push spacer sleeve (4) from the insert. Remove oil deflector (11), thrust ring (10), thrust

plate (9) and bearing (12). Remove "O" ring (8) from flinger plate insert (6) and remove both seal rings (5) from spacer sleeve and turbine shaft.

Soak all parts in Bendix metal cleaner or equivalent and use a soft brush, plastic blade or compressed air to remove carbon deposits. CAUTION: Do not use wire brush, steel scraper or caustic solution for cleaning, as this will damage turbocharger parts. Glass bead dry blast may be used for cleaning if air pressure does not exceed 40 psi and all traces of the glass beads are rinsed out before assembling.

Inspect turbine wheel and compressor wheel for broken or distorted vanes. **DO NOT** attempt to straighten bent vanes. Check bearing bore in bearing housing, floating bearing (12) and turbine shaft for excessive wear or scoring. Inspect flinger plate insert (6), flinger sleeve (4), oil deflector (11), thrust plate (9) and, if so equipped, thrust ring (10) for excessive wear or other damage. Refer to Fig. 91 and the following for specifications of new parts.

Bearing Bore in Housing
(15) . . . . 0.8750-0.8755 in.
Bearing (12) Length . . . . 1.484-1.485 in.
Flinger Sleeve (4)
Length . . . . 0.517-0.519 in.
Piston Ring Grooves . . . . 0.064-0.065 in.
Shaft (17) Concentricity–
Maximum Run-out . . . . 0.001 in.
Shaft (17) Diameter . . . 0.4800-0.4803 in.
Shaft (17) Shoulder
Length . . . . 1.595-1.596 in.
Thrust Plate (9)
Thickness . . . . 0.107-0.108 in.
Assembled Bearing Clearances–
Axial (End Play of
Shaft . . . . 0.003-0.008 in.
Radial Clearance Measured at
Exhaust Turbine
Blades . . . . 0.018-0.049 in.

Renew all damaged parts and use new "O" ring (8) and seal rings (5) when reassembling. The seal ring used on turbine shaft is copper plated and is larger in diameter than the seal ring used on spacer sleeve. Refer to Figs. 90 and 91 when reassembling.

Install seal ring on turbine shaft, lubricate seal ring and install turbine wheel and shaft in bearing housing. Lubricate I.D. and O.D. of bearing (12), install bearing over end of turbine shaft and into bearing housing. Lubricate both sides of thrust plate (9) and install plate (bronze side out) on the aligning dowels. Install thrust ring (10) and oil deflector (11), making certain holes in deflector are positioned over dowel pins. Install new seal ring on spacer sleeve (4), lubricate seal ring and press spacer sleeve into flinger plate insert (6). Position new "O" ring (8) on insert, lubricate "O" ring and install insert and spacer sleeve assembly in bearing housing, then secure with snap ring (7). Place compressor wheel on turbine shaft, coat threads and back side of nut (2) with graphite grease, or equivalent, then install and tighten nut to 156 in.-lbs. torque. Assemble bearing housing to turbine housing and align punch marks. Install clamp ring, apply graphite grease to threads, install nut and tighten to 120 in.-lbs. torque. Apply a light coat of graphite grease around machined flange of compressor cover (3). Install compressor cover, align punch marks and secure cover with cap screws, washers and clamp plates. Tighten cap screws evenly to 60 in.-lbs. torque. Fill the oil inlet with engine oil and turn turbine shaft by hand to lubricate bearing and thrust plate.

Check rotating unit for free rotation within the housings. Cover all openings until the turbocharger is reinstalled.

Use a new gasket and install and prime turbocharger as outlined in paragraph 78.

Turbocharger oil supply pressure at 2200 engine rpm should be within 10 psi of engine oil pressure but never less than 25 psi. Minimum return oil flow from turbocharger is ½-gpm at 2200 engine rpm.

Schwitzer Model 3LM

81. **OVERHAUL.** Remove turbocharger as outlined in paragraph 79. Before disassembling, place a row of light punch marks across compressor cover, bearing housing and turbine housing to aid in reassembly. Clamp turbocharger mounting flange (exhaust inlet) in a vise and remove cap screws (14 – Fig. 92), lockwashers and clamp plates (13). Remove compressor cover (3). Remove screws (24) then separate bearing housing assembly (15) from turbine housing (18).

CAUTION: Never allow the weight of the bearing housing assembly to rest on either the turbine or compressor wheel vanes. Lay the bearing housing assembly on a bench so that turbine shaft is horizontal.

Remove locknut (2) and slip compressor wheel (1) from end of shaft. Withdraw turbine wheel and shaft (17) from bearing housing. Remove back plate (20), then remove and discard all of gasket (21). Place bearing housing on bench with compressor side up. Remove snap ring (7), then using two screwdrivers, lift insert (6) from bearing housing. Push spacer sleeve (4) from the insert. Remove oil deflector (11), thrust plate (9) and bearing (12). Remove "O" ring (8) from flinger plate insert (6) and remove both seal rings (5) from spacer sleeve and turbine shaft.

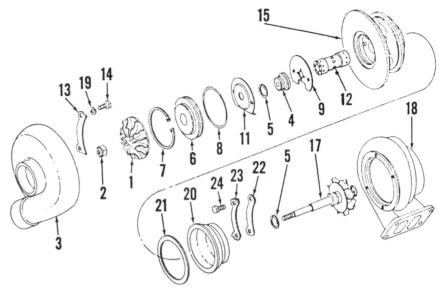

Fig. 92 – Exploded view of Schwitzer 3LM turbocharger assembly used on some models.

1. Compressor wheel	6. Insert	13. Clamp plate	19. Lockwasher
2. Locknut	7. Snap ring	14. Cap screw	20. Back plate
3. Compressor cover	8. "O" ring	15. Bearing housing	21. Gasket
4. Thrust sleeve	9. Thrust plate	17. Turbine wheel and	22. Clamp plate
5. Seal rings (not	10. Oil deflector	shaft	23. Lock plate
identical)	12. Bearing	18. Turbine housing	24. Cap screws

Soak all parts in Bendix metal cleaner or equivalent and use a soft brush, plastic blade or compressed air to remove carbon deposits. CAUTION: Do not use wire brush, steel scraper or caustic solution for cleaning, as this will damage turbocharged parts. Glass bead dry blast may be used for cleaning if air pressure does not exceed 40 psi and all traces of the glass beads are rinsed out before assembling.

Inspect turbine wheel and compressor wheel for broken or distorted vanes. **DO NOT** attempt to straighten bent vanes. Check bearing bore in bearing housing, floating bearing (12) and turbine shaft for excessive wear or scoring. Inspect flinger plate insert (6), thrust sleeve (4), oil deflector (11) and thrust plate (9) for excessive wear or other damage. Refer to Fig. 92 and the following for specifications of new parts.

Bearing Bore in Housing
(15) . . . . 0.7500-0.7505 in.
Bearing (12) Length . . . . 2.425-2.426 in.
Piston Ring Grooves . . . . 0.064-0.065 in.
Shaft (17) Concentricity –
  Maximum Run-out . . . . 0.0006 in.
Shaft (17) Diameter –
  Bearing Journal . . . . 0.4400-0.4403 in.
  Compressor wheel . . 0.3123-0.3125 in.
Shaft (17) Shoulder
  Length . . . . 2.536-2.537 in.
Thrust Plate (9)
  Thickness . . . . 0.107-0.108 in.
Thrust Sleeve (4)
  Length . . . . 0.517-0.519 in.
Assembled Bearing Clearances –
  Axial (End) Play of
  Shaft . . . . 0.002-0.005 in.

Radial Clearance Measured at Exhaust Turbine Blades . . . . 0.018-0.049 in.
Clearance Between Back of Turbine Blades and Back Plate (20) Must Be Even and Within Limits
of . . . . 0.017-0.049 in.

Renew all damaged parts and use new "O" ring (8) and seal rings (5) when reassembling. Use Fig. 92 as a guide when reassembling. Install back plate (20) using new gasket (21). Install seal ring on turbine shaft, lubricate seal ring and install turbine wheel and shaft in bearing housing. Lubricate I.D. and O.D. of bearing (12), install bearing over end of turbine shaft and into bearing housing. Lubricate both sides of thrust plate (9) and install plate (bronze side out) on the aligning dowels. Install oil deflector (11) making certain holes in deflector are positioned over dowel pins. Install new seal ring on spacer sleeve (4), lubricate seal ring and press spacer sleeve into insert (6). Position new "O" ring (8) on insert, lubricate "O" ring and install insert and spacer sleeve assembly in bearing housing, then secure with snap ring (7). Place Compressor wheel on turbine shaft, coat threads and back side of nut (2) with graphite grease, or equivalent, then install and tighten nut to 156 in.-lbs. torque. Apply graphite grease to turbine housing (18), align previously affixed punch marks and tighten screws (24) evenly to 144 in.-lbs. torque. Lock position of screws with lock plate (23). Apply a light coat of graphite grease around machined flange of compressor cover (3). Install compressor cover, align punch marks and

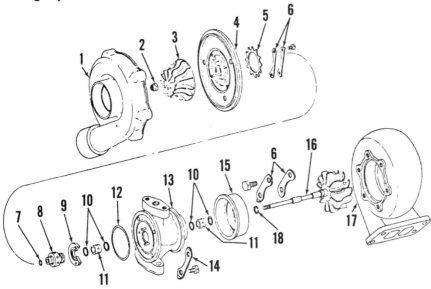

Fig. 94—Exploded view of AiResearch turbocharger assembly used on some models. See Figs. 91 & 92 for Schwitzer models.

1. Compressor housing
2. Locknut
3. Compressor impeller
4. Back plate
5. Spring
6. Clamp plate
7. Seal ring
8. Thrust collar
9. Thrust bearing
10. Bearing retainers
11. Bearings (2)
12. "O" ring
13. Center housing
14. Lock plate
15. Shroud
16. Turbine shaft & wheel
17. Turbine housing
18. Seal ring

secure cover with cap screws, washer and clamp plates. Tighten cap screws evenly to 60 in.-lbs. torque. Fill the oil inlet with engine oil and turn turbine shaft by hand to lubricate bearing and thrust plate.

Check rotating unit for free rotation within the housings. Cover all openings until the turbocharger is reinstalled.

Use a new gasket and install and prime turbocharger as outlined in paragraph 78.

Turbocharger oil supply pressure at 2200 engine rpm should be within 10 psi of engine oil pressure but never less than 25 psi.

AiResearch Turbocharger

82. INSPECTION. Remove turbocharger unit as outlined in paragraph 79. To inspect the removed turbocharger unit, examine turbine wheel and compressor impeller for blade

Fig. 94A—AiResearch model back plate and spring must be renewed as an assembly if damaged.

damage, looking through housing openings. Using a dial indicator with plunger extension, check radial bearing play through outlet oil port while moving both ends of turbine shaft equally. Radial bearing clearance should not exceed 0.006 inch. Check shaft end play with dial indicator working from either end. If end play exceeds 0.004 inch or if any of the blades are broken or damaged, overhaul or renew the turbocharger unit.

83. OVERHAUL. Mark across compressor housing (1–Fig. 94), center housing (13) and turbine housing (17) to aid alignment when assembling.

CAUTION: Do not rest weight of any parts on impeller or turbine blades. Weight of only the turbocharger unit is enough to damage the blades.

Remove lock plates and clamp plates (6) from compressor housing (1) and remove housing. Remove lock plates and clamp plates (6) from turbine (17) and remove housing. Hold turbine shaft from turning using the appropriate type of holding fixture for turbine wheel (16) and remove locknut (2).

NOTE: Use a "T" handle or double universal socket to remove locknut in order to prevent bending turbine shaft.

Lift compressor impeller (3) off, then remove center housing from turbine shaft while holding shroud (15) onto center housing. Remove backplate (4), thrust bearing (9) and thrust collar (8). Carefully remove bearing retainers (10) from ends and withdraw bearings (11). Spring (5) is available as an assembly

with backplate (4). Refer to Fig. 94A.

CAUTION: Be careful not to damage bearings or surface of center housing when removing retainers. The center two retainers do not have to be removed unless damaged or unseated. Always renew bearing retainers if removed from grooves in housing.

Clean all parts in a cleaning solution which is not harmful to aluminum. A stiff brush and plastic or wood scraper should be used after deposits have softened. When cleaning, use extreme caution to prevent parts from being nicked, scratched or bent.

Inspect bearing bores in center housing (13–Fig. 94) for scored surfaces, out of round or excessive wear. Make certain bore in center housing is not grooved in area where seal (18) rides. Compressor impeller (3) must not show signs of rubbing with either the compressor housing (1) or the backplate (4). Impeller should have 0.0002 inch tight to 0.0004 inch loose fit on turbine shaft. Make certain that impeller blades are not bent, chipped, cracked or eroded. Oil passages in thrust collar (8) must be clean and thrust faces must not be warped or scored. Ring groove shoulders must not have step wear. Inspect turbine shroud (15) for evidence of turbine wheel rubbing. Turbine wheel (16) should not show evidence of rubbing and vanes must not be bent, cracked, nicked, or eroded. Turbine wheel shaft must not show signs of scoring, scratching or overheating. Groove in shaft for seal ring (18) must not be stepped. If turbine shaft journals are damaged, undersized bearings of 0.005 and 0.010 inch may be ordered, and shaft reconditioned if so desired. Check shaft end play and radial clearance when assembling.

If the bearing inner retainers (10) were removed, install new retainers. Oil bearings and install in center housing and install outer retainers. Position the shroud (15) on turbine shaft (16) and install seal ring (18) in groove. Apply a light, even coat of engine oil to shaft journals, compress seal ring (18) with a thin strong tool such as a dental pick and install center housing (13). Install new seal ring (7) in groove of thrust collar (8), then install thrust bearing so that smooth side of bearing (9) is toward seal ring (7) end of collar. Install thrust bearing and collar assembly over shaft, making certain that pins in center housing engage holes in thrust bearing. Install new rubber seal ring (12), make certain that spring (5) is positioned in backplate (4), then install backplate making certain that seal ring (7) is not damaged. Seal ring will be less likely broken if open end of ring is installed in bore of backplate first. Install lock plates (14)

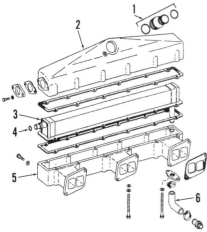

Fig. 95 — Exploded view of intake manifold and intercooler used on 4640 and 4840 models.

1. "O" rings and air intake coupling
2. Cover
3. Intercooler core
4. "O" rings
5. Manifold
6. Water inlet hose

and screws, tightening screws to 75-90 in.-lbs. torque. Install compressor impeller (3) and make certain that impeller is seated against thrust collar (8). Install locknut (2) to 18-20 in.-lbs. torque, then use a "T" handle or double universal joint socket to turn locknut an additional 90 degrees in order to stretch the turbine shaft the necessary 0.0055-0.0065 inch for proper tension.

CAUTION: If "T" handle or double universal joint socket is not used, shaft may be bent when tightening nut (2).

Install turbine housing (17) with clamp plates (6) next to housing, tighten screws to 100-130 in.-lbs., then bend lock plates up around screw heads.

Check shaft end play and radial play at this point of assembly. If shaft end play exceeds 0.004 inch, thrust collar (8) and/or thrust bearing (9) is worn excessively. End play of less than 0.001 inch indicates incomplete cleaning (carbon not all removed) or dirty assembly and unit should be disassembled and cleaned.

If turbine shaft radial play exceeds 0.007 inch, unit should be disassembled and bearings, shaft and/or center housing should be renewed, or shaft should be reconditioned and undersized bearings installed. Center housing bearing bore should not exceed 0.6228 inch and seal bore should not exceed 0.703 inch diameter. Shaft journal diameter should not be less than 0.3994 inch and seal hub diameter should not be less than 0.681 inch. Maximum permissible limits of all of these parts may result in radial play which is not acceptable.

Install compressor housing (1), but do not tighten until unit is installed, so perfect alignment can be made. Fill reservoir with engine oil and protect all openings of turbocharger until unit is installed on tractor. After intake hoses are all assembled, tighten clamp plates (6) to

110-130 in.-lbs. torque to compressor housing (1), and complete the assembly. Prime turbocharger as outlined in paragraph 78.

INTERCOOLER

4640-4840 Models

84. The intake manifold on 4640 and 4840 models contains an intercooler (3 – Fig. 95) to lower the temperature of the intake air. Coolant from the engine cooling system flows through the intercooler core and heat from the compressed (turbocharged) intake air is conducted into the engine coolant, then circulated to the engine radiator where it is permitted to cool (lose heat to the air flowing through the radiator). The intercooler can lower the temperature of the intake air as much as 80-90 degrees F. which will make the intake air more dense and permit more air to be delivered to the engine cylinders. This increased volume of air together with additional quantity of fuel can result in

the production of more power from the engine.

Since coolant from the radiator is circulated through the intercooler core, a leak in the core could cause serious damage to the engine by allowing coolant into the combustion area.

To remove the intercooler, drain cooling system, remove muffler, air stack and hood. Remove turbocharger as outlined in paragraph 79. Disconnect both water hoses at intercooler and the ether starting aid pipe. Remove both adapter plates at hose connections to intercooler. Remove the 12 cap screws from underneath intake manifold, lift off manifold cover and intercooler core.

Test and repair procedures are much the same as for a radiator. The intercooler can be pressurized with air (20-25 psi) then submerged in water to check for leaks. Repair or renew the aluminum intercooler core as necessary.

Reassemble in reverse order of disassembly using all new gaskets and prime turbocharger as outlined in paragraph 78.

FUEL SYSTEM

FUEL FILTER, PRIMARY PUMP AND LINES

All Models

90. Models without turbochargers are equipped with a single two-stage fuel filter and sediment bowl, and turbocharged models have dual filters. On all models the bowl assembly should be renewed when necessary. The primary fuel pumps on 4040 and 4240 models have no sediment bowl, with all filtering done at the single two-stage filter. Turbocharged 4440, 4640 and 4840 models use a fuel transfer pump between the tank and dual two-stage filters, that has a glass sediment bowl with a renewable filter.

The fuel tank is mounted vertically in front of radiator, and shut-off valve on

all models is accessible after removing right grille screen. The fuel tank drain valve is mounted in the lowest part of fuel tank, forward of the front axle.

91. **BLEEDING.** To bleed the system, refer to Fig. 100. Open filter bleed plug on all models. On models without turbocharger, actuate hand primer lever on fuel pump until air-free fuel flows out bleed plug opening.

NOTE: If no resistance is felt when pumping lever, turn engine so that the camshaft is not on its pump stroke.

One bleeding operation is usually all that is required. Tighten bleed plug. Make sure hand primer lever is in down position before attempting to start engine.

On models equipped with turbocharger, the fuel transfer pump is attached to the injector pump (Fig. 101) and the hand primer must be unscrewed

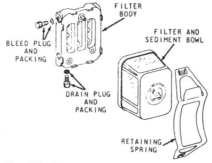

Fig. 100 — Exploded view of single renewable two-stage filter and sediment bowl element used on 4040 and 4240 models. Models with turbocharger are equipped with dual, two-stage filters.

Fig. 101 — Fuel transfer pump with hand primer as used on turbocharged models.

until it can be pulled up. Pump the primer until air-free fuel flows out the open bleed plug hole.

Loosen the pressure line connections at injector assemblies and, with throttle open, turn engine over with starter until fuel flows from all injector lines. Tighten the connections and start engine. If engine will not start or misses, repeat the above procedure until system is free of trapped air.

FUEL TRANSFER PUMP

All Models

92. The fuel transfer pump on models without turbocharger is mounted on right side of engine block, driven by the camshaft, and includes a hand primer lever. This type of fuel pump has no sediment bowl and is available as an assembly only.

INJECTOR NOZZLES

All Models

All models are equipped with Robert Bosch KDEL injector nozzles. Refer to Fig. 105 for identifying features. All threaded components are metric sizes.

94. **TESTING AND LOCATING A FAULTY NOZZLE.** If rough or uneven engine operation or misfiring indicate a faulty injector, the defective unit can usually be located as follows:

With engine running at the speed where malfunction is most noticeable (usually slow idle speed), loosen the compression nut on high pressure line for each injector in turn, and listen for a change in engine performance. As in checking spark plugs, the faulty unit is the one which, when its line is loosened, least affects the running of the engine.

If a faulty nozzle is found and considerable time has elapsed since the injectors have been serviced, it is recommended that all nozzles be removed and checked, or that new or reconditioned units be installed. Refer to the following paragraphs for removal and test procedure.

108. **REMOVE AND REINSTALL.** Refer to paragraph 94 for testing procedures to determine if injector nozzle shows indication of a malfunction before removing for service.

If nozzle to be removed is near the alternator, disconnect battery ground strap to prevent a short circuit through tools. Wash injector and surrounding area with clean diesel fuel, disconnect leak-off line and fuel pressure line and use special tool (JDE-92) to remove gland nut (9 – Fig. 105). The gland nut

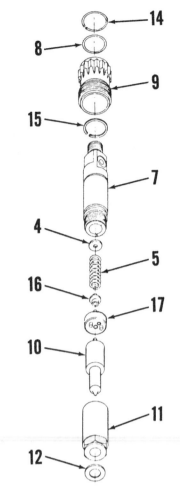

Fig. 105 – Exploded view of 21mm Robert Bosch KDEL injector typical of type used on all models.

will raise the nozzle out of cylinder head as it is removed.

When reinstalling, make sure injector and hole in cylinder head are clean and dry. Nozzle seat reamer (JDE-99) can be used to clean nozzle seat in head. Threads in head for gland nut (9) can be cleaned using a metric (M28 x 1.5) tap. This thread is different than earlier models with KDL nozzles. Threads for leak-off connectors are metric (M6 x 1). Apply anti-seize compound to bottom, inner bore and threads of gland nut and to nozzle barrel before installing. Renew nozzle gasket (12) if necessary. Use special tool JDE-92, align leak-off threaded port and tighten gland nut to 55-65 ft.-lbs. torque. Install leak-off connectors, attach leak-off line and connect delivery pipes to nozzles. Tighten fittings between delivery pipes and nozzles to 35 ft.-lbs. torque. Bleed injectors as outlined in paragraph 91.

109. **NOZZLE TEST.** A complete job of testing and adjusting an injector requires the use of special test equipment. Only clean approved testing oil should be used in tester tank. The nozzle should be tested for opening pressure, seat leakage and spray pattern. When tested,

the nozzle should open with a soft chatter, and then only when the lever is moved very rapidly. A bent or binding nozzle valve can prevent chatter. Spray will be broad and well atomized if injector is working properly.

Use the tester to check injector as outlined in the following paragraphs:

CAUTION: Fuel leaves the nozzle tip with sufficient force to penetrate the skin. Keep unprotected parts of body clear of nozzle spray when testing.

110. OPENING PRESSURE. Before conducting the test, operate tester lever until fuel flows, then attach the injector using the proper adapter. Close the valve to tester gage and pump the tester lever a few quick strokes to be sure nozzle valve is not plugged, that all spray holes are open and that possibilities are good that injector can be returned to service without overhaul.

Open valve to tester gage and operate tester lever slowly while observing gage reading. Opening pressure should be 3200 psi for 4040 and 4240 models; 3800 psi for all turbocharged models. If pressure is too low, recheck by releasing pressure and retesting. If pressure is still not correct, refer to paragraph 113, disassemble nozzle and change thickness of shim (4 – Fig. 105) until opening pressure is correct. Use only specially hardened shims. Shims are available in 0.002 inch steps from 0.039 to 0.077 inch. Each 0.002 inch increase in shim thickness varies the pressure by about 100 psi. Opening pressure should not vary more than 50 psi between nozzles. If pressure is not correct after changing shims, disassemble injector and recondition.

NOTE: When adjusting a new injector or an overhauled injector with a new pressure spring, set the pressure at 3350 psi for 4040 and 4240 models; to 4050 psi for turbocharged models. The increase in pressure is to allow for initial pressure loss as the spring takes a set.

111. SPRAY PATTERN. The finely atomized nozzle spray should be evenly distributed around the nozzle. Check for clogged or partially clogged orifices or for a wet spray which would indicate a sticking or improperly seating nozzle valve. If the spray pattern is not broad and even, and very rapid stroking of the tester handle does not cause injector to chatter softly, disassemble and overhaul the injector as outlined in paragraph 113.

112. SEAT LEAKAGE. Pump the tester handle slowly to maintain a gage pressure of 285 psi while examining nozzle tip for fuel accumulation. If nozzle is

in good condition, there should be no noticeable accumulation for a period of at least 10 seconds. If a drop or undue wetness appears on nozzle tip, renew the injector or overhaul as outlined in paragraph 113.

113. **OVERHAUL.** First clean outside of injector thoroughly and clamp flats of nozzle retaining nut (11 – Fig. 105) in a soft jawed vise. If not already removed, the leak-off connector fitting must be removed before removing upper snap ring (14), "O" ring (8), gland nut (9) and lower snap ring (15). Clamp the flats at fuel inlet end of nozzle holder (7) in a soft jawed vise, then unscrew the nozzle retaining nut (11). Be careful not to mar the polished surfaces of holder (7), intermediate plate (17) or nozzle (10). It may be necessary to soak nozzle assembly in carburetor cleaner, acetone or other commercial solvent if the valve does not fall freely from nozzle (10).

Clean all parts, then use care to keep parts clean and free from grit by submerging in a pan of clean diesel fuel, and handle only with hands that are wet with fuel. Avoid mixing of parts and other injector, and do not allow any lapped surface to come in contact with a hard object.

Valves should be cleaned of all carbon and washed in diesel fuel. Hard carbon may be cleaned off with a brass wire brush. NEVER use a steel wire brush or emery cloth on valve or tip. Use a cleaning wire 0.003 to 0.004 inch smaller than nozzle orifices to clean nozzle tips. The number and size of orifices are etched on nozzle tip such as "4 x 0,33". The "4" indicates four orifices, "0,33" indicates that each is 0.33 mm diameter. The following data applies to standard nozzles on models with KDEL nozzles.

4040 ....5 x 0,25
   Orifice diameter ....0.010 in.
   Opening pressure – New ....3350 psi
     Used ....3200 psi
4240 ....4 x 0,28
   Orifice size ....0.011 in.
   Opening pressure – New ....3800 psi
     Used ....3600 psi

4440 ....4 x 0,33
   Orifice diameter ....0.013 in.
   Opening pressure – New ....4050 psi
     Used ....3800 psi
4640 ....4 x 0,345
   Orifice diameter ....0.0136 in.
   Opening pressure – New ....4050 psi
     Used ....3800 psi
4840 ....4 x 0,36
   Orifice diameter ....0.014 in.
   Opening pressure – New ....4050 psi
     Used ....3800 psi

A stone may be used to cut flat surfaces on two sides of cleaning wire to aid in removing carbon from orifices. Finish cleaning orifices by using a wire that is 0.001 inch smaller than hole diameter. A pin vise should be used to hold cleaning wires and wire should extend only about 1/32-inch from vise to prevent breakage. Clean seat in nozzle (10) with sac hole drill furnished with cleaning kit. When held vertically, a valve that is wet with fuel should slide down to the seat in nozzle under its own weight.

Inspect all lapped and seating surfaces for excessive wear or damage. Check spindle, spring, shims and seats. Renew any parts in question and replace shims only if they are smooth and flat. Edge type filter in fuel inlet passage of body (7) can be cleaned by blowing air through passage from nozzle end of body. This will provide a reverse flushing action in filter.

Assemble in reverse order of disassembly. Submerge valve and nozzle in fuel while assembling, and make sure all other parts are wet with fuel. Do not dry parts with air or towels before assembly.

Apply anti-seize compound to bottom, inner bore and to threads of gland nut (9) and to body (7) before assembling snap rings (14 & 15), gland nut (9) and "O" ring (8). Clamp body in a soft jawed vise, then assemble shim (4), spring (5), spring seat (16), intermediate plate (17), nozzle (10) and retaining nut (11). Tighten retaining nut (11) to 44-58 ft.-lbs. torque, then retest injector as outlined in paragraphs 109, 110, 111 and 112. Use a new gasket (12) when reinstalling in engine. Bleed system as outlined in paragraph 91.

INJECTION PUMP

Roosa-Master Pumps

Roosa-Master model DM pumps are used on 4040 and 4240 models. This pump is also driven by the engine timing gears, but is located on right side of engine and is attached to a mounting flange which is part of the engine block casting.

Proper injection timing depends upon correct positioning of timing marks on pump and engine as outlined in paragraphs 115, 116 and 117.

Service to the injection pump requires use of special tools and specialized training which is beyond the scope of this manual. This service section will cover only the information required for removal, installation and field adjustments of the pump.

115. **REMOVE AND REINSTALL.** To remove the Roosa-Master fuel injection pump, first shut off fuel supply and thoroughly clean dirt from pump, lines and connections.

NOTE: Do not steam clean or pour water on a pump while it is warm or running, as this could cause pump to seize.

Remove clamps on injector lines, six connectors at pump and loosen lines at injector nozzles to avoid bending lines. Plug opening in fuel lines to prevent dirt from entering. Remove line from filter and fuel leak-off line.

Remove the crankcase vent hose and clean area around timing gear access cover at front of timing gear cover. Remove access cover, then remove the three screws which attach pump drive gear to drive hub. Clean the area around the fuel transfer pump, then disconnect fuel lines and remove transfer pump.

On all models, remove cover from timing window, install engine rotating tool and rotate engine until marks on governor weight retainer and cam ring are aligned as shown in Fig. 109. Disconnect speed control rod and fuel shut-off cable. Remove the pump mounting stud nuts, then lift pump back away from mounting. There are three stud nuts on 4040; four stud nuts on 4240 models.

To reinstall pump, reverse removal procedure. Make sure that No. 1 piston is at TDC on compression stroke and that the timing marks on cam ring and governor weight are aligned. Renew pump mounting seal if necessary. Be sure that reference mark on tang of pump shaft aligns with mark inside pump. This will prevent timing from being 180° off. The backlash in gears can allow pump timing to be off several degrees, so it is very important to recheck timing with engine running as outlined in paragraph 117. Bleed fuel system as outlined in paragraph 91. If pump drive gear must be renewed, remove timing cover (paragraph 56).

116. **STATIC TIMING.** To check injection pump static timing, proceed as outlined in paragraph 115 but without removing pump.

If adjustment is required, loosen mounting stud nuts and turn pump body until mark on cam ring aligns with mark on governor weight retainer (Fig. 109). Turn engine in normal direction two complete turns and recheck setting.

Fig. 108 – View of DM Roosa Master Injection pump, typical of all models equipped with this pump.

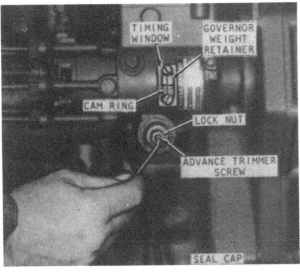

Fig. 109—View of DM Roosa Master injection pump installed showing alignment marks, timing window and advance trimmer screw typical of all models with Roosa Master pump.

117. ADVANCE TIMING. The injection pump has automatic speed advance which is factory set and will not normally need to be checked or reset. Minor adjustments can, however, be made without removal or disassembly of the pump; proceed as follows:

Shut off fuel, remove timing window cover (Fig. 109) and install timing window No. 19918. Make sure that pump has been static timed to engine as outlined in paragraph 116, and engine is up to operating temperature. Install a master tachometer on speed-hour meter drive using adapter JDE-28. Each mark on timing window is 2 degrees. Timing may be changed by removing seal cap on advance trimmer screw, loosening locknut, and with engine running, turning trimmer screw IN to advance, or OUT to retard. Specifications are as follows:

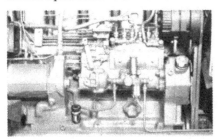

Fig. 110—View of A-Series Robert Bosch injection pump typical of installation on 4440 and 4640 models.

4040 & 4240 (Pump DM4629)
2100 rpm (FULL LOAD) . . .9° advance
Total Advance—FULL
   LOAD . . . . .9° advance
   NO LOAD . . . . .9° advance
Full load must be set using a pto dynamometer, but if a dynamometer is not available, the no load advance setting can be used, which would be the next best way to adjust the pump.

If proper advance cannot be obtained, pump must be either removed and adjusted on a test stand, overhauled or renewed.

Robert Bosch Pumps

Robert Bosch multiple plunger injection pumps are used on turbocharged models. Series A-2000 pumps are used on 4440 tractors; Series A-3000 pumps are used on 4640 tractors; and Series P-110 pumps are used on 4840 model tractors.

Series A-2000, A-3000 and P-110 pumps are all multiple plunger in-line pumps, with a governor and an aneroid control. They are equipped with an externally mounted fuel transfer pump which includes a hand primer pump, and are lubricated by engine oil pressure.

Series A-3000 pump has larger diameter plungers and other parts, than the smaller A-2000 pumps, in order to increase the amount of fuel delivered to the injectors.

Series P-110 pump is easily identified by the absence of the side cover which is used on A-series pumps. The P-series pumps are designed to support heavier loads necessary to delivering even more fuel at higher pressures than with A-series pumps.

Service to the injection pump requires use of special tools and specialized training which is beyond the scope of this manual. This service section will cover only the information required for removal, installation and field adjustments of the pump.

119. REMOVE AND REINSTALL. To remove the Robert Bosch multiple plunger fuel injection pump, first shut off fuel supply and thoroughly clean dirt from pump, lines and connections.

NOTE: Do not steam clean or pour water on a pump while it is warm or running, as this could cause pump to seize.

Cap all fittings as they are disconnected to prevent dirt entry. Remove access plate from front of timing gear cover and remove timing hole plug which is just to the rear of engine oil filler cap. Install engine rotation tool and timing pin in flywheel housing and refer to Fig. 112. Rotate engine in normal rotation direction until the mark on pump drive hub lines up with pointer mark, and timing pin enters hole in flywheel. Remove the three cap screws holding pump drive gear to pump. Disconnect the fuel and oil lines to pump and the pipe to aneroid. Disconnect the injector lines, return hose, speed control rod and fuel shut-off cable. On 4840 models, it is necessary to unbolt and remove the oil filter and valve body from engine block. On all models, remove the mounting nuts which hold pump to engine and withdraw pump.

When installing injection pump, No. 1 piston must be at TDC on compression stroke and pump timing marks aligned (Fig. 112) for proper timing. Compression stroke can be found with rocker cover removed, by turning engine until No. 1 intake valve is closing and continuing to turn until timing pin enters hole in flywheel. Renew "O" ring on front bear-

Fig. 111—View of P-110 Robert Bosch injection pump installed on 4840 model.

Fig. 112—View of Robert Bosch injection pump drive gear and timing marks as used on turbocharged models.

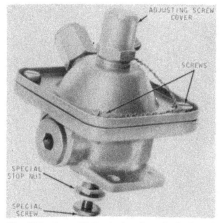

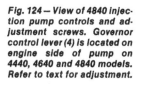

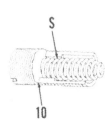

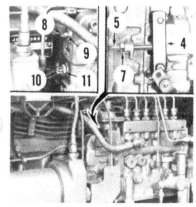

Fig. 124 — View of 4840 injection pump controls and adjustment screws. Governor control lever (4) is located on engine side of pump on 4440, 4640 and 4840 models. Refer to text for adjustment.

Fig. 113 — Aneroid Control unit typical of type used on Robert Bosch injector pumps. Refer to paragraph 121 for explanation of operation.

ing plate on pump, lubricate liberally and slide pump onto mounting studs. Make sure that the three drive gear slots are nearly centered with pump drive holes and drive hub mark is aligned with pointer mark before tightening the three cap screws. Tighten mounting stud nuts and drive gear cap screws to 35 ft.-lbs. torque. Reverse disassembly procedure and add 1 U.S. pint of engine oil through filler plug in side cover of pump. Check timing as outlined in paragraph 120 and bleed system as outlined in paragraph 91.

120. TIMING. To check injection pump static timing, proceed as outlined in paragraph 119 but without removing pump, lines or linkage.

If adjustment is required, loosen the three cap screws on pump drive gear until the drive hub can be moved in the slotted holes in the drive gear. When timing marks are aligned (Fig. 112) retighten three cap screws to 35 ft.-lbs. torque, rotate engine two complete revolutions and recheck timing marks and condition of drive gear teeth. No further timing checks are necessary with this type of pump.

121. ANEROID. Tractors with turbocharged engines are equipped with a diaphragm type control unit, which fits

on top of governor assembly on the Robert Bosch fuel injection pump (Fig. 113). This diaphragm is operated by positive pressure from the intake manifold, which results from the turbocharger producing boost pressure on a hard pull. Until positive pressure is built up by the turbocharger, the vertical shaft which extends down into governor housing provides a stop to limit the travel of the fuel control rack. This allows the engine to accelerate without producing black smoke unnecessarily. When the turbocharger builds sufficient pressure to depress the diaphragm spring in the aneroid, the aneroid shaft moves down and allows the fuel control rack to move farther open, which delivers more fuel at the time the engine can burn it.

Aneroids must be adjusted to the pump on a test stand, so if it becomes necessary to renew the diaphragm or the entire assembly, pump should be recalibrated on a test stand.

THROTTLE LINKAGE

All Models

125. LINKAGE AND SPEED ADJUSTMENT. Fast idle speed should be 2380-2420 rpm for 4040 and 4240 models; 2325-2425 rpm for all turbocharged models. Slow idle speed on all models should be set at 780-820 rpm. To adjust, proceed as follows:

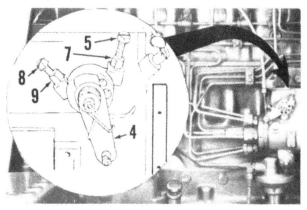

Fig. 123 — Inset shows control lever and stop screws typical of 4040 and 4240 models. Speed control lever (4) is on engine side of pump, slow speed stop screw (8) is toward front and high speed stop screw (5) is toward rear of engine and pump.

Check the force necessary to move the hand throttle lever. If force necessary is not approximately 8 lbs., remove access panel and turn the lever friction spring screw.

Move throttle lever all the way forward until it stops. Adjust the front control rod by loosening nuts on ball joint connections, or the turnbuckle if so equipped. The arm on injector pump should contact the fast idle stop, but should not depress over-ride spring. From this point, lengthen the control rod by 1½ turns of turnbuckle. This should permit full contact with fast idle stop screw on pump. The stop screws on pump limit engine speeds, but the length of control rod must permit full travel.

Move the hand throttle to the rear and adjust the stop screw and jam nut located under control panel so that hand lever is stopped by stop screw and does not damage pump linkage. Be sure that adjustment of hand lever stop nut does not change engine idle speed.

On 4440, 4640 and 4840 models, mechanical interlocking fuel control linkage is not used. Instead, these later models are equipped with hydraulic aneroid activator. Most repair and adjustment should be accomplished with pump removed from engine and attached to a test stand. Slow engine acceleration and excessive smoke when accelerating can be caused by incorrectly adjusted aneroid. Slow engine acceleration can also be caused by loose or broken inlet pipe or fitting to aneroid, cracked aneroid cover or defective aneroid diaphragm.

126. SPEED ADJUSTMENT. Refer to Figs. 123 and 124 for location of high speed adjustment screw (5) and low idle speed adjustment screw (8). Adjust the linkage as described in paragraph 125, then refer to the following:

On models with Roosa-Master pump, the low idle stop screw (8 – Fig. 123) and the high speed stop screw (5) stop positions of the pump. The pump control lever (4) is equipped with a spring loaded override to prevent damage to pump while assuring full travel to contact stop screws. Low idle speed should be 780-820 rpm and high idle should be set to 2380-2420 rpm for all models with Roosa-Master pump.

Robert Bosch Series A-2000 pumps are used on 4440 models; Robert Bosch Series A-3000 pumps are used on 4640 models; Robert Bosch Series P-110 pump is used on 4840 models. There are several differences in these pumps, but the idle speed stop screw (8 – Fig. 124) is located in similar location for all models. Slow idle speed should be approximately 780-820 rpm. To adjust idle speed, loosen locknut (11) and back screw (10) out about 3 or 4 turns. Loosen locknut (9), then turn slow idle stop screw (8) as required to set speed about 20 rpm less than desired slow idle rpm and tighten locknut (9). Turn spring loaded screw (10) in until idle rpm increases 20 rpm, then tighten locknut (11). Increasing the idle speed about 20 rpm with screw (10) should prevent slow speed surging. If surging continues at idle speed, stop engine, remove screw (10) and inspect spring for damage and freedom of movement in hollow of screw (10). Do not increase idle speed above 850 rpm in an attempt to smooth out idle.

The high speed stop screw (5) is located on engine side of pump on 4440, 4640 and 4840 models. High idle stop screw should be set to limit speed to 2325-2425 rpm. Be sure that locknut (7) is tightened after speed is correctly set.

COOLING SYSTEM

RADIATOR

All Models

135. To remove radiator, drain cooling system and remove vertical air stack, muffler, side panels, grilles, screens and hood. Remove all brackets attached to radiator, disconnect hydraulic fluid line clamps and remove the screws retaining the fan shroud. Remove air cleaner hose if necessary, and disconnect radiator hoses. Remove radiator retaining cap screws and slide radiator out right or left side of tractor after fuel return line is removed. Install by reversing the removal procedure.

FAN AND WATER PUMP

All Models

136. **REMOVE AND REINSTALL.** To remove fan and/or water pump, drain cooling system and remove vertical air stack, muffler, hood, side panels and grilles. Remove the screws attaching fan

shroud to radiator and cap screws attaching fan to pump hub; then, slide fan and shroud together out left side of tractor. Loosen fan belts, disconnect by-pass line if so equipped and lower radiator hose. Unbolt and remove the water pump. Install by reversing the removal procedure.

137. **OVERHAUL.** To disassemble the removed water pump unit, first remove hub using a suitable puller which attaches to two fan screw holes; then remove shaft, seal and impeller as an assemble from pump housing using a press. Press shaft and bearing assembly out of impeller and remove ceramic sealing inset and rubber cup from impeller bore.

All parts are available individually except that ceramic sealing insert and cup for impeller is sold as a part of seal kit.

Coat bearing with light oil and press into housing bore until bearing is flush with front edge of housing. Coat outside of seal with Permatex of a type that is resistant to heat and ethyleneglycol antifreeze and press seal into housing until it bottoms.

Apply a light coat of Permatex to shaft bore of impeller. Make sure seal lip and ceramic insert face are perfectly clean and apply a coat of light oil to insert. Insert a feeler gage between impeller blades and machined surface of housing. Clearance should be 0.015-0.025 inch.

NOTE: Support front end of shaft on bed of press when installing impeller; and impeller end of shaft when installing fan hub or pulley.

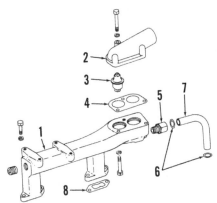

Fig. 129 — Exploded view of typical water manifold. See text for differences between models.

1. Water manifold	5. Connector
2. Thermostat cover	6. "O" ring
3. Thermostat (2)	7. By-pass pipe
4. Gasket	8. Gasket (2)

Install fan pulley and rear cover, then reinstall water pump and associated parts by reversing the removal procedure.

THERMOSTAT AND WATER MANIFOLD

All Models

138. The thermostats are contained in a thermostat housing in water manifold. Refer to Fig. 129 for view of typical manifold. All models use double thermostats. To remove thermostats, drain system, detach upper radiator hose at front and remove housing cap screws, then remove hose and housing as a unit. If a thermostat is suspected of being faulty, check temperature range of unit suspected and test in heated water with a thermometer to be sure opening temperature is correct.

ELECTRICAL SYSTEM

ALTERNATOR AND REGULATOR

All Models

141. Delco-Remy "DELCOTRON" alternator is used on most models; however, a 90 amp John Deere alternator is used on 4840 models. Delco-Remy 37 amp (1102359), 55 amp (1100491) and all of the 72 amp alternators used include an internally mounted solid state regulator. Voltage regulator for 90 amp John Deere alternator is attached to outside of alternator.

CAUTION: Because certain components of the alternator can be damaged by procedures that will not affect a D.C. generator, the following precautions MUST be observed.

a. When installing batteries or connecting a booster battery, the negative post of battery must be grounded.

b. Never short across any terminal of the alternator or regulator unless specifically recommended.

c. Do not attempt to polarize the alternator.

d. Disconnect all battery ground straps before removing or installing any electrical unit.

e. Do not operate alternator on an open circuit and be sure all leads are properly connected before starting engine.

STARTING MOTOR

All Models

142. Delco-Remy and John Deere starting motors are used on all models. Delco-Remy 1113399, 1113402, and 1113672 or John Deere AR55639 and

AR77254 starters are used on all models.

START-SAFETY SWITCH ADJUSTMENT

150. The start-safety switch is located high on right side of transmission housing, on all but Power Shift models, which has switch located in the transmission control valve.

To adjust switch, place transmission in neutral, remove switch and place enough washers under switch to close switch when reinstalled. Remove switch again, remove one washer and check selector in all positions. Switch should only close in the neutral and park posi-

Fig. 140 — Start-safety switch adjustment should include end play check of shifter camshaft.

tions. On Power Shift tractors it may be necessary to remove one additional washer. Switch will not operate properly on a Syncro Range or Quad-Range tractor if shifter camshaft end play exceeds 0.005 inch (Fig. 140).

ENGINE CLUTCH (PERMA-CLUTCH)

NOTE: This section covers only the engine clutch, which is called a PERMA-CLUTCH and is used in tractors equipped with Syncro-Range, Quad-Range and Creeper transmissions. For models equipped with Power Shift transmission, refer to paragraphs 195 through 216.

The engine clutch includes two hydraulically applied power take-off clutch plates. When engine is not running, no hydraulic pressure is available, so clutch cannot be applied and tractor can not be push-started.

Refer to the appropriate following paragraph for adjustment and to paragraph 167 for overhaul data.

LINKAGE ADJUSTMENT

All Models

160. **TRANSMISSION CLUTCH.** To adjust the transmission clutch linkage, remove left side cowl panel, and refer to Fig. 150. Correct distance (D) is 5½ inches. Be sure pedal return spring is holding operating arm against its stop. If necessary to change adjustment, remove cotter pin and washer from clutch operating rod, and change position of the hex adapter in clutch pedal to obtain correct pedal height.

161. **POWER TAKE-OFF.** To adjust the power take-off clutch linkage, remove left side cowl panel and refer to Fig. 150. Disconnect upper end of pto operating rod, then hold the rod down until it bottoms. Push the pto lever forward until front edge of lever is 3/16 inch from front edge of lever slot. Adjust length of operating rod by turning yoke at upper end of operating rod until rod can be connected to the correct hole in operating arm. The rod should be con-

nected to the lower forward hole on 4040, 4240 and 4640 models. Rod should be connected to the upper (rear) hole on 4440 models. The operating arm of some models has three holes, while others have only two holes.

PERMA-CLUTCH PRESSURE TESTS

All Models So Equipped

In order for the Perma-Clutch to function properly and give long, trouble-free operation, it is very important that it be supplied with plenty of clean oil. Before

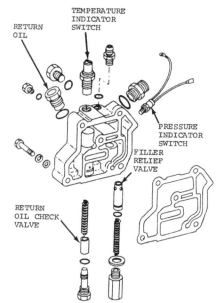

Fig. 151 — Exploded view of filter relief valve housing. Check filter relief valve to make sure it is free.

making pressure checks to locate a malfunction, make certain that the filter is not plugged, and that the filter relief valve (Fig. 151) which is located on the left side of the tractor, just behind the filter, is free and operating properly.

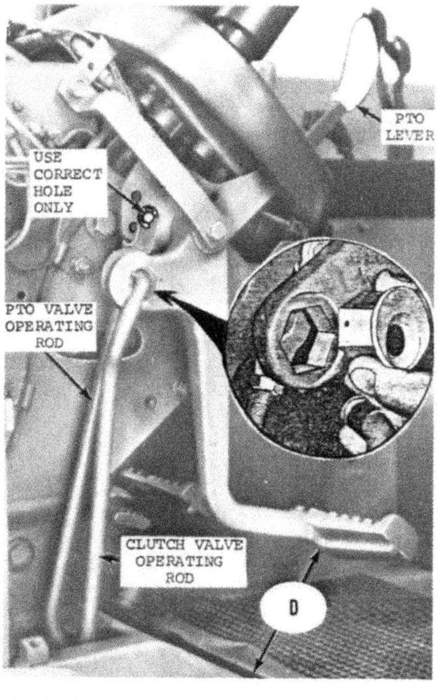

162. **CLUTCH PRESSURE CHECKS.** To make a clutch pressure check, install a 300 lb. gage in plug hole marked "clutch" (Fig. 152). With tractor in PARK, engine at operating temperature and running at recommended rpm, oil pressure should be as follows:

Model	Test Rpm	Pressure (psi)
4040	1000	125-135
	1500	130-140
	2000	135-145
4240	1000	140-150
	1500	150-160
	2000	155-165
4440	1000	160-170
	1500	165-175
	2000	170-180
4640	1000	125-135
	1500	130-140
	2000	135-145

If pressure is below specifications, test both the pto brake and pto clutch at plugs shown. If pressure is normal at these points, there is leakage in the clutch valve circuits or 2-speed planetary circuit. If pressure is low at all three check points, increase and decrease engine speed approximately 300 rpm. If pressure rises and falls with engine speed, there is an internal leak or insufficient oil flow. If pressure changes

Fig. 150 — View of clutch pedal and hex adapter linkage adjustment on all models except Power Shift.

only slightly with rpm change and all three test points show low pressure, adjust system pressure as follows: Remove pressure regulating valve plug (Fig. 152). Shims are available to adjust pressure, and each shim (19 – Fig. 160) should boost pressure approximately 5 psi.

When pressure is satisfactory, depress clutch pedal while observing gage. Pressure should fall to zero when pedal is depressed, and come up gradually as pedal is slowly released, if clutch valve is working properly.

If pressure will not come up to acceptable level, refer to transmission pump test, paragraph 275 in Hydraulic Section.

Refer to paragraphs 259 through 261 for overhaul procedures.

163. **PTO CLUTCH AND BRAKE.** If pto is suspected of having a malfunction, refer to Fig. 152 and install a 300 psi gage in the pto clutch plug hole. Start engine and engage pto clutch. Pressure should be within limits that follow:

Model	RPM	Pressure (psi)
4040	2000	135-145
4240	2000	155-165
4440	2000	170-180
4640	2000	155-165

Engage pto brake and observe clutch pressure. Pto clutch pressure should drop to zero. When pto clutch is engaged gradually, pressure should rise gradually.

To test pto brake pressure, install another 300 psi gage in pto brake plug hole. With engine again running, apply the pto brake. Pressure should equal system pressure at specified rpm. Slowly push lever forward to the engaged position. Before clutch pressure starts to rise, brake pressure should drop to zero.

If pressure cannot be brought up to specifications, test transmission pump as outlined in paragraph 275 in Hydraulic Section. If pump tests good, remove pressure regulating valve housing (Fig. 152) and refer to Fig. 159A. Use an air hose with at least 125 psi pressure and a rubber tip on air gun to test each circuit as shown. A leaking circuit will be found when air can be heard escaping in any one passage. Move the 2-Speed planetary lever and apply air in both underdrive and direct positions to check for leak in planetary circuit.

164. **LUBRICATION PRESSURE AND FLOW.** Models with Perma-Clutch use a transmission pump that flows a volume of approximately 10 gpm. Lubrication pressure can be checked by connecting a pressure gage as shown in Fig. 152A. Pressure should be checked with engine operating at 1500 rpm and with engine operating at 2000 rpm. Correct pressure is 25 psi at 1500 rpm with oil at 110 degrees F. or 17 psi with oil at 150 degrees F. Correct pressure at 2000 rpm is 43 psi at 110 degrees F. or 28 psi with oil at 150 degrees F.

Low lubrication oil pressure could be caused by: Pressure regulator to clutch valve housing gasket leaking; Adapter tube "O" rings leaking. If equipped with Quad-Range, additional possible causes are: Two speed shift valve leaking; Charge circuit malfunctioning.

The lubrication reduction valve can be checked with same gage attachment shown in Fig. 152A. Adjust engine speed until lubrication pressure indicated on gage is 10 psi, then depress clutch pedal and observe gage pressure. Proper operation of reduction valve is indicated by a dip in pressure. Check for stuck lubrication reduction valve if pressure does not dip.

165. **OIL PRESSURE SWITCH.** Remove oil pressure indicator switch (Fig. 151). Connect a hydraulic hand pump to switch and a 12 volt test light and battery across switch leads. The switch should open and close as follows:

Model	Open	Close
4040, 4240	115-125 psi	90-100 psi
4440	145-155 psi	115-125 psi
4640	115-125 psi	90-100 psi

TRACTOR CLUTCH SPLIT

All Models

166. To detach (split) engine from clutch housing for access to engine clutch and flywheel, proceed as follows:

Drain cooling system and remove air stack, hood and muffler. Remove side shields, grille screens, cowl and step plates. Remove batteries, boxes and long connecting cable between batteries.

Discharge brake accumulator by opening the right brake bleeder screw and holding brake pedal down a few moments.

If tractor is equipped with air conditioning, disconnect the two couplers above oil filter on left side. Use two wrenches, one to hold coupler body and one to loosen coupler. If gas can be heard escaping, tighten coupler and loosen again. Keep coupler ends clean while disassembled.

On tractors with Perma-Clutch, remove the hex drive shaft for transmission pump to prevent bending shaft when tractor is separated. To remove hex shaft, remove three point hitch center link attaching bracket on rear of transmission housing. Remove the large plug which contains the rear bearing for the hex shaft and draw the long hex shaft rearward with pliers.

Disconnect throttle rod, wiring, hydraulic lines, heater hoses, tachometer cable and temperature indicator sending unit. Be sure to cap all disconnected hydraulic fittings to prevent dirt entry. If tractor has a Perma-Clutch, place a drain pan under clutch housing to catch the oil from clutch as tractor is separated. If tractor is equipped with front end weights, remove the weights. Support engine and transmission securely and separately, remove the connecting cap screws and roll transmission assembly rearward away from engine.

To attach, reverse the above procedure and tighten the connecting cap

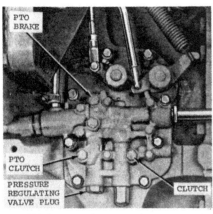

Fig. 152 – View of typical clutch regulating valve housing. Remove clutch plug and install gage to check Perma-Clutch pressure.

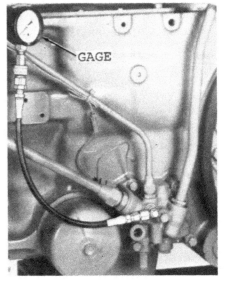

Fig. 152A – Attach pressure gage as shown to check lubrication pressure. Refer to text.

screws to a torque of 85 ft.-lbs. on ½-inch screws, or 300 ft.-lbs. on ¾-inch screws. Bleed steering system as outlined in paragraph 22.

R&R AND OVERHAUL PERMA-CLUTCH

All Models

Unless clutch has been slipping or giving some indication of trouble, DO NOT disassemble, as a properly adjusted and oil cooled assembly will give trouble free service for a very long time. Check the three transmission clutch levers for adequate clearance between the underside of levers and the clutch cover by driving wedges under the outer edge of levers as shown in Fig. 153. If some clearance does not exist with levers applied, unit should be overhauled. Transmission clutch levers will contact clutch cover before clutch discs are worn enough to damage other parts of the assembly. The clutch would then slip badly, and if transmission clutch levers were readjusted to compensate for the wear on the discs, serious damage could occur to the friction surfaces of clutch backing plate and separator plates with further wear.

167. Split tractor as outlined in paragraph 166. If clutch is to be overhauled, remove the six cover cap screws and disassemble unit. (If clutch is being removed to allow other work to be done, refer to the above procedures). Remove locknuts and adjusting nuts on the three clutch operating levers (Fig. 153). Refer to Fig. 154 and separate clutch pack, using care to avoid damaging discs and friction surfaces. Inspect all friction surfaces for scoring, cracking or sharp edges which could damage clutch discs. Smooth any raised area with crocus cloth. Original thickness of disc is 0.137-0.143 inch. Renew any disc that is not at least 0.110 inch in thickness, has less than 0.005 inch grooves in its face,

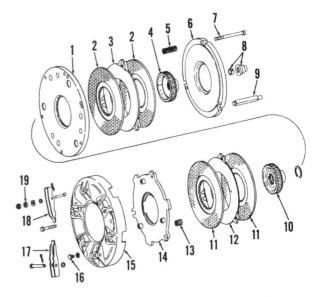

Fig. 154 — Exploded view of Perma-Clutch and associated parts typical of all models. On 4640 and 4840 models, three separators (3), four clutch discs (2), two separators (12) and three discs (11) are used.

1. Backing Plate
2. Clutch disc
3. Separator, clutch
4. Clutch hub
5. Spring
6. Pressure plate
7. Operating bolt (3)
8. Pilot adapter
9. Flywheel pin (3)
10. Pto hub
11. Pto disc
12. Separator, pto
13. Spring
14. Pto pressure plate
15. Clutch cover
16. Adjusting screw
17. Pto clutch lever
18. Clutch lever
19. Adjusting nut

or has chunks missing from friction surface. A dark color of disc lining is normal after use, and should not be considered sufficient reason to renew the disc, if thickness and groove depth are within specifications.

When reassembling, place clutch pressure plate (6 – Fig. 154) with friction surface up, insert the three clutch operating bolts (7) from the underside and place three blocks under bolts as shown in Fig. 155. Note that the transmission separator plate (3 – Fig. 154) is thicker than pto separator plate (12). Install transmission clutch hub, disc, separator plate, second disc and backing plate in that order. Three separator plates (3) and four clutch discs (2) are

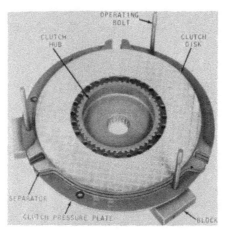

Fig. 155 — Insert blocks under operating bolts as shown.

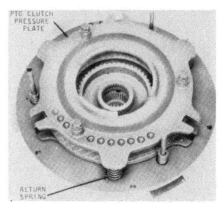

Fig. 157 — Place pto pressure plate tabs on return springs as shown.

Fig. 153 — Insert wedges under clutch operating levers as shown and check clearance under inside of lever.

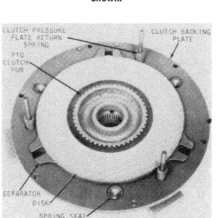

Fig. 156 — Align separator plate slots with holes in backing plate as shown.

Fig. 158 — Use a screwdriver through pin hole to align separator plates.

used on 4640 and 4840 models. Refer to Fig. 156 and install clutch pressure plate return springs, pto clutch hub, pto disc, pto separator plate, and second disc as shown. On 4640 and 4840 models, two pto separator plates (12) and three pto clutch discs (11) are used. Make sure that the three tabs on pto clutch pressure plate are installed over the three return springs (Fig. 157). Install the clutch cover, three long clutch levers, bearing bars and adjusting nuts (Fig. 158). Tighten the three adjusting nuts enough to hold the clutch backing plate snugly against the clutch cover. This will squeeze the inside pto disc so that it will not fall off pto hub when clutch pack is installed in flywheel. When clutch pack is being installed, align the slots in separator plates with pins in flywheel. It may be necessary to slightly loosen the adjusting nuts in

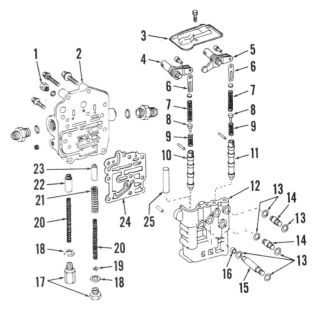

Fig. 160 — Exploded view of clutch regulating valve housing and associated parts. Flange gasket for item 2 not shown.

1. Plug
2. Outer housing
3. Shield
4. Clutch shaft
5. Pto shaft
6. Links
7. Springs
8. Pins
9. Springs
10. Clutch valve
11. Pto clutch valve
12. Valve housing
13. "O" rings
14. Adapters (4)
15. Adapter (Quad-Range)
16. Plug (Syncro-Range)
17. Plugs
18. Washers
19. Shim
20. Inner springs
21. Outer spring
22. Cooler relief valve
23. Clutch pressure valve
24. Gasket
25. Pto lock piston

Fig. 159 — JDE-78 clutch adjusting tool for adjusting transmission and pto clutch levers. Note two different adjusting surfaces for levers.

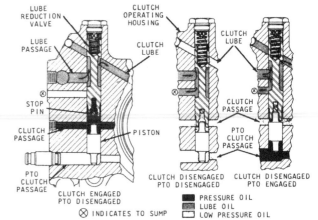

Fig. 162 — Schematic view of lubrication reduction valve used.

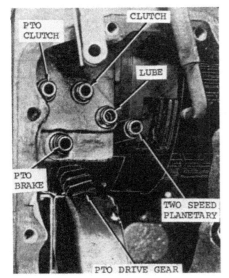

Fig. 159A — View of clutch and pto housing with pressure regulating housing (2 — Fig. 160) removed.

order to move separator plates. Install only four cap screws opposite each other in clutch cover.

168. ADJUSTMENT. It is very important that clutch adjustment be accurately performed, if uneven wear and hot spots are to be avoided. Since the Perma-Clutch is hydraulically applied, the clutch levers must fully release when apply pressure is removed from the apply piston. This will insure that the discs will not be overheated by a partially applied condition when engine is running with clutch released. The position of the clutch levers on the clutch cover is very critical, as it determines the point at which the apply piston and bearing first make contact with levers. JDE-78 clutch adjusting tool with spacer plates is required for proper adjustment (Fig. 159). The spacer plate numbers to be used for each tractor model are as follows:

Make certain that only the proper plates for each tractor are used. Loosen the adjusting nuts on the three transmission clutch levers. Loosen three locknuts on pto levers and turn the adjusting screws in. Install the two mounting studs for adjusting tool in the two remaining clutch cover holes, which should be directly opposite from each other. Install adjusting tool so that the long screw will pass through the hole in cross bar and at the same time the three adjusting tool screws will bear on a flat part of pto pressure plate, and not on the lip. Leave the three adjusting tool screws loose and tighten the bar screw to 20 ft.-lbs. torque. Tighten the three adjusting tool screws evenly to 30 in.-lbs. torque, to load the pto pressure plate.

To adjust pto levers, make a mark or number on pto lever and pivot pin so that they will be returned to the same

Model	Pto Clutch	Engine Clutch
4040, 4240, 4440	Gage 78-12 (0.070 in.) + Gage 78-3 (0.466 in.)	Gage 78-4 (0.253 in.)
4640	Gage 78-15 (0.159 in.)	Gage 78-17 (0.147 in.)

location that they were in when adjusted. Install one pto lever and pivot pin. Move lever out until it contacts outer adjusting surface, and turn adjusting screw out until it contacts the underside of lever. Carefully remove pivot pin and lever without moving adjusting screw and hold the screw from moving while locknut is tightened. Recheck setting by reinstalling pto lever and pin. If clearance is more than 0.010 inch, readjust lever to less than 0.010 inch. Pivot pin will go in tight because tightening the locknut will cause adjusting screw to stretch slightly. When all three levers have been adjusted using this procedure, loosen the three adjusting tool screws, but leave the bar screw tight.

Loosen the nuts on the three transmission clutch operating bolts and make sure the operating bolt heads are in their slots in pressure plate so that they will not turn. Move the transmission clutch levers out until they contact the **INNER** adjusting surface and finger tighten adjusting nuts. Tighten adjusting nuts in several steps (40, 70, then 90 in.-lbs. suggested) until final correct torque of 90 in.-lbs. is correct for all three nuts. Snug the locknuts onto adjusting nuts, remove the adjusting tool, and carefully tighten locknuts while making sure that adjustment does not change. Install the two remaining clutch cover cap screws and tighten to a torque of 35 ft.-lbs.

Since the transmission clutch and pto clutch are applied by hydraulic pressure

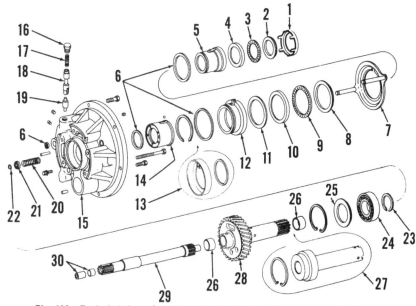

Fig. 166 — Exploded view of clutch operating housing showing associated parts.

1. Bearing retainer	8. Outer race	16. Plug
2. Outer race	9. Needle bearing	17. Spring
3. Needle bearing	10. Inner race	18. Lubrication
4. Inner race	11. Thrust washer	reduction valve
5. Clutch operating	12. Pto operating piston	19. Piston
piston	13. Oil collector (no pto)	20. Spring
6. Packing	14. Piston sleeve	21. Keeper
7. Pto bearing retainer	15. Operating housing	22. Retainer

23. Snap ring	
24. Ball bearing	
25. Oil shield	
26. Bushing	
27. Adapter (no pto)	
28. Pto drive shaft	
29. Clutch shaft	
30. Bushing (1 or 2)	

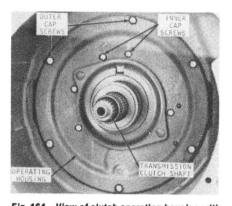

Fig. 164 — View of clutch operating housing with Quad-Range transmission, showing inner and outer cap screws. Creeper transmission has only three inner cap screws.

Fig. 165 — View of removed clutch operating housing with front side up.

through needle thrust bearings, the bearings are running any time either clutch is applied. These needle bearings and their races should be carefully inspected for wear, scoring, or flattened rollers. Renew any bearing assembly that is not in perfect condition.

Rejoin tractor as outlined in paragraph 166, and bleed steering system as outlined in paragraph 22.

LUBRICATION REDUCTION VALVE

All Models With Perma-Clutch

When the transmission oil is cold and thick, the lube oil which is being routed to the transmission clutch discs tend to keep the discs rotating after clutch is released, making shifting difficult. The lubrication reduction valve (Fig. 162), located on the backside of clutch operating housing, prevents this condition by cutting off lube oil to the discs when clutch is released. When pto clutch is engaged with transmission clutch released, the valve is only partially opened, allowing a reduced volume of lube oil to transmission clutch, with most of the oil directed to pto discs.

170. **R&R AND OVERHAUL.** Split tractor as outlined in paragraph 166, which will allow access to clutch operating housing (Fig. 164). Remove six housing cap screws on outer rim of operating housing, and remove five inner cap screws, if equipped with Quad-

Range transmission, or remove three inner cap screws if equipped with Creeper transmission. Remove operating housing, and if only the lubrication reduction valve is to be removed, no further disassembly is needed. Remove the large hex plug at the rear of the clutch operating housing (Fig. 165). Remove spring and valve. The stop pin must be driven out before piston can be removed. Check valve and piston for wear or scoring and check spring for pitting or poor tension. When reassembling, drive stop pin into housing from the rear, until pin is flush with housing. Reassemble in reverse order of disassembly. Rejoin tractor as outlined in paragraph 166, and bleed steering system as outlined in paragraph 22.

CREEPER TRANSMISSION

Models So Equipped

The creeper transmission is a mechanically shifted planetary gearset, which gives an underdrive or direct gear ratio. The unit fits in front of the 8-speed transmission as an option on the Syncro-Range only. The optional creeper transmission is **NOT** available on 4440, 4640 or 4840 models. In normal operations the unit is left in direct, but for operations where a very large speed reduction is needed, the unit can be shifted to underdrive. The underdrive may be used in the five lowest forward gears and both reverse gears. An interlock on the shift lever prevents the use of underdrive in forward gears 6, 7 and 8. The creeper shift lever must be in "Direct" in order to engage PARK position.

173. REMOVE AND REINSTALL. Split tractor as outlined in paragraph 166. Remove clutch operating housing (Fig. 164) as outlined in paragraph 170. Remove the small cap screw and retainer, which retains the planetary shifter shaft, from the right side of clutch housing. Pull shifter shaft out of housing far enough to clear the shifter assembly (Fig. 167). Pull planetary assembly and clutch shaft forward until assembly can be tilted away and removed for shifter assembly.

To reinstall, place planetary and transmission input shaft assembly (Fig. 167) into bearing quill and tilt sideways until shifter assembly can be installed with the shift yoke in REAR detent position. Use clutch shaft to guide entire assembly into splines of the transmission drive shaft. Move shifter assembly into the FRONT detent position and align assembly so that the hex shifter shaft can be pushed into the hex shaped hole in shifter arm. Shifter assembly should now be in proper position to accept three cap screws from clutch operating housing (Fig. 164). Install the cap screw and retainer which retains shift shaft in clutch housing and tighten to 20 ft.-lbs. torque. Reinstall clutch operating housing outer cap screws and be sure the three inner cap screws align with shifter assembly before final tightening.

Rejoin tractor as outlined in paragraph 166 and bleed steering system as outlined in paragraph 22.

174. OVERHAUL. Remove creeper planetary assembly as outlined in paragraph 173. Remove shifter collar (24–Fig. 168) from rear sun gear (20). Remove snap ring, collar plate (22), snap ring, rear sun gear, and transmission in-

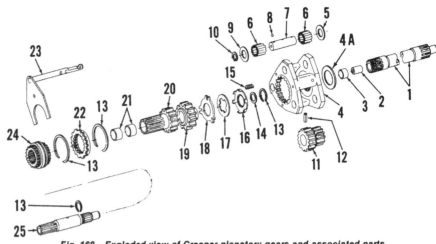

Fig. 168—Exploded view of Creeper planetary gears and associated parts.

1. Clutch shaft	7. Pinion shaft (3)	13. Snap ring	20. Rear sun gear
2. Hex bushing	8. Ball bearing (3)	14. Phenolic washer	21. Bushing (2)
3. Bushing	9. Washer	15. Spring (6)	22. Collar plate
4. Planet carrier	10. Snap ring (3)	16. Pressure plate	23. Shifter yoke
4A. Thrust washer	11. Pinion gear (3)	17. Brake disc	24. Shifter collar
5. Washer	12. Spring pin (3)	18. Backing plate	25. Transmission input
6. Bearing (6)		19. Front sun gear	shaft

put shaft (25). Remove three snap rings (10) from planet carrier (4) and remove three planet pinion shafts (7), being careful not to lose the three ball bearings (8). Remove pinions and bearings (11 and 6) and front sun gear (19). Drive out the three spring pins (12) far enough to remove snubber backing plate (18), brake disc (17), pressure plate (16) and springs (15). Inspect all parts for scoring and wear. Pressure plate (16) should be a minimum of 0.049 inch thick, and brake disc (17) should measure 0.075 to

0.085 inch. Springs should have a free length of approximately 1-1/16 inch, and should register 10.8 to 13.2 lbs. pressure when compressed to a length of 13/16-inch. Remove snap ring and clutch shaft (1) and inspect bushing (3) in shaft. If necessary to renew, press new bushing in 0.040 inch below the end of shaft.

Reassemble in reverse order of disassembly and refer to Fig. 168 as guide. First install hex bushing (2) into clutch shaft (1), install clutch shaft into carrier

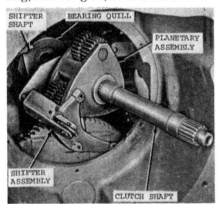

Fig. 167—View of Creeper transmission planetary and shifter assembly after clutch operating housing is removed.

Fig. 169—Compress snubber brake springs with a press as shown, to install spring pins to retain backing plate.

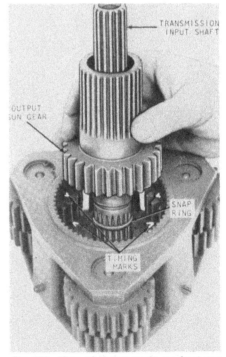

Fig. 170—View of timing marks on front sun gear and pinion gears. Marks must align as shown so that rear sun gear teeth will mesh properly.

(4) and install snap ring to retain shaft. Install thrust washer (14) on clutch shaft and place carrier in a vise with clutch shaft pointed down, or in a hole in a bench so that unit will be held steady. Assemble the springs, pressure plate, disc brake and backing plate. Use a press (Fig. 169) to compress the pieces until the three spring pins can be driven into planet carrier (4 – Fig. 168). About 13/32-inch of pin should stick into the inside of carrier to retain backing plate. Install front sun gear (19) into carrier so that the notches align with tangs on disc brake. Refer to Fig. 170 and install planet pinions so that the timing marks on all pinions align with the three marks on sun gear. This will allow the output sun gear teeth to mesh with pinions. Install snap ring on transmission input shaft and insert forward splines into front sun gear. If output (rear) pinion bushings must be renewed, press new bushings (21 – Fig. 168) to 0.055 inch below sun gear end of pinion, and 0.037 inch below splined end and install gear. Install large snap ring into inner groove in carrier, install collar plate (22) and the other snap ring into outer groove. Place shifter collar (24) onto splines of output sun gear and install assembly as outlined in paragraph 173.

SYNCRO-RANGE (8-SPEED GEAR) TRANSMISSION

(For Quad-Range Models, Refer to Paragraph 192)

(For Creeper Transmission, Refer to Paragraph 173)

(For Power Shift Models, Refer to Paragraph 195)

The "Syncro-Range" transmission is a mechanically engaged transmission consisting of three transmission shafts and a single, mechanically connected, remote mounted control lever as shown in Fig. 179. The four basic gear speeds are selected by coupling one of the differential bevel pinion shaft idler gears to the splined bevel pinion shaft, and on all models, can only be accomplished by disengaging the engine clutch and bringing the tractor to a stop. The high and low speed ranges within the four basic speeds are selected by shifting only the synchronized coupler on the transmission drive shaft and, because of the design, can be accomplished by disengaging the engine clutch and moving the control lever, without bringing tractor to a halt.

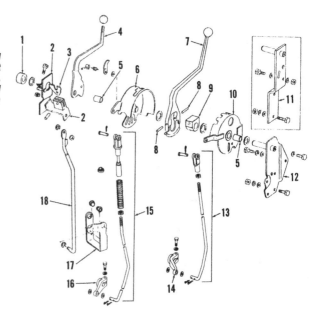

Fig. 179 — Syncro-Range and Creeper transmission shift levers and associated parts. Items 4, 17 and 18 are used only if equipped with Creeper transmission.

1. Spacer
2. Support
3. Shift latch
4. Creeper shift lever
5. Bushing (2)
6. Speed change quadrant
7. Shift lever
8. Spring pin (2)
9. Pivot
10. Speed range quadrant
11. Support
12. Support
13. Speed range rod
14. Speed range arm
15. Speed change rod
16. Speed change arm
17. Bellcrank
18. Creeper shift rod

The idler gears and bearings on the main shaft and bevel pinion shaft are pressure lubricated by a separate transmission oil pump which also serves as the charging pump for main hydraulic system pump.

INSPECTION

All Models

180. To inspect the transmission gears, shafts and shifters, first drain the transmission and hydraulic fluid and remove operator's platform. Disconnect the control support brackets, Perma-Clutch control valve operating rod linkage and Quad-Range control valve linkage, then remove transmission top cover. Examine shaft gears for worn or broken teeth and the shift linkage and cam slots for wear.

CONTROL QUADRANT

This section covers disassembly and overhaul of shifter controls mounted on shift console to the right of operator. Removal, inspection and overhaul of shift mechanism inside the transmission housing is included with transmission gears and shafts.

All Models

181. **R&R AND OVERHAUL.** To overhaul the control quadrant, remove the console sheet metal by removing the six cap screws holding the vertical panel, and three cap screws under the right fender. Disconnect the shifter rods (13 and 15 – Fig. 179), and Front Drive Rotary Switch rods on models so equipped; then unbolt and remove quadrant assembly.

If lever (7) or pivot (9) must be re-newed, proceed as follows: Clamp the lower curved portion of lever in a protected vise and slip a 5/32-inch cotter pin inside the two spring pins (8). Refer to Fig. 180. Grasp spring pin with clamping pliers and extract with a twisting motion. When reassembling, leave at least 7/16-inch of roll pin protruding from lever.

TRANSMISSION DISASSEMBLY AND REASSEMBLY

Paragraphs 182 and 183 outline the general procedure for removal and installation of the main transmission components. Disassembly, inspection and overhaul of the removed assemblies is covered in overhaul section beginning with paragraph 184, which also outlines

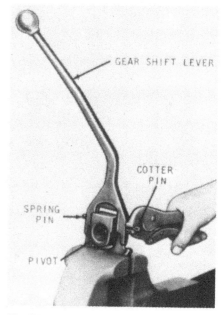

Fig. 180 — Insert a cotter pin inside spring pin to aid in removal. Refer to paragraph 181.

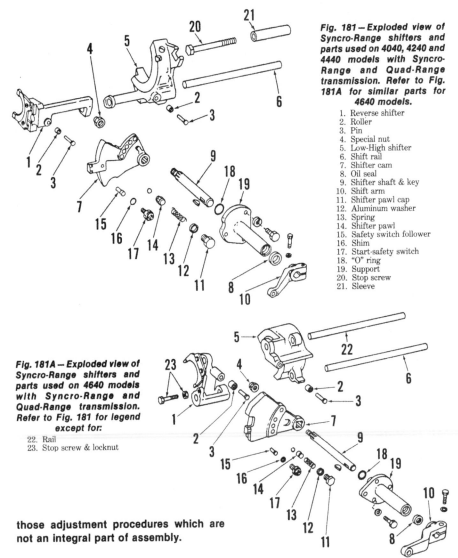

Fig. 181 — Exploded view of Syncro-Range shifters and parts used on 4040, 4240 and 4440 models with Syncro-Range and Quad-Range transmission. Refer to Fig. 181A for similar parts for 4640 models.

1. Reverse shifter
2. Roller
3. Pin
4. Special nut
5. Low-High shifter
6. Shift rail
7. Shifter cam
8. Oil seal
9. Shifter shaft & key
10. Shift arm
11. Shifter pawl cap
12. Aluminum washer
13. Spring
14. Shifter pawl
15. Safety switch follower
16. Shim
17. Start-safety switch
18. "O" ring
19. Support
20. Stop screw
21. Sleeve

Fig. 181A — Exploded view of Syncro-Range shifters and parts used on 4640 models with Syncro-Range and Quad-Range transmission. Refer to Fig. 181 for legend except for:

22. Rail
23. Stop screw & locknut

those adjustment procedures which are not an integral part of assembly.

All Models

182. DISASSEMBLY. Any disassembly of transmission gears and controls requires that tractor first be separated (split) between engine and clutch housing as outlined in paragraph 166. Dis-

connect Power Front Wheel Drive drain pipe, if so equipped. Remove the transmission top cover, and all necessary wiring and hydraulic lines.

NOTE: Tractor may be equipped with either a two gear pto drive train system, or a four gear system. A four gear system tractor may be separated between clutch

housing and transmission housing without first removing pto lower drive gear and shaft assembly. Be sure to remove the cap screws or hex nuts which must be reached from front of clutch housing and through top of transmission to rear of clutch housing, before attempting to separate the housings.

A two gear system has a large lower pto drive gear inside clutch housing which, together with pto drive shaft, must be removed BEFORE housings are separated. Remove the rear bearing quill and pto shaft as an assembly if equipped with a 1000 rpm only pto. If unit is a dual speed, remove rear bearing quill, stub shaft and 540 rpm drive gear, which is directly under bearing quill. Compress large snap ring which holds pto drive shaft bearing in transmission housing and pull drive shaft rearward out of housing. The large pto drive gear in clutch housing will remain on a sleeve in the front drive shaft bearing. Slide the large gear backward off the sleeve, being careful not to damage the oil shield which is made onto the gear, then tip gear forward and remove toward the front. If oil shield is damaged and will rub on gear, separate the welds around shield halves and renew either half. Weld around edges by tacking halves in five places. To separate clutch housing, remove transmission top cover and the two top cap screws in backside of clutch housing. Remove remaining cap screws, support the rear of tractor under front of transmission housing, lift clutch housing with a chain hoist and roll rear of tractor away. Supports should be placed under front and rear of transmission housing so that it is very stable. Front drive shaft bearing and sleeve should be inspected after clutch housing is removed, and renewed at this time if necessary. Clutch housing may be attached temporarily to engine if it is necessary to get it out of the way.

Transmission must be disassembled in the approximate sequence outlined in the following paragraphs; however

Fig. 181B — View of shifters installed on 4640 models. Refer to Fig. 181A for exploded view.

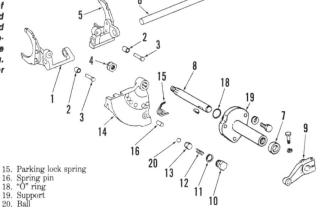

Fig. 182 — Exploded view of Speed Change shifters and parts used on 4040, 4240 and 4440 models with Syncro-Range and Quad-Range transmissions. Refer to Fig. 182A for similar parts for 4640 models.

1. Rear collar shifter
2. Roller
3. Pin
4. Special nut
5. Front collar shifter
6. Rail
7. Oil seal
8. Shifter shaft & key
9. Shift arm
10. Shifter pawl cap
11. Aluminum washer
12. Spring
13. Shifter pawl
14. Shifter cam
15. Parking lock spring
16. Spring pin
18. "O" ring
19. Support
20. Ball

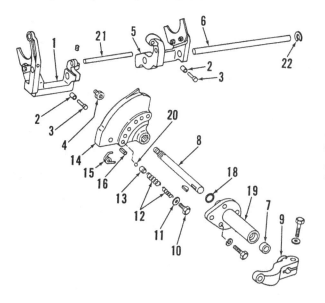

Fig. 182A — Exploded view of Speed Change shifters and parts used on 4640 models with Syncro-Range and Quad-Range transmissions. Refer to Fig. 182 for legend except for 21. Shift rail.

disassembly need not be completed once defective or damaged parts are removed.

Remove transmission top cover and rockshaft housing or transmission rear cover. Remove the detent spring caps from right side of housing, using caution, as caps are under spring pressure. Lower the upper shifter arm to its lowest position and remove special lock nut from inner end of shaft. Be careful not to drop parts in housing, as they will be difficult to remove. Refer to Fig. 181 and Fig. 181A. Withdraw shifter arm and shaft from housing. Oil seal may be

Fig. 182B — View of front of transmission case typical of all models with Syncro-Range transmission except 4640 model. Refer to Fig. 182C for 4640 model.

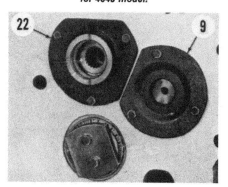

Fig. 182C — View of front of transmission case on 4640 model with Syncro-Range transmission.

renewed at this time. Remove speed range shifters and rail.

Jack up one rear wheel of tractor and turn bevel ring gear until one of flat surfaces of differential housing is toward transmission oil pump, then unbolt and remove the manifold, oil tubes and pump.

Tape or clip the synchronizer clutches together to keep them from separating while input shaft is being removed.

Unbolt and remove front bearing housing (22 – Fig. 182B or Fig. 182C), then use a brass drift to drive shaft back until gears contact gears on countershaft. Insert a special tool or spacers between bearing cone (3 – Fig. 190 or Fig. 190A) and cup (2), then finish removing cup toward rear. Tilt shaft up in front and withdraw from housing. Remove front bearing retainer using care to keep shim pack together and undamaged, then lift transmission input shaft assembly out top opening.

Remove lower (speed change) shifter cam and forks Fig. 182 and Fig. 182A. Remove differential assembly as outlined in paragraph 222. If tractor is equipped with four gear pto drive train, remove retainer plate on pto lower idler gear and use a puller to remove gear and roller bearing (front). The rear bearing cone will remain on differential pinion shaft. If tractor has the two gear pto drive train, remove retainer plate on pinion shaft. Before driving pinion shaft rearward, install a "C" clamp through hole in transmission housing (Fig. 183) and tighten against front (fourth and seventh speed) gear while using a brass drift or soft hammer to drive on shaft. As shaft begins to move, slide the parts forward, retighten "C" clamp and remove snap rings (Fig. 184) as they become exposed.

Continue to move shaft components and snap rings forward until rear snap

ring has been unseated; then bump differential drive shaft (pinion shaft) rearward, lifting gears and associated parts out top opening as shaft is removed.

NOTE: The snap rings are the same diameter but of different thickness; the thickest snap ring being in rear groove. Thus no snap ring can fall into another groove as shaft is removed or installed.

Remove countershaft front bearing retainer (9 – Fig. 189) and shim pack (8). Remove snap ring retaining rear bearing cup (1); then using a brass drift, drive the countershaft rearward until rear bearing cup is removed. Countershaft can now be lifted out top opening of transmission case.

Overhaul transmission main components as outlined in paragraphs 184 through 191; assemble as outlined in paragraph 183.

183. **ASSEMBLY.** To assemble the transmission unit, proceed as follows: Install countershaft, rear bearing cup and retaining snap ring, then install front bearing retainer using the removed shim pack. Tighten retaining cap screws to a torque of 35 ft.-lbs.; then using a dial indicator, check countershaft end play. Adjust end play if necessary, to the recommended 0.001-0.004 inch by varying the thickness of shim pack (8 – Fig. 189). Shims are available in thicknesses of 0.006, 0.010 and 0.018 inch.

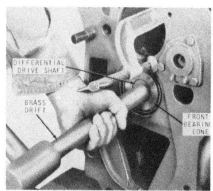

Fig. 183 — Install a "C" clamp to hold front gear forward while front bearing is driven off shaft.

Fig. 184 — Remove four snap rings as they become exposed as shaft moves rearward. Keep snap rings in order, with thickest to the rear.

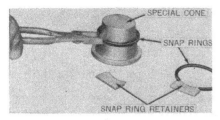

Fig. 185—Special assembly tools are needed to expand and install snap rings.

Fig. 186—Install arbor and support to hold parts while assembling differential pinion shaft gears.

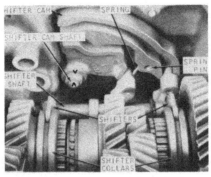

Fig. 187—Speed change shifter mechanism showing component parts. Note "V" alignment marks on shifter cam and shaft. Install shifter cam spring as shown. Refer to Fig. 187A for 4640 model.

Fig. 187A—View of speed change shift mechanism for 4640 models. Refer to Fig. 187 for identification of parts on other smaller models.

If bevel pinion shaft or transmission housing are being renewed; or if shaft bearings require renewal or adjustment; refer to paragraph 191 for adjustment procedure.

Special tools are required to reinstall bevel pinion shaft. The needed tools are 1–JDT-1, snap ring pliers; 4–JDT-3 expander plates and 1–JDT-2 expander cone. Expand the snap rings and install expander plates as shown in Fig. 185, then lay out snap rings in order according to thickness. Refer to Fig. 192 for order of reassembly of shaft components. Install arbor and support as shown in Fig. 186 (or other suitable support) to hold parts while assembling. Insert bevel pinion shaft (1–Fig. 192) through rear bore, install largest gear (5) and thickest snap ring (6) with expander plate in place. Continue to move shaft forward, installing remaining parts in proper order. Second thickest snap ring is installed next, and so on, with the thinnest at front. After fourth speed gear (14) is installed, remove expander plates and seat snap rings in their grooves; then install remainder of parts using the removed (or previously determined) shim pack (16). Tighten cap screws retaining end plate (20) to a torque of 35 ft.-lbs. on grade 5 cap screws, or 50 ft.-lbs. on grade 8 cap screws.

Install speed change shifter forks and cam, and insert shift rail through forks. Install camshaft and arm, making sure index marks are aligned as shown in

Fig. 187. Tighten locknut securely then adjust camshaft end play to 0.002-0.007 inch if necessary, by loosening clamp nut and sliding actuating arm on shaft. Too much end play in shifter camshaft can allow detent springs to force the shift cam out of proper operating position. If shift cam spring (Fig. 187) has been removed or renewed, make sure the long end of spring is installed toward the shifter cam shaft.

To install transmission drive shaft and shifter mechanism assembly, place shaft

in transmission housing and install front bearing cup, shims and retainer. Tighten cap screws securely, then install rear bearing cup and transmission oil pump. Tighten pump cap screws to 20 ft.-lbs. torque. Check transmission drive shaft end play using a dial indicator and adjust to 0.004-0.006 inch by means of shims (21–Fig. 190). Shims are available in thicknesses of 0.006 and 0.010 inch. Tighten front bearing retainer cap screws to 35 ft.-lbs. torque.

Reinstall speed range shifters and associated parts, making sure index marks are aligned and camshaft end play is within the recommended range of 0.002-0.007 inch.

OVERHAUL

The following paragraphs cover overhaul procedure of transmission main components after transmission has been disassembled as outlined in paragraph 182.

184. **SHIFTER CAMS AND FORKS.** Refer to Fig. 181 for an exploded view of speed range shifter mechanism and Fig. 182 for speed change shifters. Examine shifting grooves in cams (7–Fig. 181 and 14–Fig. 182) for wear or other damage. Parking lock spring (15–Fig. 182) is retained to cam by spring pin (16). The spring must have sufficient tension to shift the front shift coupling into engagement.

185. **TRANSMISSION PUMP.** The removed transmission pump may be disassembled by removing manifold (12–Fig. 188). Check all surfaces, bushings and bushing bores. Lubricating check valve spring (10) should test 3.2-3.8 lbs. pressure at ¾-inch length. Clearance between driven gear and bushing should be 0.0015-0.0045 inch and cover should be flat within 0.003 inch. All parts are available individually.

Fig. 188—Exploded view of Syncro-Range transmission oil pump and associated parts typical of 4040, 4240 and 4440 models.

1. Plug w/bushing
2. "O" ring
3. Hex drive shaft
4. Outlet tube
5. Intake screen
6. Clamp
7. Plug
8. "O" ring
10. Spring
11. Ball
12. Manifold
13. Tube
14. Intake tube
16. Gears
17. Dowel pin
18. Body

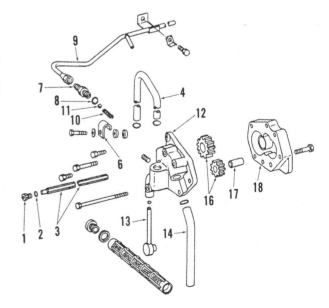

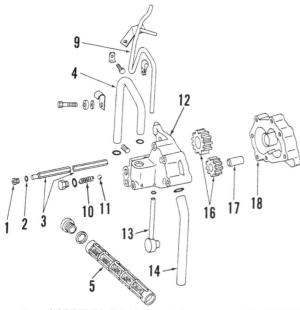

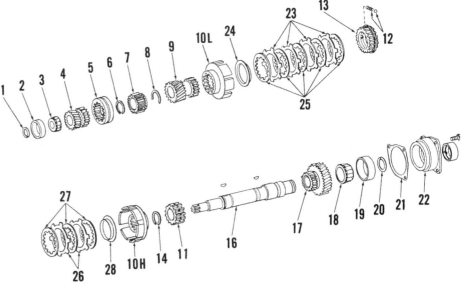

Fig. 188A — Exploded view of transmission pump used on 4640 models with Quad-Range transmission. Refer to Fig. 188 for legend.

drive tangs) to the clutch drum and blocker ring. When engine clutch is disengaged and control lever moved to change gear speeds, the first movement of shifter linkage applies contact pressure to clutch discs causing blocker to try to rotate on drive collar. Drive lugs inside the blocker ring ride up the ramps in drive collar preventing further movement of blocker ring until shaft and gear speeds are equalized. When rotative speeds are equal, thrust force on blocker ring is relieved and synchronizer drum couples gear to shaft without clashing.

189. **INSPECTION AND ASSEMBLY.** Inspect transmission drive shaft for scoring or wear in areas of range pinion rotation and make sure oil passages are open and clean. Refer to Fig. 190B for standard diameters for drive shaft. Inspect drive lugs of blocker rings and friction faces of blocker rings and synchronizer drums. Check synchronizer discs for wear using a micrometer. Renew the entire set if any disc is worn excessively. Standard thickness of internally splined friction discs (25 and 27 – Fig. 190) for all models except 4640 is 0.070 inch. Wear limit is 0.058 inch. Standard thickness of the externally splined friction discs (25 and 27 – Fig. 190A) for 4640 models is 0.078 inch except for the discs between retainers and drums (10L & 24 or 10H & 28). Standard thickness for discs in drums should be 0.120 inch. The two

186. **COUNTERSHAFT.** Refer to Fig. 189 for an exploded view of countershaft and bearings. The shaft is a one-piece unit except for high-speed gear (4). Gear is keyed to shaft and retained by snap ring (3). Countershaft should have 0.001-0.004 inch end play in bearings when properly installed.

187. **TRANSMISSION DRIVE SHAFT.** To disassemble the removed transmission drive shaft, remove snap ring (1 – Fig. 190 or Fig. 190A); then remove rear bearing cone (3) using a suitable press. Lift off reverse range pinion (4) and reverse range shift collar (5).

CAUTION: The four detent balls and springs (12) will be released when blocker is withdrawn. Use care not to lose these parts.

Remove snap ring (6) and use a press or puller to remove drive collar (7). Remove snap ring (8) and withdraw low range pinion (9); then remove high-low synchronizer observing the precautions outlined in note above. High-low range drive collar (11) can be pressed from shaft after removing snap ring (14). High range pinion (17) can be removed after removing high-low range collar (11) or bearing cone (18).

188. SYNCHRONIZER CLUTCHES. The purpose of the synchronizer clutches is to equalize the speeds of the transmission drive shaft and the selected range pinion for easy shifting without stopping the tractor. The synchronizer clutches operate as follows:

The range drive collars (7 or 11-Fig. 190 or Fig. 190A) are keyed to the shaft. Synchronizer clutch drums (10L and 10H) are splined to the range pinions. The blocker ring (13) is centered in drive collar slots by the detent assemblies (12). Synchronizer clutch discs (23 & 25 and 26 & 27) are connected alternately (by

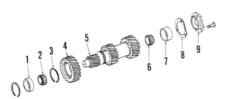

Fig. 189 — Exploded view of transmission countershaft and associated parts.

1. Bearing cup
2. Bearing cone
3. Snap ring
4. Gear
5. Countershaft
6. Bearing cone
7. Bearing cup
8. Shim
9. Bearing retainer

Fig. 190 — Exploded view of transmission drive (input) shaft and associated parts. Refer to Fig. 190A for 4640 models.

1. Snap ring	8. Snap ring	13. High-Low range blocker	21. Shim
2. Bearing cup	9. Low range pinion	14. Snap ring	22. Bearing housing
3. Bearing cone	10. High-Low range drum	16. Drive shaft	23. Plates
4. Reverse range pinion	10H. High range drum	17. High range pinion	24. Retainer
5. Shift collar	10L. Low range drum	18. Bearing cone	25. Discs
6. Snap ring	11. Drive collar	19. Bearing cup	26. Plates
7. Reverse range drive collar	12. Spring and Ball	20. Snap ring	27. Discs
			28. Retainer

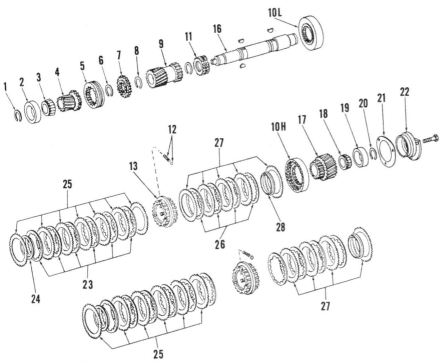

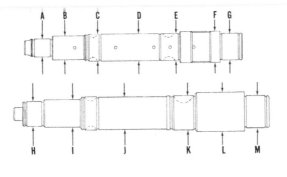

Fig. 190A — Exploded view of transmission drive shaft and associated parts for 4640 models. Refer to Fig. 190 for legend. The stack of clutch discs, separator plates, retainers and blocker shown at bottom is for late models (serial number 006828 and higher). Notice that late models use one less disc (25 and 27) in each clutch. Also, blocker (13) and drums (10H & 10L) are different.

Fig. 190B — Diameters of transmission drive shaft should be as shown. Shaft shown at top is for all models except 4640, lower shaft is for 4640 with Syncro-Range transmission.

A. 1.1260-1.1266 in.
B. 1.4895-1.4905 in.
C. 1.6243-1.6249 in.
D. 1.6805-1.6815 in.
E. 1.6870-1.6880 in.
F. 1.9195-1.9205 in.
G. 1.6260-1.6270 in.
H. 1.6259-1.6266 in.
I. 1.8745-1.8755 in.
J. 2.2021-2.2031 in.
K. 2.2695-2.2705 in.
L. 2.5932-2.5942 in.
M. 2.1265-2.1275 in.

thick discs should not be less than 0.112 inch and thin discs should not be less than 0.060 inch. Standard thickness of plates (23 and 26 – Fig. 190) for all models except 4640 should be 0.048 inch. Standard thickness of plates (23 and 26 – Fig. 190A) for 4640 models should be 0.0619-0.0621 inch. Grooves in disc facings for all models should be more than 0.006 inch deep. Minimum facing thickness should be 0.009 inch. Also check drive tangs for thickening due to peening.

Bushing in housing (22 – Fig. 190 or Fig. 190A) should be 2.255-2.258 inches inside diameter and should be 1/16 inch below front edge of housing.

Refer to Fig. 190C for assembly of discs, plates and retainer on blocker of all models except 4640. Refer to Fig. 190D for assembly of discs, plates and retainer on blocker of 4640 models. Special John Deere Tool-4A (ST-Fig. 191) is required to assemble blocker assembly on drive collar of all models except 4640. Special John Deere Tool-31 (ST – Fig. 191A) is required to assemble 4640 model.

On all models, install the thickest snap ring possible at locations (1 and 20 – Fig. 190 or Fig. 190A).

190. BEVEL PINION SHAFT. Except for bearing cups in housing and rear bearing cone on shaft, the bevel pinion unit is disassembled during removal. Refer to Fig. 192 for exploded view.

All gears should have a diametral clearance of 0.004-0.006 inch on shaft. First-third gear contains a bushing but bushing is not available for service. The bevel pinion shaft is available only as a matched set with the bevel ring gear. Refer to paragraph 225 for information on renewal of ring gear. Refer to paragraph 191 for mesh position adjust-

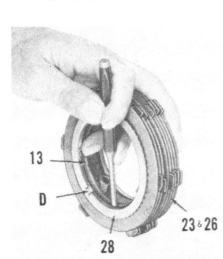

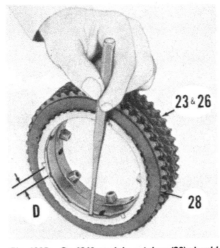

Fig. 190C — On all models except 4640, retainer (28) should be staked to blocker (13) on inside edge in at least four equally spaced locations. Dimple (D) should be aligned with slot in blocker.

Fig. 190D — On 4640 models, retainer (28) should be cleaned and coated with "Loctite" before staking to blocker on inside edge in at least twelve equally spaced locations. Be sure dimple in retainer and notch in blocker are aligned. Do not stake closer than ⅜-inch from dimple as shown at (D).

Fig. 191 — View showing special tool (ST) necessary for assembling blocker for all models except 4640.

ment procedure if bevel gears and/or housing are renewed.

191. PINION SHAFT ADJUSTMENT. The cone point (mesh position) of main drive bevel gear and pinion is adjustable by means of shims (4 – Fig. 192) which are available in thicknesses of 0.003, 0.005 and 0.010 inch. The cone point will only need to be checked if transmission housing or ring gear and pinion assembly are renewed.

The correct cone point of housing and pinion are factory determined. To make the adjustment, refer to appropriate model below.

4040 and 4240 MODELS. Subtract the number etched on end of pinion gear from 8.330. The difference will be the recommended thickness (in inches) of shim pack to be installed.

4440 MODELS. Subtract the number etched on end of pinion gear from 8.326. The difference will be the recommended thickness (in inches) of shim pack to be installed.

4640 MODEL. Subtract the number etched on end of pinion gear from 9.5345. The difference will be the recommended thickness (in inches) of shim pack to be installed.

On all models use a punch and drive out rear bearing cup (3 – Fig. 192) and add or remove shims from shim pack (4) until correct total thickness is obtained.

The bevel pinion bearings are adjusted to a preload of 0.004-0.006 inch by means of shims (16). If adjustment is required, it should be made before installing the gears as follows:

First make sure cone point is correctly adjusted as outlined and that cone (2) is bottomed on pinion shaft shoulder. Install the shaft, washer (15), the removed shim pack (16) plus one 0.010 inch shim and bearing cone (18). If tractor has a four gear pto drive train, install a

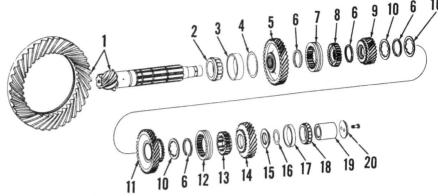

Fig. 192 – Exploded view of bevel pinion shaft and associated parts.

1. Pinion shaft & gear	6. Snap ring (4)	11. 2nd - 5th gear	16. Shim
2. Bearing cone	7. Shift collar	12. Shift collar	17. Bearing cup
3. Bearing cup	8. Rear shifter gear	13. Front shifter gear	18. Bearing cone
4. Shim	9. 6th - 8th gear	14. 4th - 7th gear	19. Spacer (no pto)
5. 1st - 3rd gear	10. Thrust washer (3)	15. Thrust washer	20. Retainer plate

suitable pipe spacer instead of the lower pto idler gear and bearings, install retainer plate (20) and the retaining cap screws. Measure shaft end play using a dial indicator. When disassembling, remove shims equal to the measured end play plus 0.005 inch. Assemble shaft and gears as outlined in paragraph 183.

QUAD-RANGE

Several production changes have occurred which effect service and, where known, these differences will be described. It is especially important to make sure that only correct parts are installed which are compatible with the tractor serial number. Also, be sure that parts for the correct model are installed, because very similar, but different, parts are sometimes used in other 4000 series tractors.

OPERATION

192. The "Quad-Range" transmission consists of an 8-speed mechanically engaged transmission which is similar to the "Syncro-Range" transmission;

however, an additional 2-speed hydraulically shifted planetary unit is installed ahead of the 8-speed transmission.

The two hand levers control speed selection at one of three locations.

The lever furthest to the right, marked A, B, C, D and P moves gears on the bevel pinion shaft to engage one of four reduction ratios or "PARK". This shift CHANGE selector lever should ONLY be moved when the tractor is stopped and the clutch pedal is depressed.

The lever located between the throttle and the shift change selector lever is the RANGE selector lever. Front to rear movement of the range selector lever moves the shift couplings located on the transmission drive shaft to provide two forward reduction ratios and one reverse. Do not attempt to change direction of tractor by moving range shift lever between 1st and Reverse until tractor is stopped and clutch pedal is depressed. The range selector lever can be moved between 2nd and 3rd with tractor moving, but be sure to depress clutch pedal.

The RANGE selector also moves from side to side as it moves from front to rear. This lateral (left to right) movement moves a hydraulic control valve located in the "Quad-Range" housing. When the lever is to the left, hydraulic pressure is used to engage a clutch which results in direct drive. When the lever is to the right, hydraulic pressure engages a brake which causes the planetary gear assembly to drive the tractor at a reduced speed. Lateral movement alone can be accomplished with the tractor moving, and without depressing the clutch pedal. Notice, that moving the range selector between 2nd and 3rd moves the lever front to rear and laterally. Shifting between 2nd and

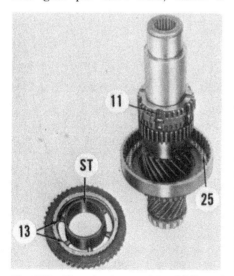

Fig. 191A – View showing special tool (ST) used to load blocker to permit assembly to drive collar for 4640 models.

Fig. 193 – Check bearing preload with dial indicator as shown. Refer to paragraph 191.

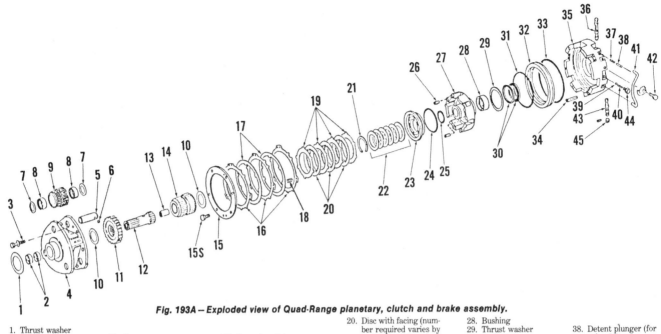

Fig. 193A – Exploded view of Quad-Range planetary, clutch and brake assembly.

1. Thrust washer
2. Bushings
3. Screws
4. Planet carrier
5. Shaft
6. Steel balls (5 mm)
7. Thrust washers
8. Needle bearings
9. Planet pinion
10. Thrust washers

11. Sun gear
12. Input shaft
13. Transmission oil pump drive shaft bushing
14. Clutch shaft sun gear
15. Brake backing plate
16. Disc with facing (number required varies by model & serial number)

17. Separator plates (number required varies by model & serial number)
18. Springs (6 used)
19. Separator plates (number required varies by model & serial number)

20. Disc with facing (number required varies by model & serial number)
21. Snap ring
22. Belleville spring washers
23. Clutch piston
24. Sealing ring
25. Packing
26. Dowel pins (3 used)
27. Clutch drum

28. Bushing
29. Thrust washer
30. Sealing rings
31. Inner packing ring
32. Brake piston
33. Outer packing ring
34. Spring pins (6 used, 5/16 x 1¾ in.)
35. Brake housing
36. Control valve
37. Detent spring

38. Detent plunger (for valve 36)
39. Detent ball (for valve 43)
40. Detent spring
41. Oil tube
42. Cap screw
43. Shift valve
44. Plug (late models)
45. Plug

SPEED SELECTION	QUAD-RANGE PLANETARY		SYNCRO-RANGE TRANSMISSION			CHANGE TRANS.			
	UNDERDRIVE	DIRECT	LOW	HIGH	REVERSE	A	B	C	D
A-1	X		X			X			
A-2		X	X			X			
A-3	X			X		X			
A-4		X		X		X			
B-1	X		X				X		
B-2		X	X				X		
B-3	X			X			X		
B-4		X		X			X		
C-1	X		X					X	
C-2		X	X					X	
C-3	X			X				X	
C-4		X		X				X	
D-1	X		X						X
D-2		X	X						X
D-3	X			X					X
D-4		X		X					X
A-1R	X				X	X			
A-2R		X			X	X			
B-1R	X				X		X		
B-2R		X			X		X		
C-1R	X				X			X	
C-2R		X			X			X	

3rd may be accomplished with tractor moving; however, **be sure** that clutch pedal is depressed.

A mechanical interlock prevents reverse from being engaged when change selector lever is in the "D" position.

The accompanying table shows position of all three sections for each of the 16 forward and 6 reverse speeds.

TESTING

192A. Several production changes have occurred and some effect assembly procedures in such a way that assembly that would be correct on certain models will cause serious malfunction on other models. If trouble occurs following service, be sure that only the correct parts were used and that parts were assembled correctly for the specific model being serviced. One example of incorrect assembly that can cause problems is with the detent ball (39 – Fig. 193B). If the ball and spring is installed under screw (42 – Fig. 193B) as it is on some earlier models, the passage will be blocked and correct operation is impossible.

If problems occur after an extended time of proper operation, first check condition of oil filter and make sure that filter relief valve (Fig. 151) is free and operating properly. Without adequate volume and pressure, the planetary cannot operate properly.

To check system pressure, refer to Fig. 152 and install a 300 psi pressure gage in the plug hole marked "CLUTCH", which is located in the clutch regulating valve housing on left side of tractor. With the engine up to operating temperature and shift lever in **PARK** position, disconnect the planetary shift rod at bellcrank. Run engine at 1900 rpm and check the system pressure in both direct and underdrive by shifting the rod back and forth by hand and observing pressure gage. Indicated pressure should be as follows:

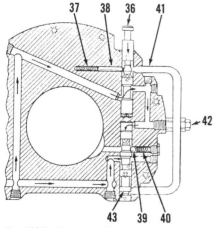

Fig. 193B— Cross-section showing location of detent ball (39) and plunger (38). Refer to Fig. 193A for legend.

4040, 4240 . . . . .175-185 psi
4440 . . . . .170-180 psi
4640 . . . . .155-165 psi

If pressure check reading is 10 psi or lower in direct than in underdrive, the direct clutch circuit is leaking. If underdrive pressure reading is 10 psi or lower than direct, the underdrive circuit is leaking. If both direct and underdrive pressures are low, the "O" rings on the 2-speed planetary adapter tube (Fig. 159A) could be leaking. Check pressure as outlined in paragraph 162 and if necessary, remove pressure regulating valve housing and perform air tests as outlined in paragraph 163.

REMOVE AND REINSTALL

193. To remove the 2-speed planetary unit, tractor must be split between engine and clutch housing as outlined in paragraph 166. Remove the hydraulic lines and operating rods from pressure regulating valve housing. Remove housing and extract all five of the oil passage adapters (Fig. 159A). Refer to Fig. 164 and remove the six outer cap screws. Remove five inner cap screws that hold the 2-speed planetary to the clutch operating housing, and remove housing. If clutch shaft did not come out with housing, pull it out of 2-speed unit. Drive spring pin (Fig. 194) from shifter arm, then remove shifter arm and shaft.

The planetary assembly can be removed without removing pto drive gear and shield. Remove the hex bushing from planetary unit if it did not come out with the clutch shaft. Make a tool to pull the transmission input shaft, using a piece of rod about 36 inches long. Flatten one end of rod and bend end over just enough to form a hook, but small enough to go through shaft. Insert rod, hook the rear of input shaft and pull forward. Move and rotate splines so that shaft will come out easily. Planetary assembly can now be removed by pulling

forward and rotating enough to clear pto gear.

Reinstall planetary assembly in reverse order of disassembly. Before installing assembly, make sure the rear thrust washer is installed. Insert hex bushing in the front of the short transmission input shaft and install shaft and bushing after the planetary assembly has been installed. Rotate gears to align splines if necessary. Complete assembly, and rejoin tractor as outlined in paragraph 166.

OVERHAUL

194. Remove 2-speed planetary unit as outlined in paragraph 193. Place assembly face down on bench and remove six cap screws (3 – Fig. 195) from planetary pinion carrier. Lift pinion carrier off dowel pins. After carrier is removed, the planet gears (9 – Fig. 196) can be removed by pushing planet gear shafts (5) out toward front and catching the retainer balls (6) as shaft starts out of carrier. The sun gear (14)

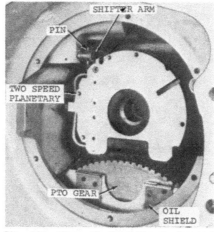

Fig. 194— View of Quad-Range 2-speed planetary with clutch operating housing removed. View is typical of all models except 4640. Refer to Fig. 194A for 4640 model.

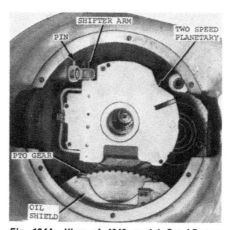

Fig. 194A— View of 4640 model Quad-Range 2-speed planetary with clutch operating housing removed. Notice different shape than other models shown in Fig. 194.

can be removed after planet gears are out of carrier. Inspect bearings, washers and shafts for excessive wear or pitting, and renew as necessary. Remove six cap screws which hold backing plate (15 – Fig. 195) to underdrive brake housing. Remove piston return springs (18 – Fig. 197), center hub and sun gear (14), brake discs (17), separator plates (16), clutch discs (20), plates (19) and direct drive clutch drum (27). Note the thrust washer (29 – Fig. 198) that fits be-

Fig. 195 -- The six screws (3) should be tightened to 20 ft.-lbs. torque, screws (15S) to 21 ft.-lbs. torque and screw (42) to 21 ft.-lbs. of torque.

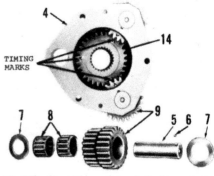

Fig. 196 – Do not lose retainer balls (6) as shafts (5) are withdrawn.

Fig. 197 – View of brake housing with backing plate removed. Discs and plates are installed alternately at (16 & 17) and (19 & 20).

tween clutch drum and brake housing. Remove brake piston (32) with its inner and outer seals. The clutch piston spring washers (22–Fig. 199) must be compressed using JDT-24 compression tool or other suitable means, before snap ring (21) can be removed. Disassemble the shift valve and control valve using Fig. 193A and Fig. 200 as a guide. Air pressure may be used to blow out the retaining plug (45) after retaining pin has been driven out. Use care to avoid losing the detent spring (39) and ball (40). A special plug (44–Fig. 200) retains the detent ball (29) and spring (40) in a separate bore.

Spring (40) should have free length of 1 7/8 inches for all models except 4640. The spring used on 4640 models is different than used on other models.

Inspect control valve, shift valve and their bores. Any nicks or scores must be smoothed on valves or in bores to insure free movement of valves. Inspect all parts for wear, and pay particular attention to the needle bearings and their retainers. Bushings should be renewed if not in excellent condition. The minimum

free height of each Belleville spring washer (22–Fig. 193A) should be 0.126 inch.

To reassemble planet gears, install sun gear in carrier, making sure the timing "V" marks point toward the three pinion shaft bores (Fig. 196). Install the three planet gears, with timing marks aligned with marks on sun gear. Thrust washer must be in place. Install the balls in pinion shafts and recheck timing marks. Install the packing ring (25–Fig. 201) in clutch drum (27) and sealing ring (24) on clutch piston (23).

Install piston in drum and assemble spring washers (22) facing each other as shown, so that the inner diameter of the end washers will contact the piston (23) and snap ring (21). Be sure that 6 Belleville spring washers are installed on all models except 4640 which should have 8 springs washers installed. Compress springs with compression tool (Fig. 199) and install snap ring. Lubricate all parts before assembling. Install brake piston seals, then use JDT-21-1 outer sleeve (Fig. 202) and JDT-21-2 inner sleeve to install brake piston on all models except 4640. On 4640 models use JDT-23-1 and JDT-23-2 sleeves to install piston.

Install shift valve (43–Fig. 200), retainer (45) and retaining pin. Insert 1-5/16 inch long spring (37) and plunger (38) through bore for tube (41) and across bore for control valve. Compress spring by pressing on plunger (38) with a small rod, then insert control valve (36) into bore with flats on valve toward detent plunger. Insert detent ball (39) and spring (40) into third hole from bottom and install plug (44).

Install tube (41–Fig. 200), clamp and screw (42). Tighten screw (42) to 21 ft.-lbs. torque.

When installing clutch drum in brake housing, be sure the sealing rings and thrust washer are in place (Fig. 198). Note that brake housing has three slots with spring pins, and three slots without pins (Fig. 202). Tangs on first separator plate (16–Fig. 193A) installed must be

installed on pins to provide a seat for piston return springs (18). Install one brake disc (17), then install remaining separator plates (16) and discs (17) alternately. Tangs on remaining separator plates go into the slots without pins.

4040

Brake separator plates (16)–
  Number used 3
  Thickness, new ....0.085-0.095 in.
Brake discs (17)–
  Number used 2
  Thickness, new ....0.087-0.093 in.
  Minimum facing groove
    depth ....0.010 in.
Brake separator springs (18)–
  Free length ....0.71 in.
  Working load ....24-30 lbs. at 0.52 in.
Clutch separator plates (19)–
  Number used 4
  Thickness, new ....0.055-0.065 in.
Clutch discs (20)–
  Number used 3
  Thickness, new ....0.071-0.075 in.
  Minimum facing groove
    depth ....0.006 in.
Spring washers (22)–
  Number used 6
  Minimum free height ....0.126 in.

4240

Brake separator plates (16)–
  Number used 3
  Thickness, new ....0.085-0.095 in.
Brake discs (17)–
  Number used 2
  Thickness, new ....0.087-0.093 in.
  Minimum facing groove
    depth ....0.010 in.

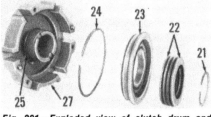

Fig. 201–*Exploded view of clutch drum and associated parts. Assemble spring washers (22) as shown, so end washers will contact piston (23) and snap ring (21) with the inner diameter.*

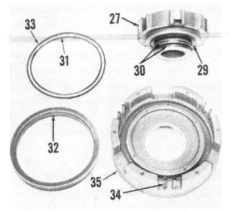

Fig. 198–*Brake housing (35) with direct drive clutch drum (27) and brake piston (32) removed. Note thrust washer (29) and pins (34).*

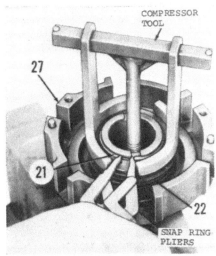

Fig. 199–*Compress Belleville spring washers (22) with suitable tool to remove snap ring (21).*

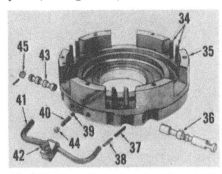

Fig. 200–*View of housing with valves and detent assemblies removed. Ball and spring (39 & 40) are located below special plug (44). Plunger and spring (37 & 38) enter bore for tube (41) and are located across bore for valve (36) as shown in cross-section Fig. 193B.*

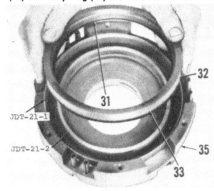

Fig. 202–*Use the proper sleeve identification number when installing brake piston. See text for model application.*

Brake separator springs (18)–
Free length . . . . 0.71 in.
Working load . . . . 24-30 lbs. at 0.52 in.
Clutch separator plates (19)–
Number used 4
Thickness, new . . . . 0.055-0.065 in.
Clutch discs (20)–
Number used 3
Thickness, new . . . . 0.071-0.075 in.
Minimum facing groove
depth . . . . 0.006 in.
Spring washers (22)–
Number used 6
Minimum free height . . . . 0.126 in.

4440
Brake separator plates (16)–
Number used 3
Thickness, new . . . . 0.085-0.095 in.
Brake discs (17)–
Number used 2
Thickness, new . . . . 0.087-0.093 in.
Minimum facing groove
depth . . . . 0.010 in.
Brake separator springs (18)–
Free length . . . . 0.71 in.
Working load . . . . 24-30 lbs. at 0.52 in.
Clutch separator plates (19)–
Number used 4
Thickness, new . . . . 0.055-0.065 in.
Clutch discs (20)–
Number used 3
Thickness, new . . . . 0.071-0.075 in.
Minimum facing groove
depth . . . . 0.006 in.
Spring washers (22)–
Number used 6
Minimum free height . . . . 0.126 in.

4640
Brake separator plates (16)–
Number used 4
Thickness, new . . . . 0.085-0.095 in.
Brake discs (17)–
Number used 3
Thickness, new . . . . 0.087-0.093 in.
Minimum facing groove
depth . . . . 0.010 in.

Brake separator springs (18)–
Free length . . . . 0.96 in.
Working load . . . . 24-30 lbs. at 0.70 in.
Clutch separator plates (19)–
Number used 5
Thickness, new . . . . 0.055-0.0065 in.
Clutch discs (20)–
Number used 4
Thickness, new . . . . 0.071-0.075 in.
Minimum facing groove
depth . . . . 0.006 in.
Spring washers (22)–
Number used 8
Minimum free height . . . . 0.126 in.

Install the six brake piston return springs (18) on pins. Tighten cap screws (15S) to 21 ft.-lbs. torque. When the planet carrier assembly is installed onto brake housing, make sure the planet pinion shafts remain in place. Tighten screws (3) to 20 ft.-lbs. torque.

Use compressed air to check action of both apply pistons. Direct approximately 50 psi to the pressure passage while moving control valve in and out. The brake port is located at (B–Fig. 202A) and brake should engage when control valve (36) is pushed in. The oil port (C) located diagonally across housing is for the clutch which should engage when control valve (36) is pulled. Make sure the rear thrust washer (1–Fig. 193A) is in place on planet gear assembly and reinstall the unit as outlined in paragraph 193.

POWER SHIFT TRANSMISSION

The power shift transmission is the only transmission available for 4840 model and is available on all other models. The power shift transmission provides 8 forward speeds and 4 reverse speeds. Gear changes are accomplished by moving a shift lever and gears can be changed without stopping tractor or operating the foot controlled feathering valve (inching pedal).

OPERATION

Power Shift Models

195. **POWER TRAIN.** The power shift transmission is a manually controlled, hydraulically actuated planetary transmission consisting essentially of a clutch pack and planetary pack as shown schematically in Fig. 203.

Hydraulic control units consist of three clutch packs (C1, C2 & C3) and

four disc brakes (B1 through B4). In addition, a multiple disc clutch (pto) is housed in the clutch pack and used in the pto drive train. All units are hydraulically engaged, and mechanically disengaged when hydraulic pressure to that unit is interrupted. The power train also contains a non-release single disc transmission clutch mounted on engine flywheel, a foot operated inching pedal, a mechanical disconnect for towing and a park pawl.

Three hydraulic control units are engaged for each of the forward and reverse speeds. In 1st speed, Clutch 1 is engaged and power is transmitted to the front planetary unit by the smaller input sun gear (C1S); Brake 1 is engaged, locking the front ring gear to housing, and the planet carrier walks around the ring gear at its slowest speed. Clutch 3 is also engaged, locking the rear planetary unit, and output shaft turns with the planet carrier. Second speed differs from 1st speed only by disengaging Brake 1 and engaging Brake 2, causing planet carrier and output shaft to rotate at a slightly faster speed.

Third speed and 4th speed are identical to 1st and 2nd except that Clutch 1 is disengaged and Clutch 2 engaged, and power enters the front planetary unit through the larger input sun gear (C2S).

Fifth speed and 6th speed differ from 3rd and 4th speeds in the rear planetary unit. Clutch 3 is disengaged and Brake 4 engaged, and the output shaft turns faster than the planet carrier through the action of the rear planet pinions and output sun gear.

In 7th and 8th speeds, both Clutch 1 and Clutch 2 are engaged, locking the in-

Fig. 202A – Views of housings showing location of ports for checking operation of brake (B) and clutch (C). Unit on left is type used on models except 4640. Housing on right is used on 4640 model.

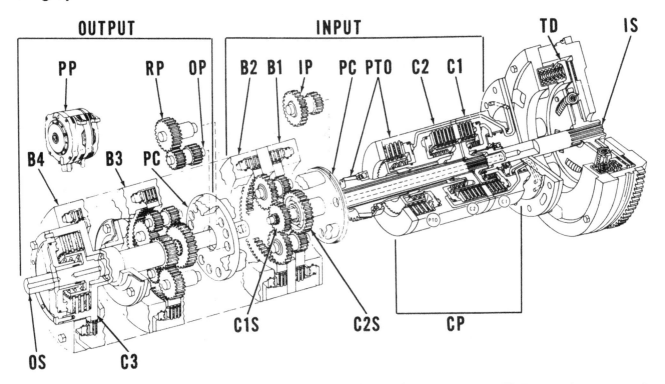

Fig. 203 — Schematic view of Power Shift Transmission showing primary function of units. Torsional damper (TD) is a non-release type clutch and pressure plate assembly. The power take-off clutch and drive gear are located in, but are not part of, the transmission power train.

B1. Brake 1	B4. Brake 4	C3. Clutch 3	CP. Clutch pack	OP. Output planet pinion	PP. Planet pack
B2. Brake 2	C1. Clutch	C1S. C1 sun gear	TD. Torsional damper	OS. Output sun gear	PTO. Power take-off drive
B3. Brake 3	C2. Clutch 2	C2S. C2 sun gear	IP. Input planet pinion	PC. Planet carrier	RP. Reverse pinions

put planetary unit, and planet carrier turns with input shaft at engine speed. Engaging the three clutch units locks both planetary units, therefore 7th speed is a direct drive, with transmission output shaft turning with, and at the same speed as, the engine. Eighth speed is an overdrive, with transmission output shaft turning faster than engine speed.

Reverse speeds are obtained by engaging Brake 3, which locks the output planetary ring gear to housing, and the output shaft turns in reverse rotation through the action of the two sets of output planetary pinions (RP & OP).

It will be noted that the front planetary unit is an input unit controlled by the two front clutch units in clutch pack and two front brake units in planetary pack. Two input control units must be engaged to transmit power, and five input speeds are obtained by selectively engaging the input brake and clutch units.

The rear planetary unit is an output unit controlled by the two rear brakes and rear clutch. One of the rear control units must be engaged to complete the power train. Two forward ranges and one reverse output range are provided, depending on which rear control unit is engaged.

The accompanying table lists the control units actuated to complete the power flow in each shift position:

	Front (Input) Control Units		Rear (Output) Control Unit
	Forward Speeds		
1st	C1	B1	C3
2nd	C1	B2	C3
3rd	C2	B1	C3
4th	C2	B2	C3
5th	C2	B1	B4
6th	C2	B2	B4
7th	C1	C2	C3
8th	C1	C2	B4
	Reverse Speeds		
1st	C1	B1	B3
2nd	C1	B2	B3
3rd	C2	B1	B3
4th	C2	B2	B3

196. **CONTROL SYSTEM.** The control valve unit consists of manually actuated speed selector and direction selector valves which operate through four hydraulically controlled shift valves to engage the desired clutch and brake units. The valve arrangement prevents the engagement of any two opposing control units which might cause transmission damage or lockup.

Power to operate the transmission system is supplied by an internal gear hydraulic pump mounted on the transmission input shaft, which also sup-plies the charging fluid for the tractor main hydraulic system. Fluid from the hydraulic pump first passes through a full flow oil filter to the main transmission oil gallery, where the pressure is regulated at approximately 175 to 195 psi for the transmission control functions. Excess oil passes through the regulating valve to the oil cooler and main hydraulic pump.

Fluid from the transmission main oil gallery is routed through the inching pedal valve to Clutches 1 and 2; and through a spring-loaded accumulator to the brake actuating pistons and to Clutch 3.

Refer to Fig. 204 for a schematic view of control circuits. The direction selector valve and shift valve 4 controls the routing of pressure to the output control units (Clutch 3, Brake 3 & Brake 4). The speed selector valve contains four pressure ports (1, 2, 3 & 4) which control the movement of the four shift valves by pressurizing the closed end (opposite the return spring) when port is open to pressure. Neutral position is provided by the selector valves or by depressing the inching pedal.

When the direction selector valve is moved to the forward detent position, system pressure is routed in Shift Valve 4. In the low range positions (1st through 4th gears), the speed selector valve charging port to the top of Shift

Valve 4 is open to pressure, Shift Valve 4 moves downward against spring pressure and Clutch 3 is actuated. In the high range positions (except 7th gear which is direct and uses all three clutch units), charging pressure is cut off to Shift Valve 4, shift valve return spring moves valve upward and Brake 4 is actuated. When the direction selector valve is moved to reverse detent position, system pressure by-passes Shift Valve 4 and is routed directly to Brake 3. In the neutral detent position, system pressure is cut off from all three output control units.

Shift Valves 1, 2 & 3 direct system pressure to the input control units (Clutches 1 & 2 and Brakes 1 & 2). Shift Valve 1 directs pressure to Clutch 2 when hydraulically actuated and to Clutch 1 when charging port to top of Valve 1 is closed.

Shift Valve 2 routes pressure to Shift Valve 3 when hydraulically actuated, and permits simultaneous engagement of Clutches 1 and 2 when charging port to top of Valve 2 is closed.

Shift Valve 3 directs pressure to Brake 2 when hydraulically actuated and to Brake 1 when charging port to top of Valve 3 is closed.

197. LINKAGE ADJUSTMENT. To adjust the shift control linkage, refer to Fig. 206 and disconnect yoke (2) from the direction selector arm (4). Move shift lever to "Reverse-N" position in the guide plate as shown at (1–Fig. 206A). Move the direction selector arm (4) down to the reverse position, then adjust cable length by turning yoke (2) until hole in yoke and arm align. Be sure that control lever (1) remains in "Reverse-N" position while adjusting. Reconnect yoke (2) to arm (4), but do not tighten yoke locknut (3). Move the control lever (1) to the center between "Reverse" and "Forward" positions and check location of the direction selector arm (4). If selector arm is correctly located at mid (neutral) position, tighten locknut (3) to retain cable yoke position. If incorrect recheck cable yoke adjustment.

Disconnect yoke (6–Fig. 206) from speed selector arm (5). Hole in yoke and hole in arm should align when arm is in uppermost ("P"–Fig. 206A) position and control lever (1) in center neutral position.

Disconnect swivel (12–Fig. 206B) from bellcrank (13). Rotate lock cam (10–Fig. 206) clockwise until tight, then adjust swivel (12–Fig. 206B) until it aligns with hole in bellcrank (B). Attach swivel to bellcrank with swivel at upper limit of hole in bellcrank. Move the speed control lever to "Park" position, then return lever to "Neutral". Check clear-

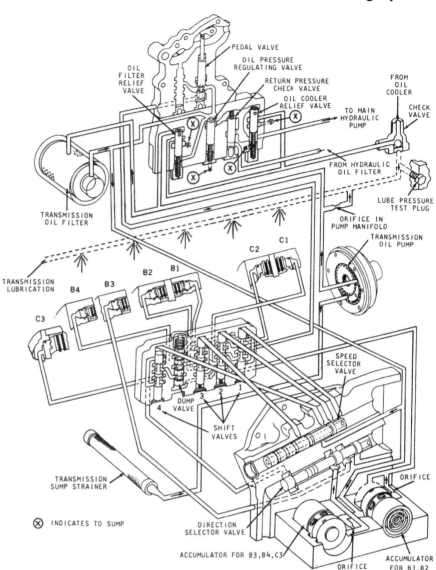

Fig. 204—Schematic view of the oil control circuits, valves and accumulators in the Power Shift transmission.

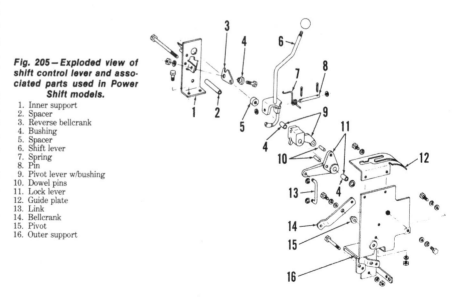

Fig. 205—Exploded view of shift control lever and associated parts used in Power Shift models.

1. Inner support
2. Spacer
3. Reverse bellcrank
4. Bushing
5. Spacer
6. Shift lever
7. Spring
8. Pin
9. Pivot lever w/bushing
10. Dowel pins
11. Lock lever
12. Guide plate
13. Link
14. Bellcrank
15. Pivot
16. Outer support

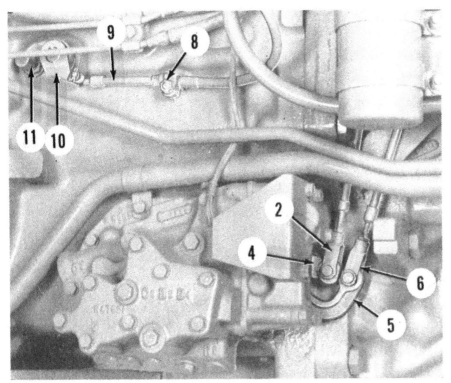

Fig. 206 — View of right side showing Power Shift adjustment points. Refer to text for adjustment procedure.

2. Cable yoke	5. Speed selector arm	8. Cable clamp	10. Lock cam arm
4. Direction selector arm	6. Cable yoke	9. Cable	11. Shifter yoke shaft end

Fig. 206B — Refer to text for adjusting position of swivel (12) on cable.

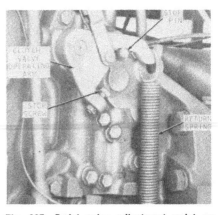

Fig. 207 — Pedal valve adjustment point on Power Shift models.

ance between lock cam arm (10–Fig. 206) and disconnect shifter yoke shaft end (11). Clearance should be less than 0.080 inch. If clearance is excessive, adjust swivel (12–Fig. 206B) to a higher position, then recheck.

After adjusting, recheck "Neutral" and "Park" positions, then check for correct operation in all gear positions.

To adjust pedal valve linkage, refer to Fig. 207. Remove clutch pedal return spring. Loosen locknut on stop screw and run screw in. Turn valve operating arm fully counter-clockwise until valve bottoms and adjust stop screw until screw head touches stop pin. Back out one-half turn and tighten locknut.

To adjust pedal height, measure from footrest to the lowest part of pedal. Distance should be 5¼-5¾ inches (Fig. 208). Adjust by removing hex adapter from pedal, rotate until the desired position is obtained and reinstall adapter.

198. PRESSURE TEST AND ADJUSTMENT. Before checking the transmission operating pressure, first be sure that transmission oil filter is in good condition and that oil level is at top of "SAFE" mark on dipstick. Place towing disconnect lever in "TOW" position, start engine and operate at 1900 rpm until transmission oil is at operating temperature. If tractor is equipped with Power Front Wheel Drive, put main hydraulic pump out of stroke by turning in the shut-off screw.

Stop engine and install a 0-300 psi pressure gage in "SYSTEM PRESSURE" plug hole (Fig. 209). Gage should register 175-195 psi with engine operating at 2000 rpm and speed control lever in any position. If pressure is not as indicated, remove plug and add or remove shims (12–Fig. 219) located between plug and spring, until pressure is correct.

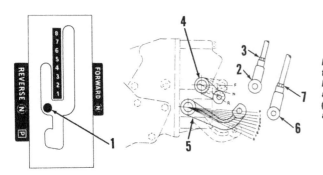

Fig. 206A — Drawing of control quadrant and control lever (1), direction selector arm (4), speed selector arm (5), connecting yokes (2 & 6) and locknuts (3 & 7). Refer to text.

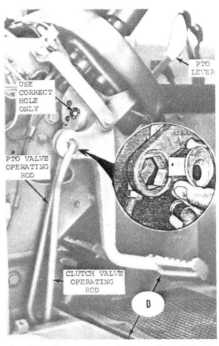

Fig. 208 — Remove hex adapter and rotate to obtain pedal height (D) of 5¼-5¾ inches.

If pressure cannot be adjusted, other possible causes are:

1. Incorrect pedal valve linkage adjustment, refer to paragraph 197.

2. Malfunctioning regulator valve, pedal valve or oil filter relief valve; overhaul as outlined in paragraphs 209 or 210.

If adjustment of operating pressure does not correct the malfunction, check pressures as outlined in paragraph 199.

TROUBLESHOOTING

Power Shift Models

199. **PRESSURE TEST.** To make a complete check of transmission hydraulic system pressures, first check and adjust operating pressure as outlined in paragraph 198, then proceed as follows:

Early housings have four test points plus the system pressure test point as shown in Fig. 209. To test, connect four 0-300 psi pressure gages in place of plugs.

Later housings have seven test points plus the system pressure test point as shown in Fig. 209A. To test, seven 0-300 psi pressure gages can be attached to test ports, or four gages can be attached as described for earlier models. The seven gage test permits easier isolation of circuits for troubleshooting.

On all models, operate engine at 2000 rpm and allow fluid to obtain normal operating temperature. Move towing disconnect lever to "TOW" position. Move shift lever to each position listed in following table and observe pressures listed.

Park			B2	
Neutral		B1		
Neutral Fwd.		B1		C3
1st Fwd.	C1	B1		C3
2nd Fwd.	C1		B2	C3
3rd Fwd.		C2 B1		C3
4th Fwd.		C2	B2	C3
5th Fwd.		C2 B1		B4
6th Fwd.		C2	B2	B4
7th Fwd.	C1 C2	*	*	C3
8th Fwd.	C1 C2	*	*	B4
Neutral Rev.		B1	B3	
1st Rev.	C1	B1	B3	
2nd Rev.	C1		B2 B3	
3rd Rev.		C2 B1	B3	
4th Rev.		C2	B2 B3	

On four gage hookup, 7th and 8th speed check shows pressure on gage at B1, B2 pressure port but elements are not engaged since pressure is stopped at shift valve.

If pressure at C1 or C2 is lower than 20 psi less than system pressure, leakage is indicated. In normal operation, C1 and C2 pressure is 15-20 psi less than system pressure. Leakage is also indicated if pressure at B1, B2, B3, B4 or C3 is lower than 10 psi less than system pressure.

With shift lever in any forward or reverse gear and engine speed at 2000 rpm, depress the inching pedal while noting C1 or C2 pressure gage reading. Gage pressure on either or both gages showing pressure should drop to zero with pedal fully depressed.

Release the pedal slowly; gage pressure should rise at a smooth, even rate until approximately 115 psi is registered with pedal ½ to 1 inch from top; then move quickly to operating pressure with further pedal movement.

Failure to perform as outlined would indicate maladjustment of pedal valve linkage (see paragraph 197) or malfunction of the valve (overhaul as outlined in paragraph 209).

200. **LUBRICATION PRESSURE TEST.** Before overhauling or repairing the Power Shift transmission, a lubrication pressure test can help isolate the problem, and prove the results after repair.

The lubrication pressure plug to be removed for the test is located in clutch housing on upper right side, just to the right of oil return line. It will be necessary to remove the footrest panel to reach the plug for this test. Install a 0-60 psi gage that does **NOT** contain a dampener orifice. In PARK position and with engine operating at 2000 rpm and oil temperature at 100 degrees F., pressure should be 26 psi. With oil temperature at 150 degrees F., pressure should be 22 psi. Excessive pressure indicates a blockage in the lube circuit. If pressure is low, put the main hydraulic pump out of stroke by turning in the shut-off screw. If pressure comes up to normal, there is a leak in the hydraulic system. Next run engine at a slow enough speed to get a reading of 10 psi. Place the towing lever in TOW position. Shift the transmission through the positions shown in table in paragraph 199 while observing pressure gage. Pressure should drop momentarily, then come back to within 3 psi of original value. If pressure fails to return at any selected speed, depress clutch pedal to cut off pressure to C1 and C2. If pressure comes back up, a clutch unit is leaking. If it does not come up with clutch depressed, the planetary pack is leaking. Test in several speeds to pinpoint leaking unit while consulting shift table.

Engage pto clutch and observe gage. If pressure fails to return to within 3 psi of original pressure, the pto clutch is leaking or the valve is out of adjustment. Refer to paragraph 206 for overhaul of pto clutch or paragraph 258 for pto valve adjustment.

201. **BEHAVIOR PATTERNS.** Erratic behavior patterns can be used to pinpoint some systems malfunctions.

ODD SHIFT PATTERN. If tractor slows down when shifted to a faster speed; speeds up when shifted to a slower speed; or fails to shift when selector lever is moved; a sticking shift valve is indicated. Refer to paragraphs 196 and 199. Overhaul the control valve as outlined in paragraph 211.

SLOW SHIFT. Possible causes are; improper regulating valve adjustment; improper pedal linkage adjustment; plugged fluid filter; malfunctioning regulating valve, pedal valve or oil filter relief valve; broken accumulator spring; sticking accumulator piston; or slipping clutch or brake unit or units.

ROUGH PEDAL ENGAGEMENT. If tractor jumps rather than starts smoothly when pedal valve is actuated, a sticking clutch pedal valve or broken pedal valve spring is indicated.

SLIPPAGE UNDER LOAD. If transmission slips, partially stalls or stalls

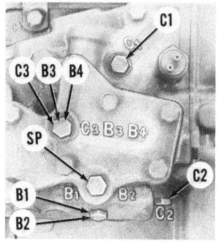

Fig. 209—View of early control valve housing showing pressure gage points. Later type housing can be checked using four gages connected to these ports. Refer to Fig. 209A for later housing.

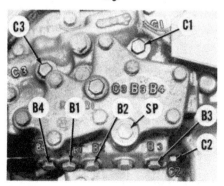

Fig. 209A—Later housing has ports for checking each of the seven points as shown. Four gages may be connected to ports shown in Fig. 209 if desired.

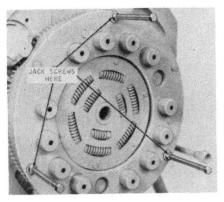

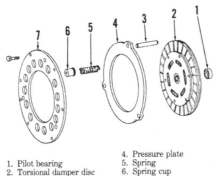

1. Pilot bearing
2. Torsional damper disc
3. Pin (3 used)
4. Pressure plate
5. Spring
6. Spring cup
7. Cover

Fig. 211 — Install jack screws ONLY in holes shown.

under full load, first check the adjustment of pedal valve as outlined in paragraph 197, then check transmission pressures as outlined in paragraphs 198 and 199. If trouble is not corrected, one of the clutch units; or torsional damper is malfunctioning. Refer to paragraph 212 for torsional damper overhaul.

If a clutch or brake unit is suspected of slipping, it will be necessary to determine which of the three units is at fault in that speed. Refer to the table in paragraph 199 to determine which three units are involved in the speed range in question. Then prove one unit at a time by choosing a speed that utilizes that particular unit, and if that speed does not slip, choose another speed that changes only one unit whenever possible. In this way the slipping unit can be isolated.

NOTE: 4th to 5th and 5th to 4th change two units at once, as do 6th to 7th and 7th to 6th.

If clutch C1 or C2 is suspected, the inching pedal can be used to determine if either clutch is bad. Since the only difference between 2nd and 4th speed is the change from C1 to C2, these two speeds can be used to isolate the faulty unit. (If either clutch is slipping, it will also slip in 7th and 8th speeds since both clutches are applied in both speeds). Place shift lever in speed desired and allow brake and clutch units time to

engage before releasing inching pedal.

If one or more units are found to be slipping in every gear in which the unit is engaged, remove and overhaul the transmission as outlined in paragraphs 204 through 216.

TRACTOR FAILS TO MOVE. If tractor fails to move when transmission is engaged, first check to see that tow disconnect is fully engaged. If tow disconnect unit is engaged, check to see that park pawl operates properly and is correctly adjusted. Park pawl is engaged by cam action and disengaged by a return spring. If spring breaks or becomes unhooked, pawl may remain engaged even though linkage operates satisfactorily. To examine or renew the park pawl return spring, remove transmission housing cover as outlined in paragraph 217.

Trouble could be caused by low pressure and/or excessive leakage. Refer to paragraph 198 for checking system pressure, paragraph 199 for checking shift pressures and paragraph 200 for checking lubrication pressures. Another possible cause is bent, broken or disconnected linkage.

TRACTOR CREEPS IN NEUTRAL. A slight amount of drag is normal in the clutch and brake units, especially when transmission oil is cold. Excessive creep can also be caused by warped (clutch or brake) plates or fragments of parts causing clutches to partially engage. Observe the following.

If tractor creeps when inching pedal is depressed and properly adjusted, either Clutch 1 or Clutch 2 is malfunctioning. Check as follows: With engine speed at 1500 rpm and transmission fluid at operating temperature shift to 2nd speed on a flat surface. Depress the inching pedal; if tractor continues to roll forward at approximately the same speed, Clutch 1 is malfunctioning, if tractor speed increases, Clutch 2 is dragging.

Place shift lever in NEUTRAL position. Disconnect yoke from direction selector arm on transmission (Fig. 206). This will leave the output section of transmission in neutral. Shift into 1st forward with shift lever, but do not depress clutch pedal. If tractor creeps forward with throttle set at 1500 rpm and transmission oil at operating temperature, Clutch 3 or Brake 4 is dragging; if tractor creeps backward, Brake 3 is malfunctioning.

NOTE: Dragging clutch or brake units, aside from causing creep, will contribute to loss of power, heat and excessive wear. Creep is merely an indication of possibly more serious trouble which needs to be corrected for best performance, or to prevent future failure.

REMOVE AND REINSTALL

Power Shift Models

202. **PEDAL AND REGULATING VALVES.** Pedal valve housing and regulating valve housing attach to left side of clutch housing using common gaskets and gasket plate. Housings may be removed separately, but both should be removed to renew the gaskets.

To remove the housings, remove left battery and battery box if necessary. Remove pedal return spring and disconnect rod from valve operating arm. Disconnect lube pipe, inlet pipe and outlet pipe from housings. Disconnect pto valve operating rod at clutch housing, then unbolt and remove the housings, gasket plate and gaskets.

Overhaul the pedal valve housing as outlined in paragraph 209 and regulating valve housing as in paragraph 210.

When installing, use light, clean grease to position gaskets and gasket plate, making sure gaskets are installed on proper sides of plate as shown in Fig. 219. Install regulator valve housing and

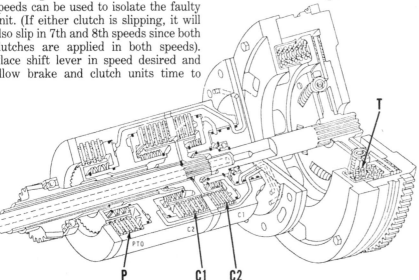

Fig. 212 — Cross-section showing early type torsion damper (T), front clutch (C1), second clutch (C2), pto clutch (P) and associated parts. Other models are similar.

retaining cap screws, then install pedal valve housing. Tighten retaining cap screws evenly and securely, and complete the assembly by reversing the disassembly procedure. Adjust as outlined in paragraphs 197 and 198.

203. CONTROL AND SHIFT VALVES. To remove the control and shift valve housing, drain transmission and remove right battery and battery box if necessary. Disconnect wiring to start-safety switch and remove the cotter pins which retain control cable yokes to control arms. Remove the retaining cap screws, then remove valve housing, being careful not to lose the detents and springs (14, 15 & 16 – Fig. 220).

Overhaul the removed unit as outlined in paragraph 211 and install by reversing the removal procedure. Make sure the inner and outer gaskets are in proper order since they are not interchangeable. Tighten retaining cap screws evenly and securely. Adjust as outlined in paragraph 197.

204. TRACTOR SPLIT. To obtain access to engine flywheel, torsional damper, pto drive gears, or power shift transmission main components, it is first necessary to detach (split) clutch housing from engine block; proceed as follows:

Drain cooling system and remove hood and muffler. Remove side shields, grille screens and right and left cowl side covers. Remove step plates and battery boxes.

Discharge brake accumulator by opening the right brake bleeder screw and holding brake pedal down for a few moments.

If equipped with air conditioning, use two wrenches to break the two connections on the lines on left side of tractor, just below cab. If gas can be heard leaking, tighten and reloosen connection until coupler will part without leaking. Cap connections to keep out dirt.

Disconnect throttle rod, wiring, hydraulic lines, tachometer cable and

temperature indicator sending unit. Be sure to cap all disconnected hydraulic fittings to prevent dirt entry. If tractor is equipped with front end weights, remove the weights. If equipped with Power Front Wheel Drive, disconnect inlet and outlet oil pipes. Support engine and transmission securely and separately, remove the connecting cap screws and roll transmission assembly rearward away from engine.

To attach, reverse the above procedure and tighten the connecting cap screws to a torque of 170 ft.-lbs. on ⅝-inch screws, or 300 ft.-lbs. on ¾-inch screws.

205. TORSIONAL DAMPER. Power Shift tractors are equipped with a non-release type clutch pressure plate, and a disc assembly with springs mounted radially around the disc hub, which absorbs torsional loads in the same manner as does a conventional dry clutch. This action softens the shock of initial engagement at a standstill, and when speed changes are made while in motion.

To remove torsional damper, detach clutch housing from engine as outlined in paragraph 204. The pressure plate has spring retainers protruding through the cover (Fig. 211) and is equipped with single springs.

Remove the three cover cap screws which are located at the center of the three spring retainer groups. Install three jack screws ONLY IN HOLES SHOWN in Fig. 211 to prevent warping the cover by uneven pressure, which could happen if jack screws were installed in holes *between* spring retainer groups. Remove remaining cap screws after locknuts are tightened on jack screws. Cover, springs and retainers can be removed from flywheel by backing off the three locknuts to relieve spring pressure. (No further disassembly is necessary). Reinstall in reverse order of removal, with a suitable aligning tool in disc and short side of hub out. Install jack screws, run the nuts down until three pressure plate cap screws can be installed, then remove jack screws and install remaining cap screws. Tighten to 35 ft.-lbs. torque.

206. CLUTCH PACK AND TRANSMISSION PUMP. The transmission pump and clutch pack can be removed as a unit after draining hydraulic system and detaching clutch housing from engine as outlined in paragraph 204.

Overhaul the removed clutch pack and pump as outlined in paragraphs 213 and 214.

To assist in easier installation of clutch pack, use alignment studs in the two side holes of housing and position gasket using light grease. Make sure oil

passages in clutch housing and gasket are properly aligned. Insert connecting shafts (71 and 72 – Fig. 222) as shown in Fig. 213. Install clutch pack with oil passages in mounting flange aligned with those of gasket and clutch housing. Tighten cap screws to 50 ft.-lbs. torque.

207. CLUTCH HOUSING. The clutch housing must be removed for access to pto drive gear train or removal of planetary unit. To remove the clutch housing, first split tractor between engine and clutch housing as outlined in paragraph 204 and remove clutch pack as in paragraph 206.

Remove Sound-Gard body, or 4 post Roll-Gard if so equipped. Remove batteries or battery boxes, differential lock pedal and front platform.

Mark and diagram location of hydraulic tubing if necessary, then remove hydraulic tubes and system control linkage. Remove rockshaft housing from tractor as outlined in paragraph 282.

Remove the snap ring (Fig. 213) securing pto clutch gear bearing to bore and withdraw clutch gear and bearing as a unit.

NOTE: A slide hammer may be required to remove the gear.

Support clutch housing and steering support assembly from a hoist and remove clutch housing flange cap screws. Two upper, center screws are accessible through inside front of clutch housing as shown in Fig. 213. Pry clutch housing from its doweled position on transmission case and swing housing away from rear unit.

Use new gasket and "O" rings when reinstalling clutch housing. "O" rings may be held in position with grease. Tighten ⅝-inch cap screws to a torque of 170 ft.-lbs. and ¾-inch cap screws to 300 ft.-lbs.

208. PLANETARY PACK. To remove the transmission planetary pack, first detach (split) engine from clutch

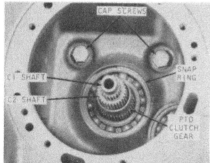

Fig. 213 – View of clutch housing with clutch pack removed. Snap ring, pto clutch gear and the two hidden cap screws must be removed before clutch housing can be separated from transmission.

Fig. 214 – Loosen cap screws to load control arm support for more wrench clearance to C3 oil pipe fitting. Refer to Fig. 215 for 4640 and 4840 models.

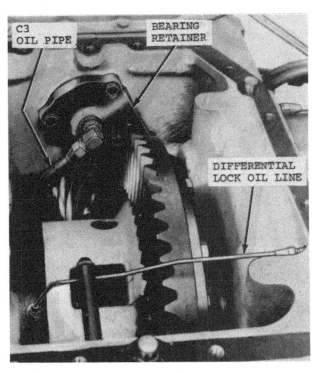

Fig. 215 — View showing location of C3 oil pipe bearing retainer and differential lock oil line for 4640 and 4840 models.

housing as outlined in paragraph 204; remove clutch pack as in paragraph 206 and clutch housing as in paragraph 207. Remove rockshaft housing as in paragraph 282 and transmission top cover plate.

Disconnect Clutch 3 pressure tube from output reduction gear rear bearing retainer and remove retainer (Fig. 214 or Fig. 215). If necessary to obtain more clearance to C3 tube fitting, loosen the cap screws holding the load control arm support at the bottom of housing of models except 4640 and 4840. On all models, use a long brass drift through center of planetary pack from front to drive reduction gear rearward. Remove bearing cup from rear of bore.

Remove the four retaining cap screws and, using a hoist and suitable lifting fixture, lift out planetary pack as shown in Fig. 216. Overhaul the removed planetary unit as outlined in paragraph 215 and 216.

Before installing planetary pack, inspect or renew the four brake-passage "O" rings in bottom of transmission housing. Reduction gear shaft should be installed with 0.001-0.002 inch bearing preload. To check bearing adjustment, make a trial installation of shaft, bearing cup and retainer, using one additional shim between bearing retainer and housing. Tighten retaining screws securely and measure shaft end play using a dial indicator. Remove the retainer and deduct from shim pack, shims equal to the measured end play plus 0.001 inch. Keep remainder of shim pack together for final installation.

Lower planetary unit straight downward being careful not to dislodge braked passage "O" rings. Tighten the four retaining cap screws alternately and evenly to 55 ft.-lbs. torque, and complete the assembly by reversing the disassembly procedure.

OVERHAUL

Power Shift Models

209. **PEDAL VALVE.** Refer to Fig. 217 for an exploded view of pedal valve and associated parts. Refer to paragraph 202 for removal and installation information.

Valve spools (10 & 13) must slide smoothly in their bores and must not be scored or excessively loose. Refer to Fig. 217 and check the pto and pedal valve springs for distortion and against the values which follow:

Pto Valve Spring (9)
　　Free Length . . . . .1.8 in.
　　Test . . . . 16.6-20.4 Lbs. at 1.4 in.
Pedal Valve Lower Spring (15)
　　Free Length . . . . .0.5 in.
　　Test . . . . .3.3-4.1 Lbs. at 0.34 in.
Pedal Valve Center Spring (17)
　　Free Length . . . . .1.3 in.
　　Test . . . . . . .15.1 at 18.3 Lbs. at 1.1 in.
Pedal Valve Upper Spring (16)
　　Free Length . . . . .2 in.
　　Test . . . . .11.7-14.3 Lbs. at 1.4 in.

210. **REGULATING VALVE.** Refer to Fig. 219 for an exploded view of regulating valve and associated parts and to paragraph 202 for removal and installation information.

Spools for all valves (6, 9, 10 and 11) are interchangeable but the springs are

Fig. 216 — Use a suitable lifting fixture and hoist to remove planetary pack.

Fig. 217 — Exploded view of pedal valve housing and component parts.

1. Housing
2. Shaft (Clutch)
3. Special pin (2)
4. Link (2)
5. Shaft (pto)
6. Arm (pto valve)
7. Clutch valve shaft
8. Shim
9. Spring (4)
10. Clutch valve
11. Spring pin (2)
12. Special plug
13. Pto valve
14. Pto valve shaft

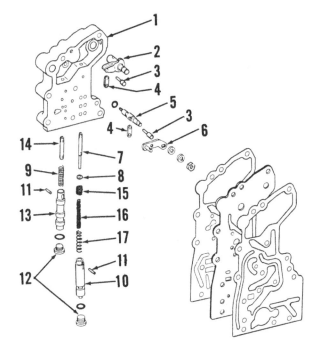

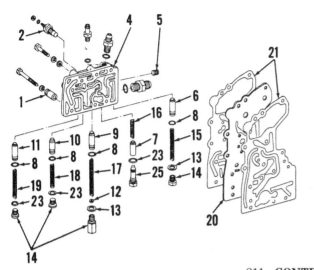

Fig. 219 – Exploded view of Power Shift regulating valve and associated parts.

1. Spacer
2. Temperature sender
4. Housing
5. Plug
6. Cooler relief valve
7. Return check valve
8. "O" ring
9. Regulating valve
10. Lube oil valve
11. Filter relief valve
12. Adjusting shim
13. Washer
14. Plugs
15. Spring
16. Spring
17. Spring
18. Spring
19. Spring
20. Plate
21. Gaskets (different)
23. "O" ring
25. Plug

Note that the accumulator pistons face opposite each other, and washer (23) is installed behind the outward facing piston only. Remove the six cap screws retaining cover (27) to housing and withdraw cover, control valves and operating linkage as a unit from control valve housing. Remove plug (2) from shift valve housing and withdraw spring (3) and dump valve spool (4). Remove the four retaining rings (10) and washers (9) then withdraw shift valve springs (8) and spools (7). The four shift valve spools and springs are interchangeable. The two replaceable accumulator charging orifices can be removed at this time.

Spring (14B) for the reverse detent ball (16) should have free length of 1.12 inch and should exert 14.4-17.6 pounds force when compressed to height of 0.80 inch. Spring (14C) for the speed detent cone (15) is similar in size, but is different. Spring (14C) should have 1.16 inch free length and should exert 5.85-7.15 pounds force when compressed to 0.83 inch.

Clean all parts in a suitable solvent and check for scoring or other damage, and for free movement of valve spools in bores. Control valve actuating mechanism need not be disassembled unless renewal of parts is indicated.

Check the valve springs for damage or distortion and against the values which follow:

Dump Valve Spring (3 – Fig. 220)
   Free Length . . . . 1.12 in.
   Test . . . . 31.5-38.5 Lbs. at 0.84 in.
Shift Valve Spring (8)
   Free Length . . . . 1.39 in.
   Test . . . . 6.8-8.4 Lbs. at 0.81 in.

not. "O" rings are used on all valves except the return pressure check valve (7). Inspect valves and bores for sticking or scoring. Valves must move freely in bore without excessive clearance. Refer to Fig. 219 and check the springs against the values which follow:

Oil Cooler Relief Valve Spring (15)
   Free Length . . . . 4.6 in.
   Test . . . . 35.7-43.7 Lbs. at 3.53 in.
Return Pressure Check Valve
Spring (16)
   Free Length . . . . 2.31 in.
   Test . . . . 10.8-13.2 Lbs. at 0.75 in.
Pressure Regulating Valve Spring (17)
   Free Length . . . . 5.09 in.
   Test . . . . 58.5-71.5 Lbs. at 4.44 in.
Transmission Lubrication Oil Valve
Spring (18)
   Free Length . . . . 2.83 in.
   Test . . . . 11.7-14.3 Lbs. at 2.13 in.
Oil Filter Relief Valve Spring (19)
   Free Length . . . . 3.71 in.
   Test . . . . 24.3-29.7 Lbs. at 2.6 in.

211. CONTROL VALVE. The shift valve housing is attached to inner face of control valve housing by a cover and six cap screws. Refer to Fig. 220 for an exploded view. Refer to paragraph 203 for removal and installation procedure.

To disassemble the removed unit, remove the six cap screws retaining shift valve housing (5) and lift off cover (1), housing, gasket plate (12) and gaskets (11 & 13). Note that the two gaskets are not interchangeable. Mark the removed gaskets "Outer" and "Inner" as they are removed to aid in the installation of new gaskets when unit is reassembled. Lift out inner detent springs (14B & 14C), plungers (15) and balls (16) to prevent loss as housing (17) is removed. Invert control valve housing (17) and remove the eight cap screws retaining accumulator cover (26), gasket (25), pistons (24) and springs (21 and 22).

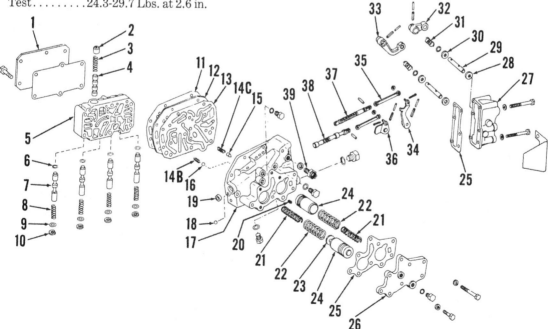

1. Cover
2. Plug
3. Spring
4. Dump valve
5. Shift valve housing
6. "O" ring
7. Shift valve
8. Spring
9. Washer
10. Retaining ring
11. Gasket
12. Plate
13. Gasket
14B. Spring for ball
14C. Spring for plunger cup
15. Detent cone (2)
16. Ball (2)
17. Control valve housing
18. Ball
19. Plug
20. Orifice (2)
21. Spring (inner)
22. Spring (outer)
23. Special washer
24. Accumulator piston
25. Gasket
26. Piston cover
27. Linkage cover
28. Oil seal (2)
29. Shaft (2)
30. Felt washer (2)
31. Spring (2)
32. Reverse arm
33. Speed control arm
34. Operating arm
35. Link
36. Operating arm
37. Speed control valve
38. Direction valve
39. Start safety switch

Fig. 220 – Exploded view of control valve assembly and associated parts used on Power Shift models.

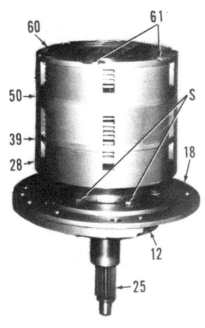

Fig. 221 — View of removed clutch assembly. The input shaft (25) and release bearing sleeve can be inserted into hole in bench or holding fixture to facilitate disassembly and reassembly. Screws (S) attach manifold plate (18) and pump housing (12) together. Refer to Fig. 222 for legend.

8. Pump body	46. High range clutch
9. Seal	piston
10. Packing ring	47. Piston outer seal
11. Pump gears	48. Piston inner seal
12. Pump housing	49. Bushing
13. Bushing	50. Rear clutch drum
14. Gasket	51. Piston inner seal
15. Spring, ball and plug	52. Piston outer seal
16. Bushing	53. Pto clutch piston
17. Steel balls (2 used)	54. Belleville spring
18. Clutch manifold	washers
19. Sealing ring (4 used)	55. Coil spring, cup and
20. Gasket	snap ring (some
21. Snap ring	models in place of
22. Ball bearing	54)
23. Thrust washer	56. Snap ring
24. Snap ring	57. Clutch hub
25. Input shaft	58. Clutch plates (flat
26. Bushing	plate next to piston
27. Snap ring (same as	53)
29)	59. Clutch friction discs
28. Front clutch drum	(alternate with 58)
29. Snap ring (same as	
27)	
30. Piston inner seal	
31. Piston outer seal	
32. Low range clutch	60. Clutch backing plate
piston	61. Through bolts
33. Belleville spring	62. Snap ring (variable
washers	thickness on models
34. Snap ring	with taper roller
35. Clutch hub	bearings at 63 and
36. Clutch plates (flat	65)
plate next to piston	63. Bearing assembly
32)	64. Pto drive shaft and
37. Clutch friction discs	gear
(alternate with	65. Bearing assembly
plates 36)	66. Roller thrust
38. Dowel pin	bearing and races
39. Clutch separator	(4640 & 4840 models)
plate	
40. Clutch plates (flat	67. Snap ring
plate next to piston	68. Thrust washer
46)	69. Bushing
41. Clutch friction discs	70. Snap ring (4640 &
(alternate with	4840 models)
plates 40)	71. Low range (C1)
42. Clutch hub	clutch shaft
43. Snap ring	72. High range (C2)
44. Belleville spring	clutch shaft
washers	
45. Coil spring and cup	
(some models in	
place of 44)	

Accumulator Valve Outer Spring (22)
Free Length . . . . 4.09 in.
Test . . . . 185-225 Lbs. at 2.91 in.
52-65 Lbs. at 3.76 in.
Accumulator Valve Inner Spring (21)
Free Length . . . . 3.78 in.
Test . . . . 100-122 Lbs. at 2.91 in.
Reassemble by reversing the disassembly procedure, using Fig. 220 as a guide. Tighten the six cap screws retaining shift valve housing to control valve evenly and alternately to a torque of 20 ft.-lbs. and tighten accumulator cover screws to 35 ft.-lbs. torque.

212. **TORSIONAL DAMPER.** To overhaul torsional damper, remove unit as outlined in paragraph 205. Check the removed disc for loose rivets, cracks or broken springs. Facing should be smooth and free of grease or oil. Renew disc if facing grooves are worn away, and determine the cause of wear, since this type coupler is never released and should not slip to any appreciable degree.

Friction surface of flywheel must not be heat-checked or scored, or have more than 0.006 inch out-of-true irregularities. Flywheel should not be machined, install new flywheel if not within specified limits.

Pins in flywheel should protrude 0.43 inch past clutch cover mounting surface of flywheel on all models except 4640 and 4840. Pins should protrude 0.46 inch on 4640 and 4840 models. On all models, tighten screws retaining flywheel to crankshaft to 85 ft.-lbs. torque.

The pilot bearing in flywheel is sealed and cannot be lubricated. Renew bearing if loose, rough or dry.

The pressure plate drive pin notches must not be excessively worn, and the friction surface must not be scored or more than 0.006 inch out-of-true. Renew pressure plate if any of these conditions are found. Check springs for rust, pitting or distortion, and against the test values which follow:
Number of Springs Used –
4040 9
4240, 4640 12
4440, 4840 15
Free Length –
4040, 4240, 4440 . . . . 2.1 in.
4640, 4840 . . . . 2.16 in.
Test –
4040, 4240 and
4440 . . . . 198-242 Lbs. at 1.6 in.
4640. 4840 . . . 247-303 Lbs. at 1.66 in.
To reassemble unit, refer to paragraph 205 and note proper holes to use when reinstalling jack screws on late type pressure plate. Tighten pressure plate retaining screws to 35 ft.-lbs. torque.

213. **CLUTCHES.** To disassemble the removed clutch pack, use a holding fix-

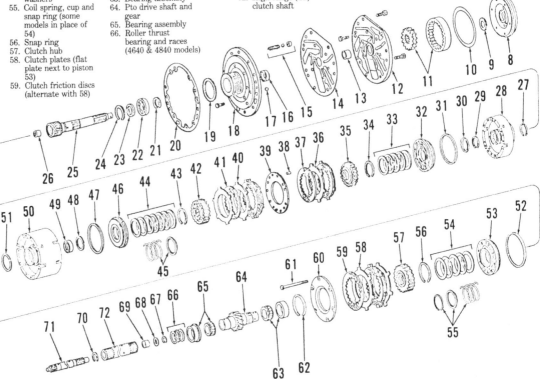

Fig. 222 — Exploded view of clutch assembly and related parts. The C1 clutch parts are (28 through 37); C2 clutch is shown at (40 through 48); the pto clutch is at (51 through 59). Coil spring, cup and snap ring shown at (45 and 55) are used in place of Belleville washer springs shown at (44 and 54) on some models. Parts shown are typical but differences may be noted.

ture with a 2-inch hole (or drill a 2-inch hole near edge of a table or bench). Insert input shaft (25 – Fig. 221) and release bearing sleeve through hole, with pto clutch pressure plate (60) up. Remove through-bolts (61) then lift off pto clutch pressure plate, pto clutch discs and clutch hub. Pto and C2 clutch drum (50), C1 and C2 pressure plate (39) and C1 clutch drum (28) can be separated after jarring slightly.

C1 clutch drum can be lifted off input shaft and manifold assembly after removing the snap ring (29) on rear splines of input shaft (25).

Clutch plates (36, 40 & 58 – Fig. 222), friction discs (37, 41 & 59) and other parts may be interchangeable between certain locations on some models; however, be sure to separate and identify by location all parts as they are separated to prevent improper assembly. Refer to Fig. 222 and check parts against the values which follow:

4040
C1 Clutch
  Piston (32) – I.D......3.125-3.126 in.
    O.D. ....6.600-6.610 in.
  Springs (33) – No. used...........4
    Minimum free height of
    one spring ....0.120 in.
  Drive plates (36) –
    No. used – Wavy ...2 (Notched tab)
    Flat ....1 (Not notched)
    Thickness – Wavy ...0.118-0.122 in.
    Flat............0.115-0.125 in.
  Friction discs (37) –
    No. used ....3 (Bronze)
    Thickness ....0.112-0.118 in.
C2 Clutch
  Piston (46) – I.D......3.125-3.126 in.
    O.D. ....6.600-6.610 in.
  Spring (45) – Free length ......2.8 in.
    Test, pounds at
    inches ....247-302 at 1.12
Drive plates (40) –
  No. used – Wavy .....4 (Notched Tab)
  Flat.............1 (Not notched)

Fig. 223 – Clutch 1 drum with piston and springs installed. Refer to paragraph 213 for disassembly procedure.

Thickness – Wavy.....0.118-0.122 in.
  Flat ....0.115-0.125 in.
Friction discs (41) –
  No. used ....5 (bronze)
  Thickness ....0.112-0.118 in.
Pto Clutch
  Piston (53) – I.D......3.125-3.126 in.
    O.D. ....6.600-6.610 in.
  Spring (55) – Free length ......2.8 in.
    Test, Pounds at
    inches ....247-302 at 1.12
  Drive plates (58) –
    No. used – Wavy ...2 (Notched tab)
    Flat ....1 (Not notched)
    Thickness – Wavy ...0.118-0.122 in.
    Flat ....0.115-0.125 in.
  Friction discs (59) –
    No. used ....3 (Fiber)
    Thickness ....0.112-0.118 in.
    Minimum groove depth....0.020 in.

4240 & 4440
C1 Clutch
  Piston (32) – I.D......3.974-3.978 in.
    O.D. ....7.377-7.387 in.
  Springs (33) – No. used...........5
    Minimum free height of one
    spring ....0.126 in.
  Drive plates (36) –
    No. used – Wavy ...2 (Notched tab)
    Flat ....1 (Not notched)
    Thickness – Wavy ...0.118-0.122 in.
    Flat............0.118-0.122 in.
  Friction discs (37) –
    No. used.............3 (bronze*)
    Thickness ....0.127-0.133 in.
C2 Clutch
  Piston (46) – I.D......3.974-3.978 in.
    O.D. ....7.377-7.387 in.
  Springs (44) – No. Used 7
    Minimum free height of one
    spring ....0.126 in.
  Drive plates (40) –
    No. used – Wavy ...4 (Notched tab)
    Flat ....1 (Not notched)
    Thickness – Wavy ...0.118-0.122 in.
    Flat............0.118-0.122 in.
  Friction discs (41) –
    No. used.............5 (bronze*)
    Thickness ....0.127-0.133 in.
Pto Clutch
  Piston (53) – I.D......3.974-3.978 in.
    O.D. ....7.377-7.387 in.
  Spring (54) – No. used............5
    Minimum free height of one
    spring ....0.126 in.
  Drive plates (58) –
    No. used – Wavy ...2 (Notched tab)
    Flat ....1 (Not notched)
    Thickness – Wavy ...0.118-0.122 in.
    Flat............0.118-0.122 in.
  Friction discs (59) –
    No. used ....3 (Fiber)
    Thickness ....0.127-0.133 in.
    Minimum groove depth....0.020 in.
*Bronze discs marked with white stripe are used in some applications and indicate harder friction material than similar discs marked by yellow stripe.

4640 & 4840
C1 Clutch
  Piston (32) – I.D......3.974-3.978 in.
    O.D. ....7.377-7.387 in.
  Springs (33) – No. used...........5
    Minimum free height of one
    spring ....0.126 in.
  Drive plates (36) –
    No. used – Wavy ...3 (Notched tab)
    Flat ....1 (Not notched)
    Thickness – Wavy ...0.118-0.122 in.
    Flat............0.118-0.122 in.
  Friction discs (37) –
    No. used.............4 (bronze*)
    Thickness ....0.127-0.133 in.
C2 Clutch
  Piston (46) – I.D......3.974-3.978 in.
    O.D. ....7.377-7.384 in.
  Springs (44) – No. used...........7
    Minimum free height of
    one spring.............0.126 in.
  Drive plates (40) –
    No. used – Wavy ...6 (Notched tab)
    Flat ....1 (Not notched)
    Thickness – Wavy ...0.118-0.122 in.
    Flat............0.118-0.122 in.
  Friction discs (41) –
    No. used.............7 (bronze*)
    Thickness ....0.127-0.133 in.
Pto Clutch
  Piston (53) – I.D......3.974-3.978 in.
    O.D. ....7.377-7.387 in.
  Spring (54) – No. used...........5
    Minimum free height of one
    spring ....0.126 in.
  Drive plates (58) –
    No. used – Wavy ...3 (Notched tab)
    Flat ....1 (Not notched)
    Thickness – Wavy ...0.118-0.122 in.
    Flat............0.118-0.122 in.
  Friction disc (59) –
    No. used ....4 (Fiber)
    Thickness ....0.127-0.133 in.
    Minimum groove depth....0.020 in.
*Bronze discs marked with white stripe are used in some later applications and indicate harder friction material than similar discs marked by yellow stripe.

Coil type springs (45 & 55 – Fig. 222) or Belleville springs (33, 44 & 54) are used to return pistons (32, 46 & 53) when hydraulic pressure is released. A suitable fixture is necessary to compress

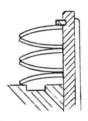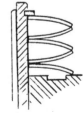

Fig. 224 – Cross-sectional schematic view showing correct method of installing five Belleville washers in clutch drum. Regardless of number used (4, 5 or 7), the inner diameter of the outer washer must always fit against snap ring as shown.

Fig. 225 – When installing assembled pump on gasket and manifold, make sure steel check balls are in place as shown.

the springs for removal or installation of snap rings (34, 43 & 56). Refer to Fig. 223 for typical compressing fixture. Pistons can be removed and new seals (30, 31, 47, 48, 51 & 52) can be installed after snap rings and springs are removed. Refer to Fig. 224 for correct back to back assembly of Belleville washers. Be sure to use correct compressing fixture to prevent damage to springs or other parts while assembling.

Inside diameter of installed bushing (49 – Fig. 222) should be 2.002-2.004 inches for 4040 model; 2.252-2.254 inches for all other models. Refer to paragraph 214 for inspection and service procedures for transmission pump, manifold plate and input shaft (items 8 through 26).

Refer to previously listed specifications for number of type of clutch plates and friction discs used. Most clutch packs use one flat plate and remaining plates (36, 40 & 58) are wavy. Plates with notch in one of the drive lugs (tabs) are wavy. Flat plates are not marked with notch. If a flat plate is used, the one flat plate should be assembled next to piston (32, 46 or 53). Alternate friction discs (37, 41 & 59) with plates (36, 40 & 58). Notches in tabs of wavy plates should be staggered (not aligned) and curve of the wavy plates should be alternated (not nested).

Tighten screws (61) to 20 ft.-lbs. torque. Screws which attach clutch assembly to clutch housing should be tightened to 50 ft.-lbs.; also, notice that two of these screws are shorter than remaining screws.

214. TRANSMISSION PUMP, MANIFOLD PLATE AND INPUT SHAFT. To overhaul the transmission pump, manifold plate and input shaft, first disassemble clutch pack as outlined

in paragraph 213. After C1 clutch drum (28 – Fig. 222) has been removed, remove the screws that hold manifold (18) to pump body (8), then slide input shaft and manifold plate assembly out of pump housing. Remove and save the two steel check balls (Fig. 225) as pump and manifold units are separated.

Remove the one remaining cap screw in rear face of pump housing (12 – Fig. 222) and lift off pump body and gears.

Input shaft and bearing assembly can be removed from manifold plate after unseating and removing the retaining ring (24) from rear of manifold plate. Press bearing from shaft if renewal is required, after removing bearing snap ring (21). Bushings (13 and 16) are not used on 4040 models; however, sealing rings are installed at these locations. On all models, chamfer of thrust washer (23) should be toward large (rear) end of shaft (25).

Examine gears and housings for scoring, wear, cracks and other damage and renew as required. Assemble by reversing the disassembly procedure. Tighten cap screws in pump housing and body to a torque of 20 ft.-lbs. Reassemble clutch pack as in paragraph 213.

215. PLANETARY PACK. Refer to paragraph 208 for removal of the planetary pack. To disassemble, place unit on bench with output end up as shown in

1. First planet piston housing
2. Pin (4 used)
3. Dowel pins
4. Piston inner seal
5. Piston outer seal
6. Brake piston
7. Brake piston return plate
8. Springs (used with washers on 4640 & 4840 models)
9. Friction discs
10. Separator plates
11. First planet ring gear
12. Brake facing plate
13. Second planet ring gear
14. Separator plate
15. Friction discs
16. Springs (used with washers on 4640 & 4840 models)
17. Brake piston return plate
18. Brake piston
19. Piston outer seal
20. Piston inner seal
21. "O" rings
22. Second planet piston housing
23. Planetary carrier assembly (Refer to Fig. 228)
24. Third planet ring gear
25. Separator plate
26. Friction discs
27. Springs
28. Brake piston return plate
29. Brake piston
30. Piston outer seal
31. Piston inner seal
32. Bushing
33. Dowel pins
34. Third planet piston housing
35. Planetary clutch drum
36. Friction discs (alternate with 37)
37. Clutch plates
38. Planetary clutch hub

39. Snap ring
40. Belleville spring washers (4640 & 4840 models)
41. Coil spring and cup (used in place of 40 on 4040, 4240 & 4440 models)
42. Planetary clutch piston
43. Piston outer seal
44. Piston inner seal
45. Ball and plug
46. Bushing
47. Ball and plug
48. Clutch piston planetary housing
49. Bushing
50. Friction discs

51. Separator plate
52. Springs
53. Brake piston return plate
54. Brake piston
55. Piston outer seal
56. Piston inner seal
57. Fourth planet brake piston housing
58. Cap screws

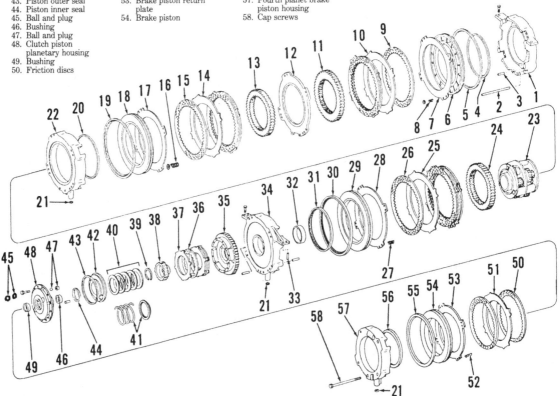

Fig. 226 – Exploded view of the planetary pack. The B1 brake includes parts (4 through 11), B2 brake is (13 through 20), B3 brake is (24 through 31), C1 clutch is parts (36 through 44), B4 brake includes parts (50 through 55). Parts shown are typical and differences may be noted.

Fig. 227. Remove the cap screws (C) and lift off C3 clutch piston housing (48). Remove the through bolts (58 – Fig. 226) then lift housing (57), drum (35), housing (34) and associated parts off until planetary assembly (23) can be removed. Refer to paragraph 216 for disassembly of planetary assembly.

Refer to Fig. 226 and check parts against the values which follow:

4040
B1 Brake
Piston (6) – I.D. ......9.258-9.262 in.
   O.D. ....11.596-11.604 in.
Piston return plate (7) –
   Thickness ....0.118-0.122 in.
Springs (8) – Free Length ....0.83 in.
   Test, pounds at inch . . 23-29 at 0.625
Friction discs (9) –
   No. used 2
   Thickness ....0.117-0.123 in.
Separator plates (10) –
   No. used 1
   Thickness ....0.118-0.122 in.
   Groove depth, Minimum . . . 0.020 in.
Brake facing plate (12) –
   Thickness ....0.851-0.881 in.

B2 Brake
Piston (18) – I.D......9.258-9.262 in.
   O.D. ....11.596-11.604 in.
Piston return plate (17) –
   Thickness ....0.118-0.122 in.
Springs (16) – Free length ....0.83 in.
   Test, pounds at inch . . 23-29 at 0.625
Friction discs (15) –
   No. used 2
   Thickness ....0.117-0.123 in.
   Groove depth, Minimum . . . 0.020 in.
Separator plates (14) –
   No. used 1
   Thickness ....0.118-0.122 in.

B3 Brake
Piston (29) – I.D......8.856-8.860 in.
   O.D. ....11.596-11.604 in.

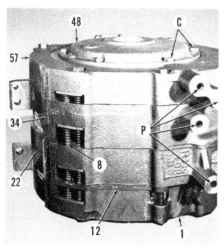

Fig. 227 – View of removed planetary pack. Refer to Fig. 226 for legend.

Piston return plate (28) –
   Thickness ....0.118-0.122 in.
Springs (27) – Free length ....1.22 in.
   Test, pounds at inch ....22-28 at 1.0
Friction discs (26) –
   No. used 3
   Thickness ....0.117-0.123 in.
   Groove depth, Minimum . . . 0.020 in.
Separator plates (25) –
   No. used 2
   Thickness ....0.118-0.122 in.
Bushing (32) –
   Installed diameter 3.7550-3.7575 in.

C3 Clutch
Piston (42) – I.D......3.125-3.126 in.
   O.D. ....6.600-6.610 in.
Spring (41) – Free length .....2.79 in.
   Test, pounds at
   inches ....247-303 at 1.12
Drive plates (37) –
   No. used 3
   Thickness ....0.118-0.122 in.
Friction discs (36) –
   No. used 3
   Thickness ....0.112-0.118 in.
   Groove depth, Minimum . . . 0.020 in.
Bushing (46) –
   Installed diameter ..1.691-1.692 in.
Bushing (49) –
   Installed diameter ..1.753-1.754 in.

B4 Brake
Piston (54) – I.D......9.258-9.262 in.
   O.D. ....11.596-11.604 in.
Piston return plate (53) –
   Thickness ....0.118-0.122 in.
Springs (52) – Free length ....0.83 in.
   Test, pounds at
   inches ....23-29 at 0.625
Friction discs (50) –
   No. used 2
   Thickness ....0.117-0.123 in.
   Groove depth, Minimum . . . 0.020 in.
Separator plates (51) –
   No. used 1
   Thickness ....0.118-0.122 in.

4240
B1 Brake
Piston (6) – I.D.......8.856-8.860 in.
   O.D. ....11.596-11.604 in.
Piston return plate (7) –
   Thickness ....0.118-0.122 in.
Springs (8) – Free length .....0.83 in.
   Test, pounds at inch . . 23-29 at 0.625
Friction discs (9) –
   No. used 2
   Thickness ....0.117-0.123 in.
Separator plates (10) –
   No. used 1
   Thickness ....0.118-0.122 in.
   Groove depth, Minimum . . . 0.020 in.
Brake facing plate (12) –
   Thickness ....0.851-0.881 in.

B2 Brake
Piston (18) – I.D......9.258-9.262 in.
   O.D. ....11.596-11.604 in.

Piston return plate (17) –
   Thickness ....0.118-0.122 in.
Springs (16) – Free length ....0.83 in.
   Test, pounds at inch . . 23-29 at 0.625
Friction discs (15) –
   No. used 2
   Thickness ....0.117-0.123 in.
   Groove depth, Minimum . . . 0.020 in.
Separator plates (14) –
   No. used 1
   Thickness ....0.118-0.122 in.

B3 Brake
Piston (29) – I.D......8.856-8.860 in.
   O.D. ....11.596-11.604 in.
Piston return plate (28) –
   Thickness ....0.118-0.122 in.
Springs (27) – Free length ....1.22 in.
   Test, pounds at inch ....22-28 at 1.0
Friction discs (26) –
   No. used 3
   Thickness ....0.117-0.123 in.
   Groove depth, Minimum . . . 0.020 in.
Separator plates (25) –
   No. used 2
   Thickness ....0.118-0.122 in.
Bushing (32) –
   Installed diameter 3.7550-3.7575 in.

C3 Clutch
Piston (42) – I.D......3.125-3.126 in.
   O.D. ....6.600-6.610 in.
Spring (41) – Free length .....2.79 in.
   Test, pounds at
   inches ....247-303 at 1.12
Drive plates (37) –
   No. used 3
   Thickness ....0.118-0.122 in.
Friction discs (36) –
   No. used 3
   Thickness ....0.112-0.118 in.
   Groove depth, Minimum . . . 0.020 in.
Bushing (46) –
   Installed diameter ..1.691-1.692 in.
Bushing (49) –
   Installed diameter ..1.753-1.754 in.

B4 Brake
Piston (54) – I.D......9.258-9.262 in.
   O.D. ....11.596-11.604 in.
Piston return plate (53) –
   Thickness ....0.118-0.122 in.
Springs (52) – Free length ....0.83 in.
   Test, pounds at
   inches ....23-29 at 0.625
Friction discs (50) –
   No. used 2
   Thickness ....0.117-0.123 in.
   Groove depth, Minimum . . . 0.020 in.
Separator plates (51)
   No. used 1
   Thickness ....0.118-0.122 in.

4440
B1 Brake
Piston (6) – I.D.......8.856-8.860 in.
   O.D. ....11.596-11.604 in.
Piston return plate (7) –
   Thickness ....0.118-0.122 in.

Springs (8) – Free length . . . . . 0.83 in.
  Test, pounds at inch . . 23-29 at 0.625
Friction discs (9) –
  No. used 2
  Thickness . . . . 0.117-0.133 in.
Separator plates (10) –
  No. used 1
  Thickness . . . . 0.117-0.123 in.
  Groove depth, Minimum . . . 0.020 in.
Brake facing plate (12) –
  Thickness . . . . 0.851-0.881 in.

B2 Brake
Piston (18) – I.D. . . . . . . 9.258-9.262 in.
  O.D. . . . . 11.596-11.604 in.
Piston return plate (17) –
  Thickness . . . . 0.118-0.122 in.
Springs (16) – Free length . . . . 0.83 in.
  Test, pounds at inch . . 23-29 at 0.625
Friction discs (15) –
  No. used 2
  Thickness . . . . 0.117-0.123 in.
  Groove depth, Minimum . . . 0.020 in.
Separator plates (14) –
  No. used 1
  Thickness . . . . 0.118-0.122 in.

B3 Brake
Piston (29) – I.D. . . . . . . 8.856-8.860 in.
  O.D. . . . . 11.596-11.604 in.
Piston return plate (28) –
  Thickness . . . . 0.118-0.122 in.
Springs (27) – Free length . . . . 1.48 in.
  Test, pounds at inches . . 23-29 at 1.12
Friction discs (26) –
  No. used 3
  Thickness . . . . 0.117-0.123 in.
  Groove depth, Minimum . . . 0.020 in.
Separator plates (25) –
  No. used 2
  Thickness . . . . 0.118-0.122 in.
Bushing (32) –
  Installed diameter 3.7550-3.7575 in.

C3 Clutch
Piston (42) – I.D. . . . . . . 3.125-3.126 in.
  O.D. . . . . 6.600-6.610 in.
Spring (41) – Free length . . . . . 2.79 in.
  Test, pounds at
    inches . . . . 247-303 at 1.12
Drives plates (37) –
  No. used 4
  Thickness . . . . 0.087-0.093 in.
Friction discs (36) –
  No. used 4
  Thickness . . . . 0.112-0.118 in.
  Groove depth, Minimum . . . 0.020 in.
Bushing (46) –
  Installed diameter 1.9425-1.9445 in.
  Installed depth . . . . 3/16-in.
Bushing (49) –
  Installed diameter . . 2.128-2.130 in.
  Inner chamfer
    toward . . . . outside (rear)

B4 Brake
Piston (54) – I.D. . . . . . . 9.258-9.262 in.
  O.D. . . . . 11.596-11.604 in.
Piston return plate (53) –
  Thickness . . . . 0.118-0.122 in.
Springs (52) – Free length . . . . 0.83 in.

Test, pounds at
  inches . . . . 23-29 at 0.625
Friction discs (50) –
  No. used 2
  Thickness . . . . 0.117-0.123 in.
  Groove depth, Minimum . . . 0.020 in.
Separator plates (51) –
  No. used 1
  Thickness . . . . 0.118-0.122 in.

4640
B1 Brake
Piston (6) – I.D. . . . . . . 9.979-9.985 in.
  O.D. . . . . 12.726-12.736 in.
Piston return plate (7) –
  Thickness . . . . 0.118-0.122 in.
Springs (8) – Free length . . . . . 0.83 in.
  Test, pounds at inch . . 23-29 at 0.625
Washer used when
  installed . . . . 0.060 in.
Friction discs (9) – No. used 2
  Thickness . . . . 0.127-0.133 in.
Separator plates (10) –
  No. used 1
  Thickness . . . . 0.118-0.122 in.
  Groove depth, Minimum . . . 0.020 in.
Brake facing plate (12) –
  Thickness . . . . 0.942-0.982 in.

B2 Brake
Piston (18) – I.D. . . . . 10.155-10.161 in.
  O.D. . . . . 12.726-12.736 in.
Piston return plate (17) –
  Thickness . . . . 0.118-0.122 in.
Springs (16) – Free length . . . . 0.83 in.
  Test, pounds at inch . . . 23-29 at 0.625
Washers used when
  installed . . . . 0.060 in.
Friction discs (15) –
  No. used 2
  Thickness . . . . 0.127-0.133 in.
  Groove depth, Minimum . . . 0.016 in.
Separator plates (14) –
  No. used 1
  Thickness . . . . 0.118-0.122 in.

B3 Brake
Piston (29) – I.D. . . . . . . 9.979-9.985 in.
  O.D. . . . . 12.726-12.736 in.
Piston return plate (28) –
  Thickness . . . . 0.118-0.122 in.
Springs (27) – Free length . . . . 1.22 in.
  Test, pounds at inch . . . . 22-28 at 1.0
Friction discs (26) –
  No. used 4
  Thickness . . . . 0.127-0.133 in.
  Groove depth, Minimum . . . 0.016 in.
Separator plates (25) –
  No. used 3
  Thickness . . . . 0.118-0.122 in.
Bushing (32) –
  Installed diameter 4.0675-4.0700 in.

C3 Clutch
Piston (42) – I.D. . . . . . . 3.974-3.978 in.
  O.D. . . . . 7.377-7.387 in.
Spring (40) –
  Free height of one . . . . 0.126 in.
  No. of spring washers used 5
Drive plates (37) –
  No. used 4

Thickness . . . . 0.118-0.122 in.
Friction discs (36) –
  No. used 4
  Thickness . . . . 0.127-0.133 in.
  Groove depth, Minimum . . . 0.020 in.
Bushing (46) –
  Installed diameter 2.0655-2.0675 in.
  Installed depth . . . . 3/16 in.
Bushing (49) –
  Installed diameter . . 2.128-2.130 in.
  Inner chamfer
    toward . . . . outside (rear)

B4 Brake
Piston (54) – I.D. . . . . 10.155-10.161 in.
  O.D. . . . . 12.726-12.736 in.
Piston return plate (53) –
  Thickness . . . . 0.118-0.122 in.
Springs (52) – Free length . . . . 0.83 in.
  Test, pounds at
    inches . . . . 23-29 at 0.625
Friction discs (50) –
  No. used 2
  Thickness . . . . 0.127-0.133 in.
  Groove depth, Minimum . . . 0.016 in.
Separator plates (51) –
  No. used 1
  Thickness . . . . 0.118-0.122 in.

4840
B1 Brake
Piston (6) – I.D. . . . . . . 9.979-9.985 in.
  O.D. . . . . 12.726-12.736 in.
Piston return plate (7) –
  Thickness . . . . 0.118-0.122 in.
Springs (8) – Free length . . . . . 0.83 in.
  Test, pounds at
    inches . . . . 23-29 at 0.625
Washers used when
  installed . . . . 0.060 in.
Friction discs (9) –
  No. used 3
  Thickness . . . . 0.127-0.133 in.
  Groove depth, Minimum . . . 0.016 in.
Separator plates (10) –
  No. used 2
  Thickness . . . . 0.118-0.122 in.
Brake facing plate (12) –
  Thickness . . . . 0.372-0.382 in.

B2 Brake
Piston (18) – I.D. . . . . 10.155-10.161 in.
  O.D. . . . . 12.726-12.736 in.
Piston return plate (17) –
  Thickness . . . . 0.118-0.122 in.
Springs (16) – Free length . . . . 0.83 in.
  Test, pounds at
    inch . . . . 23-29 at 0.625
Washers used when
  installed . . . . 0.060 in.
Friction discs (15) –
  No. used 3
  Thickness . . . . 0.127-0.133 in.
  Groove depth, Minimum . . . 0.016 in.
Separator plates (14) –
  No. used 2
  Thickness . . . . 0.118-0.122 in.

B3 Brake
Piston (29) – I.D. . . . . . . 9.979-9.985 in.
  O.D. . . . . 12.726-12.736 in.

Piston return plate (28)–
    Thickness . . . . .0.118-0.122 in.
Springs (27)–Free length . . . .1.22 in.
    Test, pounds at inch . . . .22-28 at 1.0
Friction discs (26)–
    No. used 4
    Thickness . . . . .0.127-0.133 in.
    Groove depth, Minimum . . .0.016 in.
Separator plates (25)–
    No. used .3
    Thickness . . . . .0.118-0.122 in.
Bushing (32)–
    Installed diameter 4.0675-4.0700 in.

C3 Clutch

Piston (42)–I.D. . . . . . .3.974-3.978 in.
    O.D. . . . . .7.377-7.387 in.
Spring (40)–
    Free height of one . . . . .0.126 in.
    No. used 5
Drive plates (37)–
    No. used 4
    Thickness . . . . .0.118-0.122 in.
Friction discs (36)–
    No. used 4
    Thickness . . . . .0.127-0.133 in.
    Groove depth, Minimum . . .0.020 in.
Bushing (46)–
    Installed diameter 2.0655-2.0675 in.
    Installed depth . . . . .3/16-in.
Bushing (49)–
    Installed diameter . .2.128-2.130 in.
    Inner diameter
        toward . . . . .outside (rear)

B4 Brake

Piston (54)–I.D. . . . .10.155-10.161 in.
    O.D. . . . . .12.726-12.736 in.
Piston return plate (53)–
    Thickness . . . . .0.118-0.122 in.
Springs (52)–Free length . . . .0.83 in.
    Test, pounds at inch . .23-29 at 0.625
Friction discs (50)–
    No. used 2
    Thickness . . . . .0.127-0.133 in.
    Groove depth, Minimum . . .0.016 in.
Separator plates (51)–
    No. used .1
    Thickness . . . . .0.118-0.122 in.

NOTE: Parts from the various tractor models and from different locations within the same tractor model may be similar, but not identical. DO NOT attempt to install similar, but incorrect parts at any location within this transmission.

Refer to paragraph 213 and Fig. 223 for suggested method of compressing spring (40 or 41–Fig. 226) in order to remove and install snap ring (39).

Bushing (32) should be pressed into housing (34) using a suitable piloted driver. Bushing should be flush with piston side of housing on all models except 4640 and 4840. On 4640 and 4840 models, the bushing should be 0.06 inch below flush with housing. On all models, be sure to align cut-out in bushing (32) with similar cut-out in housing.

Bushing (46 & 49) should be pressed into bores of clutch housing (48) using appropriate size pilots. Front bushing (46) should be installed 3/16-inch deep in bore. Large chamfer in inside diameter of rear bushing (49) should be toward outside (rear) of clutch housing (48).

NOTE: Coat all parts with John Deere Hy-Gard transmission and hydraulic oil or equivalent while assembling.

To assemble the planetary pack, place B1 piston housing (Fig. 234) closed end down on a bench. Install four guide studs in threaded holes and alignment dowels in remaining holes as shown.

The piston return plates (7, 17, 28 & 53–Fig. 226), which are installed next to pistons, have drilled holes in the four extended lugs to serve as seats for the brake return springs. Be sure to install correct brake return springs at locations (8, 16, 27 & 52). Refer to the preceding specifications for identification at the different locations. Washers, which are 0.060 inch thick, are used with springs at locations (8 & 16) only on 4640 and 4840 models.

NOTE: Be sure to align oil holes in brake facing plate (12–Fig. 226 & Fig. 227) with oil holes in B1 housing (1) when assembling. If not correctly assembled, B1 brake will not receive pressure to apply brake.

Refer to paragraph 216 for assembly of planetary assembly. Remainder of assembly procedure should be accomplished by reversing disassembly procedure and observing Fig. 226 and Fig. 227. Alternate separator plates and lined friction brake discs at all locations. All oil ports (P–Fig. 227) must be aligned as shown. Remove aligning studs, install through bolts (58–Fig. 226) and tighten to 35 ft.-lbs. torque. Tighten screws (C–Fig. 227) to 20 ft.-lbs. torque.

Before reinstalling planetary brake pack, apply 50-80 psi air pressure to each of the oil passage ports (P–Fig. 227) in turn, listen for air leaks and note action of brake plates. If leaks are noted or if brake return springs do not compress, recheck assembly procedure and correct the trouble before reassembling the tractor. Be sure the "O" rings (21–Fig. 226) in bottom of transmission case do not move when installing planetary pack. Tighten cap screws to 55 ft.-lbs. torque.

216. PLANETARY ASSEMBLY. The planetary assembly shown in Fig. 228 is located in the planetary brake pack as shown at 23–Fig. 226. Three different types of planetary units are used: Type A (Fig. 228) is typical of parts on 4040 and 4240 models; Type B is typical of

Model 4440; Type C is typical of planetary assembly used on 4640 and 4840 models.

Refer to paragraph 208 for removal of the planetary pack and to paragraph 215 for disassembly and removal of the planetary assembly.

On type A and type B planetary assemblies shown in Fig. 228, proceed as follows: Place unit on a bench with the rear retainer (12) up. Remove three cap screws holding retainer (12) to carrier (1). Pry cover plate (11) from dowels, remove three planet pinions (10), being careful not to lose roller bearings. All rear planet pinions are equipped with **TWO** rows of 31 bearings and one spacer each. Turn planet assembly upside down and remove three cap screws holding retainer (25) to carrier. Pry cover (24) from carrier. Remove ring gear and pinions (17), being careful not to lose roller bearings.

4040 Models

Bearing rollers in pinion (6)–
    No. used each pinion . . . . .31 x 2 rows
    Diameter of rollers . .0.1559-0.1560 in.
    Length of rollers . . . . . . .1.17-1.19 in.
Thrust washer (3)
    thickness . . . . .0.033-0.039 in.
Bearing rollers in pinion (10)–
    No. used each pinion . . . . .31 x 2 rows
    Diameter of rollers . .0.1559-0.1560 in.
    Length of rollers . . . . . . .1.17-1.19 in.
Bearing rollers in pinion (17)–
    No. used each pinion . . . . .31 x 2 rows
    Diameter of rollers . .0.1559-0.1560 in.
    Length of rollers . . . . . . .0.98-1.00 in.
Bushing (23)–
    Installed diameter . . . .2.130-2.132 in.
    Installed depth . . . . .flush to 0.020 in.
                        below flush
    Notched end toward . . .bottom of bore

4240 Models

Bearing rollers in pinion (6)–
    No. used each pinion . . . . .31 x 2 rows
    Diameter of rollers . .0.1559-0.1560 in.
    Length of rollers . . . . . . .1.17-1.19 in.
Thrust washer (3)
    thickness . . . . .0.033-0.039 in.
Bearing rollers in pinion (10)–
    No. used each pinion . . . . .31 x 2 rows
    Diameter of rollers . .0.1559-0.1560 in.
    Length of rollers . . . . . . .1.17-1.19 in.
Bearing rollers in pinion (17)–
    No. used each pinion . . . . .31 x 2 rows
    Diameter of rollers . .0.1559-0.1560 in.
    Length of rollers . . . . . . .0.98-1.00 in.
Bushing (23)–
    Installed diameter . . . .2.366-2.368 in.
    Installed depth . . . . .flush to 0.020 in.
                        below flush
    Notched end toward . . .bottom of bore

4440 Models

Bearing rollers in pinion (6)–
    No. used each pinion . . . . .31 x 2 rows
    Diameter of rollers . .0.1559-0.1560 in.
    Length of rollers . . . . .0.790-0.810 in.

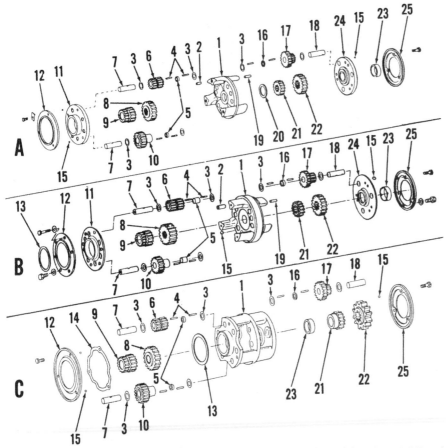

Fig. 228 — Exploded view of the three different planetary types used. Type "A" is typical of 4040 and 4240; Type "B" is typical of 4440; Type "C" is typical of 4640 and 4840 models.

1. Planet pinion and carrier housing
2. Dowel pins
3. Thrust washers (18 used)
4. Needle rollers
5. Spacer (6 used)
6. Third planet pinion (3 used)
7. Shafts (6 used)
8. Third planet sun gear
9. Fourth planet sun gear)
10. Fourth planet pinion (3 used)
11. Pinion carrier rear cover (Types "A" & "B")
12. Pinion shaft rear retainer
13. Thrust washer (Type "B")
14. Gasket (Type "C")
15. Steel balls (9 used)
16. Spacer (3 used)
17. First and second planet pinion (3 used)
18. Pinion shaft (3 used)
19. Dowel pin
20. Thrust washer
21. Second planet (C1-B2) sun gear
22. First planet (C2-B1) sun gear
23. Bushing
24. Pinion carrier front cover (Type "A" & "B")
25. Retainer

Thrust washer (3)
  thickness . . . . 0.033-0.039 in.
Bearing rollers in pinion (10)—
  No. used each pinion . . . . . 31 x 2 rows
  Diameter of rollers . . 0.1559-0.1560 in.
  Length of rollers . . . . . 0.790-0.810 in.
Bearing rollers in pinion (17)—
  No. used each pinion . . . . . 31 x 2 rows
  Diameter of rollers . . 0.1559-0.1560 in.
  Length of rollers . . . . . 0.928-0.938 in.
Bushing (23)—
  Installed diameter . . . . 2.366-2.368 in.
  Installed depth . . . . . flush to 0.020 in.
                       below flush
  Notched end toward . . . bottom of bore

Fig. 229 — View of B1 piston housing with piston withdrawn showing seal rings. Refer to Fig. 228 for legend.

4640 & 4840 Models

Bearing rollers in pinion (6)—
  No. used each pinion . . . . . 20 x 2 rows
  Diameter of rollers . . 0.1559-0.1560 in.
  Length of rollers . . . . 1.17-1.19 in.
Thrust washer (3)
  thickness . . . . 0.033-0.039 in.
Bearing rollers in pinion (10)—
  No. used each pinion . . . . . 20 x 2 rows
  Diameter of rollers . . 0.1559-0.1560 in.
  Length of rollers . . . . 1.17-1.19 in.
Bearing rollers in pinion (17)—
  No. used each pinion . . . . . 20 x 2 rows
  Diameter of rollers . . 0.1559-0.1560 in.
  Length of rollers . . . . 0.98-1.00 in.
Bushing (23)—
  Installed diameter . . . . 2.505-2.507 in.
  Installed depth . . . . . flush to 0.020 in.
                       below flush
  Notched end toward . . . bottom of bore

Models 4040 and 4240 are planetary typical of type "A" shown in Fig. 228. When assembling, refer to Fig. 230. Install the three planet pinion shafts (18) in plate (24) and locate using the steel balls (15 – Fig. 228). Position retainer (25) under cover plate to hold shafts in place. Position C2-B1 sun gear (22) and the B1 ring gear (11 – Fig. 226) on plate (24 – Fig. 230). Locate timing marks on ring gear at center of planet pinion shafts. Install the three planet pinions (17) with bearing rollers, thrust washers and spacers and with timing marks aligned with marks on C1-B2 sun gear (21). Position thrust washer (20 – Fig. 228) on sun gear, then position carrier (1) over the planet assembly being sure that dowel pins (19) properly engage holes in plate. Carefully turn assembly over and install the three cap screws through retainer (25) and plate (24). Tighten screws to 35 ft.-lbs. and bend lock plates around heads. Install six pinion shafts (7) and steel balls (15) in cover plate (11). Place retainer (12) under cover plate to hold shafts in plate. Install planet pinions (6 & 10) and B3 sun gear (8) as shown in Fig. 231. On models so equipped, be sure that thrust washer is correctly centered on B3 sun gear. On all models, position the planet carrier assembly (1 – Fig. 228) on pinion shafts correctly engaging hollow dowels (2). Turn assembly over carefully, install cap screws through lock plates, retainer (12) and plate (11). Tighten cap screws to 35 ft.-lbs. torque, then bend lock plates up around heads of cap screws.

Model 4440 uses planetary typical of type "B" shown in Fig. 228. When assembling, refer to Fig. 232 and be sure to align timing marks. Notched sides of B1 ring gear and B2 ring gear should be away from each other when assembling. Tighten the cap screws to 35 ft.-lbs. torque and lock with lock plates.

Model 4640 and 4840 planetary

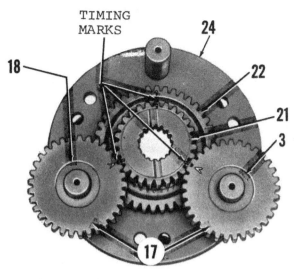

TIMING
MARKS

18

24

22

21

3

17

Fig. 230 — View of type "A" planetary assembly showing C1 and C2 sun gears and related timing marks. Refer to Fig. 228 for legend.

assembly is type "C" shown in Fig. 228. Each planet pinion shaft is retained in carrier housing by a steel ball (15) and retainer plates (12 and 25). Shaft is a slip fit in carrier housing bores. Each planet pinion contains two rows of twenty loose needle rollers (4) separated by a spacer for a total of forty bearings to each pinion. Withdraw pinions carefully to prevent loss of bearings. All planet pinion

bearing rollers are interchangeable, but should be kept in sets. Front and rear bearing spacers (5 & 16) are of different thickness, the thicker spacers being used in rear (output) planetary unit. Notice that the first and second planet pinion (17) is index-marked "1", "2" and "3" and that corresponding marks appear on planet carrier adjacent to pinion shaft bore. When assembling planet carrier place carrier on a bench front down, install first sun gear (22) with cupped out side down and install second sun gear (21) with gear side down. Special loading tools (JDT-22-1 and JDT-22-2) can be used to more easily load the roller bearings in pinion gears. When inserting loaded pinion gears, the corresponding index marks on pinion and carrier MUST be aligned as shown in Fig. 233. No indexing is required on other planetary unit, and third sun gear (8 – Fig. 228) can be installed with either side up.

REDUCTION GEARS, TOW DISCONNECT AND PARK PAWL

All Power Shift Models

217. **TRANSMISSION TOP COVER AND PARK PAWL.** If park pawl remains engaged even though linkage is properly adjusted and operates satisfactorily, the park pawl return spring may be unhooked or broken. The unit can be inspected and spring renewed after removing transmission top cover. Proceed as follows:

Remove operator's platform, steering support side panels and interfering hydraulic tubes and linkage. Unbolt and remove transmission top cover. Remove damaged or broken spring and install a new spring. Malfunction can also occur if engagement spring arm (9 – Fig. 243) is damaged or broken or if camshaft (5) binds. Pivot shaft (4) on some models is retained by a vertical dowel in transmission case.

218. **TOW DISCONNECT.** To remove the tow disconnect mechanism, first remove planetary output shaft as outlined in paragraph 208. Unscrew retaining nut (5 – Fig. 244) and right hand shaft (10) and lift out disconnect fork (7), collar (4 – Fig. 246) reduction gear (3) and cone of bearing (2). Front bearing

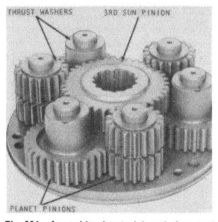

Fig. 231 — Assemble planet pinion shafts, steel balls and planet pinions in rear cover plate as shown before installing sun pinion gears on type "A" unit.

Fig. 233 — Numbers on pinion gears must be aligned with the corresponding number stamped on housing.

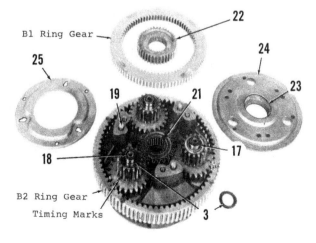

B1 Ring Gear

25

19

21

18

B2 Ring Gear

Timing Marks

22

24

23

17

3

Fig. 232 — View of front planet assembly of type "B" (4440) models showing ring gears and timing marks. Refer to Fig. 228 for legend.

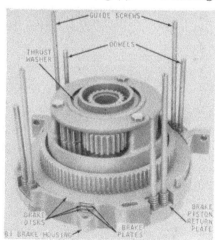

Fig. 234 — Install guide screws and dowels in place of through-bolts.

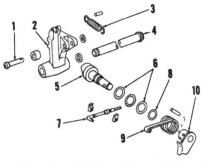

Fig. 243 — *Exploded view of transmission park lock pawl and associated parts. Type shown is used on 4640 and 4840 models. Refer to Fig. 243A for type used on other models.*

1. Pin
2. Lock pawl
3. Spring
4. Pivot shaft
5. Lock cam
6. Thrust bearing
7. Operating cable
8. "O" ring
9. Spring arm
10. Arm hub

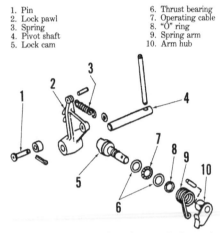

Fig. 243A — *Exploded view of transmission park lock pawl and associated parts used on 4040, 4240 and 4440. Refer to Fig. 243 for legend.*

cup can be drifted rearward out of housing after gear has been removed.

Install by reversing the removal procedure. Adjust reduction gear shaft bearings as outlined in paragraph 208. After tow disconnect parts have been installed, shift disconnect lever into forwardmost position with spring unhooked (Fig. 245). With cap screws loose, align the mark on stop plate with REAR of lever as shown in Fig. 245 for all models except 4640 and 4840. Adjustment is similar on 4640 and 4840 except mark is in **FRONT** of lever when correctly set. On all models, tighten cap

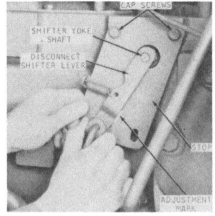

Fig. 245 — *Method of adjusting tow disconnect lever. See paragraph 218 for adjustment procedure.*

screws while holding disconnect lever fully forward and mark aligned. Attach spring, and shift lever to detent (engaged) position.

220. BEVEL PINION SHAFT. Bevel pinion shaft and bearings can be removed for service after removing planetary unit as outlined in paragraph 208 and differential assembly as in paragraph 222.

Remove the nut (24 – Fig. 246) from front end of bevel pinion shaft and drift the shaft rearward until front bearing cone, shims (26) and spacer (27) can be removed. Withdraw shaft and rear bearing cone from rear while lifting gear (28) and spacer (29) out top opening.

The bevel pinion shaft is available only as a matched set with bevel ring gear. If rear bearing cup is renewed, keep cone point adjusting shim pack (30) intact and reinstall same pack or shims of equal thickness. If gears and/or housing are renewed, check and adjust cone point as outlined in paragraph 221.

When reinstalling bevel pinion shaft and bearings, make a trial installation using the removed shim pack (26) plus one additional 0.010 inch shim. Tighten shaft nut, then measure shaft end play

using a dial indicator. Remove shims equal to measured end play plus 0.005 inch, to obtain the recommended 0.004-0.006 inch bearing preload. A preferred method of measuring bearing preload is by wrapping string around pinion shaft just in front of pinion and measuring rolling torque of shaft with a spring scale. The spring scale should indicate 2½-7½ pounds of pull required to rotate shaft one revolution per second with nut (24) tightened to 400 ft.-lbs. torque.

NOTE: If main drive bevel pinion and/or transmission housing are renewed, cone point (mesh position) of gears must be checked and adjusted BEFORE adjusting bearing preload.

Tighten nut (24) to 400 ft.-lbs. torque, then stake in at least two locations to maintain setting.

221. CONE POINT ADJUSTMENT. The cone point (mesh position) of the main drive bevel gear and pinion is adjusted by means of shims (30 – Fig. 246) which are available in 0.003, 0.005 and 0.010 inch thicknesses. The cone point will only need to be checked if the transmission housing or ring gear and pinion assembly are renewed. To make the adjustment, proceed as follows:

The correct cone point of housing and pinion are factory determined and assembly numbers are etched on rear face of pinion on all models. To determine correct thickness of shim pack (30 – Fig. 246), subtract the number etched on end of pinion from the correct guide number which follows:

Model	Guide Number
4040	8.330
4240	8.330
4440	8.3285
4640 & 4840	9.533

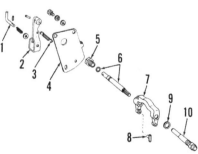

Fig. 244 — *View of 4440 model tow disconnect lever. Other models are similar.*

1. Latch
2. Shift lever
3. Spring
4. Stop plate
5. Retainer
6. L.H. shaft & "O" ring
7. Shift yoke
8. Shoe (2)
9. Aluminum washer
10. R.H. shaft

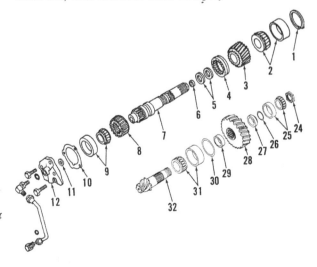

Fig. 246 — *Exploded view of output gear reduction unit, main drive bevel pinion and associated parts. Item (8) is separate from shaft (7), as shown, only on 4640 and 4840 models.*

1. Snap ring
2. Bearing
3. Reduction gear
4. Shift collar
5. Sealing rings
6. Oil seal
7. Reduction shaft
8. Drive sleeve
9. Rear bearing
10. Shim pack
11. Sealing bearing
12. Retainer
24. Nut
25. Bearing
26. Shim pack
27. Spacer
28. Gear
29. Spacer
30. Shim pack
31. Bearing
32. Pinion

The result is the correct thickness of shim pack (30). As an example, if the guide number is 8.330 and the number etched on rear of pinion is 8.315, (8.330 – 8.315 = 0.015) the correct thickness of shim pack is 0.015 inch.

Adjust pinion shaft bearings to 0.004-0.006 inch preload AFTER cone point is correctly set. Refer to paragraph 220.

DIFFERENTIAL AND MAIN DRIVE BEVEL RING GEAR

REMOVE AND REINSTALL

All Models

222. To remove the differential assembly, first drain transmission and hydraulic fluid. Remove rockshaft housing as outlined in paragraph 282, and on all models except Power Shift, remove hex shaft as outlined in paragraph 166. Remove load control arm on all models except Power Shift. The load control arm comes out with differential on these models. On Syncro-Range and Quad-Range models, remove transmission oil pump inlet and outlet lines.

Block up tractor and remove both final drive units as outlined in paragraph 230. Remove brake backing plates and brake discs, and withdraw both differential output shafts.

If two of the ring gear to housing cap screws are recessed, rotate ring gear until the two screws are horizontal, which will provide more clearance for removal.

Place a chain around differential housing as close to bevel ring gear as possible, attach a hoist and lift the differential enough to relieve the weight on carrier bearings. Remove both bearing quills using care not to lose, damage or intermix the shims located under bearing quill flanges. Differential assembly (and load control arm on power shift models) may be removed.

Overhaul the removed unit as outlined in paragraph 223.

When installing, place an additional 0.010 inch shim on bearing quill on ring gear side, tighten retaining cap screws and measure differential bearing end play using a dial indicator. Preload the carrier bearings by removing shims equal in thickness to the measured end

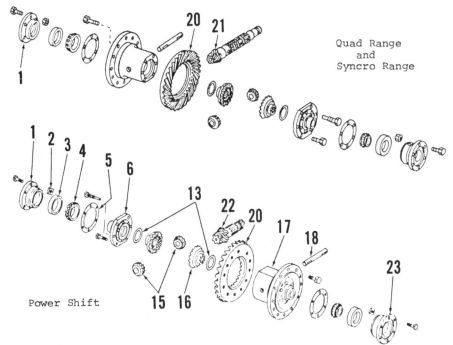

Fig. 247—Exploded view of differential assembly available on early 4040, 4240 and 4440 models without differential lock. Refer to Fig. 247A for legend.

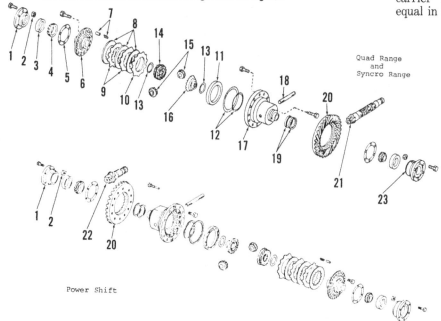

Fig. 247A—Exploded view of locking differential assembly available on early 4040, 4240 and 4440 models.

1. L.H. bearing quill
2. Square nut (2)
3. Bearing cup
4. Bearing cone
5. Shim pack
6. Housing cover
7. Spring and pin (3)
8. Drive disc (3)
9. Clutch plate (2)
10. Backing plate
11. Piston
12. "O" rings
13. Thrust washer
14. Gear (diff. lock)
15. Differential pinions
16. Axle gear
17. Differential housing
18. Pinion shaft
19. Sealing rings (2)
20. Bevel ring gear
21. Pinion (Syncro-Range)
22. Pinion (Power Shift)
23. R.H. bearing quill

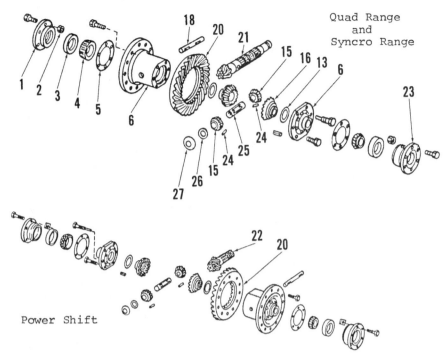

Quad Range
and
Syncro Range

Power Shift

Fig. 247B — Exploded view of non-locking differential available on late 4040, 4240 and 4440 models. Refer to Fig. 247C for legend.

play plus 0.002-0.005 inch. Shims are available in thicknesses of 0.003, 0.005 and 0.010 inch.

After the correct carrier bearing preload is obtained, attach a dial indicator, zero indicator button on one bevel ring gear tooth and check the backlash between bevel ring gear and pinion, in at least two places 180 degrees apart. Proper backlash is 0.008-0.015 inch. Moving one 0.005 inch shim from one bearing quill to the other will change the backlash by about 0.010 inch.

When bearing preload and backlash are established, tighten differential bearing quill cap screws to a torque of 85 ft.-lbs. and bend up lock plates. Tighten rockshaft housing bolts to 85 ft.-lbs. torque. Complete tractor assembly by reversing the disassembly procedure.

DIFFERENTIAL OVERHAUL

All Models

223. Differential may incorporate a hydraulically actuated locking clutch,

two, three or four pinions or other features. Refer to appropriate Fig. 247, 247A, 247B, 247C, 247D, 247E, 247F or 247G.

Differential gears and differential lock parts can be inspected after removing cover (6). Refer to paragraph 224 for service on differential lock clutch and to paragraph 225 for bevel ring gear. Tighten cover cap screws to 55 ft.-lbs. torque if equipped with differential lock, and 85 ft.-lbs. without lock.

224. **DIFFERENTIAL LOCK CLUTCH.** The multiple disc differential clutch can be overhauled after removing the unit as outlined in paragraph 222 and removing cover (6–Fig. 247A, 247C, 247E or 247G). The three internally splined clutch drive discs (8) are 0.112-0.118 inch thick. Discs should be renewed when thickness is less than 0.100 inch. Externally splined clutch plates (9) are 0.115-0.125 inch thick. The three piston return springs ride on guide dowels contained in cover (6) as shown in Fig. 248.

Remove piston (11–Fig. 247A, 247C, 247E or 247G) with air pressure or by grasping two opposing strengthening ribs with pliers. Renew sealing "O" rings whenever piston is removed. Examine sealing surface in bore of bearing quill which houses sealing rings (19), and renew quill if sealing area is damaged. Renew cast iron sealing rings (19) if broken, scored or badly worn.

225. **BEVEL RING GEAR.** The main drive bevel ring gear and pinion are available only as a matched set. Always

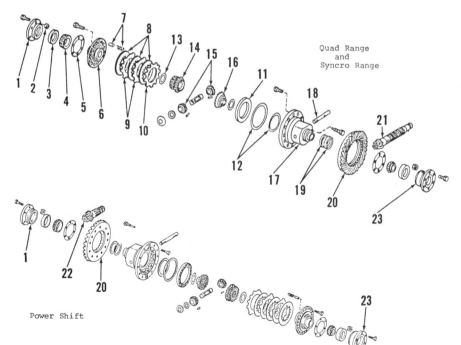

Quad Range
and
Syncro Range

Power Shift

1. L.H. bearing quill
2. Square nut (2)
3. Bearing cup
4. Bearing cone
5. Shim pack
6. Housing cover
7. Spring and pin (3)
8. Drive disc (3)
9. Clutch plate (2)
10. Backing plate
11. Piston
12. "O" rings
13. Thrust washer
14. Gear (diff. lock)
15. Differential pinions
16. Axle gear
17. Differential housing
18. Pinion shaft
19. Sealing rings (2)
20. Bevel ring gear
21. Pinion (Syncro-Range)
22. Pinion (Power Shift)
23. R. H. bearing quill

Fig. 247C — Exploded view of locking differential used on late 4040, 4240 and 4440 models.

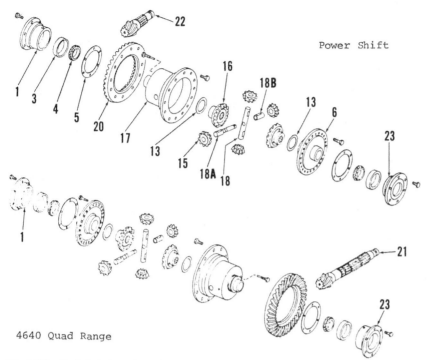

Power Shift

4640 Quad Range

Fig. 247D — Exploded view of non-locking differential used on some early 4640 and 4840 models. Refer to Fig. 247E for legend.

replace both gears. To remove ring gear, remove retaining cap screws and use a heavy drift, hammer or suitable press. Heat new gear to 300 degrees F. in an oven and position gear. Tighten retaining cap screws to 85 ft.-lbs. torque

on all models except 4640 and 4840, which should be tightened to 170 ft.-lbs. Renew pinion shaft as outlined in paragraph 183 for Syncro-Range models or paragraph 220 for Power Shift models.

DIFFERENTIAL LOCK

Tractors may be optionally equipped with a hydraulically actuated differential lock which may be engaged to insure full power delivery to both rear wheels when traction is a problem. The differential lock consists of a foot operated control and

regulating valve and a multiple disc clutch located in differential housing.

OPERATION AND ADJUSTMENT
Models So Equipped

226. Refer to Fig. 249. When pedal is depressed, pressurized fluid from the

hydraulic system is directed to clutch piston (11 – Fig. 247A, 247C, 247E or 247G) locking axle gear (14) to differential case. Elimination of the differential as a working part causes both differential output shafts and main drive bevel gear to turn together as a unit. Available power is thus transmitted to both rear wheels equally despite variations in traction.

To release the differential lock, slightly depress either brake pedal, which releases system pressure by acting through linkage.

227. **ADJUSTMENT.** The differential lock valve is applied through an operating rod from the pedal to an over-center arm. When the pedal is depressed, the over-center feature holds valve in the applied position until either brake pedal is lightly applied. The head of lock adjusting screw then contacts a clearance bar, which is mounted crosswise on operating rod (17 – Fig. 250) and pushes the rod upward to release the lock valve.

To adjust clearance, refer to Fig. 249. Depress differential lock pedal and leave pedal in the "ON" position. With brake pedals off, adjust lock adjusting screws to approximately 0.020 inch clearance between head of adjusting screw and clearance bar on operating rod. A small movement of either brake pedal will release lock valve.

Operating pressure need only be checked if valve has been overhauled or differential lock is suspected of having a malfunction. Check operating pressure by mounting a "T" fitting and a 0-1000 psi gage in the line from lock valve to differential lock. This line is located above inner end of right axle housing. Operate engine at 2000 rpm; then, when fluid temperature is 100 degrees F., depress lock pedal and observe pressure. Correct pressure is 435-485

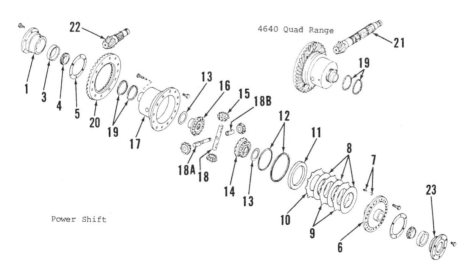

Power Shift

4640 Quad Range

1. L.H. bearing quill
2. Square nut (2)
3. Bearing cup
4. Bearing cone
5. Shim pack
6. Housing cover
7. Spring and pin (3)
8. Drive disc (3)
9. Clutch plate (2)
10. Backing plate
11. Piston
12. "O" rings
13. Thrust washer
14. Gear (diff. lock)
15. Differential pinions
16. Axle gear
17. Differential housing
18. Pinion shaft
18A. Pinion cross shaft
18B. Sleeve
19. Sealing rings (2)
20. Bevel ring gear
21. Pinion (Syncro-Range)
22. Pinion (Power Shift)
23. R.H. bearing quill

Fig. 247E — Exploded view of locking differential available on early 4640 and 4840 models.

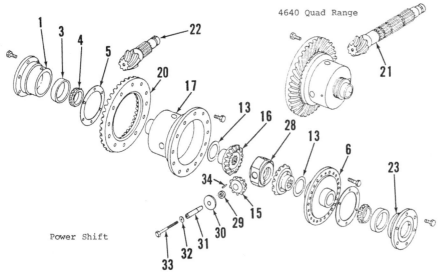

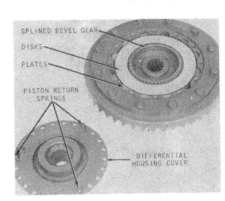

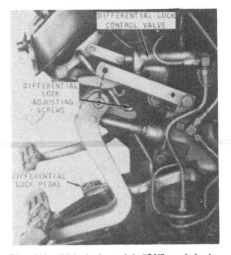

Fig. 248—Differential lock clutch partially disassembled. Piston return springs ride on guide dowels in cover and enter blank splines in clutch plate lugs.

Fig. 247F—Exploded view of non-locking differential used on some late 4640 and 4840 models. Refer to Fig. 247G for legend.

psi. If incorrect, disconnect links (15 – Fig. 250), remove plunger (1) and add or remove shims (4). One shim should change pressure 25-30 psi.

OVERHAUL

Model So Equipped

228. **CONTROL VALVE.** To remove the differential lock control valve, disconnect pin at plunger (1 – Fig. 250), rear cap screw on brace from pedals to brake valve (as shown in Fig. 249), and the oil pipes at the valve. Remove retaining cap screws and lift off valve unit and spacer(s). Items 10 through 14 (Fig. 250) must be removed before pressure control valve (8) can be removed.

All parts are renewable individually. Examine parts for wear or scoring and springs for distortion, and renew any parts which are questionable. Keep shim pack intact for use as a starting point when readjusting operating pressure. Renew "O" rings whenever valve is disassembled.

Assemble by reversing the disassembly procedure using Fig. 250 as a guide. Tighten control valve retaining cap screws securely and adjust as outlined in paragraph 227.

229. **CLUTCH.** Refer to paragraph 224 for overhaul procedures on differential lock clutch unit and associated parts.

Fig. 249—With lock pedal "ON" and brake pedals "OFF" adjust screw on brake pedals to obtain approximately 0.020 clearance.

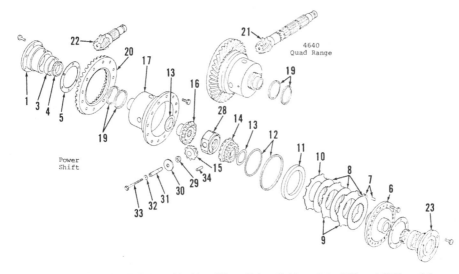

1. L.H. bearing quill
3. Bearing cup
4. Bearing cone
5. Shim pack
6. Housing cover
7. Spring and pin (3)
8. Drive disc (3)
9. Clutch plate (2)
10. Backing plate
11. Piston
12. "O" rings
13. Thrust washer
14. Gear (diff. lock)
15. Differential pinions
16. Axle gear
17. Differential housing
19. Sealing rings (2)
20. Bevel ring gear
21. Pinion (Syncro-Range)
22. Pinion (Power Shift)
23. R.H. bearing quill
28. Hub
29. Needle thrust bearing
30. Thrust washer (3 used)
31. Pinion shaft (3 used)
32. Special washer (3 used)
33. Cap screw (3 used)
34. Pinion needles (54 used)

Fig. 247G—Exploded view of locking differential available on late 4640 and 4840 models.

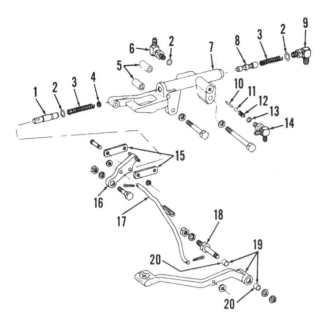

Fig. 250 — Exploded view of differential lock control valve, pedal and associated parts.

1. Plunger
2. "O" ring
3. Spring
4. Steel shim
5. Spacers
6. Oil return elbow
7. Lock valve housing
8. Pressure control valve
9. Elbow (outlet)
10. Pin
11. Steel ball
12. Spring
13. "O" ring
14. Inlet "T"
15. Links
16. Arm & pin
17. Operating rod & bar
18. Special pin
19. Pedal w/bushing
20. Bushing

REAR AXLE AND FINAL DRIVE

Some models may be equipped with high clearance rear axle drop housings, which contain the final reduction bull gear and pinion. Standard equipment on all models is a planetary reduction final drive gear located at inner ends of rear axle housings.

REMOVE AND REINSTALL

All Models

230. To remove either final drive as a unit, first drain the transmission and hydraulic fluid, suitably support rear of tractor and remove rear wheel or wheels.

On standard models, remove fenders and light wiring if so equipped. On high clearance models remove 3-point hitch lift links and draft links or entire drawbar assembly. On right final drive of models so equipped, remove differential lock pressure pipe.

On left final drive, remove rockshaft return pipe.

On models with Sound-Gard body or Roll-Gard, remove only the mount bolts on side being removed, and loosen front mount bolts on the same side for clearance on rear mount.

Support final drive assembly with a jack or a hoist, remove attaching cap screws and swing the unit away from transmission housing.

On high clearance models remove the six stud nuts securing drop housing to shaft housing, remove the two retainer plugs and thread jack screws into retainer plug holes. Tighten jack screws evenly to force housings apart.

When reinstalling, be sure the sun gear (25 – Fig. 251 or Fig. 252) on standard models stays all the way in, to prevent brake disc from falling behind teeth on gear. Tighten the retaining cap screws to transmission case to a torque of 170 ft.-lbs. on standard models or 150 ft.-lbs. on high clearance models. Complete the installation by reversing the removal procedure. Tighten the outer gear housing (11 – Fig. 253) to shaft housing (8) to a torque of 275 ft.-lbs.

OVERHAUL

All Except High Clearance

231. To disassemble the removed final drive unit, remove lock plate (24 – Fig. 251 or Fig. 252) and cap screw (23), then withdraw planet carrier (21) and associated parts.

Planet pinion shaft (17) is retained in carrier by snap ring (16). To remove, ex-

Fig. 251 — Exploded view of planetary type final drive assembly typical of most standard axle models. Refer to Fig. 252 for axle used on 4840 models.

1. Axle
2. Oil seal
3. Retainer
4. Spacer
5. Bearing cone
6. Bearing cup
7. Lube plug
8. Axle housing
9. Dowel
10. Dowel
11. Gasket
12. Oil seal
13. Bearing cone
14. Bearing cup
15. Thrust washer
16. Retaining ring
17. Pinion shaft
18. Bearing washer
19. Planet pinion
20. Bearing roller
21. Planet carrier
22. Washer
23. Cap screw
24. Lock plate
25. Sun gear

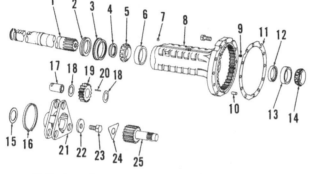

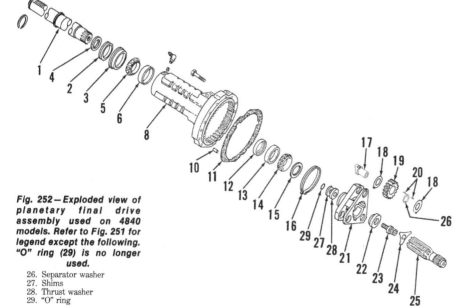

Fig. 252 — Exploded view of planetary final drive assembly used on 4840 models. Refer to Fig. 251 for legend except the following. "O" ring (29) is no longer used.

26. Separator washer
27. Shims
28. Thrust washer
29. "O" ring

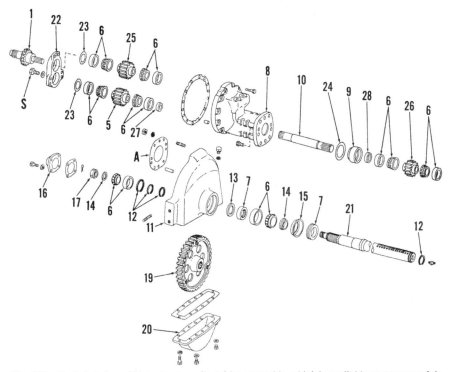

Fig. 253 — Exploded view of high clearance final drive assembly, which is available on some models.

1. Diff. gear	11. Gear housing	17. Bearing nut	24. Shims (0.002, 0.005 & 0.010 in.)
5. Drive shaft gear	12. Snap rings	19. Bull gear	25. Differential output gear
6. Bearing & cup	13. Washer	20. Cover	26. Drive shaft gear
7. Oil seal	14. Spacer	21. Axle shaft	27. Oil seal
8. Drive shaft housing	15. Bearing	22. Inner quill	28. Oil seal
9. Quill	16. Oil seal cup	23. Shims (0.002, 0.005 & 0.010 in.)	
10. Drive shaft	16. Bearing cover		

ready to install. Insert bearing and install planet carrier, washer (22) and cap screw (23).

Adjustment of bearings (5, 6, 13 & 14) is accomplished on 4040 and 4240 models by applying specified torque to screw (23), then maintaining adjustment by locking with plate (24). Correct torque should be as follows: 4040–170 ft.-lbs.; 4240–170 ft.-lbs.

Adjustment of bearings (5, 6, 13 & 14–Fig. 252) is accomplished on 4440, 4640 and 4840 models, by varying the thickness of shims installed between end of axle (1) and washer (22). The correct thickness can be determined by crushing lead balls between washer (22) and end of axle. To check, remove cap screw (23), washer (22) and all shims from end of axle. Use heavy grease to stick two lead balls which are 0.3-0.4 inch diameter (1 to 000 buck shot are approximately correct size). Be sure the two lead balls are on exactly opposite positions on end of axle, then install washer (22) and screw (23). Tighten screw (23) to 200 ft.-lbs. torque, rotate axle (or housing) several times, then retorque to 200 ft.-lbs. Remove screw (23) and washer (22). Remove and measure each of the two flattened lead balls, then average the thickness. Install shims between washer and end of axle equal to the average flattened thickness and reassemble. Upon reassembly, screw (23) should be tightened to 580 ft.-lbs. torque. Usual procedure for applying this much torque is with torque multiplying wrench such as JDST-38 or equivalent. Check rolling torque of axle using a standard torque wrench on end of screw (23) after final assembly is complete. Correct rolling torque without wheel hub is 10-23 ft.-lbs. Incorrect rolling torque should be corrected by removing screw (23) and washer (22), then adding or removing shims as required. Be sure to retorque to 580 ft.-lbs. and recheck rolling torque if any change is required. Lock position of screw (23) using plate (24). Tighten screw (23) a maximum of 8 degrees after obtaining correct (580 ft.-lbs.) torque if necessary in order to install lock plate (24).

High Clearance Models

232. OUTER HOUSING AND GEARS. If only outer housing, gears, shafts, bearings or oil seals are being overhauled, the complete final drive assembly will not need to be removed. Suitably support the tractor and remove wheel and tire unit. Remove draft link or disconnect drawbar from gear housing.

Remove the six stud nuts securing gear housing to shaft housing, remove the two retainer plugs and thread jack screws into retainer plug holes. Tighten

pand the snap ring and, working around the carrier, tap all three shafts out while snap ring is expanded. Withdraw the parts, being careful not to lose any of the loose bearing rollers (20) in each planet pinion. Examine shaft, bearing rollers and gear bore for wear, scoring or other damage and renew as indicated.

NOTE: Bearing rollers should be renewed as a set.

After planet carrier has been removed, axle shaft (1) can be removed from inner bearing and housing by pressing on inner end of axle shaft. Remove bearing cone (5) and spacer (4) if they are damaged or worn. When assembling, heat spacer and cone to approximately 300 degrees F. and install on shaft making sure they are fully seated. Heat inner bearing cone (13) to 300 degrees F. Have planet carrier

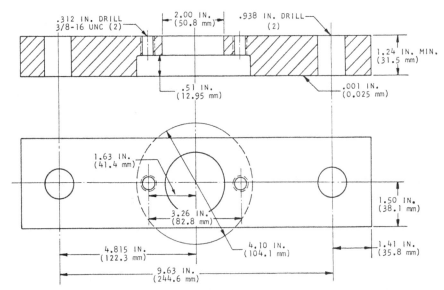

Fig. 254 — The special tool shown in Fig. 255 can be fabricated to dimensions shown.

jack screws evenly to force housings apart.

To disassemble the removed gear housing, first remove the bull gear cover and wheel axle inner bearing cover (16 – Fig. 253), then remove inner bearing nuts (17) with a spanner wrench. Unseat snap ring on inner side of bull gear, install spacers between insides of gear housing to prevent damage to housing, then press or drive out rear axle shaft. Axle shaft is equipped with two oil seals, an inner seal which is pressed into housing and a two-piece outer seal in housing and on shaft. When assembling, heat bearing cones to a temperature of 300 degrees F. to facilitate installation. Install and tighten inner axle shaft nut (17) to torque of 200 ft.-lbs. to seat bearing races. Rotate axle several times, then recheck torque. Loosen nut (17), then retighten to 50 ft.-lbs. torque. Install locking cotter pin through axle and nut. It may be necessary to loosen nut slightly before cotter pin can be installed.

Selection of shims (24) requires using a special tool which can be fabricated to dimensions shown in Fig. 254. Assemble bearings (6 – Fig. 253), gear (26), seal (28) and outer quill (9) in housing (11). Position gasket (A) over studs and install special tool over two studs as shown in Fig. 255. Tighten the two stud nuts (F) to 250 ft.-lbs. torque while making sure that screws (D) remain loose. Turn the outer drive shaft gear (26 – Fig. 253) while tightening the two screws (D – Fig. 255) evenly to a torque of 25 in.-lbs. Use a feeler gage (B) as shown to measure clearance between outer quill (C) and special tool (E). Remove special tool and install shims (24 – Fig. 253) equal to measured clearance.

Tighten stud nuts which attach drop housing (11 – Fig. 253) to drive shaft housing (8) to 445 ft.-lbs. torque. Fill each final drive housing with 4½ pints of SAE90 multi-purpose gear lubricant.

233. SHAFT HOUSING AND INNER GEARS. To overhaul the drive shaft housing and associated parts, first remove final drive assembly as outlined in paragraph 230 and outer gear housing as in paragraph 232.

Disassembly will be obvious after removing inner quill (22 – Fig. 253). When assembling, select thickness of shims (23) that will provide gears (5 & 25) with 0 to 0.004 inch end play. Be sure that bearings are firmly in place and that screws (S) are tightened to 55 ft.-lbs. torque, but **DO NOT** preload bearings by installing too many shims (23). Refer to paragraph 232 for remainder of repair and adjustment procedures. Tighten cap screws which attach drive shaft housing (8) to transmission housing to 150 ft.-lbs. torque.

BRAKES

The hydraulically actuated power disc brakes use the main hydraulic system as the power source. Discs are located on differential output shafts.

OPERATION AND ADJUSTMENT

All Models

244. Power to the wet type single disc brakes is supplied by the system hydraulic pump through foot operated control valves when engine is running. A nitrogen filled accumulator provides standby hydraulic pressure in sufficient volume to apply the brakes several times

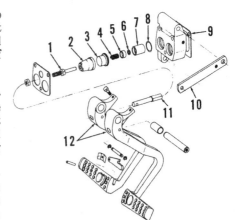

Fig. 263 – Exploded view of brake valve cover, pedals and associated parts. Operating parts shown are for left brake. Right side is identical.

1. Operating rod
2. Boot
3. Boot retainer
4. Spring
5. Stop
6. Snap ring
7. Operating rod guide
8. "O" ring
9. Cover
10. Strap
11. Extension
12. Pedals

after main hydraulic system ceases to operate. Control valve also contains master cylinders to permit manual operation when hydraulic pressure is not available.

Refer to Fig. 263. Parts 1 through 8 are duplicated for right and left brakes. In the first ¾-inch of pedal travel, operating rod (1) moves operating rod guide (7) which mechanically opens an equalizing valve pin and ball (1 and 6 – Fig. 264). This insures equal pressure to both brakes when both pedals are depressed. Further movement of operating rod guide closes brake valve plunger (10) to close escape passage. Brake valve (12) is also unseated which allows pressure oil to fill the cavity under manual brake piston (9) and continue on to unseat check valve disc (20) which allows oil to reach brake cylinders. If pressure in valve should become

Fig. 255 – Refer to text for use of special tool necessary for determining correct thickness of shims (24 – Fig. 253) required for assembly.

A. Gasket
B. Feeler gage
C. Outer quill (9 – Fig. 253)
D. Pressure screws
E. Special tool
F. Stud nuts

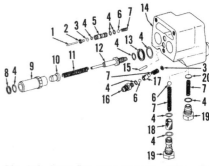

Fig. 264 – Exploded view of brake valve with cover removed. Operating parts shown are for right brake.

1. Equalizing valve shaft
2. Sleeve
3. Washer
4. "O" ring
5. Guide
6. Steel ball
7. Spring
8. Paper washer
9. Manual brake piston
10. Plunger
11. Spring
12. Inlet valve nipples (Brake valve)
13. Back-up ring
14. Housing
15. Equalizing pin
16. Inlet connector
17. Inlet guide
18. Screen
19. Plugs
20. Check valve disc

too high, plunger (10) and brake valve (12) will move up as soon as pedal is released slightly, which allows escape passage to open and inlet passage to close stabilizing pressure.

In case of pressure failure, manual braking is accomplished as follows: Pressure on the operating rod and guide closes plunger (10) and opens brake valve (12). Since oil is trapped under manual brake piston (9) and the pressure line from pump is closed by a check ball and spring, pressure exerted by manual piston can open check valve seat (20) and apply brakes. As pedal is released, the reservoir check ball is pulled off its seat, allowing oil from brake valve reservoir to be pulled into manual brake piston cavity, and continued pumping of the pedal can build pressure in the brake lines.

The only adjustment provided is at operating rod and operating rod extension (1 and 11 – Fig. 263). This adjustment is used to equalize brake pedal height. To adjust the pedals, loosen locknut and turn operating rod in or out until pedals are equal in height and are approximately 5¾ inch from top of foot pad on pedal to the floor, when measured at a right angle to the floor pan (without mat). Pedal height should be equal. For service on the hydraulic pump, refer to paragraph 279. Refer to

Fig. 265 – Manual brake pistons and brake valve plungers can be lifted out after cover is removed.

Fig. 266 – Remove guides and "O" rings from cover bore.

the appropriate following paragraphs for service on brake valve, actuating cylinders, brake discs and accumulator.

BLEEDING

All Models

245. When brake system has been disconnected or disassembled, bleed the system as follows:

Pump brakes until accumulator is discharged. Start the engine, loosen locknut on bleed screws, located on both sides of transmission housing, just above axle housings. Back out bleed screws two full turns, then tighten locknuts to prevent external oil leak. Fully depress brake pedal and hold down for 1-2 minutes. With pedal depressed, tighten bleed screw then release the pedal.

To test brake system, stop engine, discharge accumulator by holding pedal down for at least 1 minute, then depress pedal three times. Pedal travel should not exceed 4.375 inches on the third application and feel should be solid. If pedal is spongy, repeat bleeding procedure and, if trouble is not corrected, overhaul system components.

BRAKE VALVE

All Models

246. To remove the brake valve, first bleed down the pressure in accumulator, then disconnect pressure and discharge lines at valve housing. Remove special pins in pedals which actuate operating rod extensions and remove extensions through holes in pedals. Remove strap to pedals and remove right pedal. Remove cap screws and lift off control valve.

Use Figs. 263, 264 and 265 as a guide when disassembling the brake valve. Manual brake pistons, brake valve plungers and springs can be lifted out.

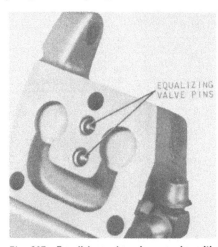

Fig. 267 – Equalizing valve pins can be withdrawn from underside of cover.

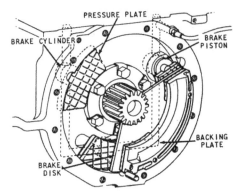

Fig. 268 – Schematic view of assembled brake unit.

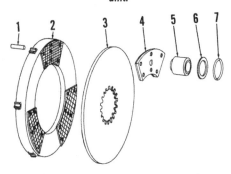

Fig. 269 – Exploded view of wet type hydraulic disc brake operating parts located on differential output shafts (final drive sun gears).

1. Dowel
2. Backing plate
3. Brake disc
4. Pressure pad
5. Piston
6. Back-up ring
7. "O" ring

Use a deep socket to remove inlet valve nipples. Clean the parts thoroughly and check against the values which follow:
Brake manual piston
    bore . . . . 0.9365-0.9375 in.
Manual piston O.D. . . . . . 0.933-0.935 in.
Plunger bore in piston . . . 0.561-0.563 in.
Plunger O.D. . . . . 0.5595-0.5605 in.

Check for free movement of operating valve guide (Fig. 266) and equalizing valve pins (Fig. 267). Binding can cause equalizing valve to fail to close and both brakes will be applied if opposite pedal is pushed. Renew "O" rings whenever unit is disassembled.

Use Figs. 263, 264, 265 and 266 as a guide when reassembling. Tighten inlet valve nipples (12 – Fig. 264) to a torque of 40 ft.-lbs. When control valve has been installed, check for equal pedal height and bleed the system as outlined in paragraph 245.

DISCS AND SHOES

All Models

247. To remove the brake discs or operating cylinders, first remove the final drive unit as outlined in paragraph 230. Remove the output shaft, backing plate and brake disc. The three stationary shoes are riveted to the backing plate (2 – Fig. 269). The three actuating pads (4) are pressed on operating

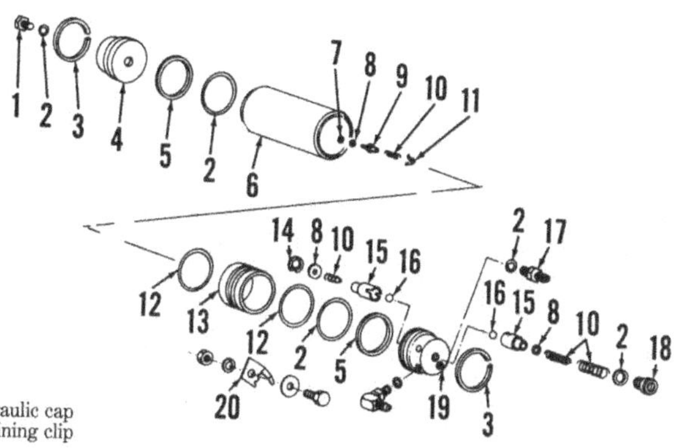

Fig. 270—Exploded view of nitrogen filled brake accumulator.

1. Plug
2. "O" ring
3. Retaining ring
4. Gas end cap
5. Back-up ring
6. Cylinder
7. Packing
8. Washer
9. Valve
10. Spring
11. Spring guide
12. Packing (U-cup)
13. Piston
14. Retaining ring
15. Guide
16. Steel ball
17. Connector
18. Plug
19. Hydraulic cap
20. Retaining clip

pistons (5) which can be withdrawn from transmission housing bores after brake disc is removed. Facings are available in sets of three and should only be renewed as a set.

Diameter of operating pistons (5) should be 2.2495-2.2505 inches for all models except 4640 and 4840. Diameter of pistons (5) for 4640 and 4840 models should be 2.6245-2.6255 inches. Pistons for all models should have 0.0025-0.0065 inch diametral clearance in cylinder bores. Refer to Fig. 268 for a schematic assembled view of brake unit.

BRAKE ACCUMULATOR

All Models

248. **R&R AND OVERHAUL.** To remove the brake accumulator, bleed fluid pressure from brake system. Disconnect pressure lines for accumulator connections, then remove retaining clip (20 – Fig. 270) and remove accumulator.

CAUTION: Bleed accumulator before attempting to disassemble. Gas side of accumulator piston is charged to 475-525 psi with NITROGEN gas.

Accumulator is discharged by removing protective plug (1) and depressing charging valve (9). With pressure removed, push cap (19) into cylinder, then unseat and remove snap ring (3). Remove gas end cap (4) by removing the other snap ring (3).

Check all parts for wear or damage, and assemble by reversing the disassembly procedure. Recharge the cylinder using approved charging equipment and DRY NITROGEN ONLY, to a pressure of 500 psi. Remove charging equipment and check by immersing the charged accumulator in water. When it has been determined that there are no leaks, reinstall plug (1) and install accumulator by reversing the removal procedure. Bleed brakes as outlined in paragraph 245. After engine has been run again and accumulator recharged, brakes should still have at least five power applications after engine is shut off.

POWER TAKE OFF

Tractors may be equipped with either a single speed or dual-speed pto. For ease of identification the pto types are divided into four groups, with the total number of drive gears in the system as the means of identification. The groups are called Two Gear, Four Gear, Five Gear, or Seven Gear systems. Refer to appropriate section for model being serviced.

TWO GEAR SYSTEM

All models with Syncro-Range or Quad-Range transmission and single speed pto.

FOUR GEAR SYSTEM

All 4640 Power Shift models and 4840 models.

FIVE GEAR SYSTEM

All 4040, 4240 and 4440 models with Syncro-Range or Quad-Range transmission and dual-speed pto.

SEVEN GEAR SYSTEM

All 4040, 4240 and 4440 Power Shift models with dual-speed pto.

Fig. 271—Exploded view of two gear drive system used on Syncro-Range or Quad-Range tractors with single speed pto. "O" ring (5) is no longer used.

1. Main drive
2. Bushing
3. Snap ring
4. Oil shield
6. Ball bearing
7. Pto gear w/shield
8. Brake piston
9. Dowel pin
10. Sleeve
11. Ball bearings
12. Bearing quill
13. Gasket
14. Oil seal
15. Ball bearing
16. Pto shaft
17. Bushing

OPERATION

All Models

249. The power take-off is driven by a hydraulically applied, independently controlled, flywheel operated, single disc clutch on Syncro-Range or Quad-Range Models. Power Shift Models use a multiple disc clutch, which is part of the transmission. See paragraph 167 for clutch overhaul on Syncro-Range or Quad-Range models and paragraph 213 for clutch pack overhaul on Power Shift models.

On all tractors the pto clutch is hydraulically engaged. The control valve on Power Shift Models is contained in the transmission pedal valve housing and service procedures are contained in paragraph 209. On Syncro-Range and Quad-Range Models a separate valve is used; refer to paragraph 259 for service procedures.

TWO GEAR SYSTEM

All Models So Equipped

250. The two gear system consists of a main drive gear (1 – Fig. 271) which is located on the clutch shaft and driven by the pto clutch hub, and the pto drive gear (7) which drives shaft (16). Bearing quill (12) and shaft can be removed as a unit, which will leave the drive gear centered on sleeve (10). Pto brake piston (8) remains applied against gear (7) when tractor is running, but pto is not in use.

251. **R&R AND OVERHAUL.** To overhaul the two gear system, split trac-

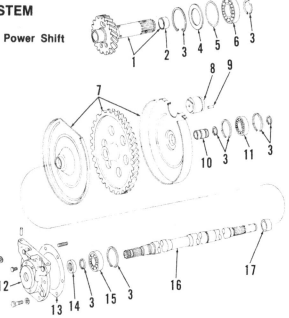

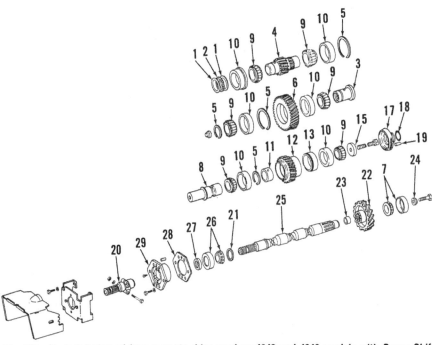

Fig. 272 – Exploded view of four gear pto drive used on 4640 and 4840 models with Power Shift transmission.

1. Race	7. Taper bearing cup and cone	15. Retainer	23. Bushing
2. Thrust bearing	8. Idler shaft (lower)	17. Pto brake shoe	24. Retainer
3. Idler shaft (upper)	9. Bearing cone	18. "O" ring	25. Pto shaft
4. Pto drive gear	10. Bearing cup	19. Dowel pin	26. Bearing cup and cone
5. Snap rings (not identical)	11. Cone spacer	20. 1⅜ inch adapter shaft	27. Seal
	12. Lower idler gear	21. Spacer	
6. Upper idler gear	13. Cup spacer	22. Pto drive gear	29. Bearing quill

tor at clutch housing as outlined in paragraph 166.

Remove clutch operating housing as outlined in paragraph 170. The large pto drive gear and oil shield must be moved rearward off sleeve (10–Fig. 271) and then lifted out toward the front, being careful not to damage oil shield. Check bearings and renew if necessary. Check bushing (2) in main drive gear and renew if worn. Reassemble in reverse order of disassembly, and renew seal (14). Tight

en rear bearing quill cap screws to 85 ft.-lbs. torque.

FOUR GEAR SYSTEM

All Models So Equipped

252. The four gear system is used only on 4640 and 4840 Power Shift models. The drive train consists of a main drive gear (4–Fig. 272) which is mounted behind the clutch pack on Power Shift models and is driven by the rear clutches. The main drive gear drives an upper

idler (6), a lower idler (12) which is held by the pto brake (17) when not in use. The pto drive gear (22) is splined to and turns shaft (25).

253. **R&R AND OVERHAUL.** The tractor must be split at the clutch housing as outlined in paragraph 204 in order to gain access to pto drive gears. Remove clutch pack as outlined in paragraph 206, remove cap screws that go into transmission housing from the front, and remove clutch housing.

Use a suitable puller to remove drive gear (22–Fig. 272) from lower shaft after cap screw and washer are removed. Before removing pto shaft toward the front, remove shield from the rear of shaft, puncture the oil seal (27); remove seal and remove snap ring inside bearing quill (29). Remove quill, and shaft can be driven out forward.

Remove upper idler (6). Remove retainer (15) and remove lower idler (12) with a suitable puller. Idler shaft (8) is retained by a set screw.

Renew any parts in question, and use Fig. 272 as a guide to reassembly. When renewing bearing cups, heat idler gears in oil (no hotter than 300°F) after installing snap ring and cup spacer. Be sure notched side of spacer (13) is seated against snap ring. Seat bearing cups against spacer (13) and snap ring (5). Heat bearing cones in oil and install on shafts while hot. Install cap screw (L–Fig. 273) against retainer (15) on lower idler gear (12). Tighten screws to 85 ft.-lbs. while bearing cones are still hot. Tighten screw (S) against retainer (24) to 170 ft.-lbs. torque. Assemble in reverse order of disassembly and tighten rear bearing quill cap screws to 85 ft.-lbs. torque.

Fig. 274 – Exploded view of drive gears and associated parts used in five gear system. Dual speed gears shown in Fig. 275 complete the system. "O" ring (5) is no longer used.

1. Main drive gear
2. Bushing
3. Snap ring
4. Oil shield
5. "O" ring
6. Ball bearing
7. Pto gear w/shield
8. Pto brake piston
9. Sleeve
10. Roller bearing
11. Pto shaft
12. Snap ring
13. Bushing

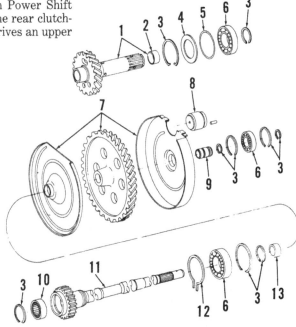

Fig. 273 – View of four gear pto installed. Refer to Fig. 272 for legend except special screws (L and S).

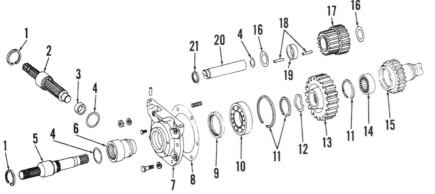

Fig. 275 — Exploded view of reduction gears and associated parts used with five gear system. Item 15 is the same as Item 11 — Fig. 274.

1. Snap ring	6. Pto pilot	11. Snap ring
2. 540 rpm shaft	7. Bearing quill	12. Spring washer
3. Bearing race	8. Gasket	13. 540 rpm gear
4. "O" ring	9. Oil seal	14. Roller bearing
5. 1000 rpm shaft	10. Ball bearing	15. Pto shaft

16. Thrust washer
17. Countershaft gear
18. Roller bearing (48)
19. Spacer
20. Idler shaft
21. Snap ring

FIVE GEAR SYSTEM

All Models So Equipped

254. The five gear system is used on 4040, 4240 and 4440 models equipped with Syncro-Range or Quad-Range, and dual-speed pto. The drive train consists of a main drive gear (1 – Fig. 274), pto shaft drive gear (7), pto shaft with gear (11), countershaft gear (17 – Fig. 275) and 540 rpm gear (13). Pto brake piston (8 – Fig. 274) is applied against gear (7) when tractor is running but pto is not in use. The 1000 rpm shaft (5 – Fig. 275) goes through the 540 rpm gear and splines into pto shaft which uses only the main drive gear and pto shaft drive gear. The drive train is then like the two gear system.

When 540 rpm is desired, the 540 rpm rear shaft (2) is inserted through bearing quill and splines into 540 gear (13) which uses all five gears to transmit power to rear shaft at the reduced speed.

255. To overhaul the reduction gear assembly at rear of tractor, drain hydraulic system, remove snap ring (1 – Fig. 275) from rear shaft and withdraw shaft. Remove rear bearing quill (7) and remove 540 rpm gear (13). Squeeze snap ring (12 – Fig. 274) and pull pto shaft (11) from tractor, which will leave the large drive shaft gear on its sleeve (9). Countershaft gear (17 – Fig. 275) can be removed by pulling shaft and snap ring (20 and 21) from case while holding countershaft gear to prevent losing the 48 roller bearings (18) and thrust washers (16). Inspect all parts for wear or other damage. When installing countershaft gear use a nonfibrous grease to hold bearings in place. Reassemble in reverse order of disassembly and tighten bearing quill cap screws and nuts to 85 ft.-lbs. torque.

If main drive gear and large pto drive gear must be removed, refer to para-graph 251 and overhaul the forward section using the same procedures as with the two gear system.

SEVEN GEAR SYSTEM

256. All 4040, 4240 and 4440 Power Shift Models with dual speed pto use this system. It consists of a four gear drive system between the planetary pack and the clutch pack in the transmission. Fig. 276 shows the main drive gear (2), upper idler gear (7), lower idler gear (10), which drives pto drive gear (12) and pto shaft (22). Pto shaft and gear (15 – Fig. 275) drives countershaft gear (17),

which drives 540 rpm gear (13) at the reduced speed when 540 rpm shaft (2) is used in bearing quill (7). When the 1000 rpm shaft (5) is used, the reduction gear system is by-passed, and the four gear system in the center of transmission drives shaft (5) at 1000 rpm.

To overhaul the reduction gears, refer to paragraph 255 and Fig. 275. Pto shaft and gear (15) on this system cannot be removed from the rear unless tractor is split, since the pto drive gear (12 – Fig. 276) is held to the shaft by special cap screw (14).

257. To overhaul the forward four gear drive train of the seven gear system, refer to paragraph 204 to split tractor, and to paragraph 206 and 207 to gain access to pto gears. If only the pto main drive clutch gear (2 – Fig. 276) is to be removed, it is not necessary to remove clutch housing. Remove snap ring in front of front bearing (15) on main drive gear, remove C1 and C2 clutch shafts (Fig. 213) and pull main

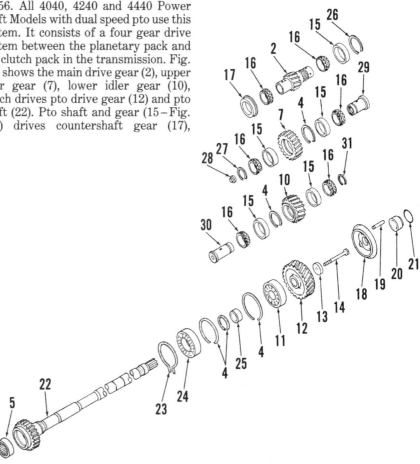

Fig. 276 — Exploded view of pto drive gears of seven gear system. Reduction gear portion of drive train is shown in Fig. 275. Install thickest snap ring (26, 27 and 31) possible to correctly adjust bearings.

2. Pto main drive	13. Washer	19. Pin	25. Bushing
4. Snap ring	14. Special cap screw	20. Brake piston	26. Snap ring
7. Upper idler	15. Bearing cup	21. "O" ring	27. Snap ring
10. Lower idler	16. Bearing cone	22. Pto shaft & gear	28. Shaft retainer
11. Ball bearing	17. Flanged bearing cup	23. Snap ring	29. Upper idler gear shaft
12. Pto drive gear	18. Brake shoe	24. Ball bearing	30. Lower idler gear shaft
			31. Snap ring

drive forward. If other pto gears require service, remove clutch housing as outlined in paragraph 207. Use a puller to remove pto drive gear (12–Fig. 276) after removing special cap screw and washer. Pto shaft may then be removed rearward by squeezing snap ring (23) and driving on pto shaft. Remove lower idler gear (10), bearings and shaft assembly. Upper gear (7) is supported in two taper roller bearings. Brake piston (20) can be forced out as follows: To remove brake piston (20), remove test plug in pedal valve housing marked "PTO BR" and use air pressure in the hole to force brake piston out.

Reassemble in reverse order of disassembly. Drive pto shaft front bearing (11) into case. Heat ball bearing (24) in oil no hotter than 300 degrees F. and install on pto shaft after making sure snap ring (23) is installed first. Seat bearing against shoulder on shaft. If upper idler has a ball bearing, heat bearing before installing. Snap rings (26, 27 & 31) are available in several thicknesses for later models. Thickest snap ring possible to install should be used to reduce bearing play to near zero. Tighten screw (14) to 120 ft.-lbs. torque and rear bearing quill cap screws and nuts to 85 ft.-lbs. torque.

PTO CLUTCH VALVE

Power Shift Models

258. The Power Shift pto clutch valve service is covered in paragraph 209 as part of the pedal valve.

To adjust the pto operating linkage on Power Shift models, remove the left cowl for access to clutch valve operating rod. Disconnect yoke from upper end of rod and push pto lever to 3/16-inch from front of lever slot (front edge of lever). Push down on operating rod until valve bottoms and adjust yoke so pin hole aligns with lower hole in lever.

Syncro-Range and Quad-Range Models

The pto clutch valve is included as part of the Perma-Clutch valve and oil pressure regulating valve housings, located on the left side of clutch housing. The inner section of the housing (12–Fig. 160) contains the clutch and pto operating valves and the pto lock piston (25), which is applied hydraulically after pto valve has been engaged. This keeps pto lever engaged firmly until lever is disengaged manually. The outer section of housing (2) contains the Perma-Clutch oil pressure regulating valve and the oil cooler relief valve.

259. **REMOVE AND REINSTALL.** To remove outer and inner sections of oil pressure regulating valve housing,

Fig. 287—View of pressure regulating valve housing and operating linkage typical of all models with Perma-Clutch.

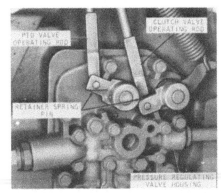

Fig. 288—View of operating linkage on pressure regulating valve housing for tractors with Perma-Clutch.

remove operating rods for both clutch valve and pto valve (Fig. 287). Loosen locknut on pto operating rod to allow the yoke pin retainer to be lifted out of pin. On some models, it may be necessary to drive the retainer spring pins out of arms and remove rods. Remove oil

pipes, clutch pedal return spring and remove only the eight outermost flange cap screws and pull assembly straight outwards to get free of adapters (See items 14 and 15 Fig. 160). After the outer and inner sections are removed as a unit, the inner cap screws can be removed to separate the halves. Inspect the "O" rings on adapters and renew if necessary. Assemble in reverse order of disassembly.

260. To overhaul the pto clutch valve and Perma-Clutch valve in the inner housing, separate inner and outer housings (Fig. 160). The pins in valves (10 and 11) must be removed before springs and valves will come out. The pto valve, springs, pin and spacers are identical with those of the Perma-Clutch, but it is better to return the valves to the bores from which they were removed. Inspect valves for nicks or burrs and make sure valves slide freely when reinstalled. Renew "O" rings on adapters if necessary and be sure pto lock piston (25) is free in its bore. Tighten cap screws to outer housing to 20 ft.-lbs. torque.

261. The oil pressure regulating valve housing (2–Fig. 160) contains the oil pressure regulating valve (23) and oil cooler relief valve (22). Valve (23) maintains approximately 140 lbs. pressure to the power train, and opens above that pressure to allow oil to be routed to the main pump and oil cooler. Make sure valves move freely in their bores and that they are free of nicks and burrs. "O" rings are used on valves to prevent sticking. Shims (19) must be in place before assembly. Refer to paragraph 162 for clutch pressure check.

HYDRAULIC SYSTEM

The closed center hydraulic system provides standby pressure to all tractor hydraulic components, with a maximum available flow of 23 gpm on models without Power Front Wheel Drive, or 26.5 gpm on models with Power Front Wheel Drive. Models with Power Front Wheel Drive have a 4.0 cubic inch main pump, while all other models are equipped with a 3.0 cubic inch pump. Working fluid for all hydraulic functions is available at a standby pressure of 2200-2300 psi.

MAIN HYDRAULIC SYSTEM

All Models

272. **OPERATION.** The main hydraulic system pump is mounted in front of radiator and coupled to engine crankshaft. This variable displacement, radial piston pump provides only the fluid necessary to maintain system pressure. When there are no demands on the system, pistons are held away from the pump camshaft by fluid pressure and no flow is present. When pressure is lowered in the supply system by hydraulic demand or by leakage, the stroke control valve in the pump meters fluid from the camshaft reservoir, permitting the pistons to operate and supply the flow necessary to maintain system pressure.

The transmission pump provides pressure lubrication for the transmission gears and shafts, and has a small priority valve in the pump manifold, which closes off the lubricating pressure until the clutch and main pump receive a preset pressure. On Power Shift Models, the pump also supplies the operating fluid for transmission operation. On all models, excess fluid flow from transmission pump passes through the full flow system filter to the inlet side of the main hydraulic system pump. If no fluid is demanded by the main pump, the fluid passes through the oil cooler then back to reservoir in transmission housing.

The oil cooler is mounted in front of tractor radiator and on air-conditioned models is an integral part of the air-conditioning condenser.

Return oil from the different functions is routed through a second filter on Power Shift models.

273. RESERVOIR AND FILTERS. The hydraulic system reservoir is the transmission housing and the same fluid provides lubrication for the transmission gears, differential and final drive units. The manufacturer recommends that only John Deere Hy-Gard Transmission and Hydraulic Oil or its equivalent be used in the system. Reservoir capacity in U.S. Gallons is as follows:

Power Shift Models
    4040 . . . . 13½
    4240 . . . . 13½
    4440 . . . . 13½
    4640 & 4840 20
Syncro-Range and Quad-Range Models
    4040 . . . . 15½
    4240 . . . . 15½
    4440 . . . . 15½
    4640 Quad-Range 25

Approximately 5 U.S. Gallons must be added to above capacities on all tractors equipped with Power Front Wheel Drive.

Fig. 289 – Hydraulic system and transmission fluid filler cap and dipstick are located at rear of tractor as shown.

Fig. 290 – Typical Power Shift dual filter arrangement showing return function filter behind transmission filter. Other models have only a transmission filter.

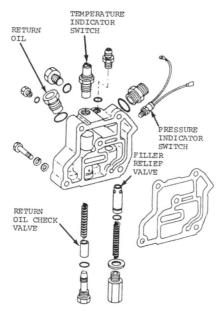

Fig. 291 – Exploded view of oil filter relief valve housing and associated parts used on early model tractors with Perma-Clutch. Late models have "O" ring grooves and "O" rings on valves, but do not have manual by-pass valve. Corresponding valves on Power Shift units are shown in Fig. 219.

To check the fluid level, stop tractor on level ground and check to make sure that fluid level is in "SAFE" range on dipstick (Fig. 289).

The oil filter element (or elements) may be renewed without draining fluid reservoir, by removing filter cover and extracting element. Filters are located on left side of transmission housing as shown in Fig. 290.

All filters are provided with a by-pass valve which opens to allow oil to flow when cold or with filter plugged. On Syncro-Range models, the by-pass valve is located in oil filter relief valve housing (Fig. 291). To service filter relief valve, remove front plug and withdraw spring and valve plunger. The housing also contains a return check valve, and the early model housing contains a manual by-pass valve. Late model housings had "O" rings on the valves, but some return oil check valves had a groove without an "O" ring and the manual by-pass valve was discontinued.

On Power Shift models the relief valve for front (transmission) filter is located in the Power Shift Regulating Valve housing and is shown at (11 – Fig. 219). The rear (hydraulic) filter by-pass valve is located in filter base housing as shown in Fig. 292. To renew or check the valve it is necessary to drain the system and remove rear filter; then unbolt and remove filter housing.

274. SYSTEM TESTS. Efficient operation of the tractor hydraulic units requires that each component operates properly. A logical procedure for testing

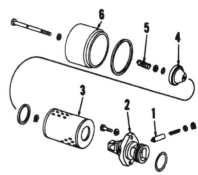

Fig. 292 – Rear hydraulic filter and associated parts used on Power Shift models.

1. By-pass valve	4. Retainer
2. Base housing	5. Spring
3. Filter element	6. Cover

the system is therefore needed. The indicated tests include Transmission Pump Flow Test, System Pressure Tests and Leakage Tests as outlined in the following paragraphs. Unless the indicated repair of hydraulic units is obvious because of breakage, these tests should be performed before attempting to repair the individual units.

275. TRANSMISSION PUMP FLOW TEST. A quick test of transmission pump operation can be performed by removing the fluid filter (front filter on Power Shift Models) and turning engine over with starter. A generous flow of fluid will be pumped into filter housing if pump is operating satisfactorily.

To more thoroughly test pump condition, connect a flow meter into main hydraulic pump supply line on left side of clutch housing (see Fig. 293 or Fig. 294). With unit at operating

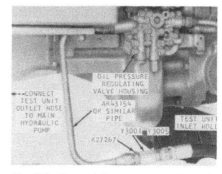

Fig. 293 – Connect flow meter as shown for transmission pump test on Syncro or Quad-Range models.

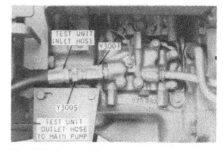

Fig. 294 – Connect flow meter as shown to test transmission pump on Power Shift models.

temperature, engine at 1900 rpm and flow unrestricted, output should be 10 gpm for all Syncro-Range or Quad-Range Models, or 12 gpm for all Power Shift Models.

Slowly close test unit pressure valve while observing flow, which should remain relatively constant to 90 psi. At approximately 100 psi, relief valve should start to open and flow decrease; if it does not, overhaul Oil Cooler Relief Valve (22 – Fig. 160) on Syncro-Range Models or (6 – Fig. 219) on Power Shift Models.

276. MAIN PUMP PRESSURE AND FLOW. To check the main pump pressure and flow, bleed off hydraulic pressure by opening the right brake bleeder screw and holding pedal down a few moments. Remove the line leading from pressure oil manifold (Fig. 295) to

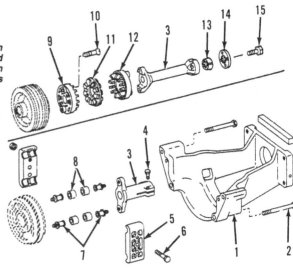

Fig. 298—Exploded view of main hydraulic pump support and drive coupling. Drive shaft shown at top of illustration is for models with front-wheel drive.

1. Support
2. Cap screw
3. Drive shaft
4. Cap screw
5. Coupler half
6. Cap screw
7. Coupler stud
8. Bushing
9. Coupler half, rear
10. Cap screw
11. Coupler cushion
12. Coupler half, front
13. Drive collar
14. Keeper
15. Cap screw

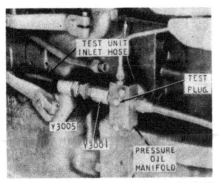

Fig. 295 – Install tester inlet hose on manifold after removing line to pressure control valve.

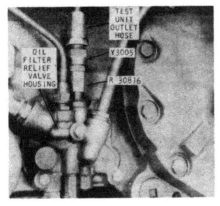

Fig. 296 – Install tester outlet hose for Syncro and Quad-Range models in filter relief valve housing as shown.

Fig. 297 – Install outlet hose for Power Shift models in pressure regulating valve housing as shown.

pressure control valve. Connect inlet side of flow meter to manifold and connect outlet line from flow meter as shown in Fig. 296 or Fig. 297 on left side of tractor. Start and run engine at 1800 rpm.

NOTE: If flow meter is not the inline type, install outlet hose in the transmission filler tube. Run engine for short periods only while testing with this type meter.

Close the test unit control valve until a pressure of 2000 psi is registered; then check fluid flow which should be 23 gpm for all models without Power Front wheel Drive; 26.5 gpm for models with Front Wheel Drive.

Slowly close flow meter control valve until flow stops; pressure should be 2250-2300 psi for models with Power Front Wheel Drive or 2200-2300 psi for other models. Adjust standby pressure if necessary, by turning the stroke control adjusting screw (2 – Fig. 299) clockwise to increase standby pressure or counterclockwise to decrease pressure.

277. PRESSURE CONTROL (PRIORITY) VALVE CHECK. Be sure hydraulic pressure is bled off. Remove flow meter and reconnect line to pressure control (priority) valve. Refer to Fig. 295 and install a 3000 psi pressure gage in test plug hole. Install a jumper hose in a breakaway coupler so that hydraulic fluid can flow through the coupler and back to the reservoir, and turn metering valve arm to "Fast" position.

Start engine and operate at 800 rpm (slow idle), then move selective control valve lever to pressurize breakaway coupler. Gage reading will be the minimum pressure which is maintained by priority valve to insure that steering and brakes will always be pressurized, even if other functions receive no pressure. If pressure is not within recommended

range of 1650-1700 psi, refer to Fig. 32 and disassemble and inspect pressure control valve. BE SURE hydraulic pressure is bled off before removing any oil lines. Shims control the pressure at which control valve starts to restrict oil flow to functions.

278. LEAKAGE TEST. To check for leakage at any of the system valves or components, move all valves to neutral and run engine for a few minutes at 1900 rpm. Check all of the hydraulic unit return pipes individually for heating. If the temperature of any return pipe is appreciably higher than the rest of the lines, that valve is probably leaking. Disconnect that return line and measure the flow for a period of one minute. Leakage should not exceed ½-pint; if it does, overhaul the indicated valves as outlined in the appropriate sections of this manual.

279. **MAIN HYDRAULIC PUMP.** When external leaks or failure to build or maintain pressure indicates a faulty hydraulic pump, remove the unit for service as follows:

Relieve hydraulic pressure. The oil cooler and radiator must both be drained and then removed. Disconnect main supply line, oil cooler line, pressure line and oil seal drain tube from pump. Remove drive shaft coupler halves and loosen clamp screws in pump half of coupler. Suitably support the pump, remove screws securing pump to support and remove the pump.

Install by reversing the removal procedure. Tighten pump mounting bolts to a torque of 85 ft.-lbs. Other applicable torques are as follows: Note: Numbers in bolt description refer to identification symbols in Fig. 298.

Pump support to cyl.
    block (1) . . . . 85 ft.-lbs.

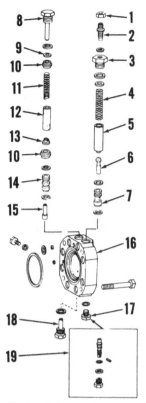

Fig. 299 — Exploded view of stroke control valve housing showing component parts. Item 19 might not be used.

1. Nut	
2. Adjusting screw	11. Spring
3. Bushing	12. Filter
4. Spring	13. Guide
5. Spring guide	14. Valve sleeve
6. Stroke control valve	15. Outlet valve
7. Valve sleeve	16. Housing
8. Plug	17. Plug
9. Shim washer	18. Plug
10. Packing	19. Lockout valve

Pump drive clamp screws (4) 35 ft.-lbs.
Drive coupler cap screws (6). 35 ft.-lbs.
Drive coupler studs (7) . . . . . 35 ft.-lbs.
Coupler cap screws (10) . . . . 35 ft.-lbs.
Front-wheel drive

keeper screws . . . . 20 ft.-lbs.
There should be a gap of 0.098-0.178 inch between fromt and rear couplers (9 and 12—Fig. 298) used on front-wheel drive shaft. Use "Loctite" when installing coupler studs (7).

280. **OVERHAUL.** Before disassembling the pump, check pump shaft end play using a dial indicator, and record the measurement for convenience in reassembling. End play should be 0.001-0.003 inch for either (3.0 or 4.0 cubic inch) pump. End play is adjusted by adding or removing shims (20—Fig. 300) which are available in thickness of 0.006 and 0.010 inch. Bearing wear, or wrong number of adjusting shims can cause excessive end play.

To disassemble the pump, remove the four cap screws retaining the stroke control valve housing (16—Fig. 299) to front pump, and remove the housing. Withdraw discharge valve plugs, guides,

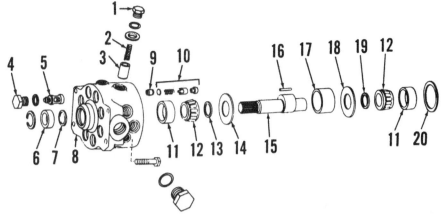

Fig. 300 — Exploded view of main hydraulic pump body, shaft and associated parts. Two additional spacers are used between bearing cones and thrust washers on 4.0 cid pumps.

1. Plug	6. Oil seal	11. Bearing cup	16. Roller bearing
2. Spring	7. Packing	12. Bearing cone	17. Race
3. Piston	8. Body	13. Spacer	18. Thrust washer
4. Plug	9. Seat	14. Thrust washer	19. Spacer
5. Inlet valve	10. Discharge valve	15. Pump shaft	20. Shim

springs and valves (10 – Fig. 300). Remove all piston plugs (1), springs (2) and pistons (3); then carefully withdraw pump shaft (15) together with bearing cones, thrust washers (14), cam race (17) and the loose needle rollers (16). Thrust washers (14) are 0.1235-0.1265 inch thick for 3.0 cubic inch pump (models without Power Front Wheel Drive); 0.0422-0.0452 inch thick for 4.0 cubic inch pump used on models with Power Front Wheel Drive. There are 36 bearing rollers (16) used on 3.0 cubic inch pumps; 25 bearing rollers (16) used on 4.0 cubic inch pumps.

Remove the plugs retaining inlet valve assemblies (5) and check inlet valve lift using a dial indicator. Lift should be 0.060-0.080 inch. If lift exceeds 0.080 inch, spring retainers are probably worn and valves should be renewed. Also check for apparent excessive looseness of valve stem in guide. Do not remove inlet valve assembly unless renewal is indicated or discharge valve seat (9) must be renewed. To remove the inlet valve, use a small pin punch and drive valve out, working through discharge valve seat (9). If inlet valve is to be re-used, place a flat disc on inlet valve head from the inside, so that all the driving force will not strike valve head in the center. Be sure the disc will drive through the hole without touching. Discharge valve seat can be driven out after inlet valve is removed. Be sure to reinstall pistons, springs, valves and seats to their own respective bores. The piston bores in all pumps are lined with a Teflon sheath, so all bores should be carefully inspected. Scored pistons or bores could cause pistons to stick.

The manufacturer recommends that the eight piston springs (2) test within 1½ lbs. of each other and within range of 34-40 lbs. at 1.62 inches for 3.0 cubic inch pump; 47-53 lbs. at 1.78 inches for

Fig. 301 — Special adjusting tool (JDH 19) can be used to determine stroke control valve setting. Refer to paragraph 280.

4.0 cubic inch pump. Install seal (6) only deep enough to allow snap ring to enter groove, to avoid blocking the relief hole in body.

Valves located in stroke control valve housing control pump output as follows: The closed hydraulic system has no discharge except through the operating valves or components. Peak pressure is thus maintained for instant use. Pumping action is halted when line pressure reaches a given point by pressurizing the camshaft reservoir of pump housing, thereby holding pistons outward in their bores.

The cutoff point of pump is controlled by pressure of spring (4 – Fig. 299) and can be adjusted by turning adjusting screw (2). When pressure reaches the standby setting, valve (6) opens and meters the required amount of fluid at reduced pressure into crankcase section of pump. Crankcase outlet valve (15) is held closed by hydraulic pressure and blocks the outlet passage. When pressure drops as a result of system demands, crankcase outlet valve is opened by the pressure of spring (11) and a temporary hydraulic balance on both ends of valve, dumping the pressurized crankcase fluid and pumping action resumes. Stroke control valve spring (4) should test 125-155 lbs. pressure when compressed to 3.3 inches, and crankcase outlet spring (11) should test 45-55 lbs. at 2.2 inches.

Cutoff pressure is regulated by the setting of adjusting screw (2) and adjustment procedure is given in paragraph 276. Cut-in pressure is determined by the thickness of shim pack (9) and/or pressure of spring (11). A special tool (JDH-19) is available to determine shim pack; refer to Fig. 301 and proceed as follows:

Assemble outlet valve units (8 through 15–Fig. 299), using existing shim pack (9). Install special tool (JDH-19) in place of plug (18), using one ⅛-inch thick washer as shown in Fig. 301. If adjusting shim pack thickness is correct, scribe line on tool plunger should align with edge of tool plug bore as shown; if it does not, remove top plug (8–Fig. 299) and add or remove shim washers (9) as required. Shims (9) are available in

0.030 inch thickness. If special tool is not available use shim washers of same thickness as those removed, then add shims to raise cut-in pressure, or remove shims to lower pressure.

When installing stroke control housing, add or remove shims (20–Fig. 300) as necessary to obtain specified pump shaft end play of 0.001-0.003 inch. Always use new "O" rings, packings and seals. Oil all parts liberally with clean hydraulic system oil. Tighten stroke control valve housing retaining cap screws to 85 ft.-lbs. and tighten piston cap plugs to 100 ft.-lbs. torque. Adjust standby pressure as outlined in paragraph 276 after tractor is reassembled.

281. PRIORITY VALVE. The Pressure Control (Priority) valve is mounted on right side of rockshaft housing, just ahead of rockshaft. Refer to paragraph 23 in Steering Section for data on the valve unit.

ROCKSHAFT HOUSING AND COMPONENTS

All Models So Equipped

282. REMOVE AND REINSTALL. To remove the rockshaft housing, first re-

move seat and operator's platform. Disconnect the three point lift links on tractors so equipped. Disconnect and remove hydraulic control rods, interfering wiring and hydraulic lines. Remove the attaching bolts and lift the housing from tractor using a hoist.

When reinstalling the unit on Models 4040, 4240 and 4440, place load selector lever in "MAX" position and make sure that linkage roller is to rear of cam follower located on load control arm as housing is lowered. Do not bend draft linkage. Load control arm is not used on Models 4640 and 4840. On all models, complete installation by reversing the removal procedure and tighten the retaining cap screws to a torque of 85 ft.-lbs.

Models 4040-4240-4440

283. CONTROL VALVE HOUSING OVERHAUL. Control valve housing (1–Fig. 302), can be removed without removing rockshaft housing from tractor. Fully lower the lift arms, then disconnect hydraulic lines and operating linkage. Unbolt and remove selective

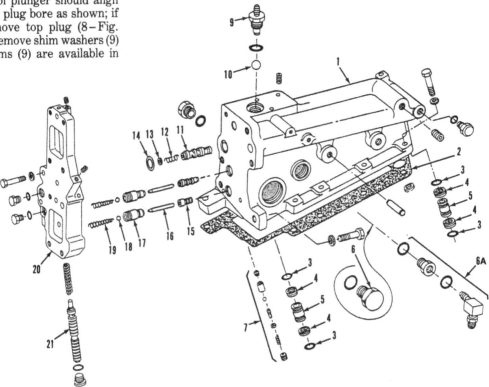

Fig. 302—Exploded view of rockshaft control valve housing, valves and associated parts used on Models 4040, 4240 and 4440.

1. Housing	6A. Fittings (w/lift	10. Check ball	15. Valve guide
2. Gasket	assist cylinder)	11. Flow control valve	16. Metering shaft
3. "O" rings	7. Thermal relief	12. Spring	17. Valve seat
4. Back-up rings	valve assy.	13. Shim	18. Ball
5. Oil inlet pipe	9. Fitting (w/lift assist	14. "O" ring	19. Spring
6. Plug (w/o lift assist | cylinder)		20. Selective control valve mounting plate
cylinder)			21. Directional valve

control valve from rear of control valve housing. Unbolt and remove control valve housing from rockshaft housing.

Remove selective control valve mounting cover (20—Fig. 302), and withdraw rockshaft operating valves (16 through 19) and flow control valve (11, 12 and 13). Remove thermal relief valve assembly (7) and plug (6).

To disassemble control linkage, remove load selector arm (22—Fig. 303), load selector control shaft (23), selector control arm (25) and links with roller (24). Remove lower operating arm (20), quill (18) and shaft (11). Remove valve camshaft (1). Disconnect spring (4) from control valve adjusting cam (5), remove retainer ring holding linkage to control valve adjusting link (6) and remove linkage from housing.

Check all linkage, springs, valves and housing for wear, scoring, or other damage and renew any parts in question. Valves may be lapped to their seats, if necessary, using fine lapping compound. Inspect thermal relief valve assembly and spring, which should have 8 to 10 pounds pressure at a compressed length of 15/32 inch.

When reassembling, install thermal relief valve, spring and shims as required. Hook cam spring (4) to pin (3) in valve body. Assemble valve adjusting cam (5), adjusting link (6), special nut (7) and adjusting screw (2). Assemble operating links (8 and 10) and negative signal limit link (9) to cam link. Position this assembly in valve housing, then install camshaft (1). Hook return spring (4) to ad-

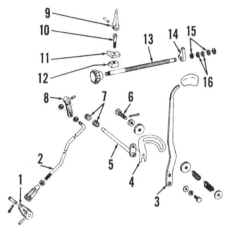

Fig. 304—Exploded view of typical rockshaft control lever and depth adjusting screw.

1. Lower operating arm	9. Cam
2. Control rod	10. Height stop screw
3. Control lever	11. Height stop
4. Friction plate	12. Special nut
5. Shaft	13. Adjusting screw
6. Friction screw	14. Lever stop
7. Bushing	15. Special washers
8. Upper operating arm	16. Spring washers

justing cam. Install lower shaft (11), quill (18) and operating link (20). Assemble load selector control arm (25) and links (24) and connect load selector links to negative signal limit link (9) and cam (12). Install load selector shaft (23) through housing bore and inner selector arm (25). Be sure that outer selector arm (22) is pointing upward.

After control valve housing is reinstalled, turn adjusting screw (2) counterclockwise until bottomed, then clockwise 1/2-turn. This will adjust the rockshaft only enough to allow operation. Refer to paragraph 284 for further adjustments.

Models 4640-4840

283A. **CONTROL VALVE HOUSING OVERHAUL.** Control valve housing (1—Fig. 305), can be removed without removing rockshaft housing from tractor. Fully lower the lift arms, then disconnect hydraulic lines and operating linkage. Unbolt and remove selective control valve from rear of control valve housing. Unbolt and remove control valve housing from rockshaft housing.

Remove selective control valve mounting cover, and withdraw rockshaft operating valves (16 through 19), flow control valve (11, 12 and 13), thermal relief valve assembly (7) and plug (6). Unbolt and remove load control valve assembly (Fig. 306) from front of control valve housing.

NOTE: Load control valve (4—Fig. 306) was adjusted at factory to an overall length of 8.36 inches. Do not disassemble or readjust valve unless absolutely necessary.

To disassemble control linkage, drive out spring pins and remove outer load selector arm shaft (22—Fig. 307), and inner arm (24). Disconnect spring (40) and remove special plug (39), then remove servo cam follower (37) and link (38). Remove valve operating link (35). Remove operating link (10), operating arm (20), quill (18) and shaft (11). Disconnect spring (4) and remove valve camshaft (1), valve operating cam (5) and adjusting link (6).

Check all linkage, springs, valves and housing for wear, scoring or other damage and renew any parts in question. Valves may be lapped to their seats, if necessary, using fine lapping compound. Inspect thermal relief valve assembly and spring, which should have 12.5 to 15 pounds pressure at a compressed length of 0.450 inch. Load control valve is available only as a complete and preadjusted assembly.

To reassemble, reverse the disassembly procedure while noting the follow-

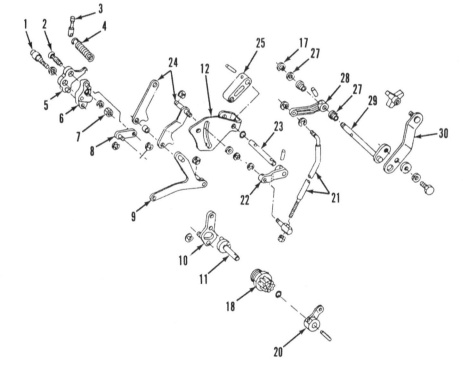

Fig. 303—Exploded view of rockshaft control linkage used on Models 4040, 4240 and 4440. Refer to Fig. 304 for rockshaft control lever and depth adjusting screw.

1. Valve camshaft	9. Negative signal limit		23. Shaft
2. Adjusting screw	link	18. Quill	25. Load selector inner
3. Anchor pin	10. Operating link	20. Lower operating	arm
4. Spring	11. Rockshaft valve	arm	27. Bushings
5. Valve operating cam	operating shaft	21. Load selector rod	28. Upper operating
6. Cam adjusting link	12. Negative signal limit	22. Load selector outer	arm
7. Adjusting screw nut	cam	arm	29. Shaft
8. Link w/pin	17. Washer		30. Load selector lever

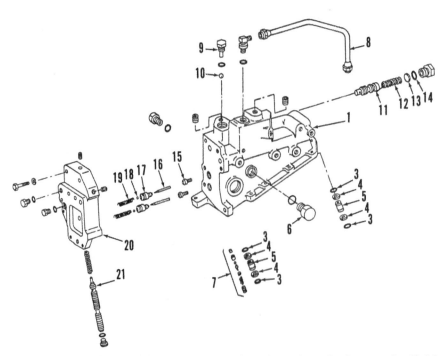

Fig. 305—Exploded view of rockshaft control valve housing, valves and associated parts used on Models 4640 and 4840.

1. Housing
3. "O" rings
4. Back-up rings
5. Oil tubes
6. Plug
7. Thermal relief valve assy.
8. Oil inlet tube
9. Special plug
10. Check ball
11. Flow control valve
12. Spring
13. Shim
14. "O" ring
15. Valve guide
16. Metering shaft
17. Valve seat
18. Ball
19. Spring
20. Selective control valve mounting plate
21. Directional valve

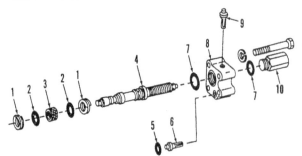

Fig. 306—Exploded view of load control valve used on Models 4640 and 4840, which are equipped with hydraulic draft load sensing system. Load control valve (4) is preadjusted at factory and is available as an assembly only.

1. Back-up rings
2. "O" rings
3. Filter screen
4. Load control valve assy.
5. "O" ring
6. Orifice & filter screen
7. "O" rings
8. Valve retainer
9. Filter
10. Special plug

ing special instructions: When assembling valve adjusting cam assembly (5 and 6—Fig. 307), tighten nut (7) so gap between nut and washer does not exceed 0.450 inch. Tighten operating shaft quill (18) to a torque of 100 ft.-lbs. Install thermal relief valve and flow control valve using same number of shims as removed.

All Models

284. **CONTROL VALVE ADJUSTMENT.** While making any of the following adjustments, no load should be on the hitch.

With engine off and rockshaft fully lowered, move console load selector knob from "MIN" to "MAX" position. The selector knob should not contact either end of console slot. If necessary, adjust length of load selector rod (21—Fig. 303 or 307) to provide 1/16 inch clearance at both ends of slot.

NOTE: Before adjusting rockshaft valve clearance on tractors equipped with lift assist cylinder, lower hitch and disconnect lift assist cylinder oil pressure line and surge relief valve return line at rockshaft piston cover. Cap fittings on piston cover. Assist cylinder oil line should be left open to allow free movement of hitch.

To adjust rockshaft valve clearance, place load selector lever in "MIN" position and disconnect control rod from rockshaft control lever arm. Clamp locking pliers to lever arm as shown in Fig. 308, and affix a reference point 10 inches from centerline of lever shaft to be used to measure valve opening clearance.

CAUTION: Avoid lift arm and lift link while making this next adjustment, as they may move unexpectedly.

Run engine at slow idle and carefully measure movement of reference mark required to change rockshaft direction. Movement should be 3/16 to 3/8 inch as shown in Fig. 308. If distance is greater than 3/8 inch, remove plug from side of housing and reaching through plug hole, insert a screwdriver in slotted end of adjusting screw. Turn adjusting screw clockwise to decrease free movement.

Reconnect control linkage rod and raise rockshaft about halfway and shut off engine. If rockshaft starts to drop, insufficient valve clearance exists or valves are leaking.

To adjust rockshaft control lever, remove plug from rear of transmission housing (Fig. 309) and turn control arm extension screw clockwise as far as it will go. With engine running at slow idle and load selector lever set at "MIN" position, move rockshaft control lever forward to lower hitch. Move rockshaft control lever rearward to raise rockshaft. Rockshaft should not be fully raised when REAR edge of lever is still in front of "O" mark on console, but hitch must be fully raised when lever is halfway past "O" mark. Decreasing length of control rod (2—Fig. 304) raises rockshaft, and increasing length of rod lowers rockshaft.

To adjust load control arm screw on Models 4040, 4240 and 4440 (Models 4640 and 4840 do not have a load control arm), place load selector lever in "MAX" position and run engine at slow idle. Move rockshaft control lever forward to lower hitch, then move control lever rearward until REAR edge of control lever is even with "1" mark on console. Turn load control arm extension screw (Fig. 309) counterclockwise until rockshaft starts to raise.

285. **CONTROL LEVER FRICTION ADJUSTMENT.** The rockshaft control lever friction adjusting screw is located near the load selector lever (Fig. 310). With valve operating rod disconnected at the lower operating arm on rockshaft valve housing, the control lever should require a 7 to 8 pound pull to move the lever. Loosen jam nut and adjust friction screw to obtain desired resistance.

286. **ROCKSHAFT HOUSING OVERHAUL.** Rockshaft piston can be removed for inspection or renewal without removal of rockshaft housing. Remove cylinder end cover (8—Fig. 311) and force piston out by pushing down on rockshaft arms with short, jerky motions.

NOTE: Be careful not to damage open end of cylinder with connecting rod or ram (crank) arm. Tighten cylinder cover retain-

ing cap screws to a torque of 170 ft.-lbs. when reinstalling.

To disassemble the rockshaft, remove housing and lower cover (6–Fig. 314). Remove dog point set screw (11–Fig. 315) and right hand lift arm (2), then slide rocker shaft (9) out left side of housing, removing crank arm (8) and servo cam (10) as shaft is withdrawn.

When installing rockshaft bushings, use suitable drivers and make sure oil holes are aligned. Add as many spacers (12) as will fit in housing, to eliminate end play. Install servo cam (10) with ramp up as shown and make sure dog point set screw (11) enters locating hole in rockshaft (9). Splines on crank arm (8) and shaft (9) are indexed for proper alignment during assembly. Tighten lift arm cap screws to 300 ft.-lbs., strike arm with hammer and retighten.

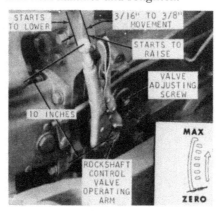

Fig. 308—With load selector lever set at "ZERO", a 10-inch lever should move the distance shown. Valve adjustment is through plug hole in housing.

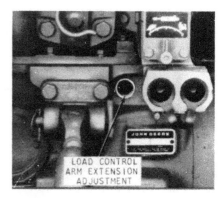

Fig. 309—Remove plug for access to load control arm adjusting nut.

Fig. 310—Refer to paragraph 285 for adjustment of friction device.

LOAD CONTROL ARM AND SHAFT

Models 4040-4240-4440

287. **OPERATION.** When the load selector lever is moved to "MAX" position, the operating depth of the three-point hitch is controlled by the draft of the attached implement acting in conjunction with the position of the control lever.

The amount of draft is transmitted by the lower links to the drawbar support (Fig. 319), then to the control valve by the load control shaft and arm. The spring steel load control shaft is anchored in each side of the transmission housing and the drawbar or draft link support frame affixed to outer ends of shaft. Positive or negative draft causes the center of the load control shaft to deflect a predetermined amount according to the load encountered. The center arc of the flexing shaft moves the straddle mounted lower end of the load control arm while the upper (follower) end transmits the required signal to the control valve. Adjustment is made as outlined in paragraph 289.

288. **REMOVE AND REINSTALL.** Remove rockshaft housing as outlined in paragraph 282 and differential as outlined in paragraph 222. Remove load

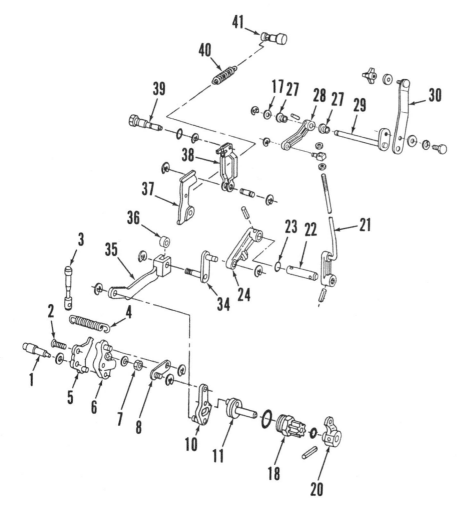

Fig. 307—Exploded view of rockshaft control linkage used on Models 4640 and 4840. Refer to Fig. 304 for rockshaft control lever and depth adjusting screw linkage.

1. Valve camshaft
2. Adjusting screw
3. Anchor pin
4. Spring
5. Valve operating cam
6. Cam adjusting link
7. Adjusting screw nut
8. Link w/pin
10. Valve operating link
11. Rockshaft valve operating shaft
17. Washer
18. Quill
20. Lower operating arm
21. Load selector rod
22. Shaft
23. "O" ring
24. Load selector inner arm
27. Bushings
28. Upper operating arm
29. Shaft
30. Load selector lever
34. Link w/pins
35. Valve operating link
36. Roller
37. Rockshaft servo cam follower
38. Cam follower link
40. Spring
41. Anchor pin

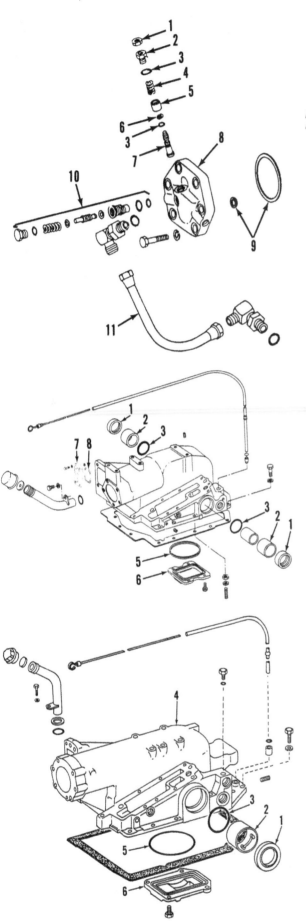

Fig. 311—Exploded view of rockshaft cylinder cover and associated parts.
1. Nut
2. Bushing
3. "O" rings
4. Spring
5. Throttle valve
6. Back-up ring
7. Throttle valve shaft
8. Cylinder cover
9. Seal rings
10. Surge relief valve assy.
11. Relief valve to selective control valve pipe

Fig. 314—Exploded view of typical rockshaft housing and associated parts typical of 4040, 4240 and 4440 models. Refer to Fig. 314A for 4640 and 4840 rockshaft housing.
1. Oil seal
2. Bushing
3. "O" ring
4. Housing
5. "O" ring
6. Cover
7. Cover
8. "O" ring

Fig. 314A—Exploded view of rockshaft housing and associated parts for 4640 and 4840 models. Refer to Fig. 314 for legend.

control arm by removing cap screws to support (12–Fig. 316). All parts are available individually.

To remove the load control shaft or draft link support, drain transmission and hydraulic fluid and remove snap rings (Fig. 319) and retainers. Suitably support draft link support, then with a brass drift and working from right side of tractor, bump load control shaft to the left and out. Bushings in draft link support and in transmission housing can be renewed at this time. Renew transmission housing bushings only if necessary. Chill bushings before installation. Inside diameter of bushings is tapered to provide a small bearing area for the flexing load control shaft. Install bushings in transmission housing with small I.D. to inside and bushings in support with small I.D. to outside, away from transmission. Refer to Fig. 321. Heat sealing ring in 160 degrees F water to soften. Drive bushing into transmission case with groove to the outside. Collapse sealing ring, and install in groove in bushing with plenty of lubrication after "O" ring is installed. Press sealing ring round so that load control shaft will go through and install as many selective washers as necessary to provide minimum clearance between transmission case and support. Carefully install load control shaft with adequate lubrication and install retainers and snap rings. Adjust as outlined in paragraph 289.

289. **ADJUSTMENT.** To adjust the control mechanism, remove the adjusting plug shown in Fig. 309. Move selector lever to "MAX" and start and idle engine.

Reaching through plug port with a socket and extension, turn slotted nut (1—Figs. 316 or 317) as necessary until rockshaft starts to raise when rear edge of rockshaft control lever aligns with "1" mark at rear of slot on console scale. Check operation of rockshaft and adjust as outlined in paragraph 284.

DRAFT SENSING CYLINDER

Models 4640-4840

289A. **OPERATION.** Hydraulic sensing of draft load is used on Models 4640 and 4840. The draft sensing system consists of a load sensing cylinder, mounted on right side of transmission case and attached to the draft link pivot arm; and a load control valve, located in rockshaft control housing, which actuates rockshaft control valves. Load control valve is actuated by varying hydraulic sensing pressure at the load control valve piston by means of a fixed orifice in load

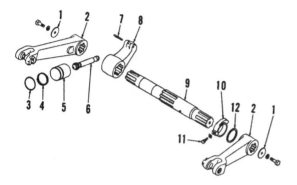

Fig. 315 — Exploded view of typical rockshaft, piston, lift arms and associated parts.
1. Washer
2. Lift arms
3. "O" ring
4. Back-up ring
5. Piston
6. Piston rod
7. Spring pin
8. Crank (ram) arm
9. Rockshaft
10. Servo cam
11. Set screw
12. Spacer washer

control valve housing and a variable orifice located in sensing cylinder.

When the draft load increases, the sensing cylinder piston (5—Fig. 318) and valve (9) are pulled rearward, which enlarges the variable orifice opening. This produces an increase in sensing pressure at the load control valve piston, which results in rearward movement of the piston. This movement is transmitted by mechanical control linkage to the rockshaft control valves, opening rockshaft pressure valve to raise rockshaft and reduce the draft load.

When draft load decreases, sensing cylinder piston moves forward and variable orifice opening is reduced. Sensing pressure at load control valve piston is reduced, and rockshaft lowers until hydraulic pressure between variable orifice and fixed orifice equalizes.

289B. **R&R AND OVERHAUL.** The draft load sensing cylinder is attached to draft link arm on right side of rear axle center housing. To remove cylinder, first lower hitch and relieve hydraulic system pressure. Disconnect hydraulic lines. Remove cylinder anchor pins from pivot arm and anchor bracket and remove cylinder.

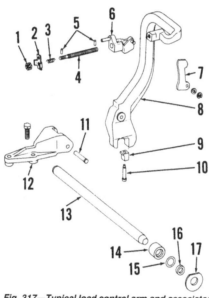

Fig. 317—Typical load control arm and associated parts used on 4040, 4240 and 4440 models with Power Shift. See Fig. 316 for legend.

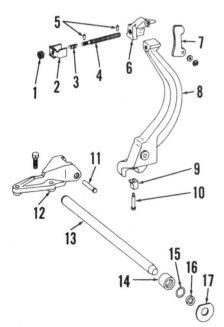

Fig. 316—Typical load control arm and associated parts used on all 4040, 4240 and 4440 models with Syncro-Range and Quad-Range.

1. Nut
2. Lock plate
3. Spring
4. Adjusting screw
5. Spring pin
6. Extension
7. Follower
8. Arm
9. Follower block
10. Pin
11. Pin
12. Support
13. Load control shaft
14. Bushing
15. "O" ring
16. Oil seal
17. Washer

Fig. 318—Exploded view of draft load sensing cylinder used on Models 4640 and 4840.

1. End cover
2. "O" ring
3. Nut
4. Seal spring
5. Piston
6. "O" ring
7. Back-up ring
8. Rod
9. Variable orifice valve
10. Spring
11. Cylinder
12. Plug
13. Wiper seal
14. Springs
15. Retainer
16. Clevis

To disassemble, remove top bracket (1—Fig. 318). Clamp cylinder rod end (16) in a vise, then loosen rod nut (3) several turns.

IMPORTANT: Do not remove rod nut at this time as spring pressure is against piston.

Clamp cylinder horizontally in a vise to relieve spring tension on rod nut, then remove nut. Release spring tension and separate components from cylinder body. Double nut piston end of rod and remove clevis (16).

Inspect piston, rod and cylinder for scoring, wear or other damage. Renew all seals and "O" rings. Note that early models used an "O" ring and back-up ring on the piston, while late models are fitted with a quad seal ring without back-up ring.

Install wiper seal (13) in cylinder with lip facing outward and drive in until flush with cylinder surface. Lubricate all parts with clean hydraulic oil, then insert rod into cylinder from piston end of cylinder. Install springs (14), retainer (15) and clevis (16). Tighten clevis about two turns. Install spring (10) and orifice valve (9). Install piston and nut on rod. Secure clevis in a vise, then tighten clevis and nut to torque of 185 ft.-lbs. Install top bracket with new "O" ring and tighten cap screws to a torque of 35 ft.-lbs.

Reinstall sensing cylinder on tractor. Attach cylinder rod end to upper hole in draft link pivot arm on 4640 tractors and to middle hole on 4840 tractors.

SELECTIVE (REMOTE) CONTROL VALVES

All Models So Equipped

290. **OPERATION.** Tractors are optionally equipped with one, two or three

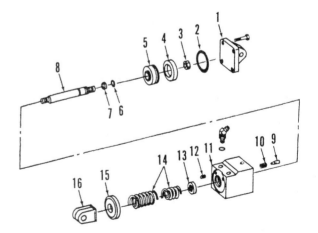

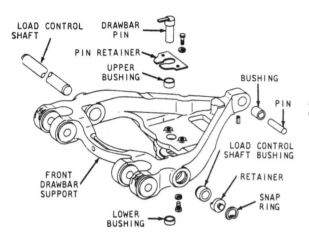

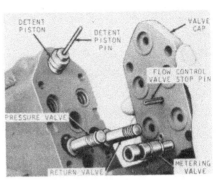

Fig. 319 — Exploded view of draft link and front drawbar support used.

Fig. 324 — Clamp housing in a vise to remove rear cap.

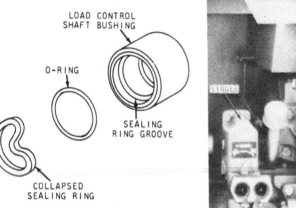

Fig. 321 — Collapse sealing ring for ease of installation.

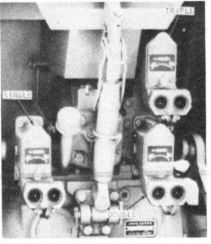

Fig. 322 — Rear view of tractor showing location of remote valves.

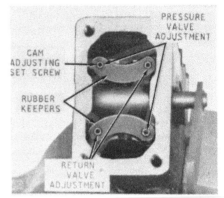

Fig. 325 — Front view of housing showing rocker assembly, adjustment screws and associated parts.

selective (remote) control valves for operation of remote cylinders. Mounting positions of valves are shown in Fig. 322.

As with all other units of the hydraulic system, pressure is always present at the valves but no flow exists until the valve is moved. Refer to Fig. 323 for an exploded view of valve mechanism. Each breakaway coupler is equipped with two return valves (20) and two pressure valves so arranged that one of each is opened when control lever is moved off center in either direction. Detent piston (16) is actuated by pressure differential across metering valve (26) and released by pressure equalization when flow stops at end of piston stroke. Flow control valve (25) maintains an even flow with varying pressure loads.

291. **OVERHAUL.** Refer to Fig. 323 for an exploded view of the selective control valve. Clamp the unit in a vise and unbolt and remove cap (23) and associated parts carefully as shown in Fig. 324. Identify parts as required for later assembly, then remove valves, springs and guides.

Rotate valve body in vise so that rocker assembly is up as shown in Fig.

325. Rocker arm can be disassembled by driving out spring pin, and removing control arm and shaft. Remove screws holding cam (3 and 7 – Fig. 323) and remove rocker (5). Notice how parts are assembled, to aid in reassembly. Inspect

all bores, valves and valve seats. Seats are non-renewable, but can be reconditioned by using NJD 150 Valve Seat Repair Kit (Use exactly as directed). Some tractors were equipped with a spring pin in cover (23) to prevent "O" ring (24) from being pulled into the passage under sudden oil surges.

Assemble by reversing the disassembly procedure. If actuating cam was dis-

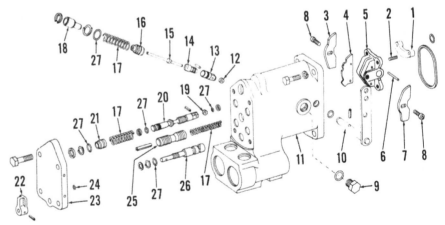

Fig. 323 — Exploded view of typical Selective (Remote) Control valve showing component parts.

1. Rubber keeper (2)	8. Cam clamp screws	15. Detent pin
2. Adjusting screw (4)	9. Plug (2)	16. Detent piston
3. Regular cam	10. Rocker shaft	17. Spring
4. Detent cam	11. Housing	18. Outer detent guide
5. Rocker	12. Detent roller	19. Cam roller (4)
6. Pin (2)	13. Detent follower	20. Poppet valve (4)
7. Float cam	14. Inner detent guide	21. Valve guide (4)
		22. Lever
		23. Cover
		24. "O" ring
		25. Flow control valve
		26. Metering valve
		27. "O" ring

assembled, refer to Fig. 326 and note that the float cam (7) is shorter than the regular cam (3), and is installed on numbered side of housing as shown. Install detent cam (4) in rocker (5) with 2 pins (6) and install rocker assembly and rocker shaft (10).

Adjusting the valve requires use of special adjusting cover (JDH-15C) and a dial indicator. Remove the two adjusting plugs (9) and loosen the two cam locking screws (8). Back out the four adjusting screws (2) at least two turns.

Install pressure and return valves, detent follower, piston, guides and retaining snap ring. Be sure detent follower roller properly rides on detent cam. Also make sure that operating valve rollers are turned to ride properly on ramps of cams. Back out all adjusting screws on special plate and install the plate with the angled screw pointing at detent pin (15–Fig. 323). Carefully FINGER TIGHTEN the four screws contacting operating valves until valves are seated; then while holding operating lever in center position FINGER TIGHTEN the detent locking screw until detent roller is seated in neutral detent on detent cam (4–Fig. 326). With operating lever in neutral position, refer to Figs. 325 and 328. Turn in the two diagonally opposite Pressure Valve Cam Adjusting Screws until screws, cams and follower rollers are in contact. Install rubber keeper then back out ¼-turn as shown. Turn in the two diagonally opposite Return Valve Cam Adjusting Screws until screws, cams and follower rollers are in contact. Install rubber keeper, then back out ⅛-turn. Move the two cams (3 & 7–Fig. 326) into contact with adjusting screws and tighten screws (8) securely.

To double-check the adjustment, install a dial indicator 3 inches from center of shaft on operating arm as shown in

Fig. 328. Zero dial indicator while locked in neutral detent, then back out the detent locking screw on adjusting cover. Back out the two adjusting cover screws which contact operating valves on lever side and measure rocker movement which should be 0.021 inch toward return valve or 0.060 inch toward pressure valve as shown. Valves contacting cam (3–Fig. 326) opposite lever side are being checked. Tighten the two adjusting

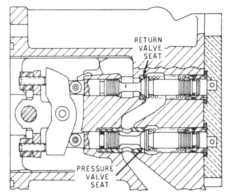

Fig. 327 – Valve seats can be reconditioned using special tool NJD 150.

Fig. 328 – Use a dial indicator to measure rocker arm movement as shown. Refer to paragraph 291 for procedure.

Fig. 329 – Exploded view of breakaway coupler and associated parts. All removable parts are duplicated in adjacent bore.

1. Snap ring
2. Snap ring
3. Back-up ring
4. "O" ring
5. Ball
6. Back-up ring
7. Receptacle
8. Ball
9. Spring
10. Plug
11. Snap ring
12. Sleeve
13. Operating lever
14. Cam
15. Lockwasher
16. Washer
17. Expansion plug

cover screws on lever side and loosen the other two screws, then check adjustment of valves on lever side. Readjust as necessary for correct rocker movement. This procedure will allow return valves to open before pressure valves, when selective control valve is used.

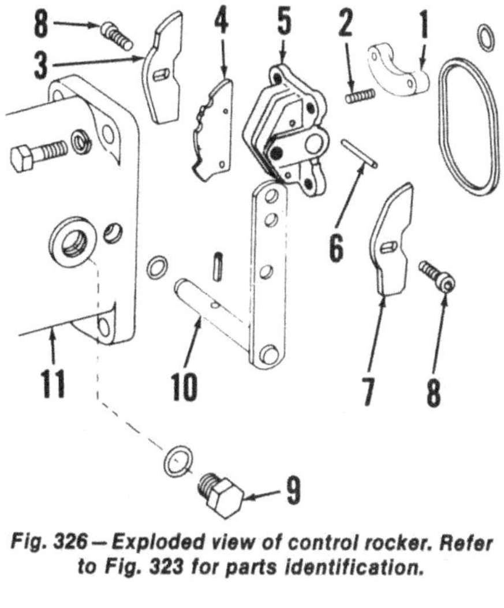

292. BREAKAWAY COUPLER. Drive a punch into the expansion plugs (17–Fig. 329) and pry out of housing. Remove retainer rings and springs. Operating levers can then be removed. Drive receptacle assembly from housing. Check steel balls, springs, and all parts for wear and replace as necessary. Renew "O" rings and back-up washers. Reassemble in reverse order of disassembly.

POWER WEIGHT TRANSFER VALVE

All Models So Equipped

293. OPERATION. The power weight transfer hitch uses a special coupler, a power weight transfer control valve, pressure gage, a special rockshaft piston cover, a double acting remote cylinder (used as a retracting cylinder only) and a transfer link. The remote cylinder takes the place of the center link on a three point hitch, and becomes a telescoping center link.

When using the power weight transfer hitch, only a drawn implement should be used and operation is in the rockshaft "MAX" position. Excessive load causes the load control arm to direct pressure oil through the control valve to the remote cylinder. As the cylinder retracts, it tilts the coupler forward at the top, which pulls on the transfer link to the implement being used, causing the weight of the implement to bear down on the draft links. This has the effect of using the rear wheels as a pivot point to unload the front wheels and give added traction to the rear when needed, without having to add ballast to rear wheels.

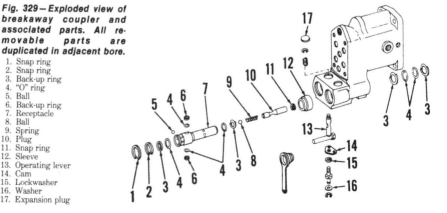

Fig. 326 – Exploded view of control rocker. Refer to Fig. 323 for parts identification.

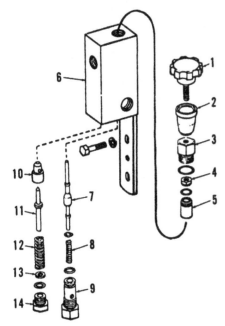

Fig. 330 — Exploded view of power weight transfer valve.

1. Valve screw	8. Spring
2. Boot	9. Plug
3. Bushing	10. Relief valve seat
4. Nut	11. Relief valve
5. Shaft	12. Spring
6. Housing	13. Shim
7. Diverting valve	14. Plug

The console mounted pressure gage shows when weight is being transferred.

Refer to Fig. 330 for an exploded view of valve unit.

The valve is primarily a switch valve which diverts fluid from the rockshaft cylinder to the control cylinder and rockshaft lever is used for the control lever.

294. OVERHAUL. Refer to Fig. 330 for an exploded view of valve unit. Diverting and relief valves can be removed after removing ports plugs (9 & 14). One seat for diverting valve (7) is on upper surface of plug (9); the other seating surface is in bore of body (6). Diverting valve moves upward by pressure of spring (8) and inward flow of oil when knob (1) is backed out for rockshaft operation; and diverter valve seals against upper seat to close off passage to remove cylinder. Turning control knob (1) clockwise mechanically moves diverter valve into contact with seat on plug (9), closing return passage to rockshaft cylinder and opening passage to remove cylinder. Relief valve spring (12) should test 180-220 lbs. when compressed to a height of 1 5/8 inches. Shims (13) may be added if necessary to increase release pressure of relief valve. Renew any parts which are worn, broken or damaged.

REMOTE CYLINDER

All Models

295. Refer to Fig. 332 for exploded view of the double acting, hydraulic stop remote cylinder. To disassemble, remove end cap (1), stop rod spring (3) and valves (4 and 5), using care not to lose ball (6), if cylinder is equipped with override provision. Fully retract the cylinder and remove nut from piston end of piston rod (19). To remove the stop rod and springs, drive the groove pin from stop rod arm (23).

Fig. 332 — Exploded view of the high pressure remote cylinder used on all models so equipped.

1. Cap	
2. Gasket	
3. Spring	
4. Stop valve	
5. Bleed valve	
6. Ball	
7. Spring	
8. Stop rod	
9. Washer	
10. Spring	
11. Cylinder	
12. Spring	
13. "V" packing	
14. Piston	
15. Back-up ring	
16. "O" ring	
17. Lever	
18. Stop screw	
19. Piston rod	
20. Stop	
21. Wiper seal	
22. Back-up ring	
23. Arm	
24. Guide	

Install new wiper seal (21) in guide (24) with lip to outside, assemble rod end of cylinder, rod packing and piston rings, and have piston fully inserted in cylinder before installing rod nut. Tighten nut securely. Be sure the piston rod stop (20) is located so that stop lever (17) is opposite the stop rod arm (23). Tighten cap screws securing piston rod guide (24) to a torque of 35 ft.-lbs. and cap screws retaining piston cap (1) to 85 ft.-lbs.

Fig. 333 shows parts identification on cylinders with mechanical stop.

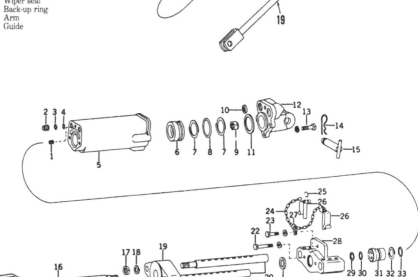

Fig. 333 — Exploded view of remote cylinder with mechanical stop. Item 6 may be equipped with an "O" ring.

1. Plug	9. Nut	17. Jam nut (2)	25. Link
2. Plug	10. Gasket (2)	18. Lockwasher	26. Stop pin
3. Back-up ring	11. Gasket	19. Piston rod stop	27. Clamp
4. "O" ring	12. Cap	20. Stop rod (2)	28. Stop rod guide
5. Cylinder	13. Cap screw	21. Oil seal	29. Back-up ring
6. Piston	14. Lock pin	22. Cap screw	30. "O" ring
7. Back-up ring (2)	15. Attaching pin	23. Cap screw	31. Piston rod guide
8. "O" ring	16. Piston rod	24. Chain	32. Back-up ring
			33. "O" ring

NOTES

NOTES

NOTES

NOTES

JOHN DEERE

SHOP MANUAL JD-202

Models	■ 2510	■ 2520			
Model	■ 2040				
Models	■ 2240	■ 2440	■ 2630	■ 2640	
Models	■ 4040	■ 4240	■ 4440	■ 4640	■ 4840

I&T
SHOP MANUALS

Information and Instructions

This shop manual contains several sections each covering a specific group of wheel type tractors. The Tab Index on the preceding page can be used to locate the section pertaining to each group of tractors. Each section contains the necessary specifications and the brief but terse procedural data needed by a mechanic when repairing a tractor on which he has had no previous actual experience.

Within each section, the material is arranged in a systematic order beginning with an index which is followed immediately by a Table of Condensed Service Specifications. These specifications include dimensions, fits, clearances and timing instructions. Next in order of arrangement is the procedures paragraphs.

In the procedures paragraphs, the order of presentation starts with the front axle system and steering and proceeding toward the rear axle. The last paragraphs are devoted to the power take-off and power lift systems. Interspersed where needed are additional tabular specifications pertaining to wear limits, torquing, etc.

HOW TO USE THE INDEX

Suppose you want to know the procedure for R&R (remove and reinstall) of the engine camshaft. Your first step is to look in the index under the main heading of ENGINE until you find the entry "Camshaft." Now read to the right where under the column covering the tractor you are repairing, you will find a number which indicates the beginning paragraph pertaining to the camshaft. To locate this wanted paragraph in the manual, turn the pages until the running index appearing on the top outside corner of each page contains the number you are seeking. In this paragraph you will find the information concerning the removal of the camshaft.

More information available at haynes.com
Phone: 805-498-6703

J H Haynes & Co. Ltd.
Haynes North America, Inc.

ISBN-10: 0-87288-366-3
ISBN-13: 978-0-87288-366-6